Geophysical Monograph Series

Geophysical Monograph Series

Geophysical Monograph 284

Noisy Oceans

Monitoring Seismic and Acoustic Signals in the Marine Environment

Gaye Bayrakci
Frauke Klingelhoefer

Editors

This Work is a co-publication of the American Geophysical Union and John Wiley and Sons, Inc.

AGU
ADVANCING EARTH AND SPACE SCIENCES

WILEY

This edition first published 2024

Published under the aegis of the AGU Publications Committee

Matthew Giampoala, Vice President, Publications
Carol Frost, Chair, Publications Committee

For details about the American Geophysical Union visit us at www.agu.org.

The right of Gaye Bayrakci and Frauke Klingelhoefer to be identified as the editors of this work has been asserted in accordance with law.

Registered Office
John Wiley & Sons, Inc., 111 River Street, Hoboken, NJ 07030, USA

Editorial Office
111 River Street, Hoboken, NJ 07030, USA

For details of our global editorial offices, customer services, and more information about Wiley products visit us at www.wiley.com.

Wiley also publishes its books in a variety of electronic formats and by print-on-demand. Some content that appears in standard print versions of this book may not be available in other formats.

Library of Congress Cataloging-in-Publication Data Applied for:
Hardback ISBN: 9781119750895

Cover Design: Wiley
Cover Image: Courtesy of Olivier Dugornay

Set in 10/12pt Times New Roman by Straive, Pondicherry, India

CONTENTS

LIST OF CONTRIBUTORS

Chastity Aiken
IFREMER
Centre Bretagne
Plouzané, France

Megan Baker
Departments of Earth Sciences, and Geography
University of Durham
Durham, UK

David Barclay
Department of Oceanography
Dalhousie University
Halifax, Nova Scotia, Canada

Gaye Bayrakci
National Oceanography Centre
Southampton, UK

Mohammed Belal
National Oceanography Centre
Southampton, UK

Jonathan M. Bull
School of Ocean and Earth Science
University of Southampton
Southampton, UK

Emmy T.Y. Chang
Institute of Oceanography
National Taiwan University
Taipei, Taiwan (R.O.C.)

Mark Chapman
School of Geosciences
University of Edinburgh
Edinburgh, UK

Chien-Chih Chen
Department of Earth Sciences
National Central University
Taoyuan City, Taiwan (R.O.C.)

Ying-Nien Chen
Department of Earth and Environmental Sciences
National Chung Cheng University
Chiayi, Taiwan (R.O.C.)

Michael A. Clare
National Oceanography Centre
Southampton, UK

David Dellong
Département Acoustique Sous-marine
Service Hydrographique et Oceanographique
de la Marine
Brest, France

Jessica Fisher
School of Ocean and Earth Science
University of Southampton
Southampton, UK

Ángel F. García-Fernández
Department of Electrical Engineering and Electronics
University of Liverpool
Liverpool, UK

Yung-Cheng Gung
Department of Geosciences
National Taiwan University
Taipei, Taiwan (R.O.C.)

Timothy J. Henstock
School of Ocean and Earth Science
University of Southampton
Southampton, UK

Finnigan Illsley-Kemp
School of Ocean and Earth Science
University of Southampton
Southampton, UK
and
School of Geography, Environment and Earth Sciences
Victoria University of Wellington
Wellington, New Zealand

Bazile G. Kinda
Département Acoustique Sous-marine
Service Hydrographique et Oceanographique
de la Marine
Brest, France

Frauke Klingelhoefer
IFREMER
Centre Bretagne
Plouzané, France

Aude Lavayssiere
School of Ocean and Earth Science
University of Southampton
Southampton, UK
and
Institut de Physique du Globe de Paris
Paris, France

Florent Le Courtois
Département Acoustique Sous-marine
Service Hydrographique et Oceanographique
de la Marine
Brest, France

Timothy G. Leighton
Institute of Sound and Vibration Research
University of Southampton
Southampton, UK

Jing-Yi Lin
Department of Earth Sciences, and
Center for Environmental Studies
National Central University
Taoyuan City, Taiwan (R.O.C.)

Gwyn Lintern
Natural Resources Canada
Geological Survey of Canada
Sidney, British Columbia, Canada

Calum Macdonald
School of Geosciences
University of Edinburgh
Edinburgh, UK

Timothy A. Minshull
School of Ocean and Earth Science
University of Southampton
Southampton, UK

Chung-Hsiang Mu
Institute of Earth Sciences
Academia Sinica
Taipei, Taiwan (R.O.C.)

Edward Pope
Department of Earth Sciences, and
Department of Geography
University of Durham
Durham, UK

Adam H. Robinson
School of Ocean and Earth Science
University of Southampton
Southampton, UK

Ben Roche
School of Ocean and Earth Science
University of Southampton
Southampton, UK

Sean Ruffell
Department of Earth Sciences, and
Department of Geography
University of Durham
Durham, UK

Vera Schlindwein
Section Geophysics
Alfred Wegener Institute Helmholtz Centre
for Polar and Marine Research
Bremerhaven, Germany
and
Faculty of Geosciences
University of Bremen
Bremen, Germany

Steven Simmons
Earth and Environment Institute
Hull, UK

Brendan Smith
Department of Oceanography
Dalhousie University
Halifax, Nova Scotia, Canada

Peter J. Talling
Department of Earth Sciences, and
Department of Geography
University of Durham
Durham, UK

Jean Baptiste Tary
Department of Geosciences
University of the Andes
Bogotá, Colombia

Chuen-Teyr Terng
Center Weather Bureau
Taipei City, Taiwan (R.O.C.)

Bai Tian
Institute of Sound and Vibration Research
University of Southampton
Southampton, UK

Victoria L.G. Todd
Ocean Science Consulting Ltd.
Dunbar, UK

Morelia Urlaub
GEOMAR Helmholtz Centre for Ocean Research
Kiel, Germany

Paul R. White
Institute of Sound and Vibration Research
University of Southampton
Southampton, UK

Wei Yi
School of Information and Communication Engineering
University of Electronic Science and Technology of China
Chengdu, China

Boxiang Zhang
School of Information and Communication Engineering
University of Electronic Science and Technology of China
Chengdu, China

Mohammad Zaki Zulkifli
National Oceanography Centre
Southampton, UK

PREFACE

The oceans are far from a "silent world"—they harbor rich and variable acoustic signals. The ocean soundscape includes natural sound sources with biological origins, such as fish and marine mammals, and environmental sound sources from earthquakes, volcanoes, and rain. In recent times, anthropogenic sound sources, such as marine engineering and construction, seismic experiments, and marine traffic, have been introduced into the oceans. The marine soundscape affects life in the oceans as well as human societies.

Many past studies using marine seismometers and hydrophones have studied earthquakes and the Earth's deep interior. Similarly, biological noise, such as marine mammal calls, has been studied by acousticians to gather information about the behaviors of varied species. Although the ocean soundscape was recorded, these marine signals were often discarded, as it was impossible to identify the source of the different signals.

With the growing demand for marine resources and recent technological developments such as increased computational power and data storage, marine scientists have begun to show interest in the marine soundscape—previously discarded as noise—as a tool to study environmental processes and the human effect on them. Although there is a clear increase in nonconventional seismology exploring unknown sound sources, no work has been conducted to compile a set of signals frequently encountered in marine seismometer or hydrophone data.

This book attempts to fill this gap and compile a comprehensive set of nontectonic marine noise commonly observed by ocean-bottom seismometers or hydrophones, providing a literature review of different signals and state-of-the-art examples of research in which these signals are studied via existing methods or new methods developed for this purpose.

After an introductory chapter about marine soundscapes and their sources, this book presents 13 chapters discussing different types of noise: of environmental origins, such as from volcanoes, icebergs, landslides, and wind; of biological origins, such as from whales; and anthropogenic sounds, such as marine traffic. In the concluding chapter, we put the book into its broader context, show its possible impact and implications, and propose future actions.

This book includes review chapters and case studies from specific geographic locations, such as the North Sea and Taiwan. The authors, most of whom are internationally recognized scientists from academia and government institutions, work in subjects ranging from geology and geophysics to biology and risk assessment.

This work aims to deepen the understanding of marine soundscapes and the identification of signal sources. It is relevant to a broad community of students, teachers, and researchers studying the marine environment, including geologists, geophysicists, volcanologists, and biologists. It will also be useful for those who develop laws and regulations concerning the impact of sound on marine life.

We would like to thank all the contributing authors for their valuable chapters. We would also like to thank the reviewers of all the chapters for their work, which helped immensely to improve the book. We also thank the publishing teams at the American Geophysical Union and John Wiley & Sons, Inc. for their guidance throughout the editing of this book, including Noel McGlinchey, Jenny Lunn, Lesley Fenske, Layla Harden, and Keerthana Govindarajan. Warmest thanks to Rituparna Bose for inviting us on this great adventure of editing a book on submarine noise and Lihini Aluwihare from the Scripps Oceanographic Institute and AGU Books Editorial Board for editorial support.

Gaye Bayrakci
National Oceanography Centre, UK
Frauke Klingelhoefer
IFREMER (French National Institute for Ocean Science), France

1

An Introduction to the Ocean Soundscape

Gaye Bayrakci[1] and Frauke Klingelhoefer[2]

ABSTRACT

The ocean soundscape is composed of sound from natural origins related to geological processes (geophony) and marine life (biophony) and of manmade origin (anthrophony). Early seismologic studies focused on earthquakes and classified most other signals as "noise." Recent studies have shown that geological processes originating from the seafloor and subseafloor (submarine volcanoes, landslides, etc.) and the water column (e.g., microseisms and icebergs) are also recorded by seismoacoustic instruments and provide important information about how our planet works. Sound is the primary way that many marine species gather information and understand their environment, and it is a valuable tool for us to study their behavior. There has been a substantial increase in anthropogenic noise in the oceans since the Industrial Revolution, and assessing the human impact on the oceans is an emerging topic aiming to leave a healthy ocean for future generations. In this chapter, we briefly present the types of seismic waves and noise sources in the oceans. We explain the seismoacoustic tools used to record noise and give an overview of standard data-processing methods to familiarize the reader with the concepts discussed in this book. We then summarize the following chapters and discuss future directions for seismoacoustic noise research.

1.1. INTRODUCTION

Early seismoacoustic submarine recording involved instruments towed behind a ship or installed on the seafloor, primarily to record earthquakes or seismic shots. Other sounds were classified as undesired "noise" and either cut out or filtered from the data. An earthquake is the shaking of the Earth due to movement along a fault or discontinuity within the Earth's subsurface (Aki, 1972). The sudden displacement along the fault generates seismic waves corresponding to elastic waves. These waves are studied extensively by earthquake seismologists because of their impacts on human life and because they provide insight into the properties of the media in which they travel (Agnew et al., 2002). The Oxford English Dictionary defines noise as a loud or unpleasant sound, or irregular fluctuations that accompany a transmitted signal but are not part of it and tend to obscure it. Since seismology is the branch of science concerned with earthquakes and related phenomena, signals of biological or anthropogenic origin are considered noise in earthquake seismology.

Studies of ocean noise began in the 1960s with the advent of more affordable data storage and better instrument performance. They showed that the oceans are far from silent and are filled with noises from different origins: human, biological, and tectonic. In this book, we present many of these nonearthquake-related ocean noises and explain various methods to interpret them.

The ambient sound field in the oceans is composed of sounds from natural and manmade processes.

[1]*National Oceanography Centre, Southampton, UK*
[2]*IFREMER, Centre Bretagne, Plouzané, France*

Noisy Oceans: Monitoring Seismic and Acoustic Signals in the Marine Environment, Geophysical Monograph 284, First Edition.
Edited by Gaye Bayrakci and Frauke Klingelhoefer.

DOI:10.1002/9781119750925.ch01

Natural noise sources in the oceans are of abiotic/geological (geophony) and biological (biophony) origin. Examples of geophony include microseisms related to the interactions of wind with oceans, volcanic events, landslides, icebergs, hydrothermal noise, rain, breaking waves, and gas bubbles, whereas biophony includes marine mammal vocalizations and noise from other marine species (e.g., crabs and shrimp). Ocean noise also includes manmade noise (anthrophony), which results from resource extraction (e.g., underwater mining), coastal or marine construction (e.g., pipeline and wind turbines), explosions, coastal and marine traffic, seismic exploration, navigation tools, etc.

Human audible sound is limited to 20 Hz to 20 kHz frequencies, but the term sound refers to more than sounds audible to humans. It means an oscillation in pressure corresponding to a particle displacement: that is, back-and-forth movement of particles caused by a passing wave. Within the oceans, as elsewhere, this expands over a wider range of frequencies than human audible sound.

Often, the origin of repeatedly recorded signals in the ocean remains unknown because there is no visual proof of their origin. For example, short-duration events (SDEs) with durations of a few seconds have been observed on seismic records since the early 1980s; however, their origin remains unknown (see Chapter 8 of this book). At first, these events were explained as fish colliding with the instruments due to an observed decrease in the number of observations with increasing instrument depth (Buskirk et al., 1981). Later studies proposed different origins related to instruments settling into the sediment (Ostrovsky, 1989), microearthquakes (Sohn et al., 1995), oscillating clouds of methane bubbles in the water column (Pontoise & Hello, 2002), and release of gas from subsurface sediments (Bayrakci et al., 2014; Diaz et al., 2007; Sultan et al., 2011). In an attempt to define the origin of SDEs and other seismoacoustic signals, Batsi et al. (2019) deployed ocean-bottom seismometers in front of a submarine observatory, monitoring the environment with an underwater camera. They found that marine species such as crabs and octopuses frequently interact with the instruments and leave signals in the seismic records (Fig. 1.1); however, no fish were observed colliding with the ocean-bottom seismometer (OBS) spheres. During a different experiment in a fish tank filled with sediments and water, Batsi et al. (2019) also recorded gas bubbles leaving the sediments. They concluded that SDEs are signals resulting from gas expulsions from the subsurface. Although there is growing evidence for the relationship between SDEs and seafloor gas expulsion, the exact mechanism that generates these events (collapse of seafloor gas migration conduits, vibration of the conduit walls, migration pathways opening via hydraulic fracturing, etc.) is still debated, illustrating the challenge of clearly associating each signal with its origin.

This book aims to provide a comprehensive list of nontectonic seismic signals recorded in the ocean (Fig. 1.2). It includes review and case-study chapters describing the characteristics of different signals, explaining the methods used for identifying and interpreting these signals and their wider significance. For each type of signal, peer-reviewed publications by domain experts can be found in the literature. Since human impact on the oceans is a subject of growing importance, recent review papers on this topic (Duarte et al., 2021; Williams et al., 2015) are also available. However, to our knowledge, no book introduces all or most known noise sources in the oceans. Policy frameworks (e.g., the United Nation's Convention on Biological Diversity or the European Commission's Marine and Coastal Environment Policy) with recommendations and goals to reduce human impact on the ocean also introduce noise sources encountered in the oceans, but they usually lack the seismoacoustic methodology related to noise identification and analysis as this is not part of their communication goals. Because some ocean noises originate within the water column and are studied by acousticians, and some originate from the subseafloor and are studied by seismologists, different terminology can act as a barrier to knowledge transfer between two very close branches of science. Here, our goal has been to produce a homogenized scientific and educative document that summarizes various types of signals for the curious reader, policymaker, student, or researcher.

In this introductory chapter, we briefly present the types of seismic waves and different noise sources within the oceans. We then explain the tools for recording ocean noise and give an overview (nonexhaustive) of common data processing methods to familiarize the reader with the concepts discussed in the following chapters. We also offer short summaries of each chapter and finish with a subsection on the future directions of seismoacoustic noise research.

1.2. SEISMIC WAVES

"A seismic wave is an elastic wave generated by an impulse such as an earthquake or an explosion" (https://www.usgs.gov/media/images/what-was-richter-scale). Seismic waves can travel within the Earth's subsurface, where they are called body waves, or along Earth's surfaces, where they are called surface waves.

1.2.1. Body Waves

Body waves include primary waves (P-waves) and secondary waves (S-waves). P-waves are compressional waves that travel and displace particles longitudinally, parallel to the direction of the propagating wave. They travel faster than other waves and can propagate in all types of material, including liquids. P-waves are also called acoustic waves because they travel as pressure fluctuations in fluids. S-waves, also called shear waves, travel and displace particles transversely, perpendicular to the direction of propagation. Their speed is related to the

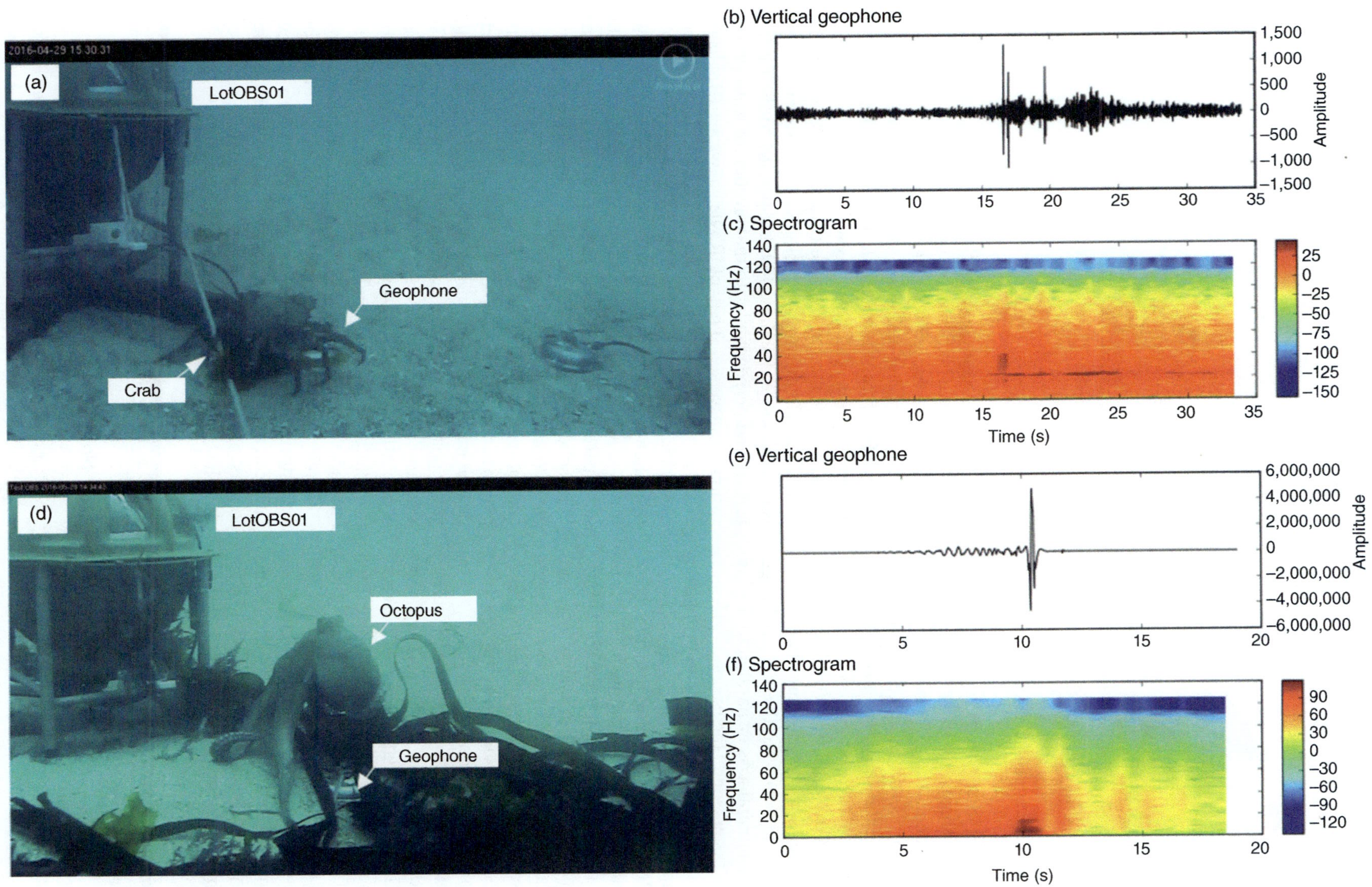

Figure 1.1 (a, d) A crab and an octopus interacting with the geophones of an ocean-bottom seismometer. (b, e) The associated waveforms. (c, f) The spectrograms. Batsi et al. (2019) / with permission from John Wiley & Sons.

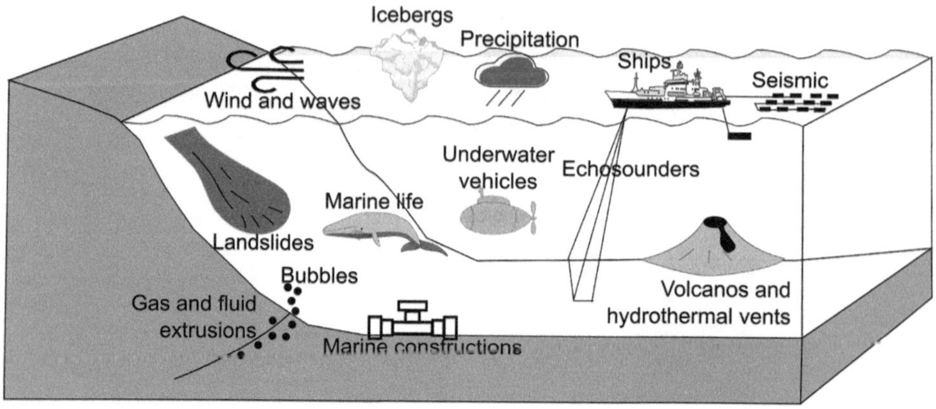

Figure 1.2 Cartoon illustrating the ocean noise sources.

medium's shear modulus; therefore, they do not travel in liquids, as liquids do not have any shear strength. In an anisotropic medium where, for example, the mineral crystals have varying properties in different directions (i.e., lattice-preferred orientation) or the medium is made of thin layers of contrasting properties (i.e., shape-preferred orientation), S-wave splitting (or S-wave birefringence) occurs. Here, S-waves split into two phases: the first phase is polarized parallel to the preferred alignment (e.g., the longer mineral axis or parallel to the layering) and travels faster, and the second phase is polarized perpendicular to the alignment and propagates more slowly (Crampin, 1985).

In addition to P- and S-waves, Biot's theory (Biot, 1956), which describes wave propagation in a porous medium fully saturated with a fluid (e.g., a solid frame and the fluid), predicts the presence of a slow compressional wave due to a 180-degree out-of-phase movement of the fluid and the solid phase. The presence of these slow compressional waves is shown theoretically, but they have not been observed on seismograms.

In data from seismometers located on islands or coastal locations or moored hydrophones in the ocean, a third seismic phase called the T- (tertiary) phase is observed when the seismic energy from earlier phases is converted to acoustic energy. T-phases can propagate long distances within the oceans in a horizontal layer called the SOFAR (sound fixing and ranging) channel. It acts as a waveguide because its sound speed is lower than that of the waters above and below. T-phases may also convert back to elastic waves when they encounter a continental slope (Buehler & Shearer, 2015).

1.2.2. Surface Waves

Surface waves are the waves trapped at a surface marked by a change in density and/or elastic properties, such as the surface of the Earth, an interface within the subsurface, or the walls of a fluid-filled subsurface crack (Takeuchi & Saito, 1972). Surface waves have larger amplitudes than body waves and travel more slowly. They are dispersive; longer wavelengths penetrate deeper and travel with higher velocities than shorter wavelengths. Types of surface waves include Love, Rayleigh, Scholte, Stoneley, Lamb, and Krauklis waves.

The particle motion (i.e., the back-and-forth movement of the particle due to the passage of the wave) of Love waves (Sheriff & Geldart, 1995) is parallel to the Earth's surface and perpendicular to the direction of wave propagation (transverse). That of Rayleigh waves (Rayleigh, 1885) is elliptical, with vertical and longitudinal components (Vidale, 1986). Love and Rayleigh waves travel along a vacuum–solid interface like the Earth's surface. Scholte waves and Stoneley waves (Stoneley, 1924) exhibit the same particle motion as Rayleigh waves but travel along fluid–solid and solid–solid interfaces, respectively. Lamb waves (Lamb, 1881, 1917) exhibit the same particle motion and propagate within thin plates (i.e., isolated elastic plates) surrounded by a contrasting material, in a symmetric (the particle motions at both surfaces of the thin plate are in phase) or asymmetric mode (Korneev et al., 2012). Similarly, Krauklis waves (Korneev, 2011; Korneev et al., 2012) propagate around a fluid layer bounded by two elastic half spaces, and they are polarized along the interfaces (e.g., the walls of a fluid-filled conduit).

Gravity waves are a type of surface wave produced by the restoring force of gravity on the distorted ocean surface. They can also occur within a body of water where a density contrast exists between two superposed fluids, where they are called internal waves (Garrett & Munk, 1979).

1.3. NOISE SOURCES IN THE OCEANS

1.3.1. Noise from Geological Origins (Geophony)

Microseisms are seismic noise arising from the vertical oscillations of solid earth due to gravity waves (Ardhuin et al., 2015). They form the second-lowest-frequency noises recorded on seismometers, after the tides resulting from the gravitational forces of the Moon and the Sun on the oceans and the Earth, which have a semidiurnal period (two low and two high tides in 24.50 hours).

Microseismic noise usually displays two period peaks. The weaker peak is produced in shallow coastal environments by gravity waves with periods of up to ~20 seconds and amplitudes of up to several meters because of pressure fluctuations caused by breaking waves. These are called primary microseisms (or single-frequency microseisms) and are observed at the period of the wave initiating them (0.05–0.1 Hz). The stronger peak displaying shorter periods is also caused by gravity waves. These are called secondary microseisms (or double-frequency microseisms), resulting from ocean waves (gravity waves) with the same frequency traveling in opposite directions. Their period is usually half the period of the wave pair generating them (0.1–0.5 Hz). Ocean infragravity waves (a subcategory of gravity waves with periods >30 seconds; Webb et al., 1991) also create secondary microseisms with periods up to 50 seconds; they are referred to as "Earth's hums" (Bertin et al., 2018).

Microseisms can be recorded anywhere on Earth and are a useful monitoring tool for climate change, as they allow the observation of ocean storm characteristics such as storm frequency or intensity. In this case, increasing microseismic noise indicates increased ocean storminess, likely due to climate change (Bromirski, 2009).

Another source of low-frequency seismic noise is volcanic very-long-period (VLP) events of 0.01–0.5 Hz frequencies, considered to be the manifestation of magma migration and resonant fluid excitation in the volcanic system (Arciniega-Ceballos et al., 1999; Chouet, 2022). When relocated, they reveal the locations of cracks, dikes, and sills in the volcanic system. Volcanos also produce monochromatic long-period (LP) signals, usually with frequencies <5 Hz and durations of ~10 seconds. When LP energy is sustained over minutes to days, the signals are called volcanic tremors and are attributed to the dynamics of magmatic and hydrothermal fluids (Chouet & Matoza, 2013). Tremors can have dominant energy in the 1–9 Hz band and can be harmonic when related to conduits that are filling or emptying. In addition to nontectonic events, volcano-tectonic (VT) events originating from shear failure-like earthquakes occur in volcanic systems and act as a gauge that maps the stress conditions around the magma conduit (Chouet & Matoza, 2013). Note that the terminology for volcano-seismologic signals is still evolving due to the ongoing discovery of new signals.

An interesting example of a developing marine volcano is the Mayotte volcano. The region became seismically active in 2018, with intense seismic swarms. In November 2018, a very low-frequency tremor was recorded worldwide, confirming that the seismic crisis was likely of magmatic origin. Other nonearthquake signals, such as hydroacoustics probably associated with the volcano, were discovered during the monitoring of the volcano (Foix et al., 2021; Saurel et al., 2021).

Submarine landslides commonly occur at continental slopes, submarine canyons, volcanic islands, and ridges. They produce both a long-period noise (20–50 seconds), related to the elastic rebound after the mass failure, and a large-amplitude noise with frequencies from 1 to 10 Hz (1–0.1 second period) due to the impact of the mass on the base of the slope. They are associated with tsunamigenic hazards, and their hydroacoustic detection is suggested to be a powerful tsunami monitoring and modeling tool (Caplan-Auerbach et al., 2001).

Noise from ice is the dominant noise source in regions covered by ice sheets. Iceberg bursts attributed to disintegration processes produce signals of uniform spectral power between 1 and 8 Hz lasting a few minutes. Monochromatic signals of <10 Hz that do not show multiple harmonic overtones (only one or two) are called *iceberg tremors* and are thought to be related to the resonance of fluid-filled cracks within the ice. Iceberg harmonic tremors of a few Hz fundamental frequency with overtones occupying the frequency band below 200 Hz are also commonly recorded in locations covered by ice sheets, and they are attributed to the stick slip at the contact between icebergs or an iceberg and the ground (see Chapter 6 of this book).

Seabed fluid flow is a widespread natural process with consequences for seabed geological features, marine biology, and the compositions of the oceans (Judd & Hovland, 2009), which also contributes to the ocean soundscape. Compaction, petroleum, fresh water, and magma-controlled systems differentiate the mechanisms through which fluids migrate in permeable sediments (Berndt, 2005). Faults, seismic pipes, and chimneys are sedimentary vertical fluid migration pathways; pockmarks and mud volcanoes are the seafloor expression of fluid emission in cooler environments; and hydrothermal vents driven by the presence of a heat source are mostly observed in regions with magmatic or geochemical control.

Hydrothermal vents occur when heated subsurface fluid is discharged at the seafloor. They are found near seafloor spreading centers and hot spots, often in water depths between 2,000 and 4,000 m. Usually the heat source is magma-related, but it can also be sourced from exothermic metamorphosis, such as serpentinization, which is the alteration of the mantle peridotite by water (Früh-Green et al., 2003). Hydrothermal vents generate seismic noise due to the interaction of turbulent flow with the vent structure and cavities (Fig. 1.3), the jet discharge into the cold ocean, and phase transitions of seawater, driven by rapidly changing temperature and pressure (see Chapter 7 of this book). In general, noise from

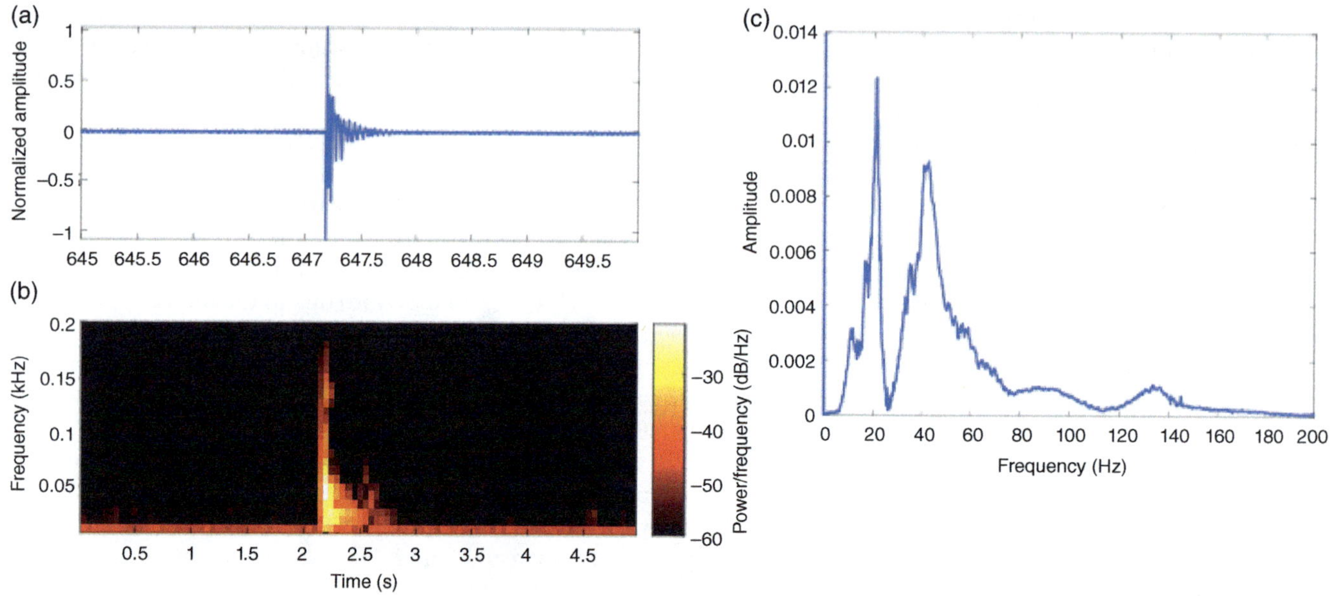

Figure 1.3 A short-duration event recorded by an ocean bottom node around the SEM1 seafloor massive sulfide deposit located on top of the Semyenov Oceanic Core Complex (OCC) at 13°N along the Mid-Atlantic Ridge. (a) Waveform, (b) spectrogram, and (c) frequency spectrum of the signal. Note the monochromatic and harmonic characteristics of the signal. Courtesy of Bramley Murton.

hydrothermal activities exhibits frequencies less than 500 Hz. Passive soundscape monitoring of hydrothermal vents is currently new. Since sound can travel long distances within the ocean, it has been suggested that passive monitoring may allow the remote detection of vent sites and vent characterization. It has also been suggested that some biological organisms use vent noise as an acoustic cue to aid settlement (Crone et al., 2006). Since hydrothermal vents are proposed to be the catalysts for life on Earth (Martin et al., 2008), acoustic monitoring of hydrothermal vents may be a valuable tool in the search for extraterrestrial life exploration (Dziak et al., 2020).

Like hydrothermal vents, active fluid discharge in cooler sedimentary environments generates seismoacoustic noise. Monochromatic (4–30 Hz frequencies and harmonics) SDEs (Fig. 1.3) a few seconds in length are recorded worldwide within sedimentary basins (see Chapter 8 of this book; Tary et al., 2012). They are best observed on geophones as only large-amplitude events are detected by hydrophones, suggesting a seafloor origin for these events. The frequency range and waveforms of SDEs can be similar to VLP events, and they may be related to the resonance of sedimentary fluid conduits. Besides the seismic noise within the sediments, fluid emission from the sediments into the water column also creates bubbles and bubble clouds (with lower-frequency oscillations), contributing to the acoustic noise.

A *gas bubble* is a volume of gas surrounded by a liquid (Leighton, 1994). Bubbles can be found throughout the water column and are produced in many ways, from volcanic emissions to breaking waves. The sound of a single bubble is insignificant compared to other marine noise sources; however, natural processes (rainfall, gas seeps, breaking waves) usually generate millions of them. At the seafloor, a single bubble forms when a small volume of gas from a larger reservoir intrudes into the water, and the two volumes of gas are connected by a thin neck (see Chapter 10 of this book). When the neck eventually breaks and releases the bubble into the water column, the snapping of the neck results in a jet of water being propelled into the bubble, triggering an oscillation at the natural frequency of the bubble that creates its acoustic signature. This phenomenon is called *Minnaert resonance* (Minnaert, 1933); the frequency of oscillation of the bubble is a function of the bubble radius, the ambient pressure, and the density of the liquid and spans several orders of magnitude (~0.1–100 kHz). *Bubble clouds*, groups of independent bubbles, also oscillate with lower oscillation frequencies than each individual bubble, comparable to frequencies of single, larger bubbles (Leighton, 2012).

Similarly, when a rain droplet falls on the sea surface, it forms an impact crater and may entrain a bubble into the water. An initial sound (a compressional wave) results from the impact of the droplet on the sea surface, followed by the simple harmonic motion of the bubble, which is again the Minnaert resonance phenomenon. The acoustic signature of rainfall is caused by this entrainment of

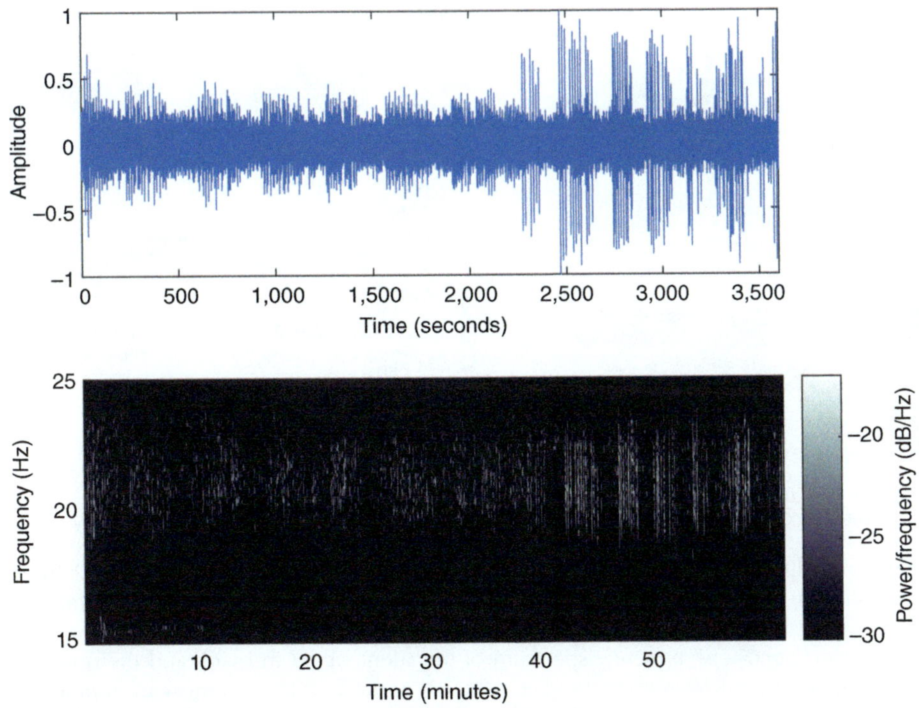

Figure 1.4 Whale calls recorded during the Galicia 3D seismic survey at the Deep Galicia rifted margin offshore Spain and Portugal. The signal's waveform is shown in the top panel and the spectrogram in the bottom panel. Whale calls are the short-duration signals observed after 35 minutes with a frequency range of 18–24 Hz. Tim Minshull (2022) / PANGAEA / CC-BY-4.0.

bubbles, so the frequency spectrum of rainfall depends on the sizes of bubbles entrained by the rain droplets (e.g., 13–25 kHz for drizzle of 1.2–2 mm droplets; Ma & Nystuen, 2005).

The acoustic signature of breaking waves also results from bubble-cloud generation. The largest sound intensity occurs during the cavity collapse in the wave's life cycle (see Chapter 10 of this book). The frequency of the acoustic noise is also related to the size of the created bubbles (e.g., 300 Hz for a bubble size distribution of 2 mm to >10 mm radius; Deane & Stokes, 2002).

1.3.2. Noise from Biological Origins (Biophony)

Sound is a highly efficient means of communication underwater and is the primary way that many marine species gather and understand information about their environment (Fig. 1.4). Cetaceans vocalize for social and echolocation purposes (Au & Hastings, 2008). Vocalizations can be periodic or nonperiodic and are a mixture of frequency- and amplitude-modulated signals (Zimmer, 2011). The 7 Hz to 180 kHz frequency band is thought to contain all frequencies within cetacean vocalizations and hearing ranges (Barker & Lepper, 2012; Bittle & Duncan, 2013). Generally, larger species generate lower-frequency sounds, and the frequency range of a given species is dictated by its anatomy. As the sound attenuates through geometric spreading, lower-frequency songs of larger species propagate over longer distances. Long-period recording of these calls is useful for gathering information about the species' migration patterns.

1.3.3. Noise from Anthropogenic Origins (Anthrophony)

There has been a substantial increase in anthropogenic noise in both terrestrial and marine environments since the Industrial Revolution. Given the rapid pace of change in the ocean soundscape, there is an urgent need to assess the impact of anthropogenic noise on marine life and mitigate further disturbances for a healthy ocean in the future (Duarte et al., 2021). The anthropogenic soundscape in the oceans includes noise from fishing, mining, drilling (Fig. 1.5), resource harvesting and related underwater construction, geophysical exploration such as seismic surveys (10–500 Hz), and other active echosounders producing higher-frequency noise to map the seafloor, among other purposes. Vessel noise with frequencies ranging from 10 Hz to tenths of kilohertz encompasses the frequency ranges of other anthropogenic noises, and it is by far the most studied. The International Maritime Organization adopted regulations for commercial shipping to reduce onboard noise to protect the ship

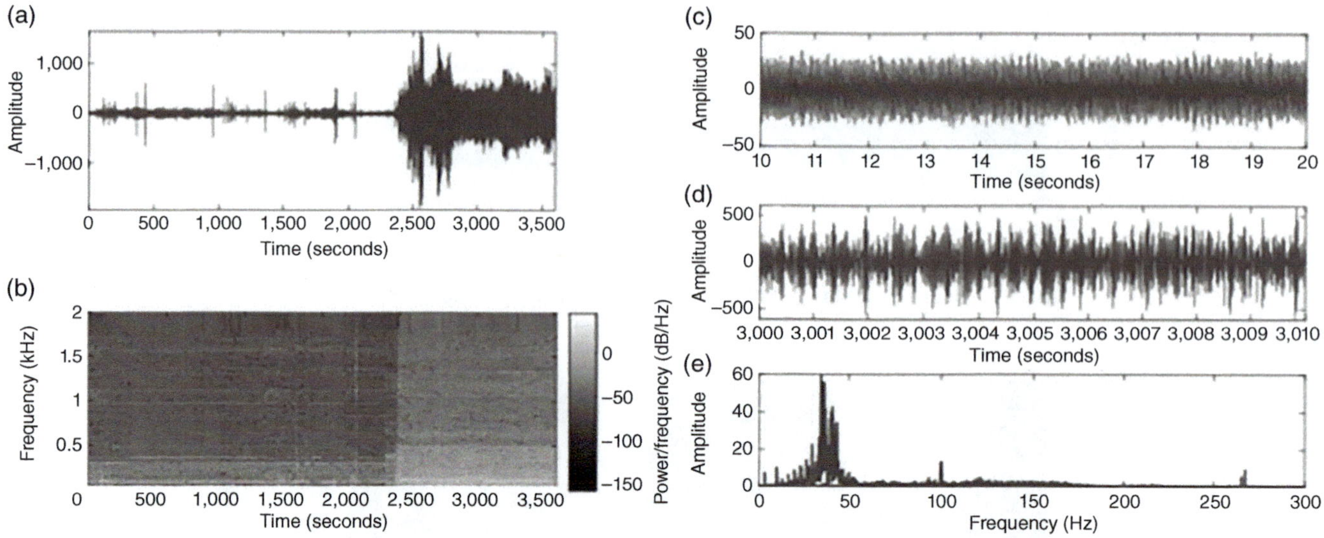

Figure 1.5 Seafloor drilling noise recorded by an ocean bottom node mounted on the RD2 (British Geologic Survey) seafloor drill. (a) Waveform and (b) spectrogram of the signal. Note that drilling starts at the fortieth minute of the recording. (c, d) Two 10-minute windows from the data shown in (a) with a silent and noisy (drilling) period, respectively. (e) Frequency spectrum of the silent signal in black and the noisy signal in grey. Note that the drilling produces low-frequency noise centered around 30 Hz. Courtesy of Bramley Murton.

personnel, along with a set of guidelines for reducing underwater noise.

1.4. TOOLS FOR RECORDING MARINE NOISE

Hydrophones, geophones, and accelerometers usually detect seismic waves traveling within the ocean or the Earth's subsurface. Other instruments, such as cameras filming bubbles emitted from the seafloor or current-meters measuring the velocity of water currents (acoustic Doppler current profiler, ADCP), are used to infer the origin of noise sources.

A hydrophone (Fig. 1.6) records pressure (absolute or differential) in the water using a piezoelectric transducer that generates an electric potential when subjected to a pressure change. It is therefore very similar to a microphone, which measures pressure variations in the air; the difference is the increased acoustic resistance of water, which is 3,750 times that of air. A hydrophone or microphone senses P-waves, which are pressure transients. At low frequencies, the pressure variations are omnidirectional, and a hydrophone with a single cylindrical ceramic transducer is used to achieve omnidirectional reception. More sophisticated directional hydrophones containing focused transducers or transducer arrays can enhance the sensitivity of the signal coming from a given direction.

A modern geophone consists of a coil of wire wrapped around a mass suspended by a spring over a magnet. As the mass moves, the magnet moves the electrons through the wire coil, producing an electrical voltage. The geophone thus converts the ground velocity caused by a seismic wave into voltage. An accelerometer is similar to a geophone, but it uses a strain gauge or a capacitance sensor and a feedback mechanism that outputs the force required to make the mass follow the case. Since the force is the product of the mass and the acceleration, the output is proportional to the acceleration. The latest technology used in ocean-bottom accelerometers is MEMS (micro-electro-mechanical system), which provides

Figure 1.6 RS Aqua single-channel Porpoise acoustic recorder/real-time streaming device with a hydrophone. Photo Credit: RS Aqua Ltd.

recordings of the signal uninfluenced by parameters such as the water temperature or the tilt of the instrument.

Three orthogonally mounted geophones are used in a seismometer to record the particle motion created by the Earth's vibrations. The first seismoscope, the ancestor of the seismometer, was created in China during the second century (Sleeswyk & Sivin, 1983). Seismoscopes detected motion of the Earth and indicated the direction of motion. Only in the nineteenth century were pendulum seismographs invented: they detect and record the Earth's movements using a pencil attached to an inertial mass that moves with respect to a fixed case, and thus record the horizontal particle motions. Initially, the word *seismograph* referred to a device detecting and recording the Earth's movements simultaneously, whereas a *seismometer* only detected the movement without recording it. Today the words *seismometer* and *seismograph* are used interchangeably.

Modern seismometers use electronic sensors, amplifiers, and recorders (data loggers). Seismic sensors (geophones and accelerometers) are defined by their bandwidth, sensitivity, self-noise, resolution, and dynamic range. The frequency of natural seismic sources spans a wide range from 10^{-5} Hz for tidal movements to hundreds of kilohertz for single bubble oscillations. However, it is difficult (if not impossible) for a single sensor to cover the entire frequency range of natural phenomena. This is why short-period, long-period, and broadband sensors have been developed to tackle analyzing different processes. The sensitivity controls how precisely the volts can be converted to displacement, velocity, or acceleration. Ideally, a given ground velocity should produce the same voltage regardless of the frequency. This is called *sensor linearity*. Seismic sensors are usually defined by their low-frequency limits (e.g., 4.5 Hz geophones), a frequency below which the sensor's response is not linear and the output voltage is less than 50% of the correct voltage. Self-noise resulting from thermomechanical processes is unavoidable in an electric circuit. It defines the resolution of a seismic sensor, which is the smallest measurable value the instrument can reliably indicate. The seismic sensors can either be passive and not require electrical power to operate or active and use electronic circuits to improve sensitivity at certain frequencies.

Seismic amplifiers are used to record seismic signals without distortion. They consist of preamplifiers that filter out unwanted signals using low- or high-cut filters and main amplifiers that reproduce the exact waveshape of the seismic pulse. The dynamic range of a seismic amplifier is the ratio between the largest and the lowest input intensities. Large dynamic ranges are required for handling the considerable variation in the amplitude of the input signal, which quickly decays in time after the initial peak.

Data loggers or digitizers record signals received via seismic sensors. They usually contain several channels to record data from multiple sensors, such as three orthogonal geophones in the case of a seismometer or three orthogonal geophones, and a hydrophone in the case of an ocean-bottom seismometer. The sensor signals recorded on different channels usually share the same timing information. The signals are recorded with a user-defined sampling frequency, which varies with respect to the frequency content of the studied phenomena. In an autonomous instrument, the maximum record length is related to the battery life or is a function of the data logger's sampling frequency and disk space.

1.4.1. Ocean-Bottom Seismometers

Ocean-bottom seismometers (Weatherall et al.) and ocean-bottom hydrophones (OBHs) are the most commonly used instruments to record seismic signals in the oceans (Fig. 1.7). Depending on their use, they are equipped with either only an OBH or a hydrophone and a three-channel seismometer (three geophones; OBS), which can be broadband or short period. These instruments can be deployed from a vessel, sink to sit on the seafloor for their deployment duration, and are released from their anchor through an acoustic release signal or at a given time chosen before deployment. To enable deployment, OBSs have anchors that ensure the coupling of the seismometer with the seafloor. Acoustic releases detach the anchor from the instrument for recovery and floatation so it can rise to the sea surface once it is released (Bialas & Flueh, 1999). A flashing light and a flag are usually placed on top of the floatation as visual aids for spotting the instruments once they rise to the sea surface.

To ensure good coupling between the geophone sensors and the seafloor, the geophones are located at the bottom of an OBS, and the hydrophone is placed higher. Newer-generation OBSs send automatic identification system (AIS) signals to facilitate retrieving the seismometer on the sea surface and can record and adjust their tilt while on the sea bottom. Some new-generation instruments also contain compasses to help orient the geophones on the seafloor.

1.4.2. Ocean-Bottom Nodes

Modern industrial applications propose deploying several hundred instruments linked together by a chain, called an ocean-bottom node (OBN) network. OBNs are smaller instruments than OBSs: they do not contain buoys, anchors, or acoustic releases (Fig. 1.8). They are deployed and recovered by remotely operated vehicles, which also reduces the location uncertainty.

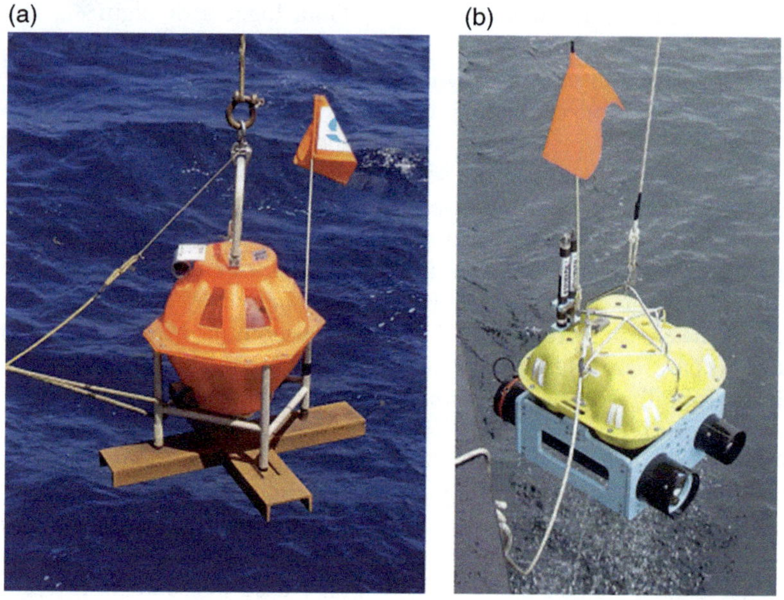

Figure 1.7 (a) Ifremer's LotsOBS. (b) An OBS from the UK Ocean Bottom Instrumentation Facility (OBIF) (Minshull et al., 2005).

Figure 1.8 An ocean-bottom node manufactured by Geospace. Courtesy of Geospace.

1.4.3. Ocean-Bottom Observatories

Semipermanent cabled underwater observatories (e.g., Ocean Observatories Initiative, US; and Ocean Networks Canada, Canada) measure seismic signals via geophones, hydrophones, and, more recently, fiber-optic cables via distributed acoustic sensing (see Section 1.7). Simultaneously, other sensors provide observations of ocean conditions. These long-term recording stations can provide near-real-time observations via cables or acoustical data transfer.

1.4.4. Acoustic Doppler Current Profilers

An acoustic Doppler current profiler (ADCP) is an active acoustic device that transmits sound signals of a constant frequency into the water using a piezoelectric transducer. The transmitted signal reflects from the particles in the water and is received by the same transducer. The frequency of the received signal is modified by the speed of the particles; if a particle is moving toward the transducer, the received frequency is higher than the emitted one, and vice versa. This phenomenon is known as the *Doppler effect*. The resulting *Doppler shift*—the difference between the emitted and received frequencies—is used to calculate the speed of the particles in the water and, therefore, the speed of the currents. ADCPs can be either autonomous and bottom mounted (i.e., deployed on the sea bottom) or ship mounted, in which case the ship speed must be considered in the current speed calculations.

1.4.5. Echosounders

Echosounders are active acoustic tools that transmit acoustic pulses into the water and receive them once reflected from an obstacle. They are used for ranging obstacles (e.g., measuring the seafloor depth or navigating by underwater vessels) and detecting fish, gas bubbles, plankton, or anything else that provides a density contrast with the surrounding water. Common types of echosounders are (i) single-beam echosounders that transmit a sound pulse of a narrow cone (i.e., beam, usually 1–12° wide), (ii) split-beam echosounders that transmit more than one beam (usually four) to position the target (e.g., fish) within the insonified region, and (iii) multibeam echosounders that map the seabed using multiple beams and via beamforming (Simmonds & MacLennan, 2008). Most echosounders can transmit more than one frequency or a range of frequencies to allow their operability in a range of depths, as lower frequencies are less susceptible to attenuation.

1.4.6. Drifters and Floats

Drifters and floats are autonomous, unmoored equipment for measuring ocean currents, temperature, and salinity. They can also host additional instruments to measure barometric pressure, wave height, and wind speed and direction, as well as optical and biochemical sensors and hydrophones. Drifters sample the surface of oceans, whereas floats can sink, so they can acquire vertical profiles within the oceans.

With contributions from more than 30 countries, the international Argo program (named after the mythical Greek ship) has been operating since the early 2000s and has nearly 4,000 floats covering the world's oceans, providing real-time data for oceanographic research. The Argo floats, containing conductivity, temperature, and depth (CTD) sensors, sink to a depth of 1,000 m and drift for 10 days, after which they change buoyancy and sink to 2,000 m, allowing vertical profiling of the ocean (Riser et al., 2016). *Mermaids*, also used by the Argo program, are floats equipped with CTD sensors and a hydrophone to measure the ocean soundscape (Sukhovich et al., 2015). The collected data helps scientists understand how the physical state of the ocean is changing over time.

1.5. COMMON DATA-PROCESSING METHODS

Here, we provide a nonexhaustive overview of common data-processing methods frequently used in seismoacoustics to familiarize the reader with the concepts discussed in the following chapters.

1.5.1. Time-Drift Correction

Since seismologic methods often use the travel time of seismic waves generated by geological, biological, or anthropogenic sources—for example, to locate the events—the seismic amplitudes need to be recorded with accurate timing. To ensure this, seismoacoustic instruments usually contain high-accuracy quartz clocks. Before deploying a network of seismic or acoustic instruments on the seafloor, the instruments are synchronized using GPS (global positioning system) time, with the recording often started at a round minute. The time synchronization with GPS is repeated after the instruments are recovered, again usually at a round minute. At the second time synchronization, the internal clocks of the instruments usually show slight deviations (usually on the order of milliseconds) from the GPS time due to clock drift. Clock drift can be either nonlinear, caused by the temperature difference between the air (while the instruments are on the vessel) and the seawater at the deployment site, or linear, which is a function of the deployment duration. Linear drift is corrected using the total drift calculated after the recovery time synchronization.

1.5.2. Data Reduction

Continuous analog sensor data is filtered to remove all energy above the Nyquist frequency, and band-limited data is then digitized using a given sampling frequency (e.g., 200 Hz: 5 ms time interval between samples). The data is windowed with a finite-length window, which can be either the entire duration of the record or a convenient duration selected by the user (e.g., the entire record duration divided into hourly time series with or without overlap, a fixed time window starting at the origin time of an earthquake, etc.). Data is usually stored as single-station and single-component files, which contain headers describing the file's contents, and the discrete time series, which consists of data points (counts) indexed against time.

1.5.3. Instrument Relocation through Travel-Time Analysis

In marine seismology, instruments are often deployed on the seafloor by free fall from a vessel. The instrument's position on the seafloor may differ from the vessel's, as it may drift due to currents traveling in the water column. The difference between the deployment position (i.e., the position of the vessel) and the seafloor position is a function of the seafloor depth and the currents. Additional drift in the position of the instruments may result from the seafloor topography. The instrument's position on the seafloor is usually estimated by a triangulation-like procedure using the travel times of controlled source shots or pings (e.g., airgun shots) or distant earthquakes with known origin times and positions.

Travel-time analysis is the basis of modern seismology. It consists of identifying a seismic event on the seismoacoustic record and estimating the travel time of this event between where it originated and where it was recorded. Travel times are used for relocating instruments on the seafloor and estimating the location of a seismic source and its origin time. Locating a seismic source and finding its origin time is usually done by solving the inverse problem, the process of inferring the parameters of a system using the observations resulting from that system (Tarantola & Valette, 1982; Tarantola, 2005). Solving the inverse problem requires multiple observations—at least more than the number of unknown parameters. Note that the travel times of different phases are often identified after processing (geophone rotation, DC correction, filtering, etc.) to increase the precision of the travel-time observations.

1.5.4. Rotation for Geophone Reorientation

Due to the free-fall deployment, the orientations of the three geophones of a seismometer on the seafloor are also unknown. To locate noise sources precisely and identify the types of waves (e.g., body waves, surface waves, etc.) via their particle motion, the orientations of the sensors (geophones or accelerometers) are estimated via controlled source shots or distant earthquakes of known location and origin time.

Another common practice in marine seismology is to rotate the data recorded by horizontal geophones into radial and transverse components. The radial plane is the direction in which the compressional P-waves travel. It is oriented along the azimuth between the noise source and the receiver, also called the *sagittal* azimuth. The transverse plane is perpendicular to the radial plane.

1.5.5. Converting from Counts to Physical Units

The seismoacoustic data recorded by hydrophones or geophones/accelerometers must be processed to convert the raw machine counts in volts to a physical unit (pressure change for hydrophones and displacement, velocity, or acceleration for seismometer data). Each recording instrument has its own calibration factor for this conversion. Hydrophone data is converted to sound pressure, which gives the pressure deviations from ambient pressure caused by the sound wave. Sound pressure is expressed in Pascal (Pa). Complementary to the sound pressure is the particle velocity, representing the velocity of the particles when they are displaced by the wave. The particle velocity is derived from geophone measurements and expressed in meters per second (m/s). Sound pressure is often expressed as sound pressure level, which is the effective pressure relative to a reference value and is a logarithmic value measured in decibels (Collett et al.).

1.5.6. Removing the Mean from the Data Set

Processing is applied to the time series to isolate signals from the noise. The long-period variations in the mean amplitude of the data over time are removed by the detrending process, as these slow variations in the mean over time may have significant energy and mask the signals of interest with higher frequencies.

1.5.7. Frequency Spectrum, Spectrogram, and Power Spectral Density

Time series can be analyzed in the time or frequency domain (Figs. 1.3–1.5). The Fourier series breaks down a periodic function into the sum of multiple sinusoidal functions, called components. The Fourier transform (and fast Fourier transform [FFT], a Fourier transform type applied to discrete time series) converts a function of time into a function of frequency called a *frequency spectrum* (Figs. 1.3c and 1.5e). The frequency *spectrum* (plural, *spectra*) of a signal is the range of frequencies contained by the signal, and it is a one-dimensional function of amplitude and frequency. Frequency spectra are sometimes shown with a logarithmic amplitude axis in decibels (dB). A spectrogram shows the variations in the frequency content of a signal over time. It is a two-dimensional function of amplitude (usually represented with a color scale), frequency, and time. For most frequency resolutions (sampling rate divided by the signal length), the frequency spectrum accurately represents the periodic frequencies contained within the signal. However, the amplitude of the random (nonperiodic) signals within the signal varies with the frequency resolution. Power spectral density is the measure of the signal's power versus frequency, and it allows the comparison of random components of the signal independent of the frequency resolution.

1.5.8. Frequency Filtering

Filtering can be applied to enhance the spectral power of chosen frequencies and remove the rest. It involves convolving the data set with a second, usually shorter time series. Convolution is a mathematical operation on two functions, and the resulting function defines how the shape of one part is modified by the other. For example, in a musical concert, the reverberations of the concert hall caused by the music are a convolution and, therefore, a filtering. Similarly, seismometers convolve the ground motion with their internal reverberations, which are the impulse response of the seismometer. If this impulse response is known, it can be inverted to an inverse filter and then convolved with the data set to remove its signature and acquire the original signal. This process is known as *deconvolution*. Time-series data can be represented in time or frequency domains, and filtering can be applied in both. *Filtering* is a convolution in the time domain, corresponding to *multiplication* in the frequency domain.

Aside from deconvolution, common filtering operations consist of high-pass, low-pass, or band-pass filters. High-pass (or low-cut) filters attenuate the energy of the frequencies below a user-set cut-off frequency, whereas low-pass filters do the opposite and attenuate the frequencies higher than the cut-off frequency. Band-pass filters attenuate the frequencies outside a user-defined frequency range called the *pass band.* A common application of high-pass filters in marine seismoacoustic data is to remove swell noise, which has low frequencies (2–10 Hz) and high amplitudes.

Note that filtering may create phase shifts at some frequencies. A zero-phase filter is one for which the phase shift is equal to zero at all frequencies. Minimum-phase filters, on the other hand, focus the energy of the impulse response toward the beginning of the signal. Minimum-phase filters are generally used to identify the first arrival phases (e.g., P-wave arrival of an earthquake), whereas zero-phase filters are used to enhance the later phases (e.g., S-wave arrival of an earthquake).

1.5.9. Polarization Analysis

Polarization analysis quantitatively describes the particle motion of a seismic wavefield. It is the fundamental vector-processing technique applied to multicomponent seismic data. It is a single-trace time-domain method carried out using *hodograms*, which are graphs or curves that display time versus distance of motion: for example, a crossplot of two components of particle motion over a time window. Hodograms are used in borehole seismology to determine the arrival directions of waves and detect shear-wave splitting. Data recorded along two geophone axes can be displayed as a function of time. Polarization analysis can facilitate extracting pure P- and S-wave sections, removing unwanted noise events, and recovering information relating to fracturing, porosity, and lithology from multicomponent data (Hendrick & Hearn, 1999).

1.6. SUMMARY OF CHAPTERS

Following this chapter, in Chapter 2, Chang and colleagues introduce nonearthquake-related noise sources in Taiwan. They conclude that the dynamic sea is the primary noise generator, with regular secondary short-period microseism generation. Longer-period primary microseisms are also observed in shallow water due to ocean storms and typhoons. Additionally, ambient noise is generated in deep-water environments by advection from internal waves that propagate at the ocean bottom. Gas hydrate-related SDEs are also recorded with frequencies of 4 and 8 Hz. The authors relate the SDEs to seafloor fluid emission. Similarly, in Chapter 3, Lin and colleagues explore the use of microseismic noise of 0.04–1 Hz frequency recorded on land stations in Taiwan to derive significant wave height.

In Chapter 4, Aiken gives a review of signals generated by submarine volcanoes and classifies them into high-frequency events (>5 Hz) related to volcano tectonics; low-frequency events, which can have very long periods (0.01–0.5 Hz) or long periods; and volcanic tremors, which are successions of long-period events of 1–9 Hz. This is followed by an introduction to the methodologies available for monitoring or analyzing these signals.

In Chapter 5, Clare and colleagues give an overview of geohazards related to submarine landslides, introduce the challenges of recording them using small-scale geophysical surveys, and present alternative passive acoustic monitoring techniques.

In Chapter 6, Schlindwein reviews studies of iceberg-generated noise recorded in the oceans and gives an overview of signal characteristics, their spatial and temporal distribution, and source mechanisms.

In Chapter 7, Smith and Barclay review the passive sensing of sound generated by hydrothermal vents. They summarize the challenges in interpreting the sound recorded around hydrothermal vents and recommend future field experiments.

In Chapter 8, Tary presents the characteristics of SDEs and compares them to more common low-magnitude earthquakes (microearthquakes, also called microseismicity in the chapter). The chapter provides a comprehensive literature review on the locations and tectonic settings where SDEs are observed, summarizes different origins proposed by various authors, and suggests future directions for SDE studies. Similarly, in Chapter 9, Bayrakci and colleagues provide a case study of SDEs recorded at an active gas vent at Scanner pockmark, North Sea. They discuss the spatial and temporal distribution of SDEs and point out the correlation between SDE occurrence and tidal height. They suggest a hydrostatic control on the initiation of these events.

In Chapter 10, Roche and colleagues discuss bubble formation within liquids, including at underwater gas seeps, rainfall, and breaking waves.

In Chapter 11, White and colleagues give an overview of the sounds produced by baleen whales and other frequently recorded sounds of biological origin and suggest that seismoacoustic instrument deployments for geophysical purposes are a useful source of information on the behavioral studies of cetaceans. They also suggest that cetacean vocalizations may offer a route to obtain information on the subsurface structure. Similarly, in Chapter 12, Fisher and colleagues present a case study of automatic identification of whale vocalizations (~20 Hz) on an array of 72 ocean-bottom seismometers/hydrophones offshore Spain, deployed for geophysical purposes, and use this data set for tracking fin whale movements. They identify six whales with swim speeds ranging from 2 to 15 km/hr over 1.5 to 15 km.

In Chapter 13, Dellong and colleagues review the marine noise produced by shipping activities in two categories: the characterization of signals from individual ships useful for the naval industry and, more globally, the contribution of shipping noise to the ocean soundscape, as part of environmental impact assessments and regulatory frameworks.

In Chapter 14, Yi and colleagues present a signal-processing chain to detect and track underwater vessels using passive sonar systems.

Finally, in Chapter 15, Klingelhoefer and Bayrakci present the conclusions that can be drawn from this book and the impact of different marine noise on the environment. They discuss the shortcomings of ocean noise research and future research directions on the subject.

1.7. FUTURE DEVELOPMENTS OF ACOUSTIC MEASUREMENTS IN THE OCEAN

Most of the existing studies presented in this book are based on recordings of ocean-bottom seismometers deployed temporarily at the seafloor. These measurements mostly provide limited data coverage in space and time. However, only long-term observations give information about the seasonality of noise levels in the sea. Permanently installed and cabled observatories, such as the Nemo-SN1 observatory in the Gulf of Catania (NEutrino Mediterranean Observatory–Submarine Network 1), record the noise level over periods of several months (Viola et al., 2017). Nonseismic signals monitored by the observatory include volcanic tremors and landslides (Sgroi et al., 2014). By comparing to recordings from AIS of passing ships, noise levels can be predicted for future ship routes (Viola et al., 2017). Even airplane noise from airports near the coast can be recorded by submarine observatories (Erbe et al., 2018). On a smaller scale, variations in signals from gas emissions from cold seeps were monitored for a year by the NEPTUNE-cabled observatory offshore Vancouver Island, British Columbia (Römer et al., 2016). Images taken by a cabled observatory compared to acoustic recordings allow improved constraints on the origin of short events commonly recorded in the marine environment (Batsi et al., 2019). Although more cabled observatories will hopefully be installed in the near future, their cost is the primary limiting factor for dense deployment in the global ocean.

An emerging technique called *distributed acoustic sensing* (DAS) has been developed, which allows fiber-optic cables, commonly employed by the telecommunication industry, to be used as a densely spaced geophone array. This technology uses an optoelectronic interrogator that repeatedly sends a laser pulse through a fiber-optic cable and measures optical phase changes in Rayleigh backscattered light. The changes in the backscatter are proportional to the strain the fiber is experiencing. Using this technique, seismic and other signals can be recorded up to several tens of kilometers along a fiber-optic cable. At first, this technique was primarily used to record seismic signals in wells and monitor reservoirs of the hydrocarbon industry (Mateeva et al., 2014). Since then, it has been shown that existing telecommunication optical fiber cables can detect earthquakes (Marra et al., 2018). The rapidly growing total length of submarine telecommunication cables increases the potential for their use as seismological networks. Several recent studies used DAS techniques in marine telecommunication cables (Lindsey et al., 2019; Sladen et al., 2019; Williams et al., 2019). The 2018 M_w 8.2 Fiji earthquake was recorded on a buried cable in the North Sea using DAS measurements (Williams et al., 2019). Regional seismicity and ocean surface gravity waves were detected on a cable deployed offshore Toulon, France (Sladen et al., 2019). Along the same cable, a study of ocean noise compared to AIS recordings showed the potential of DAS measurements to record ship noise (Rivet et al., 2021). The interrogation of a cable of the Monterey Accelerated Research System with DAS showed noise mostly from ocean surface waves and microseisms (Lindsey et al., 2019). However, the large data quantities and the scarcity of DAS instrumentation in the scientific community have prevented the large-scale use and testing of DAS recording in the marine environment today.

REFERENCES

Agnew, D. C., Lee, W., Kanamori, H., Jennings, P., & Kisslinger, C. (2002). History of seismology. *International Handbook of Earthquake and Engineering Seismology*, *81*(A), 3–11.

Aki, K. (1972). Earthquake mechanism. *Tectonophysics*, *13*(1–4), 423–446.

Arciniega-Ceballos, A., Chouet, B. A., & Dawson, P. (1999). Very long-period signals associated with vulcanian explosions at Popocatépetl volcano, Mexico. *Geophysical Research Letters*, *26*(19), 3013–3016.

Ardhuin, F., Gualtieri, L., & Stutzmann, E. (2015). How ocean waves rock the Earth: Two mechanisms explain microseisms with periods 3 to 300 s. *Geophysical Research Letters*, *42*(3), 765–772.

Au, W. W., & Hastings, M. C. (2008). *Principles of marine bioacoustics* (Vol. *510*). Springer.

Barker, P., & Lepper, P. A. (2012). Development of a versatile platform for long-term underwater acoustic monitoring. In *Proceedings of Meetings on Acoustics ECUA2012*. Acoustical Society of America.

Batsi, E., Tsang-Hin-Sun, E., Klingelhoefer, F., Bayrakci, G., Chang, E. T., Lin, J. Y., et al. (2019). Nonseismic signals in the ocean: Indicators of deep sea and seafloor processes on ocean-bottom seismometer data. *Geochemistry, Geophysics, Geosystems*, *20*(8), 3882–3900.

Bayrakci, G., Scalabrin, C., Dupré, S., Leblond, I., Tary, J.-B., Lanteri, N., et al. (2014). Acoustic monitoring of gas emissions from the seafloor. Part II: a case study from the Sea of Marmara. *Marine Geophysical Research*, *35*(3), 211–229.

Berndt, C. (2005). Focused fluid flow in passive continental margins. *Philosophical Transactions of the Royal Society A: Mathematical, Physical and Engineering Sciences, 363*(1837), 2855–2871.

Bertin, X., de Bakker, A., Van Dongeren, A., Coco, G., André, G., Ardhuin, F., et al. (2018). Infragravity waves: From driving mechanisms to impacts. *Earth-Science Reviews, 177*, 774–799.

Bialas, J., & Flueh, E. R. (1999). Ocean bottom seismometers. *Sea Technology, 40*(4), 41–46.

Biot, M. A. (1956). Theory of propagation of elastic waves in a fluid-saturated porous solid. II. Higher frequency range. *The Journal of the Acoustical Society of America, 28*, 179–191.

Bittle, M., & Duncan, A. (2013). A review of current marine mammal detection and classification algorithms for use in automated passive acoustic monitoring. Paper presented at Proceedings of Acoustics 2013, Victor Harbor, Australia.

Bromirski, P. D. (2009). Earth vibrations. *Science, 324*(5930), 1026–1027.

Buehler, J. S., & Shearer, P. M. (2015). T phase observations in global seismogram stacks. *Geophysical Research Letters, 42*(16), 6607–6613.

Buskirk, R. E., Frohlich, C., Latham, G. V., Chen, A. T., & Lawton, J. (1981). Evidence that biological activity affects ocean bottom seismograph recordings. *Marine Geophysical Researches, 5*, 189–205.

Caplan-Auerbach, J., Fox, C. G., & Duennebier, F. K. (2001). Hydroacoustic detection of submarine landslides on Kilauea volcano. *Geophysical Research Letters, 28*(9), 1811–1813.

Chouet, B. (2022). Volcanoes, non-linear processes. In Tilling, R. I. (Ed.), *Complexity in tsunamis, volcanoes, and their hazards* (pp. 469–502). Springer.

Chouet, B. A., & Matoza, R. S. (2013). A multi-decadal view of seismic methods for detecting precursors of magma movement and eruption. *Journal of Volcanology and Geothermal Research, 252*, 108–175.

Clemens, S. C., Kuhnt, W., LeVay, L. J., & Expedition Scientists (2014). Indian monsoon rainfall. *Proceedings of the International Ocean Discovery Program, 353*.

Collett, T., Bahk, J.-J., Frye, M., Goldberg, D., Husebo, J., Koh, C., et al. (2013). *Historical methane hydrate project review*. United States Geological Survey.

Crampin, S. (1985). Evaluation of anisotropy by shear-wave splitting. *Geophysics, 50*, 142–152.

Crone, T. J., Wilcock, W. S., Barclay, A. H., & Parsons, J. D. (2006). The sound generated by mid-ocean ridge black smoker hydrothermal vents. *PLoS One, 1*, e133.

Deane, G. B., & Stokes, M. D. (2002). Scale dependence of bubble creation mechanisms in breaking waves. *Nature, 418*(6900), 839–844.

Diaz, J., Gallart, J., & Gaspà, O. (2007). Atypical seismic signals at the Galicia Margin, North Atlantic Ocean, related to the resonance of subsurface fluid-filled cracks. *Tectonophysics, 433*(1–4), 1–13.

Duarte, C. M., Chapuis, L., Collin, S. P., Costa, D. P., Devassy, R. P., Eguiluz, V. M., et al. (2021). The soundscape of the Anthropocene ocean. *Science, 371*(6529), eaba4658.

Erbe, C., Williams, R., Parsons, M., Parsons, S. K., Hendrawan, I. G., & Dewantama, I. M. I. (2018). Underwater noise from airplanes: an overlooked source of ocean noise. *Marine Pollution Bulletin, 137*, 656–661.

Foix, O., Aiken, C., Saurel, J.-M., Feuillet, N., Jorry, S. J., Rinnert, E., & Thinon, I. (2021). Offshore Mayotte volcanic plumbing revealed by local passive tomography. *Journal of Volcanology and Geothermal Research, 420*, 107395.

Früh-Green, G. L., Kelley, D. S., Bernasconi, S. M., Karson, J. A., Ludwig, K. A., Butterfield, D. A., et al. (2003). 30,000 years of hydrothermal activity at the Lost City vent field. *Science, 301*(5632), 495–498.

Garrett, C., & Munk, W. (1979). Internal waves in the ocean. *Annual review of fluid mechanics, 11*, 339–369.

Hendrick, N., & Hearn, S. (1999). Polarisation analysis: what is it? Why do you need it? How do you do it? *Exploration Geophysics, 30*(4), 177–190.

Judd, A., & Hovland, M. (2009). *Seabed fluid flow: the impact on geology, biology and the marine environment*. Cambridge University Press.

Korneev, V., Goloshubin, G., Kashtan, B., Troyan, V., Shigapov, R., Bakulin, A., et al. (2012). Krauklis wave-half a century after. Paper presented at Saint Petersburg 2012–5th Saint Petersburg international conference & exhibition 2012: making the most of the earth's resources.

Korneev, V. A. (2011). Krauklis wave in a stack of alternating fluid-elastic layers. *Geophysics, 76*(6), N47–N53.

Lamb, H. (1881). On the vibrations of an elastic sphere. *Proceedings of the London Mathematical Society, 1*, 189–212.

Lamb, H. (1917). On waves in an elastic plate. *Proceedings of the Royal Society of London. Series A, Containing Papers of a Mathematical and Physical Character, 93*(648), 114–128.

Leighton, T. (1994). The freely-oscillating bubble. In *The acoustic bubble* (pp. 129–286). Academic Press.

Leighton, T. (2012). *The acoustic bubble*. Academic Press.

Lindsey, N. J., Dawe, T. C., & Ajo-Franklin, J. B. (2019). Illuminating seafloor faults and ocean dynamics with dark fiber distributed acoustic sensing. *Science, 366*(6469), 1103–1107.

Ma, B. B., & Nystuen, J. A. (2005). Passive acoustic detection and measurement of rainfall at sea. *Journal of Atmospheric and Oceanic Technology, 22*(8), 1225–1248.

Marra, G., Clivati, C., Luckett, R., Tampellini, A., Kronjäger, J., Wright, L., et al. (2018). Ultrastable laser interferometry for earthquake detection with terrestrial and submarine cables. *Science, 361*(6401), 486–490.

Martin, W., Baross, J., Kelley, D., & Russell, M. J. (2008). Hydrothermal vents and the origin of life. *Nature Reviews Microbiology, 6*, 805–814.

Mateeva, A., Lopez, J., Potters, H., Mestayer, J., Cox, B., Kiyashchenko, D., et al. (2014). Distributed acoustic sensing for reservoir monitoring with vertical seismic profiling. *Geophysical Prospecting, 62*(4), 679–692.

Minnaert, M. (1933). On musical air-bubbles and the sounds of running water. *The London, Edinburgh, and Dublin Philosophical Magazine and Journal of Science, 16*(104), 235–248.

Minshull, T., Sinha, M., & Peirce, C. (2005). Multi-disciplinary, sub-seabed geophysical imaging. *Sea Technology, 46*(10), 27–31.

Ostrovsky, A. (1989). On the nature of microshocks recorded by ocean bottom seismographs. *Marine Geophysical Researches, 11*, 113–118.

Pecher, I. A., Barnes, P. M., LeVay, L. J., & Edpedition Scientists (2019). Creeping gas hydrate slides. *Proceedings of the International Ocean Discovery Program, 372A.*

Pontoise, B., & Hello, Y. (2002). Monochromatic infra-sound waves recorded offshore Ecuador: possible evidence of methane release. *Terra Nova, 14*(6), 425–435.

Rayleigh, L. (1885). On waves propagated along the plane surface of an elastic solid. *Proceedings of the London Mathematical Society, 1*, pp. 4–11.

Riedel, M., Collett, T. S., J., M. M., & Expedition Scientists (2006). Stages of gas hydrate evolution on the Northern Cascadia margin. *Proceedings of the Integrated Ocean Drilling Program, 311.*

Riser, S. C., Freeland, H. J., Roemmich, D., Wijffels, S., Troisi, A., Belbéoch, M., et al. (2016). Fifteen years of ocean observations with the global Argo array. *Nature Climate Change, 6*, 145–153.

Rivet, D., de Cacqueray, B., Sladen, A., Roques, A., & Calbris, G. (2021). Preliminary assessment of ship detection and trajectory evaluation using distributed acoustic sensing on an optical fiber telecom cable. *The Journal of the Acoustical Society of America, 149*(4), 2615–2627.

Römer, M., Riedel, M., Scherwath, M., Heesemann, M., & Spence, G. D. (2016). Tidally controlled gas bubble emissions: A comprehensive study using long-term monitoring data from the NEPTUNE cabled observatory offshore Vancouver Island. *Geochemistry, Geophysics, Geosystems, 17*(9), 3797–3814.

Saurel, J.-M., Jacques, E., Aiken, C., Lemoine, A., Retailleau, L., Lavayssière, A., et al. (2021). Mayotte seismic crisis: building knowledge in near real-time by combining land and ocean-bottom seismometers, first results. *Geophysical Journal International, 228*, 1281–1293.

Sgroi, T., Monna, S., Embriaco, D., Giovanetti, G., Marinaro, G., & Favali, P. (2014). Geohazards in the western Ionian Sea: Insights from non-earthquake signals recorded by the NEMO-SN1 seafloor observatory. *Oceanography, 27*, 154–166.

Sheriff, R. E., & Geldart, L. P. (1995). *Exploration seismology.* Cambridge University Press.

Simmonds, J., & MacLennan, D. N. (2008). *Fisheries acoustics: theory and practice.* John Wiley & Sons.

Sladen, A., Rivet, D., Ampuero, J. P., De Barros, L., Hello, Y., Calbris, G., & Lamare, P. (2019). Distributed sensing of earthquakes and ocean-solid Earth interactions on seafloor telecom cables. *Nature Communications, 10*, 5777.

Sleeswyk, A. W., & Sivin, N. (1983). Dragons and toads. The Chinese seismoscope of AD 132. *Chinese Science*, 1–19.

Sohn, R. A., Hildebrand, J. A., Webb, S. C., & Fox, C. G. (1995). Hydrothermal microseismicity at the megaplume site on the southern Juan de Fuca Ridge. *Bulletin of the Seismological Society of America, 85*(3), 775–786.

Stoneley, R. (1924). Elastic waves at the surface of separation of two solids. *Proceedings of the Royal Society of London. Series A, Containing Papers of a Mathematical and Physical Character, 106*(738), 416–428.

Sukhovich, A., Bonnieux, S., Hello, Y., Irisson, J.-O., Simons, F. J., & Nolet, G. (2015). Seismic monitoring in the oceans by autonomous floats. *Nature Communications, 6*(1), 8027.

Sultan, N., Riboulot, V., Ker, S., Marsset, B., Geli, L., Tary, J.-B., et al. (2011). Dynamics of fault-fluid-hydrate system around a shale-cored anticline in deepwater Nigeria. *Journal of Geophysical Research: Solid Earth, 116*(B12).

Takeuchi, H., & Saito, M. (1972). Seismic surface waves. *Methods in Computational Physics, 11*, 217–295.

Tarantola, A. (2005). *Inverse problem theory and methods for model parameter estimation.* SIAM.

Tarantola, A., & Valette, B. (1982). Inverse problems = quest for information. *Journal of Geophysics, 50*, 159–170.

Tary, J.-B., Geli, L., Guennou, C., Henry, P., Sultan, N., Çağatay, N., & Vidal, V. (2012). Microevents produced by gas migration and expulsion at the seabed: a study based on sea bottom recordings from the Sea of Marmara. *Geophysical Journal International, 190*, 993–1007.

Vidale, J. E. (1986). Complex polarization analysis of particle motion. *Bulletin of the Seismological Society of America, 76*, 1393–1405.

Viola, S., Grammauta, R., Sciacca, V., Bellia, G., Beranzoli, L., Buscaino, G., et al. (2017). Continuous monitoring of noise levels in the Gulf of Catania (Ionian Sea). Study of correlation with ship traffic. *Marine Pollution Bulletin, 121*(1–2), 97–103.

Weatherall, P., Marks, K. M., Jakobsson, M., Schmitt, T., Tani, S., Arndt, J. E., et al. (2015). A new digital bathymetric model of the world's oceans. *Earth and Space Science, 2*(8), 331–345.

Webb, S. C., Zhang, X., & Crawford, W. (1991). Infragravity waves in the deep ocean. *Journal of Geophysical Research: Oceans, 96*(C2), pp. 2723–2736.

Williams, E. F., Fernández-Ruiz, M. R., Magalhaes, R., Vanthillo, R., Zhan, Z., González-Herráez, M., & Martins, H. F. (2019). Distributed sensing of microseisms and teleseisms with submarine dark fibers. *Nature Communications, 10*, 5778.

Williams, R., Wright, A. J., Ashe, E., Blight, L. K., Bruintjes, R., Canessa, R., et al. (2015). Impacts of anthropogenic noise on marine life: Publication patterns, new discoveries, and future directions in research and management. *Ocean & Coastal Management, 115*, 17–24.

Zimmer, W. M. (2011). *Passive acoustic monitoring of cetaceans.* Cambridge University Press.

2

Seismic Ambient Noise: Application to Taiwanese Data

Emmy T.Y. Chang[1], Yung-Cheng Gung[2], and Ying-Nien Chen[3]

ABSTRACT

This chapter explores the various nonearthquake noises recorded at Taiwan's seismometric stations, utilizing land-based and ocean-bottom observations to examine their frequency and generation mechanisms. The primary source of microseisms for the Taiwan area is the dynamic sea, with short-period microseisms regularly originating from the Taiwan Strait. Moreover, annual intense ocean storms expand the ambient noise's frequency range to include longer-period microseisms. These longer-period microseisms are typically generated in shallow water and can be attributed to energetic events like typhoons and ocean swells. In addition to shallow-water noise, internal waves generate noise detected by deep-sea seismic stations (ocean-bottom seismometers [OBSs]). OBS records from southeast Taiwan reveal water advection caused by deep-sea internal waves, affecting the seafloor stations. Furthermore, OBSs in gas-hydrate-rich areas record burst-type short-duration events (SDEs) characterized by rapid vibrations (~4–8 Hz) and exponential decay, indicating seafloor fluid emissions. This chapter quantifies SDEs to pinpoint seafloor fluid conduits' locations. In summary, the research highlights that ambient seismic noise is not solely linked to ocean surface waves but also stems from geological processes and converted ocean waves in the deep sea. This information is valuable for probing deep-sea environments using ambient seismic noise as a proxy.

2.1. INTRODUCTION

In seismology, "noises" in seismometric records generally mean signals that are not earthquake-related. Recent studies with seismic noise have focused on seismic phase-based analyses. It has further extended the discussion to the seismic structure of the interior Earth (e.g., Huang et al., 2015; Ritzwoller & Levshin, 1998; Sabra et al., 2005; Shapiro et al., 2005) and the Earth's surface environment (e.g., Brooks et al., 2009; Bromirski et al., 2013; Tanimoto, 2007; Hilmo & Wilcock, 2020; Nishida, 2017; Webb, 1998). Ambient seismic noise has been deduced to be the nonlinear coupling of atmospheric and ocean conditions to the solid Earth to perform the energy partition among seismic wave types (e.g., Nishida, 2017; Webb, 1998).

Of all ambient seismic noise, microseisms, which are excited by oceanic gravity waves at the sea surface, are among the most energetic components of noise for seafloor and coastal site seismometers (Bromirski, 2001; Dorman et al., 1993; Friedrich et al., 1998; Rhie & Romanowicz, 2004). Two major spectral categories are analyzed with the noise driven from wind and

[1]*Institute of Oceanography, National Taiwan University Taipei, Taiwan (R.O.C.)*
[2]*Department of Geosciences, National Taiwan University, Taipei, Taiwan (R.O.C.)*
[3]*Department of Earth and Environmental Sciences, National Chung Cheng University, Chiayi, Taiwan (R.O.C.)*

Noisy Oceans: Monitoring Seismic and Acoustic Signals in the Marine Environment, Geophysical Monograph 284, First Edition.
Edited by Gaye Bayrakci and Frauke Klingelhoefer.

DOI:10.1002/9781119750925.ch02

waves: (1) the direct coupling between ocean waves and topography, which is responsible for primary microseisms (PMs) in a frequency range of 0.05 to ~0.1 Hz; and (2) nonlinear coupling to elastic waves, which is responsible for secondary microseisms (SMs) in a frequency range of approximately 0.1 to ~0.5 Hz (Hasselmann, 1963; Longuet-Higgins, 1950). The depth-dependent pressure fluctuations generate PMs; therefore, PMs are generally sourced from coastal or shallow-water areas. SMs are converted from nonlinear interactions between ocean waves with nearly the same frequency and travel in opposite directions of standing ocean swells at the sea surface (Longuet-Higgins, 1950). Therefore, SMs are depth-independent and sourced at the sea surface in both pelagic and coastal regions. Ocean swells generate relatively long-period SMs (LPSMs, typically 0.1–0.2 Hz), and the local sea winds primarily respond to short-period SMs (SPSMs, typically 0.2–0.4 Hz) (Stephen et al., 2003).

Below the frequency of PMs and SMs, pressure fluctuations caused by nonlinear interactions of long-period ocean swells and atmospheric disturbances can generate long-period infragravity waves (<0.05 Hz) and "earth hum" (<0.01 Hz, Webb, 2007). These are the surf beat and low-frequency resonance oscillation as the background noise of the Earth. However, all origins of this seismic ocean noise correspond to wind-wave climatic conditions. Intense storms such as extra-tropical storms and typhoons (i.e., tropical cyclones) are the strong driving force for the excitation of all types of ambient seismic noise (Gutenberg, 1931; Bromirski & Gerstoft, 2009).

In this chapter, we demonstrate the ambient seismic ocean noise detected in the Taiwan area. Located at the western edge of the Pacific Ocean, Taiwan island is on the pathway of typhoons and at the western boundary of the Kuroshio Current (Fig. 2.1). The dynamics of the ocean significantly affect the background level of the ambient seismic noise around Taiwan. Furthermore, seafloor bathymetry abruptly changes from the deep ocean basin to the east to the shallow continental shelf to the west of Taiwan (Fig. 2.1). Therefore, the predominant frequency content of ambient noise varies for eastern and western offshore Taiwan.

Such a sharp change in bathymetry surrounding Taiwan is due to the plate convergence of two subduction systems that join at the Taiwan island: the Philippine Sea plate moves northwest against the southeast China continental shelf, whereas the South China Sea, which is a subplate plate moving with southeast mainland China, subducts eastward in southern offshore Taiwan at the Manila trench (e.g., Tsai et al., 1977). The active plate subduction and collision in and surrounding Taiwan generate significant seismic activity. Therefore, in addition to the effect of tropical storms, Taiwan is prone to earthquakes. A dense seismic network has been installed throughout Taiwan since the 1970s (Tsai et al., 1977; Wang & Shin, 1998), and the use of OBSs began in the 2000s (Chang et al., 2008; Lin et al., 2010).

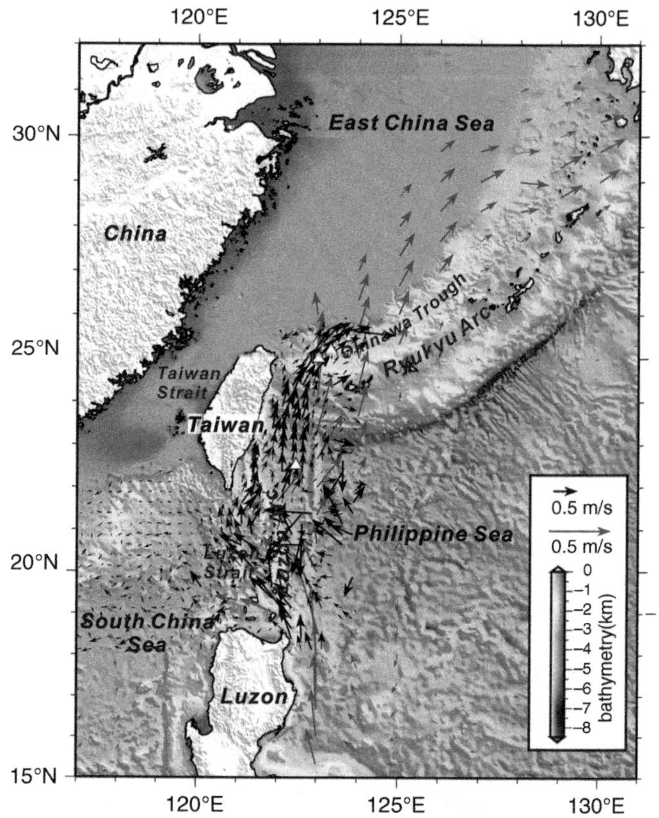

Figure 2.1 Seafloor topography around Taiwan. The black and gray arrows show the mean Kuroshio Current speeds in water depths of 200 and 750 m. The three white triangles in eastern offshore Taiwan are OBS stations. The data is discussed in Section 2.2. From the Ocean Data Bank of the Ministry of Science and Technology, Taiwan, www.odb.ntu.edu.tw.

This chapter reviews the sparse literature on the origins and predominant patterns of ambient seismic noise with Taiwanese data. The energy levels of PMs and SMs are discussed with island-based and ocean-bottom seismic observations. Using seismic interferometry, which forms the cross-correlation functions (CCFs) with the paired stations, Chen et al. (2011) identified the directivity of asymmetric noise propagation with the land stations of Taiwan. Their study proceeded with energy migration to image the topographic effect in water-topography coupling surrounding Taiwan. In addition to microseisms caused by the dynamic sea state, OBSs deployed in the Taiwan area detect deep-sea internal waves created from the combined effect of local ocean tides and kinematic energy of the Kuroshio Current (Chang et al., 2016). The recorded signals were the to-and-fro motion of the ambient ocean water. In addition to the seismic waves converted from the coupling between the ocean and solid Earth, OBSs deployed southwest offshore Taiwan, along the continental slope in the northern South China Sea, systematically record burst-type SDEs (usually less than 2 seconds). Their waveforms exhibit a monotone vibration with rapid exponential decay. Based on recent studies (e.g., Batsi et al., 2019; Géli et al., 2018; Tary et al., 2012), such signals are associated with gas emissions at the seafloor and marine natural resources.

The generators of the ambient seismic noise are multiple and compounded. Moreover, the ambient seismic energy can be converted from different wave types via transport over different media. Therefore, we consider that the study of ambient seismic noise can advance the comprehension of the sea state in an entire water regime and provide a tool to probe the marine geological environment. This chapter is a compendium of ambient seismic noise sources resolved for the offshore area of Taiwan. It is intended to serve as a reference to interpret OBS background signals.

2.2. BACKGROUND AMBIENT SEISMIC NOISE IN TAIWAN

Similar to other observations of anthropogenic seismic noise (e.g., McNamara & Buland, 2004), significant cultural noise at high frequencies (>1 Hz) is detected in near-populated areas in Taiwan, which shows diurnal and weekly variations as expected (Liang et al., 2008). In addition to cultural noises sourced from human activity, microseisms related to the dynamic ocean are significant for the Taiwan area. Ocean-surface agitation driven by wind and waves has long been accepted as a major source of microseisms on land and sea (Knudsen et al., 1948; Longuet-Higgins, 1950). Taiwan is surrounded by the tropical-to-subtropical ocean regime, and oceanic weather conditions make the background noise level much higher than the global average, particularly for periods less than ~8 seconds (>0.125 Hz) (Lin et al., 2010).

Undoubtedly, the average noise level at the central mountain stations is the quietest in Taiwan at all frequencies. In Liang et al. (2008), estimates of power spectral density with island-based broadband stations (Broadband Array in Taiwan for Seismology) and mobile cross-island seismic arrays in the TAIGER project (the Taiwan-US cooperative research project Taiwan Integrated GEodynamic Research, from April 2006 to April 2008) were employed to quantify ocean-generated noise. Their results showed that SMs with peak spectra at ~0.2 Hz dominated ocean noise, and the western coastal plain had ~20 dB higher noise than eastern Taiwan. Seismicity in Taiwan is much more active along the east coast than the west (e.g., Wu & Rau, 1998). Hence, the transient energy release of earthquakes contributes little to the background noise levels at the SM frequency band.

Based on the analysis of coherent ambient noise, Chen et al. (2011) investigated the mechanism that caused a higher SM level on the west coast of Taiwan. Chen et al. conducted CCFs from 86 pairs of vertical components (Z-Z) and 79 pairs of transverse components (T-T) with coastal and selected mountain stations. The primary energy of the resulting CCFs fell within the period range of approximately 2–6 seconds, or short-period secondary microseisms (SPSMs). In a fully diffusive wave field, the causal (time-positive portion) and acausal (time-negative portion) signals of the derived CCFs are identical in both phases and amplitudes (e.g., Shapiro & Campillo, 2004; Snieder, 2004; Weaver & Lobkis, 2001). However, the asymmetric energy response at the causal and acausal time lags was most reported, which reveals inhomogeneity in the spatial distribution of noise sources (e.g., Brzak et al., 2009; Gu et al., 2007). The amplitude asymmetry of the CCFs illustrated in Chen et al. (2011) indicates the landward transportation of SPSM energy from coastal to inland stations (Fig. 2.2). Furthermore, the SPSM excitations from the Taiwan Strait and northern offshore Taiwan were more substantial than those from the eastern and southern coasts. The SPSM intensity resulting from noise source migration (method: Brzak et al., 2009) illustrates that the strength of SPSM is linked to offshore bathymetric features, and the shallow water of the continental shelf is an excellent condition for SPSM excitation (Fig. 2.2). Furthermore, the temporal variation in the CCF strength exhibits correlations with wind speed and wave height, which explicitly indicates that the observed SPSM is generated by local wind-driven ocean waves around offshore Taiwan.

Seismic noise levels between ocean-bottom and island-based stations were compared with the broadband instruments by Lin et al. (2010). The Taiwan broadband stations (BATS, http://bats.earth.sinica.tw) near the coastline of eastern Taiwan and two F-net stations are employed for comparison with the OBS stations in the Okinawa trough (at a water depth of ~2000 m) and Huatung basin in eastern offshore Taiwan (~4700 m water depth) (see OBS locations in Fig. 2.1) in their study. The results show that while the spectrum patterns were similar for the ocean-bottom and land stations, the ocean environment was noisier than the land environment for most frequencies. The noise notch between high-frequency microseisms and infragravity waves is located in the frequency range of 0.02–0.05 Hz for the vertical components of OBSs, but it shifts to lower frequencies for the land stations. Based on power spectral density (PSD) analysis from the selected stations in Lin et al. (2010), the predominant noise level for both land-based and ocean-bottom observations was at the frequency

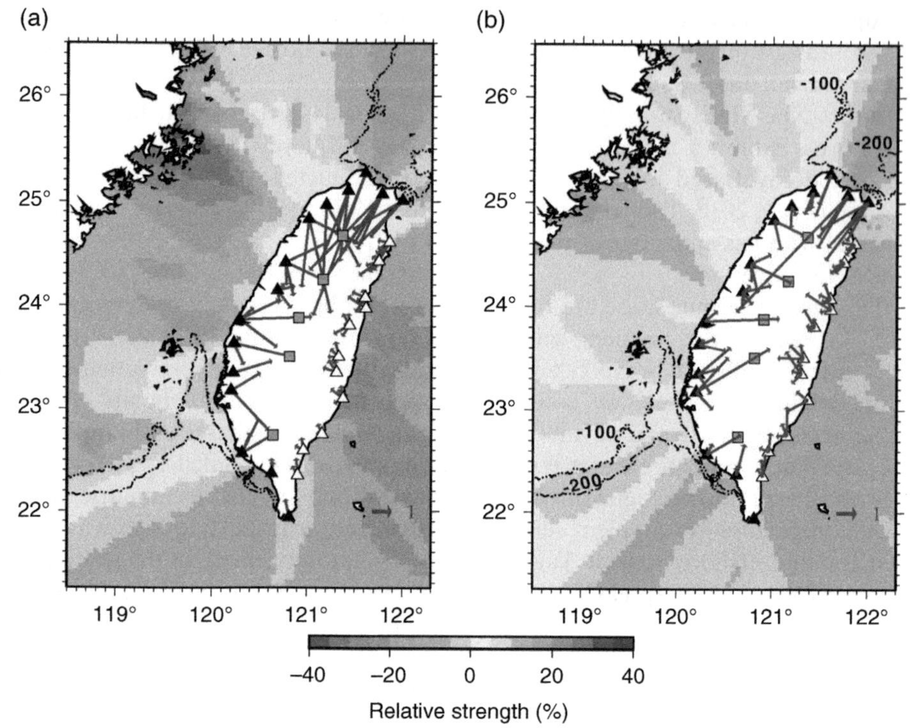

Figure 2.2 Results of migration imaging of SPSM strengths and relative peak amplitudes of annual CCF stacks for the (a) Z-Z and (b) T-T components. The relative strengths of CCFs are shown as vectors on the locations of coastal stations; the vector lengths are proportional to the amplitude ratios (scale shown in the lower-right corner of each panel), and their directions indicate the energy flow along the line of station pairs (see text for details). The results of migration images are expressed as perturbations with respect to the regional average. The dash contour lines indicate depths of –100 and –200 meters, which approximate the break of the continental shelf from shallow water to the deep sea. Adapted from Chen et al. (2010).

band of SMs: i.e., approximately 0.2–0.5 Hz. This observation is similar to the study of Chen et al. (2010). PM energy was observed at only two OBS stations (at water depths of 1700 and 2000 meters) in the Okinawa Trough. These OBS stations were deployed along the path of the Kuroshio Current. We consider that the kinematic energy may be coupled with the topography of the Okinawa Trough and generate seismic waves in the PM.

According to Lin et al. (2010), infragravity waves were mainly detectable at OBS stations in the Okinawa Trough. Following the energy-dispersion relation raised in Webb et al. (1991), the infragravity waves at the OBS stations in eastern offshore Taiwan, where the water depth surpasses 5 km, were detected at periods below 160 seconds (0.006 Hz). They may already exceed the detection capacity of OBSs at the seafloor.

2.3. AMBIENT SEISMIC NOISE GENERATED BY INTENSE STORMS

Intense ocean storms called *typhoons* in the western Pacific Ocean form large and complex ocean waves at the sea surface. A portion of the dynamic energy of a typhoon can propagate to the seafloor and be transferred into seismic waves as microseisms. These microseisms propagate as both surface waves (mainly Rayleigh-type) and compressional (P) waves, typically with frequencies of 0.05–0.4 Hz (Gerstoft et al., 2008; Zhang et al., 2010). In general, broadband seismometers can record these seismic waves from far away (Chi et al., 2010; Gerstoft et al., 2008; Kedar & Webb, 2005; Kedar et al., 2008; Longuet-Higgins, 1950; Sutton, 1996; Webb, 2007; Wilson & Makris, 2008).

In the Taiwan area, the seismic noise level is augmented when typhoons approach. Chi et al. (2010) demonstrated that ground motions induced by Typhoon Shanshan (category 4, 18–20 September 2006) lit up the seismic stations along its way, crossing eastern offshore Taiwan (Fig. 2.3). The typhoon agitated ocean water and generated microseisms on the seafloor at frequencies of PM and SM. Chi et al. (2010) showed that distant storms excited microseisms at the splitting peaks of LPSM (0.085–0.2 Hz) and SPSM (0.2–0.45 Hz) frequencies recorded by the land and ocean-bottom seismic stations in eastern Taiwan. When the typhoon approached the stations, the frequency content of microseisms even extended to the PM band (0.01–0.085 Hz) and formed a spindle-shaped PSD energy

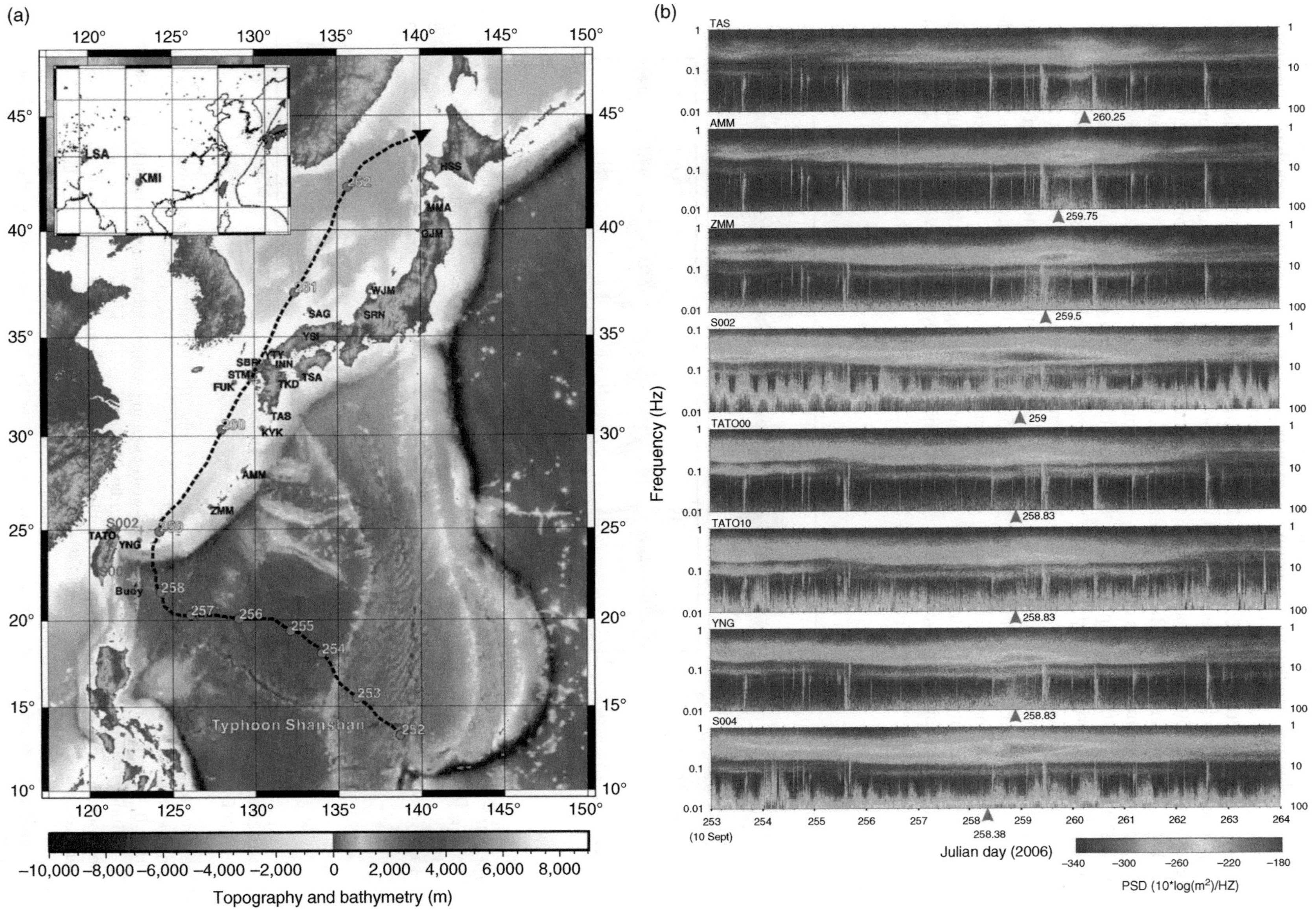

Figure 2.3 (a) Moving track of Typhoon Shanshan across the western Pacific Ocean. The dots and the nearby numbers depict the typhoon locations and time by Julian day. S002 and S004 mark the locations of the two OBSs. All other stations are land stations, primarily from the F-net of Japan. Stations KMI and LSA are part of the Global Seismic Network (GSN; see inset). (b) Spectrograms of typhoon-induced ground motions. The triangle marks the time when the typhoon was closest to the station. Chi et al. (2010) / Springer Nature/ CC by 2.0.

envelope (Fig. 2.3). However, the maximum energy was seen a few hours after the typhoon center had passed, not when the wall of the highest-wind-speed typhoon eye was closest to the station (Chi et al., 2010). This delay correlated with the wave–wave interaction region to the rear of a moving cyclone, which was first theoretically proposed by Longuet-Higgins (1950). Chi et al. (2010) also demonstrated that Typhoon Shanshan affected the seismic noise level of far-distant seismic stations in China, which ranged up to 2,800 km from the typhoon center.

Another template study on typhoon effects at Taiwanese stations is given by Lin et al. (2017). They investigated the temporal evolution of the microseisms excited by the super-typhoon Megi (13–24 October 2010). Their observations about the frequency evolution of SMs are similar to those of Chi et al. (2010). Their study detected the LPSM (0.12–0.2 Hz) when the typhoon moved to the South China Sea, a relatively shallow water zone. They also stressed the significant energy dispersion in the deep basins of the Philippine Sea and strong nonlinear interaction of ocean waves at the coastal area during the lifespan of typhoons.

In addition to the excitation of microseisms, typhoon energy compounds with the specific seafloor topography and generates standing oscillation for a deep-sea basin to southeast offshore Taiwan. In Chang et al. (2016), the passage of Typhoon Sungda (Category 5, 27–28 May 2011) agitated ocean water from the surface to approximately 1,600 meters deep. Based on the measurement of the ambient temperature of the seawater at the OBS station, a strong oscillation of up to 0.5 °C at the O1 tidal period (25.82 hours) lasting a day or longer was detected at the seafloor thermal probe in the eastern offshore region (Taitung Trough in Fig. 2.4). This caused the ambient seismic noise level to vary with the diurnal tidal periods.

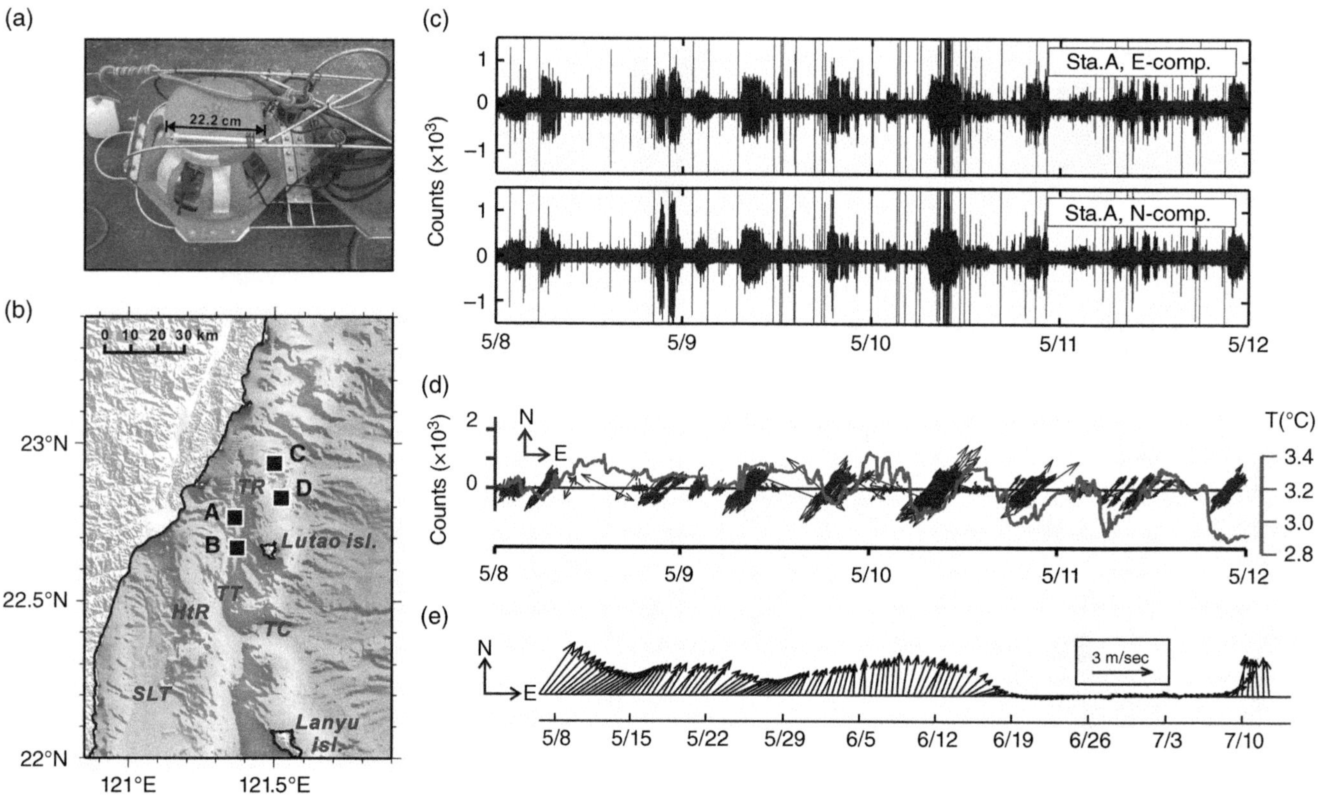

Figure 2.4 Offshore campaign OBS experiment from 5 May to 15 July 2011 in eastern offshore Taiwan. (a) Each OBS instrument mounted a high-sensitivity CHTL thermoprobe. (b) Seafloor topography and four instrument stations labeled A, B, C, and D, relative to the geological units Hautung Ridge (HtR), Lanyu Ridge (LR), Taitung Canyon (TC), Taitung Trough (TT), and South Longitudinal Trough (SLT). TT and SLT are longitudinal semiclosure deep-sea troughs. (c) Four-day time series (8–11 May) of the horizontal (north and east) motion vector of water agitation at OBS station A, synthesized from channels H1 and H2, given their geographical orientations. (d) Motion vectors overlaid with the temperature record at station A during the four days of (c). (e) Estimated horizontal velocity vector of the M2 plane wave flows over the station array during the timeline of the experiment. Reproduced from Chang et al. (2016) / John Wiley & Sons.

2.4. DEEPSEA INTERNAL WAVES SOUTHEAST OF OFFSHORE TAIWAN

In east offshore Taiwan, the Kuroshio Current flows to the north at approximately 130 cm/s (e.g., Chung et al., 1993; Liang et al., 2003; Tang et al., 2000). The center stream of the Kuroshio Current is approximately 100 km east of Taiwan island within a vertical range from the sea surface to approximately 1,000 meters deep. To date, the background seismic level affected by the Kuroshio Current is not yet precise. However, nonlinear energy conversion from the motion of the Kuroshio Current drives the internal waves and has been detected at deepsea OBS stations in eastern offshore Taiwan (Chang et al., 2016). Internal tides are baroclinic internal wave trains generated at tidal intervals by barotropic tidal currents flowing over steep ocean-bottom topography (e.g., Garrett, 2003; Phillips, 1977; Wunsch, 1975). In contrast to the more familiar wind-driven waves at the surface, these internal waves propagate along interfaces such as the thermocline in the density-stratified ocean and are considered able to enhance vertical water mixing and warming in the deep sea (e.g., Garrett, 2003; Jayne et al., 2004; Levitus et al., 2012; St. Laurent et al., 2003).

Chang et al. (2016) provided deep-sea internal waves recorded at seafloor seismic stations in eastern offshore Taiwan. The east and southeast offshore of Asia, which includes Taiwan island, is one of the world's most active "hot spots" of internal wave generation (Alford et al., 2015; Egbert & Ray, 2000; Farmer et al., 2011; Liu et al., 1998). Recent studies reveal that internal tides radiated westward into the northern South China Sea and eastward into the Philippine Sea/western Pacific (e.g., Alford et al., 2015; Farmer et al., 2011; Simmons et al., 2011) have been routinely detected by satellite sea-surface height and in situ oceanographic monitoring (e.g., Alford et al., 2011, 2015; Farmer et al., 2011; Hsu et al., 2000; Ramp et al., 2012).

In Chang et al. (2016), four stations were situated on the west flank inside the narrow, semiclosed Taitung Trough and the east flank bordering the Philippine Sea basin (Fig. 2.4). The local bathymetries of these stations are in the range of 1,250 to ~1,600 m. Each station consists of an OBS and a compact high-sensitivity temperature logger (CHTL; Chang & Shyual, 2011). These four stations were employed to monitor the elastic waves of ground motion and measure the perturbation of the ambient water temperature at high precision. In their experiment, the seismograms recorded intermittent "tremor"-like vibrations, where the amplitudes are abruptly raised several times above the background noise level. These intermittent vibrations emerge for as long as 2–5 hours each time and are monotonic in their frequency content, with energy being predominantly within the 2–4 Hz band. The tremor signals were probably the resonant motion of the OBS/geophone devices that were physically agitated by external forcing since the dominant vibration frequencies matched the natural frequency of the geophones (at ~4 Hz). Therefore, the external agitating force at OBS stations likely came from ocean-bottom currents, not actual ground motions. The particle motions composed of the horizontal components of the OBS station showed systematic shaking as the waves flowed by (Fig. 2.4).

Meanwhile, the change in ambient water temperature was synchronized with the emergency of the tremor signals (Fig. 2.4). An apparent time periodicity in the tremor signals and ambient temperature variation was observed, with semidiurnal periodicities; however, it was out of phase with the local sea-surface tides. With observations from OBSs and the ambient water temperature from the CHTL, Chang et al. inferred that the high-frequency, intermittent emergence of tremor-like signals was from the motions of deep-sea internal waves that propagated over seafloor OBS stations. The deep-sea internal waves caused vertical fluctuations in the stratified ocean water and generated ambient temperature variations in the semidiurnal period. The simulation in Chang et al. (2016) demonstrated that deep-sea internal waves resulted from the Kuroshio Current's motion coupled with the local topographic relief.

In terms of ambient seismic noise, there are two aspects from Chang et al. (2016): (1) deep-sea internal waves converted from barotropic tides are the kinetic transformation and dissipation of ocean tidal energy, which influences the vertical mixing of stratified water layers and generates fluctuations in water layers at the seafloor; and (2) the passage of internal waves generates advection motions of ocean water over the OBS stations. This observation helps seismologists understand or eliminate sensor recordings on the seafloor.

2.5. GAS EMISSIONS AT THE SEAFLOOR AND "BUBBLE" SDES IN SW OFFSHORE TAIWAN

Over the last two decades, numerous studies have been performed to recognize the importance of submarine gas emissions in many submarine environments, including at the West Spitzbergen continental margin (Hustoft et al., 2009; Knies et al., 2004; Mienert et al., 2005), in the North Sea (Hovland, 2007; Hovland & Sommerville, 1985), in the Santa Barbara Basin (Fischer, 1978; Leifer & Clark, 2002), in the Nile deep-sea fan (Bayon et al., 2013; Dupré et al., 2008), and in the Marmara Sea (Gasperini et al., 2012; Géli et al., 2008). Gases emitted at the seafloor are mainly composed of methane, whereas fluid seeps include hydrocarbons. Seep activities at the seafloor correspond to various signal patterns such as continuous, transient (periodic or sporadic), or eruptive, and temporal variations on scales ranging from tidal to subhourly periods have been documented (e.g., Leifer et al., 2004; Hsu et al., 2013). In this study, we specifically investigate methane degassing signals recorded at OBS stations. Recent studies reveal that gas migration followed by the collapse of fluid-filled cavities or conduits can generate burst-type SDEs at OBS stations (e.g., Géli et al., 2018; Tary et al., 2012). The experiment of Vidal et al. (2009) testified that the bubble acted as a resonator when bursting at the free surface. We consider that the degassing process at the seafloor presents a similar pressure-release mechanism as methane gas escaping through a subvertical conduit generates a "bubble waveform" with a high-frequency burst followed by resonant attenuation. Bubble waveforms are one type of SDE with no identified P- or S-phase arrivals. Based on the character of quickly resonant attenuation, we can describe the attenuating section of the bubble waveform $B(t)$ with a harmonic cosine wave chain shaped by exponential decay:

$$B(t) = \mathbf{A}\cos(2\mathrm{pi}ft)\exp(at) \qquad (2.1)$$

where $\mathbf{A}$, f, and a are the maximum amplitude, resonant vibration frequency, and attenuation rate of the bubble, respectively. Figure 2.5 shows examples of SDE waveforms detected in the northern South China Sea and their mathematical fittings. This result shows that harmonic vibration and exponential decay can restore their attenuation sections well.

Southwest offshore Taiwan and the northern South China Sea are well known to be rich in gas hydrate (e.g., Huang et al., 2021; Liu et al., 2006). OBS data acquired from these areas contain many bubble SDEs. As demonstrated in Fig. 2.5, the SDEs lasted less than 2 seconds. Similar to the observation raised in Tary et al. (2012), bubble SDEs detected in southwest offshore Taiwan are better identified at the geophone components than the hydrophone. This observation again proves that the energy of bubble SDEs was transported from underground layers instead of in water layers. With a quick waveform search along the OBS seismograms, bubble SDEs in the northern South China Sea can be roughly described by waveforms at resonant frequencies of approximately 4.5 to ~6.5 Hz with attenuation rates of –0.8 to ~ –2.0 count/sec.

Traditional energy-based signal detectors, such as a short-term average (STA)/long-term average (LTA), identify events by impulsive, high signal-to-noise ratios. However, these detectors are prone to high false-positive and false-negative rates at low magnitudes, especially for noisy backgrounds (Gibbons & Ringdal, 2006). Template

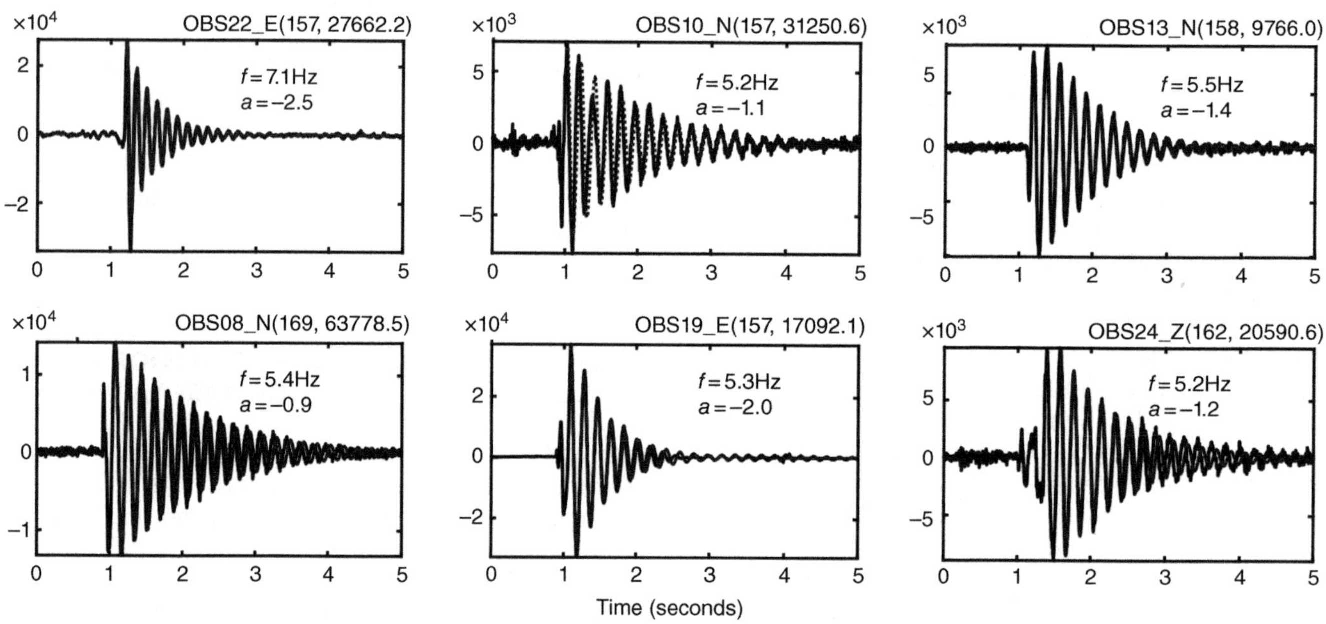

Figure 2.5 Six examples of bubble waveforms recorded by OBS stations deployed in the northern South China Sea in 2011 (see Fig. 2.7 for station locations). The station's name and recording component (E, N, or Z) are indicated in the title of each plot. The numbers in parentheses are the Julian day in 2011 and the beginning time (in seconds) counted from the day. The black curves are the observed bubble SDE waveforms. The gray dashed curves are the synthetic waveforms from equation 2.1 using parameters f and a, which are the dominant frequency and attenuation rate, respectively.

matching, or waveform cross-correlation with template waveforms of known signals, has proven more effective for detecting known seismic signals in noisy data (Bobrov et al., 2014; Schaff & Waldhauser, 2010). However, the method relies on template waveforms that cannot handle bubble signals well because bubble waveforms do not exhibit identical resonant frequencies or fixed attenuation rates.

We apply convolutional neural network (CNN) supervised machine learning to screen out the numerous bubble SDEs in the OBS data acquired from southwest offshore Taiwan (located along the continental slope of the northern South China Sea) in 2011. Each input signal is a 5-second data window that contains 1×1,251 data points (sampling rate: 250 Hz). The CNN structure comprises three convolutional layers, a max-pooling layer, and two fully connected layers with a two-way softmax function at the network's end. The three convolutional layers contain 8, 16, and 32 kernels. The size of the max-pooling kernel is 1×35 with a stride of 5×5. The ReLU function is used as the activation function for the results of the three convolutional layers. The algorithm adopted in this model is the stochastic gradient descent with momentum (SGDM) optimizer (Fig. 2.6).

We iteratively modify the model parameters for up to 20 runs. In each run, the newly found bubbles are labeled and integrated into the input data for the next iteration. The number of detected bubble SDEs rapidly increases in the first 10 runs and gradually converges to approximately 5,000 s over 18 OBSs in 25 days. The spatial and temporal variations in the detected numbers are presented in Fig. 2.7. It shows that the seafloor degassing activity was concentrated on certain days and certain stations in the experimental period: most bubble SDEs are on Julian day 158 and at stations located at the feet of the continental slope (Stations 23, 24) and along the topographic ridge (Stations 04–08). The localized distribution of the bubble SDEs reflects the structures of seafloor fractures and chimneys, which correspond to regional tectonic forces and fluid migration.

2.6. CONCLUSION

Taiwanese seismic observations enforce the concepts that topographic coupling and the dynamics of air–sea conditions are two main factors for the regular excitation of microseisms. These two factors interact to produce different frequency responses for ambient seismic levels. Without intense storms, the predominant microseism in the Taiwan area is at the frequency band of SPSMs for both land and ocean-bottom seismic stations. The coastal feature can enhance ambient energy diffusion, whereas the greatest excitation of ambient seismic energy is the SPSM from the shallow-water Taiwan Strait. PM signals are most evident for OBS stations in the Okinawa Trough, where the Kuroshio Current pathway is located. Meanwhile, intense typhoon events significantly excite

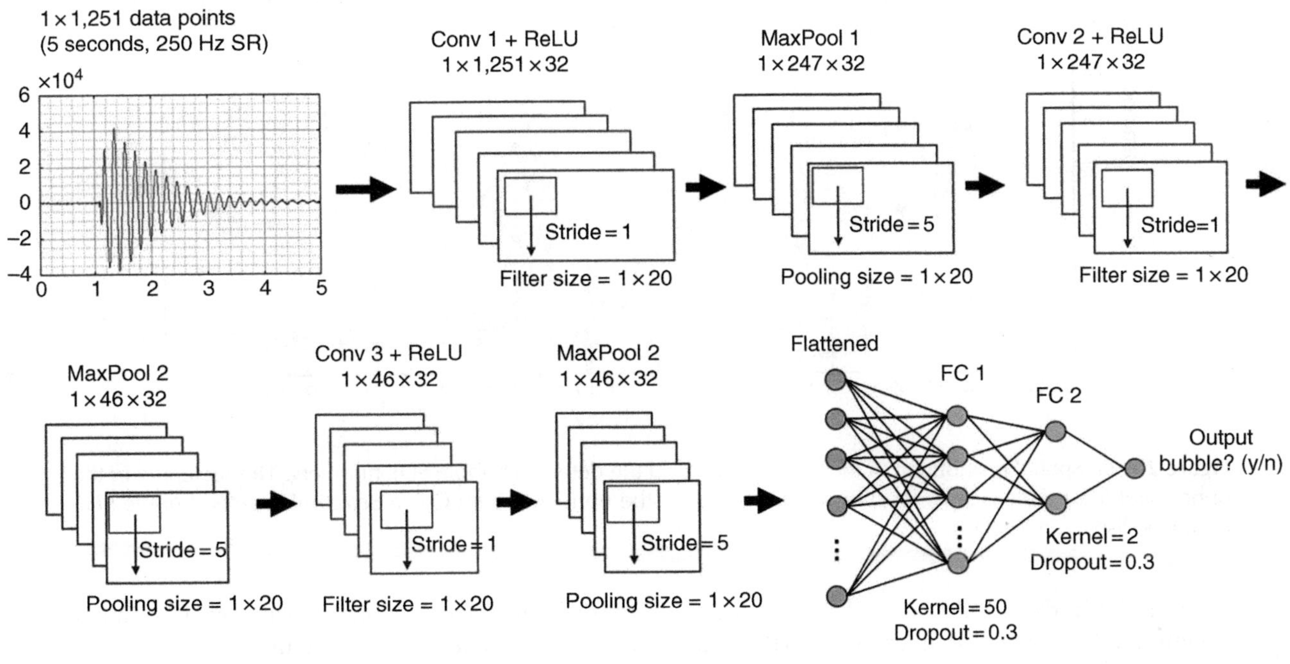

Figure 2.6 CNN model structure constructed in this study. This model consists of three pairs of convolutional layers and max-pooling layers, followed by two fully connected layers to predict the classification.

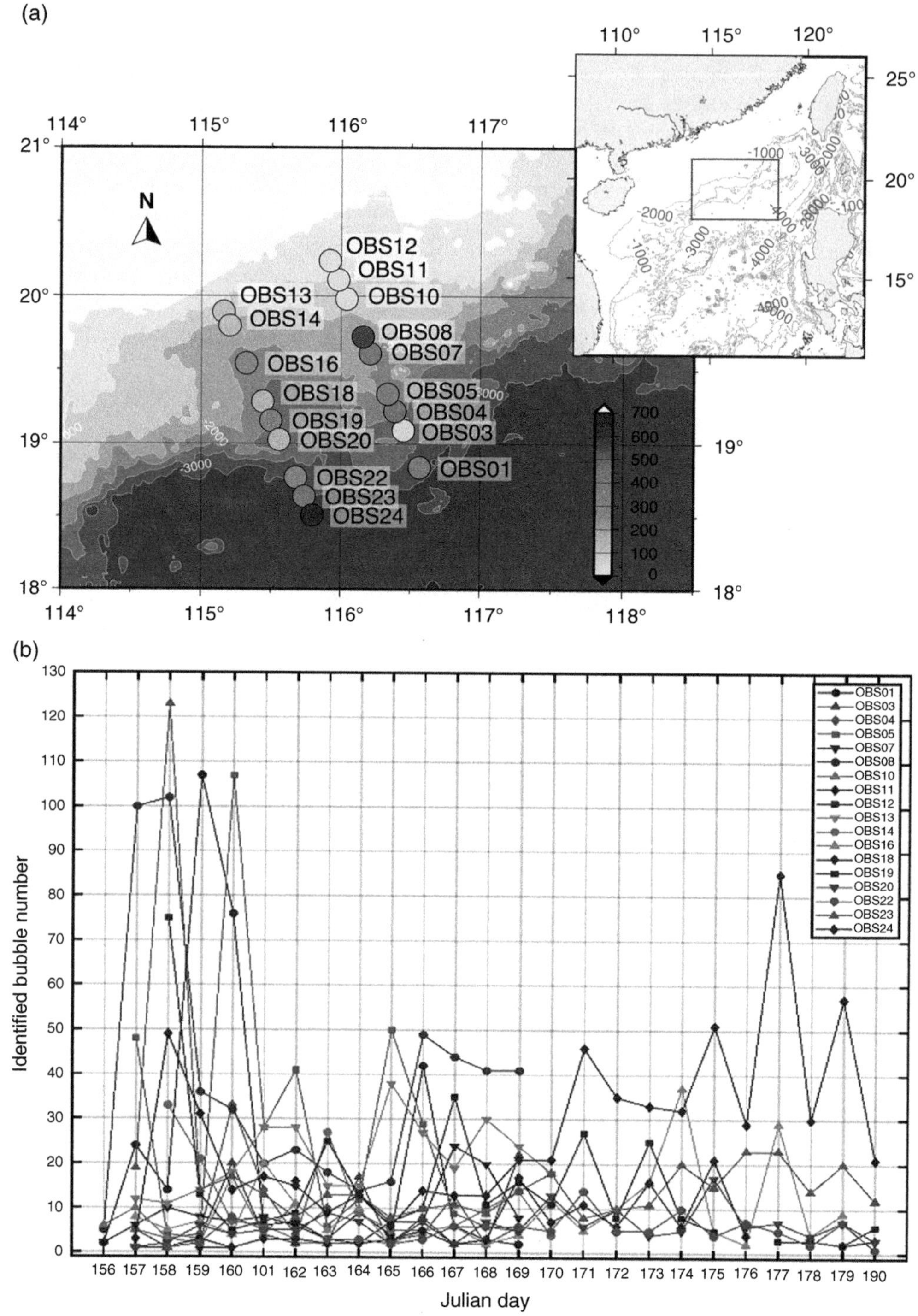

Figure 2.7 (a) Spatial distribution of the OBS stations and the detected bubble SDE numbers. The rectangle in the upper-right inset indicates the location of OBS arrays in the northern South China Sea. (b) Detected bubble SDE numbers displayed by days.

microseisms at both the LPSM and SPSM frequency bands. Relatively longer-period microseisms (PMs and LPSMs) are primarily detected when typhoons are in shallow-water areas. These observations stress that the shallow-water Taiwan Strait and southeast continental shelf are the "birthplace" of microseismic excitation in the Taiwan area. In addition, the excitation of relatively longer-period microseisms requires a significant driving force, such as typhoons or the kinetic energy of strong ocean currents.

The energy of air–sea interactions dissipates in deep water. However, internal waves converted from Kuroshio Current kinematics and tidal energy have been detected in the deep sea in eastern offshore Taiwan. Furthermore, SDEs that exhibit discernible signal amplitudes in a resonance vibration and rapid attenuation are noticeably detected from the OBS data acquired from SW offshore Taiwan. Seafloor gas emissions are likely candidates to explain their origin. The essential geological background to generate seafloor bubble events is seafloor fractures, fluid migration within shallow strata, and gas hydrate deposits in the ocean.

Based on the study with Taiwan seismic data, ambient seismic noise can be converted from a diverse mechanism and involve signals from the geological-related marine environment.

ACKNOWLEDGMENTS

The seismic data presented in this chapter was assembled from Taiwan Ministry of Science and Technology (MOST) projects. Excellent collaboration with W.C. Chi and W.T. Liang created the discussion on the background level of microseisms and typhoon excitation. We thank Ziyi Lee for assistance with the machine learning modeling. This research was supported by MOST projects 110-2611-M-002-012 and 110-2116-M-194-006.

REFERENCES

Alford, M. H., MacKinnon, J. A., Nash, J. D., Simmons, H., Pickering, A., Klymak, J. M., et al. (2011). Energy flux and dissipation in Luzon Strait: two tales of two ridges. *Journal of Physical Oceanography*, *41*(11), 2211–2222.

Alford, M. H., Peacock, T., MacKinnon, J., Nash, J., Buijsman, M. C., Centurioni, L. R., et al. (2015). The formation and fate of internal waves in the South China Sea. *Nature*, *521*, 65–69. https://doi.org/10.1038/nature14399

Batsi, E., Tsang-hin-sun, E., Klingelhoefer, F., Bayrakci, G., Chang, E., Lin, J. Y., et al. (2019). Nonseismic signals in the ocean: indicators of deep sea and seafloor processes on ocean-bottom seismometer data. *Geochemistry Geophysics Geosystems*, *20*(8), 3882–3900. https://doi.org/10.1029/2019GC008349

Bayon, G., Dupré, S., Ponzevera, E., Etoubleau, J., Chéron, S., Pierre, C., et al. (2013). Formation of carbonate chimneys in the Mediterranean Sea linked to deep-water oxygen depletion. *Nature Geoscience*, *6*(9):755–760. https://doi.org/10.1038/ngeo1888

Bobrov, D., Kitov, I., & Zerbo, L. (2014). Perspectives of cross-correlation in seismic monitoring at the international data centre. *Pure and Applied Geophysics*, *171*(3):439–468.

Bromirski, P. D. (2001). Vibrations from the "perfect storm." *Geochemistry, Geophysics, Geosystems*, *2*(7), 1030. https://doi.org/10.1029/2000GC000119

Bromirski, P. D., & Gerstoft, P. (2009). Dominant source regions of the Earth's "hum" are coastal. *Geophysical Research Letters*, *36*. https://doi.org/10.1029/2009GL038903

Bromirski, P. D., Stephen, R. A., & Gerstoft, P. (2013). Are deep-ocean-generated surface-wave microseisms observed on land? *Journal of Geophysical Research, Solid Earth*, *118*, 3610–3629. https://doi.org/10.1002/jgrb.50268

Brooks, L. A., Townend, J., Gerstoft, P., Bannister, S., & Carter, L. (2009). Fundamental and higher-mode Rayleigh wave characteristics of ambient seismic noise in New Zealand. *Geophysical Research Letters*, *36*, L23303. https://doi.org/10.1029/2009GL040434

Brzak, K., Gu, Y. J., Ökeler, A., Steckler, M., & Lerner-Lam, A. (2009). Migration imaging and forward modeling of microseismic noise sources near southern Italy. *Geochemistry Geophysics Geosystems*, *10*, Q01012. https://doi.org/10.1029/2008GC002234

Chang, E., Chao, B. F., Chen, G. Y., & Liau, J. M. (2016). Internal tides recorded at ocean bottom off the coast of Southeast Taiwan. *Journal of Geophysical Research: Ocean*, *121*(5), 3381–3394.

Chang, E., Hsu, S. K., & Lee, C. S. (2008). Earthquake swarm recorded by an ocean bottom seismic array in southwest offshore of Taiwan in October, 2005. *Terrestrial, Atmospheric and Oceanic Sciences*, *19*(6), 717–728

Chang, H. I., & Shyu, C. T. (2011). Compact high-resolution temperature loggers for measuring the thermal gradients of marine sediments. *Marine Geophysical Research*, *32*, 465–479

Chen, Y. N., Gung, Y. C., You, S. H., Hung, S. H., Chiao, L. Y., Huang, T. Y., et al. (2011). Characteristics of short period secondary microseisms (SPSM) in Taiwan: the influence of shallow ocean strait on SPSM, *Geophysical Research Letters*, *38*, L04305. https://doi.org/10.1029/2010GL046290

Chi, W., Chen, W., Kuo, B., & Dolenc, D. (2010). Seismic monitoring of western Pacific typhoons. *Marine Geophysical Research*, *31*(4), 239–251

Chuang, W. S., Li, H. W., Tang, T. Y., Wu, C. K. (1993). Observation of the countercurrent on the inshore side of the Kuroshio northeast of Taiwan. *Journal of Oceanography*, *49*, 581–592.

Dorman, L. M., Schreiner, A. E., Bibee, L. D., & Hildebrand, J. A. (1993). Deep-water seafloor array observations of seismo-acoustic noise in the eastern Pacific and comparisons with wind and swell. In B. R. Kerman (Ed.), *Natural physical sources of underwater sound* (pp. 165–174). Springer.

Dupré, S., Buffet, G., Mascle, J., Foucher, J. P., Gauger, S., Boetius, A., et al. (2008). High-resolution mapping of large gas emitting mud volcanoes on the Egyptian continental margin (Nile Deep Sea Fan) by AUV surveys. *Marine Geophysical Research*, *29*(4):275–290. https://doi.org/10.1007/s11001-009-9063-3

Egbert, G. D., & Ray, R. D. (2000). Significant dissipation of tidal energy in the deep ocean inferred from satellite altimeter data, *Nature*, *405*, 775–778.

Farmer, D. M., Alford, M. H., Lien, R. C., Yang, Y. J., Chang, M. H., & Li, Q. (2011). From Luzon Strait to Dongsha Plateau: stages in the life of an internal wave. *Oceanography*, *24*(4), 64–77.

Fischer, P. J. (ed) (1978). Natural gas and oil seeps, Santa Barbara Basin, California. In: *California offshore gas, oil, and tar seeps* (pp 1–62). Sacramento: California State Lands Commission.

Friedrich, A., Kruger, F., & Klinge, K. (1998). Ocean generated microseismic noise located with the Grafenberg array. *Journal of Seismology*, *2*(1), 47–64.

Garrett, C. (2003). Internal tides and ocean mixing, *Science*, *301*, 1858–1859. https://doi.org/10.1126/science.1090002

Gasperini, L., Polonia, A., Del Bianco, F., Etiope, G., Marinaro, G., Favali, P., et al. (2012). Gas seepages and seismogenic structures along the North Anatolian Fault in the eastern Marmara Sea. *Geochemistry Geophysics Geosystems*, *13*, Q10018. https://doi.org/10.1029/2012GC004190

Géli, L., Henry, P., Zitter, T., Dupré, S., Tryon, M., Çağatay, M. N., et al. (2008). Gas emissions and active tectonics within the submerged section of the North Anatolian Fault zone in the Sea of Marmara. *Earth and Planetary Science Letters*, *274*, 34–39. https://doi.org/10.1016/j.epsl.2008.06.047

Géli, L., Henry, P., Grall, C., Tary, J. B., Lomax, A., Batsi, E. (2018). Gas and seismicity within the Istanbul seismic gap. *Scientific Reports*, *8*(1), 6819. https://doi.org/10.1038/s41598-018-23536-7

Gerstoft, P., Shearer, P. M., Harmon, N., & Zhang, J. (2008). Global P, PP, and PKP wave microseisms observed from distant storms. *Geophysical Research Letters*, *35*, L23306. https://doi.org/10.1029/2008GL036111

Gibbons, S. J., & Ringdal, F. (2006). The detection of low magnitude seismic events using array-based waveform correlation. *Geophysical Journal International*, *165*(1), 149–166.

Gu, Y. J., Deblanko, C., Lener-Lam, A., Brzak, K., & Steckler, M. (2007). Probing the source of ambient seismic noise near the coasts of southern Italy. *Geophysical Research Letters*, *34*, L22315. https://doi.org/10.1029/2007GL03967

Gutenberg, B. (1931). Microseisms in North America. *Bulletin of the Seismological Society of America*, *21*, 1–24.

Knudsen, V. O., Alford, R. S., & Emling, J. W. (1948). Underwater ambient noise. *Journal of Marine Research*, *7*, 410–429.

Hasselmann, K. A. (1963). Statistical analysis of the generation of microseisms. *Reviews of Geophysics*, *1*(2), 177–210.

Hilmo, R., & Wilcock, W. S. D. (2020). Physical sources of high-frequency seismic noise on Cascadia Initiative ocean bottom seismometers. *Geochemistry, Geophysics, Geosystems*, *21*, e2020GC009085. https://doi.org/10.1029/2020GC009085

Hovland, M. (2007). Discovery of prolific natural methane seeps at Gullfaks, northern North Sea. *Geo-Marine Letters*, *27* (2–4), 197–201. https://doi.org/10.1007/s00367-007-0070-6

Hovland, M., & Sommerville, J. H. (1985). Characteristics of two natural gas seepages in the North Sea. *Marine and Petroleum Geology*, *2*, 319–326. https://doi.org/10.1016/0264-8172(85)90027-3

Hsu, M. K., A. K. Liu, A. K., & Liu, C. (2000). A study of internal waves in the China Seas and Yellow Sea using SAR. *Continental Shelf Research*, *20*, 389–410. https://doi.org/10.1016/S0278-4343(99)00078-3

Hsu, S.K, Wang, S., Liao, Y. C., Yang, T. F., Jan, S., Lin, J. Y., & Chen, S. C. (2013). Tide-modulated gas emissions and tremors off SW Taiwan. *Earth and Planetary Science Letters*, *369–370*, 98–107.

Huang, T. Y., Gung, Y., Kuo, B. Y., Chiao, L. Y., & Chen, Y. N. (2015). Layered deformation in the Taiwan orogen. *Science*, *349*(6249), 720–723. https://doi.org/10.1126/science.aab1879

Huang, Y. S., Hsu, SK., Su, CC., Lin, A. T. S., Yu, P. S., Babonneau, N., et al. (2021). Shallow gas hydrates off southwest Taiwan and their mechanisms. *Marine Geophysical Research*, *42*, 7. https://doi.org/10.1007/s11001-021-09429-x

Hustoft, S., Bünz, S., Mienert, J., & Chand, S. (2009). Gas hydrate reservoir and active methane-venting province in sediments on < 20 Ma young oceanic crust in the Fram Strait, offshore NW-Svalbard. *Earth and Planetary Science Letters*, *284* (1–2), 12–24. https://doi.org/10.1016/j.epsl.2009.03.038.s

Jayne, S. R., St. Laurent, L. C., & Gille, S. T. (2004). Connections between ocean-bottom topography and Earth's climate. *Oceanography*, *17*, 65–4.

Kedar S, & Webb, F. (2005). The ocean's seismic hum. *Science*, *307*(5710):682–683. https://doi.org/10.1126/science.1108380

Kedar, S., Longuet-Higgins, M., Webb, F., Graham, N., Clayton, R., & Jones, C. (2008). The origin of deep ocean microseisms in the North Atlantic Ocean. *Proceedings of the Royal Society of London, Series A*, *464*, 777–793. https://doi.org/10.1098/rspa.2007.0277

Knies, J., Damm, E., Gutt, J., Mann, U., & Pinturier, L. (2004). Near-surface hydrocarbon anomalies in shelf sediments off Spitsbergen: evidences for past seepages. *Geochemistry Geophysics Geosystems*, *5*, Q06003. https://doi.org/10.01029/02003GC000687

Leifer, I., & Clark, J. F. (2002). Modeling trace gases in hydrocarbon seep bubbles: application to marine hydrocarbon seeps in the Santa Barbara Channel. *Geologiya i Geofizika*. *43*(7), 613–621.

Leifer, I., Boles, J. R., Luyendyk, B. P., & Clark, J. F. (2004). Transient discharges from marine hydrocarbon seeps: spatial and temporal variability. *Environmental Geology*, *46*:1038–1052. https://doi.org/10.1007/s00254-004-1091-3

Levitus, S., Antonov, J. I., Boyer, T. P., Baranova, O. K., Garcia, H. E., Locarnini, A. V., et al. (2012). World ocean heat content and thermosteric sea level change (0–2000 m), 1955–2010, *Geophysical Research Letters*, *39*, L10603. https://doi.org/10.1029/2012GL051106

Liang, W. D., Tang, T. Y., Yang, Y. J., Ko, M. T., & Chuang, W. S. (2003). Upper-ocean currents around Taiwan. *Deep-Sea Research II*, *50* (2003), 1085–1105.

Liang, W. T., Liu, C. C., Chen, R. Y., Huang, B. S., Wu, F. T., & Wang, C. Y. (2008). Ambient noise levels in the Taiwan region. Paper presented at American Geophysical Union, Fall Meeting.

Lin, C. R., Kuo, B. Y., Liang, W. T., Chi, W. C., Huang, Y. C., Collins, J., & Wang, C. Y. (2010). Ambient noise and teleseismic signals recorded by ocean-bottom seismometers offshore eastern Taiwan. *Terrestrial, Atmospheric and Oceanic Sciences*, *21*, 743–755.

Lin, J., Lin, J., & Xu, M. (2017). Microseisms generated by super typhoon Megi in the western Pacific Ocean. *Journal of Geophysical Research: Oceans*, *122*, 9518–9529. https://doi.org/10.1002/2017JC013310

Liu, A. K., Chang, Y. S., Hsu, M. K., & Liang N. K. (1998). Evolution of nonlinear internal waves in the East and South China Seas, *Journal Geophysical Research*, *103*, 7995–8008. https://doi.org/10.1029/97JC01918

Liu, CS, Schnürle, P., Wang, Y., Chung, S. H., Chen, S. C., & Hsiuan, T. H. (2006). Distribution and characters of gas hydrate offshore of southwestern Taiwan. *Terrestrial, Atmospheric and Oceanic Sciences*, *17*, 615–644. https://doi.org/10.3319/TAO.2006.17.4.615(GH)

Longuet-Higgins, M. S. (1950). A theory of the origin of microseisms. *Philosophical Transactions of the Royal Society of London. Series A, Mathematical and Physical Sciences*, *243*, 1–35.

McNamara, D., & Raymond, P. B. (2004). Ambient noise levels in the continental United States. *Bulletin of the Seismological Society of America, 2004*; *94* (4): 1517–1527. https://doi.org/10.1785/012003001

Mienert, J., Bünz, S., Guidard, S., Vanneste, M., & Berndt, C. (2005). Ocean bottom seismometer investigations in the Ormen Lange area offshore mid-Norway provide evidence for shallow gas layers in subsurface sediments. *Marine and Petroleum Geology*, *22*(1–2), 287–297. https://doi.org/10.1016/j.marpetgeo.2004.10.020

Nishida, K. (2017). Ambient seismic wave field. *Proceedings of the Japan Academy. Series B, Physical and Biological Sciences*, *93*(7), 423–448. https://doi.org/10.2183/pjab.93.026

Phillips, O. M. (1977). *The dynamics of the upper ocean*. Cambridge Univ. Press.

Ramp, S. R., Yang, Y. J., Reeder, D. B., & Bahr, F. L. (2012). Observations of a mode-2 nonlinear internal wave on the northern Heng-Chun Ridge south of Taiwan. *Journal of Geophysical Research*, *117*, C03043. https://doi.org/10.1029/2011JC007662

Rhie, J., & Romanowicz, B. (2004). Excitation of Earth's continuous free oscillations by atmosphere-ocean-seafloor coupling. *Nature*, *431*, 552–556.

Ritzwoller, M. H., & Levshin, A. L. (1998). Surface wave tomography of Eurasia: group velocities. *Journal Geophysical Research*, *103*, 4839–4878.

Sabra, K. G., Gerstoft, P., Roux, P., Kuperman, W. A., & Fehler, M. C. (2005). Surface wave tomography from microseism in southern California, *Geophysical Research Letters*, *32*, L14311. https://doi.org/10.1029/2005GL023155

Schaff, D. P. & Waldhauser, F. (2010). One magnitude unit reduction in detection threshold by cross correlation applied to Parkfield (California) and China seismicity. *Bulletin of the Seismological Society of America*, *100*(6), 3224. https://doi.org/10.1785/0120100042

Shapiro, N. M., Campillo, M., Stehly, L., & Ritzwoller, M. H. (2005). High resolution surface wave tomography from ambient seismic noise. *Science*, *307*, 1615–1618.

Shapiro, N. M., & Campillo, M. (2004). Emergence of broadband Rayleigh waves from correlations of the ambient seismic noise. *Geophysical Research Letters*, *31*, L07614. https://doi.org/10.1029/2004GL019491

Simmons, H., Chang, M. H., Chang, Y. T., Chao, S. Y., Fringer, O., Jackson, C. R., & Ko, D. S. (2011). Modeling and prediction of internal waves in the South China Sea, *Oceanography*, *24*(4), 88–99, https://doi.org/10.5670/oceanog.2011.97

Snieder, R. (2004). Extracting the Green's function from the correlation of coda waves: A derivation based on stationary phase. *Physical Review E*, *69*, 046610. https://doi.org/10.1103/PhysRevE.69.046610

St. Laurent, L. C., Stringer, S., Garrett, C., & Perrault-Joncas, D. (2003). The generation of internal tides at abrupt topography, *Deep-Sea Research I*, *50*, 987–1003. https://doi.org/10.1016/S0967-0637(03)00096-7

Stephen, R. A., Spiess, F. N., Collins, J. A., Hildebrand, J. A., Orcutt, J. A., Peal, K. R., et al. (2003). Ocean seismic network pilot experiment, *Geochemistry Geophysics Geosystems*, *4*(10), 1092. https://doi.org/10.1029/2002GC000485

Sutton, G. H. (1996). Ocean bottom microseisms from a distant supertyphoon. *Geophysical Research Letters*, *23*(5):499–502.

Tang, T. Y., Tai, J. H., & Yang, Y. J. (2000). The flow pattern north of Taiwan and migration of the Kuroshio. *Continental Shelf Research*, *20*, 349–371.

Tanimoto, T. (2007). Excitation of microseisms. *Geophysical Research Letters*, *34*, L05308. https://doi.org/10.1029/2006GL029046

Tary, J. B., Géli, L., Guennou, C., Henry, P., Sultan, N., Çağatay, N., & Vidal, V. (2012). Microevents produced by gas migration and expulsion at the seabed: a study based on sea bottom recordings from the Sea of Marmara, *Geophysical Journal International*, *190*(2), 993–1007.

Tsai, Y. B., Teng, T. L., Chiu, J. M., & Liu, H. L. (1977). Tectonic implications of the seismicity in the Taiwan region. *Memoir of the Geological Society of China*, *2*, 13–41.

Vidal, V., Ichihara, M., Ripepe, M., & Kurita, K. (2009). Acoustic waveform of continuous bubbling in a non-Newtonian fluid. *Physical Review E*, *80*(6), 066314. https://doi.org/10.1103/PhysRevE.80.066314

Wang, C. Y., & Shin, T. C. (1998). Illustrating 100 years of Taiwan seismicity. *TAO*, *9*(4), 589–614.

Weaver, R. L., & Lobkis, O. I. (2001). On the emergence of the Green's function in the correlations of a diffuse field. *The Journal of the Acoustical Society of America*, *110*, 3011–3017. https://doi.org/10.1121/1.1417528

Webb, S. C. (1998). Broadband seismology and noise under the ocean. *Reviews of Geophysics*, *36*(1), 105–142.

Webb, S. C. (2007). The earth's 'hum' is driven by ocean waves over the continental shelves. *Nature*, *445*, 754–756.

Webb, S. C., Zhang, X., & Crawford, W. (1991). Infragravity waves in the deep ocean. *Journal Geophysical Research*, *96*, 2723–2736.

Wilson, J. D., & Makris, N. C. (2008). Quantifying hurricane destructive power, wind speed, and air-sea material exchange with natural undersea sound. *Geophysical Research Letters*, *35*, L10603. https://doi.org/10.1029/2008 GL033200

Wu, F. T., & Rau, R. J. (1998). Seismotectonics and identification of potential seismic source zone in Taiwan. *Terrestrial, Atmospheric and Oceanic Sciences*, *9*(4), 739–754.

Wunsch, C. (1975). Internal tides in the ocean. *Review of Geophysics*, *13*, 167–182.

Zhang, J., Gerstoft, P., & Bromirski, P. D. (2010). Pelagic and coastal sources of P-wave microseisms: Generation under tropical cyclones. *Geophysical Research Letters*, *37*, L15301. https://doi.org/10.1029/2010GL044288

3

Seasonal and Geographical Variations in the Quantified Relationship Between Significant Wave Heights and Microseisms: An Example From Taiwan

Jing-Yi Lin[1,2], Chuen-Teyr Terng[3], Chien-Chih Chen[1], and Chung-Hsiang Mu[4]

ABSTRACT

Investigating significant wave heights (SWHs) predicted from microseisms can complement climate observations and provide information about the interaction of sea storms with coastal environments. By comparing the spectrogram calculated from the data of inland seismic stations and SWHs recorded by buoys around Taiwan, we discern seasonal and geographical differences while estimating wave height with an empirical law. A monthly analysis based on nine seismic stations and buoy pairs shows an approximately linear relationship between the square root of the seismic energy and the wave heights during summer for most stations and a nonlinear relationship in winter. The high correlation coefficient of over 0.8 between the seismic energy and SWHs in winter indicated that direct effects from weather conditions are the primary factor influencing microseism behavior. Although the station in southernmost Taiwan revealed microseism energy distribution similarly to that of other stations, the wave-height fluctuation was distinct, showing slight variations in winter. This observation could relate to the shielding effect of the land topography, which prevents infiltration by strong winds and decreases the ramifications of wave height. This work assesses the possibility of estimating significant sea wave heights around Taiwan and the importance of seasonal and geographical variation while selecting parameters.

3.1. INTRODUCTION

Taiwan experiences a subtropical and tropical climate heavily influenced by the ocean, the movement of air currents controlling the East Asian monsoon, and numerous typhoons. Changes in the marine environment generally significantly affect the formation of extreme climates, which threaten people's life and property. Therefore, it is crucial to understand the marine environment and its interaction with the land. Traditionally, ocean and climate change observations have been primarily realized based on satellite measurements, aerial reconnaissance missions, and in situ monitoring using land-based radars or sea buoys (Kalra et al., 2005; Reul et al., 2012). Although these methods generate abundant data and results, the observation conditions have limitations that are likely to affect the quality and scale of the collected data. For example, satellite images may not have sufficient resolution for rapidly changing climate phenomena. The weather or flight path could restrict the performance

[1]*Department of Earth Sciences, National Central University, Taoyuan City, Taiwan (R.O.C.)*

[2]*Center for Environmental Studies, National Central University, Taoyuan City, Taiwan (R.O.C.)*

[3]*Center Weather Bureau, Taipei City, Taiwan (R.O.C.)*

[4]*Institute of Earth Sciences, Academia Sinica, Taipei, Taiwan (R.O.C.)*

Noisy Oceans: Monitoring Seismic and Acoustic Signals in the Marine Environment, Geophysical Monograph 284, First Edition.
Edited by Gaye Bayrakci and Frauke Klingelhoefer.

DOI:10.1002/9781119750925.ch03

of aerial reconnaissance missions. Land and sea observations are subject to station distribution, etc. Periodic data-transmission and power-supply difficulties affecting sea buoys often lead to discontinuities or data loss for both near-shore and offshore measurements. This situation obstructs the efficiency of the oceanographic data-collecting system. In addition, deploying and maintaining sea buoys is more costly and complex than inland stations. Therefore, it is of great importance to use inland geophysical data to complement investigations of the marine environment such as wave height, wave period, and energy.

Climate change, especially large-scale extreme events, usually involves extensive interactions between the atmosphere, ocean, and solid earth. Pressure, waves, and wind fields interact closely with solid earth (Longuet-Higgins, 1950), generating continuous background seismic noise called *microseisms* in the frequency band between 0.04 and 1 Hz (Peterson, 1993). The two energy peaks in the seismic spectrum located in the vicinity of the 0.04–0.1 Hz and 0.1–1 Hz frequency band consist of the primary microseism with lower amplitude and the secondary microseism with higher amplitude (Bromirski, 2009; R. Haubrich et al., 1963; R. A. Haubrich and McCamy, 1969). Previous studies have analyzed the spectral properties of microseisms, making a close correlation between their power spectral density spectrum and sea weather conditions (e.g., Aster et al., 2008; Lin et al., 2014). Several empirical laws were built using the power spectral energy estimated from seismic stations and sea wave heights measured from buoys (Ardhuin et al., 2011; Bromirski, 2009; Ferretti et al., 2013; Tindle & Murphy, 1999). Longuet-Higgins (1950) suggested that standing waves generated along the coast within a narrow region could quantitatively account for the generation of microseisms. Based on his theory, Bodvarsson (1975) found a roughly linear relationship between the root mean square amplitudes of microseisms and the perfect square of local ocean wave heights. Then Ardhuin et al. (2011) presented the first comprehensive numerical modeling of microseisms by considering the generation of random ocean waves and their coastal reflections, which is valid for the global ocean. Recently, Ferretti et al. (2013) confirmed the basic model for significant wave height evaluation and refined it with a lognormal distribution to develop an empirical, data-driven model for the Mediterranean area; they concluded that estimating wave height depended strongly on the spectral energy of microseisms. Their study found the optimal parameters for the inverse problem using the Markov chain Monte Carlo algorithm.

In an island country like Taiwan, the coastline is characterized by high-density infrastructure and dynamic sediment evolution. Monitoring onshore wave heights is important for the management of coastal development. Thus the ability to estimate and predict onshore wave heights is of significant public interest. The studies mentioned have recently demonstrated a high feasibility of estimating wave height based on land-based seismic stations. However, these studies considered one undiversified geographical location and/or short-term information, which may prevent comprehensive conclusions regarding the relative efficacy of the proposed methods. Taiwan has a complex bathymetry context and is regularly lashed by typhoons. The related study performed in the area should provide a new perspective for estimating wave height via the energy of microseisms.

This chapter is focused on mapping offshore wave heights at particular onshore locations and attempts definitive testing of the method of Ferretti et al. (2013) for modeling near-shore wave height variations. A comprehensive evaluation of the empirical method is performed by considering nine seismic station–buoy pairs across five geographical regions distributed across the island of Taiwan (Fig. 3.1) for 1 to 2 years. Towards quantification of ocean wave heights around Taiwan using land-based seismic data, we hope to explore the primary factors that should be noticed when the microseism energy is used for wave height forecasting with an empirical law.

3.2. METHOD AND DATA PROCESSING

3.2.1. Data

The significant wave height ($H_{1/3}$, hereafter SWH) is a commonly used ocean wave parameter to quantify sea wave height. It is defined traditionally as the mean wave height of the highest third of the waves measured over a period T (Sverdrup & Munk, 1947).

Marine meteorological observation in Taiwan is conducted by the Marine Meteorology Center of the Central Weather Bureau (CWB). The Meteorological Buoy Network comprises several marine buoys providing near-real-time monitoring of coastal waves, ocean currents, sea-level variation, and seawater temperature in the vicinity of Taiwan. The monthly values of the collected data from several years are published on the CWB website (http://www.cwb.gov.tw/V7/climate/marine_stat/wave.htm). Every hour, the buoys report SWHs, wave directions, and corresponding periods using data from 10 minutes before the specific observation time (i.e., the data recorded between 00:50 and 1:00 are used for the value of 1:00). The data is recorded via a sampling interval of 0.5 seconds. Notably, the wave heights and their periods are computed through signal filtering and energy density spectrum analysis.

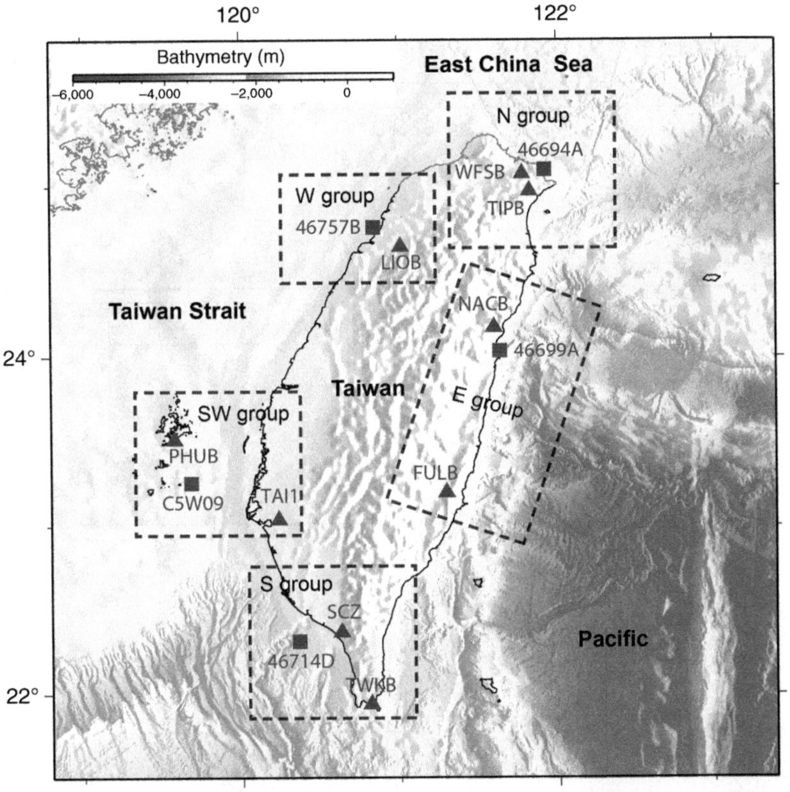

Figure 3.1 General map for the distribution of seismic stations (triangles) and buoys (squares). The gray dashed rectangles framed the five different geographic areas (E, N, W, SW, and S groups) discussed in the text.

Figure 3.2b shows the wave-height distribution of three buoys, 46699A (E group in Fig. 3.1), 4657B (W group in Fig. 3.1), and 46714D (S group in Fig. 3.1) for the year 2014. The wave height range is between 0 and about 748 cm. Relatively larger SWHs occur in winter for stations located in east and west Taiwan, such as 46699A and 46757B. However, no significant changes occurred for the station in southernmost Taiwan (46714D). The largest wave heights occur in the summertime and are suggested to be due to the passing of typhoons (Fig. 3.2).

In this study, we use broadband seismic data with a sampling frequency of 100 Hz from selected stations of the CWB and Broadband Array in Taiwan for Seismology (BATS) network (http://bats.earth.sinica.edu.tw) to estimate spectral energy. Continuous vertical component seismic data were analyzed by resampling the signal at a frequency of 2 Hz. After removing the offset and linear trend, the Fourier transform algorithm was applied to the seismic waveform data with a one-hour time window corresponding to one half-hour before and after the observation time for the significant wave height record. The spectral power density was calculated using a 0.025 Hz frequency sample rate. Thus, we obtained the microseism energy for the different frequency bands.

To avoid the influence of occasional large-amplitude transients (e.g., earthquakes), outbursts of high-amplitude signals were excluded from the calculation of spectral energy. Since the duration of earthquake-related event is not long for teleseisms or low-frequency earthquakes with small magnitude, their presence should make only a negligible contribution to our results. Some data gaps exist due to power failures or discontinuous data transmission caused by bad weather (e.g., data of the SCZ station in Fig. 3.2c). The spectrogram of microseisms at different seismic stations generally displays similar patterns. However, relatively higher energy was observed for frequencies higher than 0.5 Hz for stations located to the west of Taiwan, such as LIOB. Overall, apart from a few individual peaks related to individual storms (Figs. 3.2a,c), noise levels during September–March (winter) are slightly higher than during the summer months, when the noise levels remain lower, particular in the low-frequency band.

To comprehensively evaluate spectral data, we also considered nine seismic station–buoy pairs across five geographical regions in Taiwan. Each region has distinct properties in terms of bathymetry and wave climate. We divided our station pairs into these five groups for convenience (Fig. 3.1). The first group, which consists of the

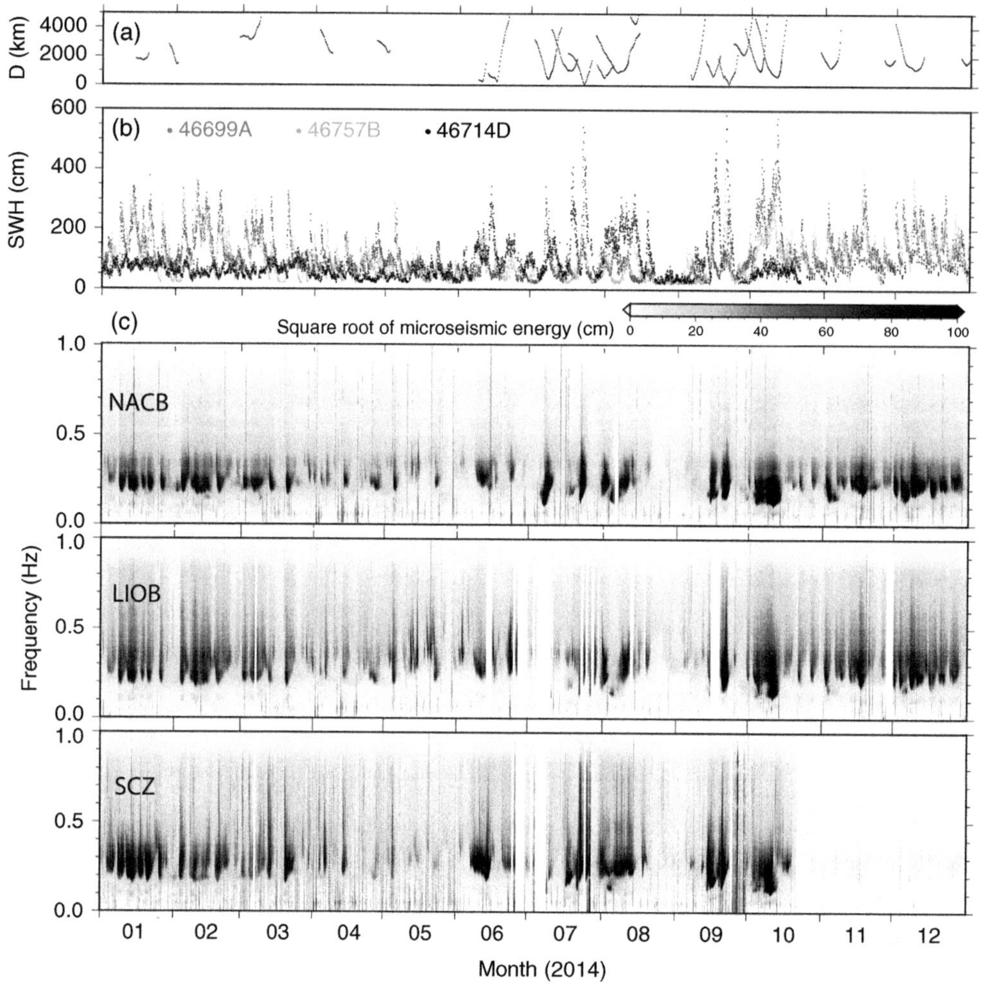

Figure 3.2 General data inspection of the wave heights and seismic data from 2014 for three buoys and seismic stations. (a) The distance between the typhoon eye and the NACB seismic station. (b) Records of SWHs of the three different buoys. (c) Spectrogram of three seismic stations located in different geographic area. The location of the buoys and the seismic stations is shown in Fig. 3.1.

NACB and FULB seismic stations, is located along the eastern coast, facing the deep and open sea (E group). The WFSB and TIPB stations belong to the N group, where the bathymetry transfers from the deep sea to shallow water. Only one station in western Taiwan, LIOB, is available for the analysis; it defines the W group and is characterized by very shallow water (about 200 m). The SW group is composed of two stations with very different coastline configurations. PHUB is located on a small islet, and TAI1 is on the southwestern side of the main Taiwan Island. TWKB and SCZ form the southernmost group (S group), where the bathymetry also changes rapidly, as in the N group.

3.2.2. Method

Previous studies have established a series of empirical laws for forecasting wave height. Ferretti et al. (2013) used one year's data to calibrate an empirical predictive relation for sea wave height. Its functional form is a modified version of the classical definition of SWH ($H_{1/3}$). Thus, the wave height estimated from the energy of microseisms $\mathrm{H}_{1/3}^{\mathrm{calc}}$ can be described as

$$H_{\frac{1}{3}}^{\mathrm{calc}} = 4c\sqrt{\int_{f_{min}}^{f_{max}} \frac{2}{T}|S_m(f)|^2 \mathrm{d}f} \tag{3.1}$$

where c is a scaling coefficient that adjusts the level of the seismic signal to the sea wave height, $\frac{2}{T}|S_m(f)|^2$ is the spectral energy of the microseism, and f_{min} and f_{max} are the minimum and maximum frequencies of integration. Then they modified the equation based on the lognormal function for the simulation and refined the equation as follows:

$$H_{\frac{1}{3}}^{\mathrm{calc}} = \exp\left(a + b\ \ln\left(\sqrt{\int_{f_{min}}^{f_{max}} \frac{2}{T}|S_m(f)|^2 \mathrm{d}f}\right)\right) \tag{3.2}$$

where a, b, f_{min}, and f_{max} are the four unknowns of the model, and the $\frac{2}{T}|S_m(f)|^2$ are calculated at different frequencies based on the broadband seismometer data. Therefore, the seismic energy at the frequencies located between f_{min} and f_{max} was summed to get the total seismic energy to estimate wave heights. Then we can obtain the square root of the microseismic energy and proceed to the following calculation. To facilitate the following discussion, we modified the equation as follows:

$$H_{\frac{1}{3}}^{\mathrm{calc}} = \exp(a) \times \left(\sqrt{\int_{f_{min}}^{f_{max}} \frac{2}{T} |S_m(f)|^2 \mathrm{d}f} \right)^b \tag{3.3}$$

In this equation, exp(a) is the slope(gradient) of the relationship between the estimated wave height and the square root of the seismic energy, which describes how the estimated wave height changes regarding the square root of seismic energy. b expresses the exponent term of the seismic energy. If b equals 1, there are no powers of variables, and the relationship between the estimated wave heights and the square root of the seismic energy should be linear.

As shown by the equation, four unknowns exist (a, b, f_{min}, and f_{max}). Thus, Ferretti et al. (2016) selected the "best-fitting model" through the Markov chain Monte Carlo algorithm. Their result demonstrates a good ability to reproduce observed sea wave heights using predictive law. However, to evaluate the relationship between the four parameters and the fitting curve, we used the iterative method to find the the root mean square (rms), measured in *cm*, of the observed SHWs (from buoys) and the predicted SHWs (from microseisms). Using this method, we can observe the evolution of the fitting curves by varying the four unknown parameters. The intervals we used in the iteration for a, b, f_{min}, and f_{max} are 1, 0.25, 0.05, and 0.05, respectively.

3.3. TESTING AND DETERMINING PARAMETERS

To study the effectiveness of applying the empirical law proposed by Ferretti et al. (2013) to the Taiwan area, we first analyze the data of the NACB seismic station and its nearest buoy, 46699A. We calculated the wave height estimated from the seismic energy by varying the four parameters, a, b, f_{min}, and f_{max}, and estimated the rms (Fig. 3.3). The result showed that we could find a local minimum for the rms that is generally smaller than 35 cm (Fig. 3.3a). However, two minimum peaks were observed in some months (May, June, and July 2014). In addition, we found that the values of a and b at the minimum rms point varied from month to month. This distribution suggests that the monthly parameter sets for predicting wave height may be different, probably inferring a time-dependent variation. The changes of f_{min} and f_{max} also do not correlate with the rms. The brightness of the symbols in Fig. 3.3 denotes the f_{max} used for the simulation, and no prevalent distribution pattern was observed. In other words, the fitting result is not sensitive to the frequency band used to calculate seismic energy. Otherwise, although no absolute correlation between the four parameters and the least rms results could be determined, the trade-off pattern between a and b was clearly illustrated when the rms was less than a certain value (50 cm in Fig. 3.3b). This phenomenon is understandable based on equations 3.2 and 3.3. As the observed wave height is fixed, when a is larger, b should be smaller to fit the observed data.

Moreover, a positive correlation exists between SWHs measured by buoys (black dots in Fig. 3.4) and seismic energy at frequencies between 0.1 and 0.9 Hz (Fig. 3.4). Despite the relatively lower seismic energy level at the 0.9 Hz frequency compared to other frequencies, a strong correlation persists. This observation suggests that wave activities can have extensive influence in the frequency bands between 0.1 and 0.9 Hz, and no matter which frequency band is used, SWHs can always be estimated with their corresponding a and b values. Consequently, if the frequency bands used for estimating seismic energy do not change the simulation result dramatically, we suggest simplifying the calculation by fixing a frequency band for evaluating seismic energy. The seismic energy located at the 0.01–1 Hz frequency band is considered the primary zone affected by microseisms, as identified by many previous studies. Therefore, this frequency range (0.01–1 Hz) was chosen for the posterior analyses. After the frequency band was fixed, the seismic energy could be calculated and compared to the observed SWHs by varying a and b.

As only two unknown parameters remain, the regression method was used to find the best-fitting curve and the correlation coefficient between the data points (Fig. 3.5). Figures 3.5c–f demonstrate the relationship between SHWs and the square root of the seismic energy between 0.01 and 1 Hz for the NACB seismic station and its corresponding buoy, which shows a visible seasonal variation. The data distribution has a bisected shape in autumn: that is from September to November (Figs. 3.5f,g). The higher branch has a pattern similar to that observed in winter and spring, while the lower branch distribution is close to that in summer. In addition, an approximately linear correlation was observed during the summertime and fall (June to November) (Figs. 3.5e,f), whereas a nonlinear distribution dominates the wintertime (December to February) (Fig. 3.5c). Fig. 3.5b shows the SWHs and the simulated SHWs by using the a and b obtained based on an entire year of data. The result indicates that the fitting parameters obtained based on annual data cannot provide a reliable estimation for the SHW variation for the entire year. Even though the fit seems good for the summertime, the wave height is often

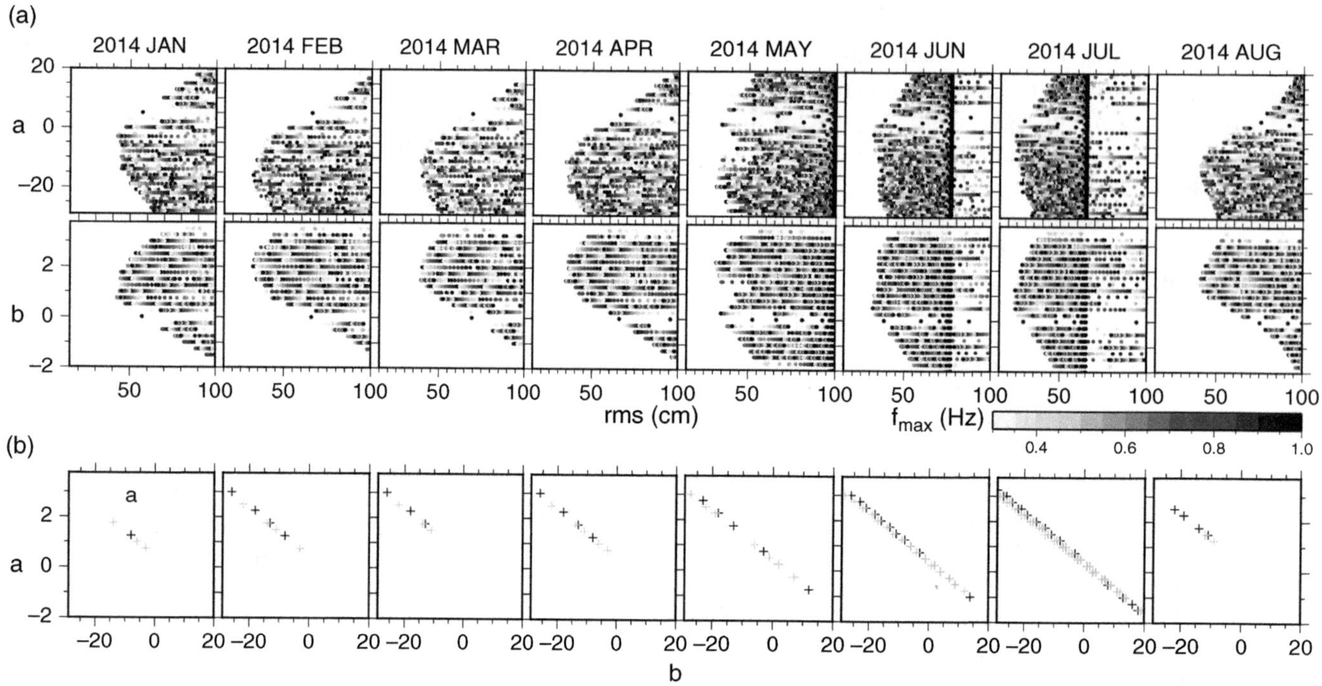

Figure 3.3 Distribution of estimated values for a, b, and f_{max} in relation to rms using data from the NACB stations pair for the first eight months of 2014. (a) The variations in a and b concerning changes in rms, with shades indicating f_{max}. Notably, rms values appear to be insensitive to the frequency band used for estimating seismic energy. (b) a versus b distribution when rms is less than 50 cm. Symbols in black and gray represent parameters obtained from seismic energy estimations in the frequency bands 0.01–0.95 Hz and 0.3–0.35 Hz, respectively, revealing a trade-off pattern in the distribution of a and b.

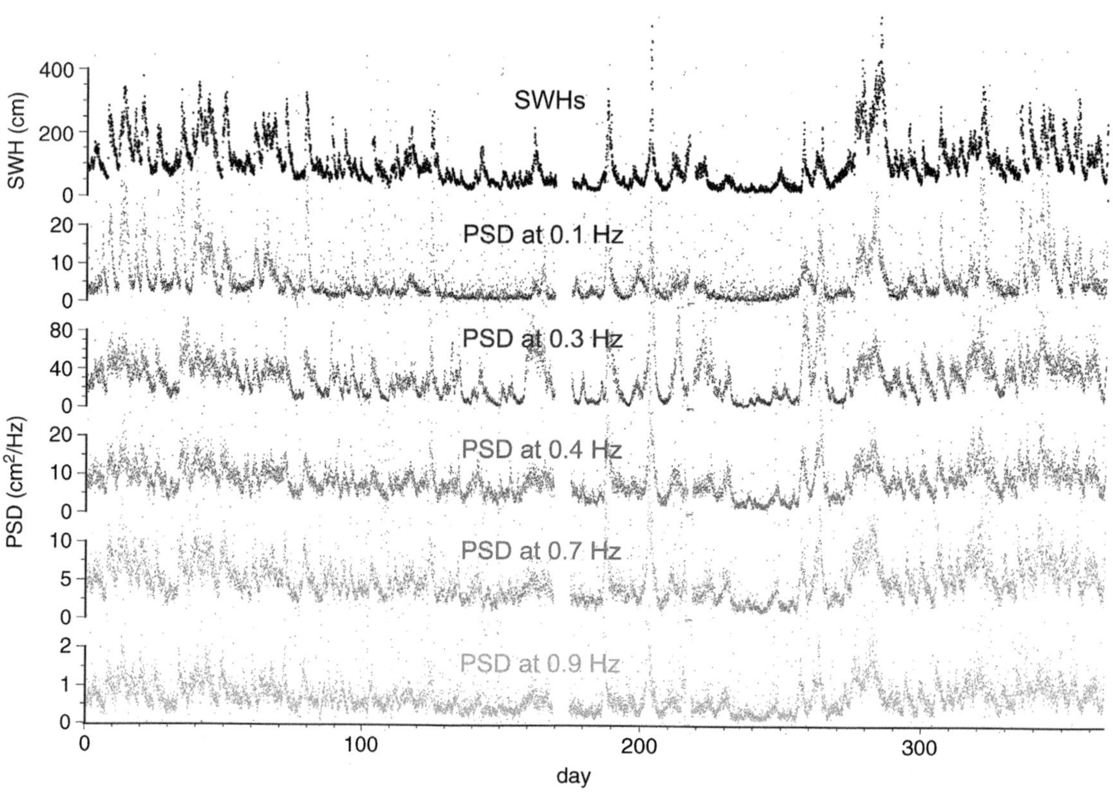

Figure 3.4 Observed SWHs alongside seismic energy, represented as the power spectral density extracted at various frequencies from the data collected at NACB station pairs. The x-axis of the graph represents the day of the year.

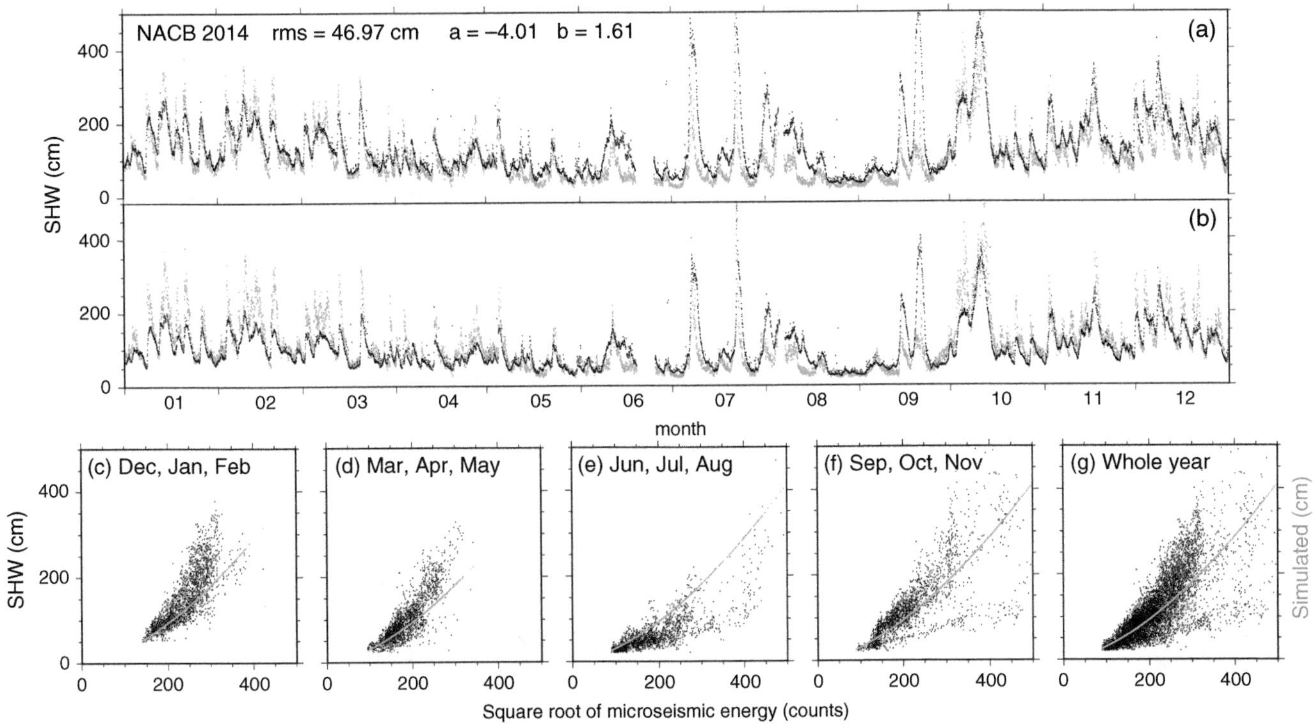

Figure 3.5 (a, b) Distribution of observed SHWs (black dots) and simulated SHWs (gray dots) for the NACB station pair. These were calculated using *a* and *b* values obtained from monthly data and a year's worth of data, respectively. (c–g) Seasonal distribution of SHWs (black dots) in relation to the square root of microseismic energy within the frequency band 0.01–1.00 Hz. Notably, the relationship between SHWs and microseismic energy varies significantly across different seasons. Gray lines represent the relationship between simulated SHWs and the square root of microseismic energy. These simulated SHWs were computed based on the best-fitting line derived from the entire year's dataset.

underestimated in winter. This distribution pattern indicated a clear time-dependent relationship between SWHs and seismic energy. Therefore, we perform the relative simulation monthly in the following analyses.

3.4. RESULTS AND DISCUSSION

The best-fitting results obtained from *a* and *b* of the nine seismic station–buoy pairs estimated every month are plotted in Fig. 3.5a. The related parameters are plotted in Fig. 3.6. We observed a significant improvement in the prediction of winter wave heights compared with those estimated on whole-year data (Figs. 3.5a,b). In the results, *b* is between –0.46 and 2.95, whereas *a* is from –13.33 to 6.49. The average rms of each evaluated month is between 0.00 cm (with the inclusion of NaN, which means instances where the value is not a number) and 147.02 cm, and *R* is in the range from 0 to 0.95. The extremely small or large rms values (about less than 10 cm or more than 120 cm) are generally related to a lack of data, which could be due to a very low *R*. In most cases, *R* is larger than 0.5, which infers a good correlation between the square root of seismic energy and observed SWHs. However, for the months where a smaller or nan correlation coefficient was obtained, bad data quality during that recording period or a slight variation of SWHs seemed to be a possible explanation, such as in September 2015. After a close check of the data, some extremely low energy records appeared in the seismic spectrogram during certain periods, such as the spectrogram of the SCZ station shown in Fig. 3.2c. The lack of wave height records or seismic data could be recurrent due to bad weather conditions or transmission problems. In these cases, the estimated *a* and *b* were shown as outliers characterized by values far from the average, such as the parameters for the SCZ station pair during November and December in both 2014 and 2015.

Based on equation 3.3, when *b* is close to 1, the square root of the seismic energy and the observed wave heights show a linear correlation. However, we found that *b* is larger than 1 in most cases, suggesting that a nonlinear relationship occurs more often. We found that when *b* is minor, *a* becomes more prominent, which could also be determined based on equation 3.3, as discussed previously. Notably, the station pairs in the same geographic

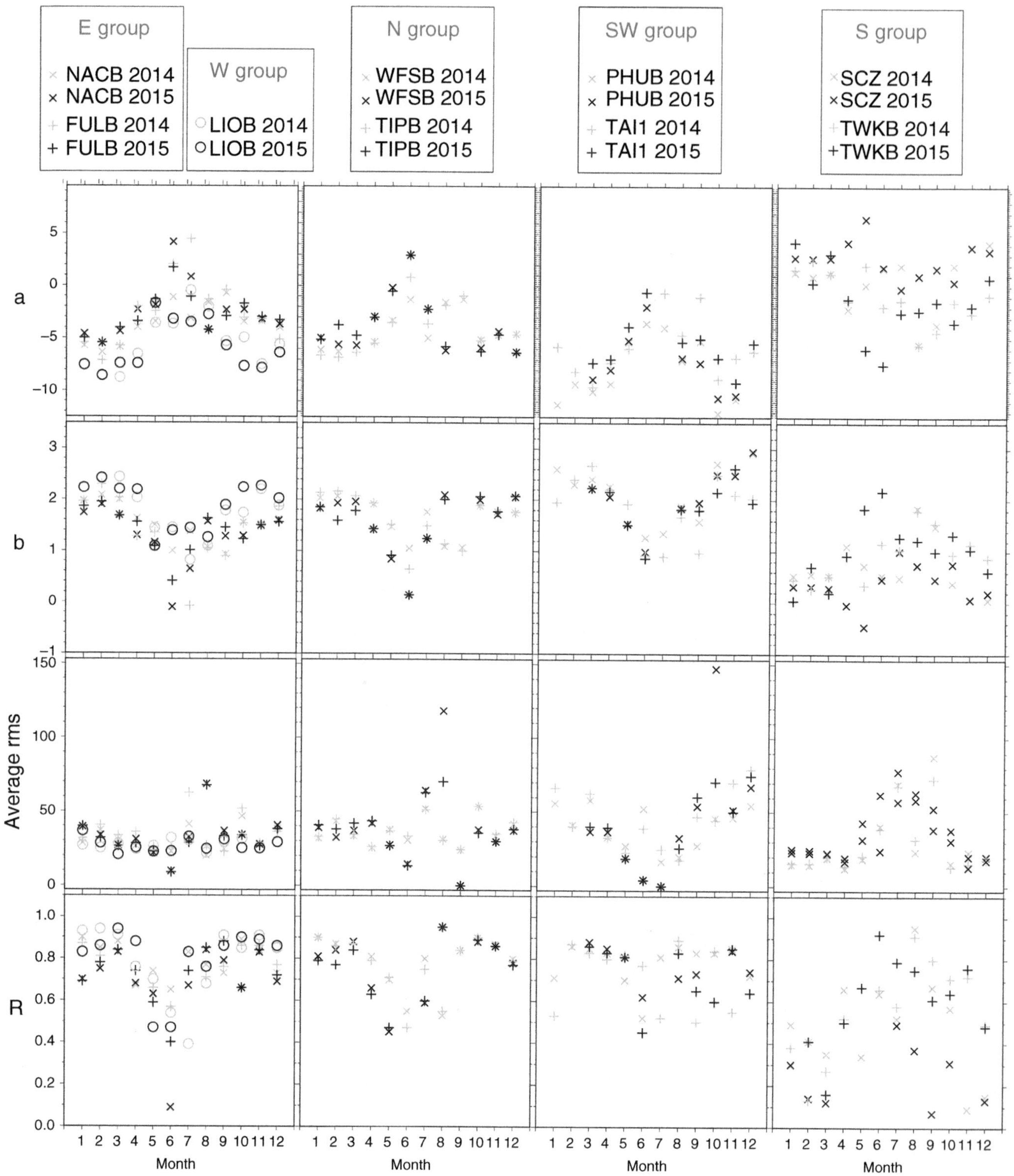

Figure 3.6 Monthly distribution of *a*, *b*, average rms, and the correlation coefficient (*R*) derived from data collected from the nine station pairs depicted in Fig. 3.1.

area show similar results (Fig. 3.6). This phenomenon suggests that seismic stations close to each other should have very similar microseismic responses regarding wave activity. Therefore, the following discussion is based on different geographical groups.

3.4.1. Seasonal Variation

Generally, the weather conditions do not change significantly for the same period of a year. Therefore, the estimated a and b for the same month in different years should be similar. In our results, this hypothesis was proven, but only for wintertime. For example, the parameters obtained in November for almost all the station pairs showed very similar values in 2014 and 2015 (Fig. 3.6). This consistency infers that the weather conditions during the same annual period, particularly in winter, cause similar wave activity, affecting the seismic stations with a comparable way. However, the parameters obtained for the summertime were less persistent. The correlation coefficient (R) showed similar seasonal variation for these station pairs: a stronger correlation in winter and a weak one in summer. These phenomena may be caused by the diversification of typhoon trajectories and their impacts on wave activity in the summertime. In general, b is greater in wintertime, except for the stations in the S group (Fig. 3.6). In the summer season, the estimated b is closer to 1, which suggests a linear relationship between the square root of seismic energy and the observed SWHs. On the other hand, a higher b in the winter reveals a nonlinear relationship, which suggests that when the wave heights are greater, the seismic energy does not increase proportionally. Although the S group does not have the same patterns as discovered in other station pairs (Fig. 3.6), seasonal variation also exists. Both b and the correlation coefficient have larger values in summer and smaller values in winter. A very high correlation coefficient appears for summer: 0.95 in August 2014 for the SCZ station pair and 0.92 in June 2015 for the TWKB station pair. However, the value is usually less than 0.5 in wintertime. This characteristic is the opposite of the results of other station pairs.

Seasonal variations in microseismic energy are observed worldwide (Rhie & Romanowicz, 2004; Schimmel et al., 2011; Stutzmann et al., 2009; Traer et al., 2012). These studies mentioned that the most energetic sources in the Northern and Southern Hemispheres during their respective winter seasons created relatively more significant microseismic noise levels. This observation agrees with what we observed for the middle-northern part of Taiwan, which showed a more coherent seismic energy density and a much better correlation coefficient in winter. This consistency suggests that the northeastern monsoon could be the most critical factor affecting the area's microseismic energy distribution. Thus, the general increase of b for almost all the station pairs located in middle-northern Taiwan indicates that most of Taiwan is influenced by this strong weather system and wave activity in winter. On the other hand, no specific intensive weather system affects the Taiwan area in summer except the irregular lashing of typhoons. This may be why the estimated a and b are so different from those in wintertime. For the southernmost area of Taiwan, the S group, the low observed SWHs and irregular a and b values in winter denote that the area is outside the influence range of the northeastern monsoon. Meanwhile, a good correlation in summer appears when typhoons pass near southern Taiwan.

Regarding the correlation coefficient, most previous studies showed that measured wave heights were in closer agreement during winter wave conditions than in summer (Howell, 1990; Thompson & Vincent, 1985). However, the result of Tillotson and Komar (1997) showed a conflicting observation with a poor agreement in winter. Our results demonstrate a high correlation coefficient in winter for most station pairs, including E, N, W, and SW groups, and in summer for the S group. This phenomenon suggests that a better-fitting result should be related to intensive weather conditions, which can produce a coast-wide effect and increase the SWHs. Meanwhile, the local weather is not energetic enough to cause larger waves, and the relation between seismic energy and wave heights is not as evident.

Based on these observations, we summarized that strong weather systems are the primary factor controlling the correlation between seismic energy and SWHs. Under these weather conditions, a relatively larger b, inferring a nonlinear relationship, also prevails. Thus, the so-called seasonal effect for microseism generation depends principally on when extreme weather conditions happen.

3.4.2. Geographical Variation

Figure 3.2c shows that the SCZ station has a microseismic energy distribution pattern similar to the other stations. However, the much smaller b (less than 1) in winter (Fig. 3.6) suggests that the seismic energy cannot be reflected in the wave height variation. This discrepancy can be explained by analyzing the source area where the microseismic energy was generated. As mentioned in the previous section, microseismic energy in winter is provided by the northeastern monsoon, which comes from the north of Taiwan. The middle and northern parts of Taiwan confront this weather system and are affected directly by the related strong winds. This microseismic source distribution is also confirmed by Chen et al. (2011) using the cross-correlation functions (CCFs) between pairs of stations. Thus, strong wind generates significant wave heights in the corresponding offshore areas, and the stations receive microseismic energy. The high mountains in the middle of Taiwan provide a sheltering effect due to the high topography that may protect southernmost Taiwan from the influence of the fierce wind. Nevertheless, microseismic energy can be transmitted through the land from northern Taiwan to the south. In a small surface area like Taiwan, microseisms may be received everywhere, but this fact may not necessarily be reflected in the wave height changes. This may be why the SCZ station demonstrated high microseismic energy in winter, but no large wave heights were observed. The similarity of the microseismic energy distribution of different seismic stations shown in Fig. 3.2c is a good demonstration.

When typhoons strike the Taiwanese coasts during the summer, some of them near southernmost Taiwan, they can significantly and directly impact the southern shoreline. Thus, tremendous seismic energy and a better correlation between seismic energy and wave heights for the S group occurred in the summer. Ferretti et al. (2013) used the microseism polarization method and revealed two prevailing orientations for the origin of the microseismic energy. Powers in the range 0.10–0.25 Hz occurred when intense storms were present far from the observed station, which did not affect the local wave heights. In contrast, storms near the station generated microseismic energy in the frequency range of 0.25–0.50 Hz, which caused significant wave heights. However, based on the results observed from the S group in our study, we cannot observe any discrepancy in the energy distribution at different frequency bands. In the study of Ferretti et al. (2013), the seismic station was far from the source area of the microseisms. Thus, high energy appeared only in a relatively low-frequency band after traveling a long distance, and no corresponding wave heights were caused locally. But in our study, as the seismic station of the S group is not far from the stations located to the north of Taiwan, microseismic energy was not transmitted through a long distance and did not have significant attenuation. This may explain why the seismic stations in the S group can receive considerable energy, but the observed SHWs show slight variation. Consequently, the seismic station's position and inland topography affect the relationship between wave activities and microseismic energy.

Besides the source area difference, other studies showed that offshore bathymetry can affect microseism generation (Kedar et al., 2008). However, in our study, the bathymetry effect does not influence microseismic patterns. It is true that the LIOB station pair (W group) is characterized by a very shallow water depth and displayed a higher value of b then other areas (Fig. 3.2). However, no significant difference in parameters existed between the station pairs in the E and N groups, which have distinct bathymetry characteristics. Therefore, we cannot verify the bathymetry effect based on our analyses.

We should consider the related factors while practicing wave height prediction. To resolve this problem, other criteria may be helpful for determining distant sources for microseisms, such as the frequency of maximal seismic energy and local wind speed data. However, further analyses concerning the frequency variation of microseisms and other oceanographic data would be needed to obtain more information.

3.4.3. Residual Distributions of the SHW Simulation

Based on the residual distributions, we noticed that a rms of about 30 cm was estimated in winter for most stations, including the N, W, and E groups (Fig. 3.6). In addition, R is relatively high compared to the results obtained from other geographic groups. Therefore, this method can be applied to northern Taiwan with a 30 cm uncertainty. This value is less than one-tenth of the maximum SHWs observed in winter. Even though the value of rms is less than 30 cm for the S group in winter, the small R infers a bad correlation between the observed and estimated SHWs. This suggests that while an area is only affected by minor wave activities, a smaller rms can be obtained. For the SW group, R in winter is generally higher than 0.8. However, we obtained an average rms of about 50 cm, slightly higher than that in the groups relatively closer to the northern part of Taiwan. This result shows that the SHWs simulation in southwestern Taiwan should be less accurate than in the northern part.

A significant variation in the rms and R is observed in the summertime for all the stations. The former ranges from about 20 to 70 cm, while the latter ranges from 0.4 to 0.7. This distribution suggests inconstant weather conditions in summer that are strongly linked to the occurrence of typhoons. In these cases, typhoon trajectories and intensity should be considered when simulating wave heights.

3.5. CONCLUSIONS

Knowledge of the wave climate at near-shore sites is crucial for coastal research and marine engineering. However, observation conditions can limit investigations of the ocean and climate change based on traditional approaches such as satellite measurements, aerial missions, and marine buoys. Using inland geophysical data can complement investigations of the marine environment. In the present study, wave measurements recorded by ocean buoys around the coast of Taiwan were compared with microseism observations at inland broadband seismometers. Evaluating near-shore wave heights in several study regions allows us to consider the reasons for differential simulation performance in each study area. Most regions of Taiwan, including the northern, eastern, and western parts, are strongly influenced by the northeastern monsoon, which results in a good simulation result with relatively higher correlation coefficients. In addition, a clear nonlinear relationship between the square root of the seismic energy and the observed SWHs exists in winter. During summer, except for the irregular lashing of typhoons, no specific weather system affects the area, and no significant wave heights are detected. Thus, the correlation between the wave heights and microseismic energy is less evident, and the value of b is close to 1. For southernmost Taiwan (the S group), the relationship between offshore wave heights and the seismic energy at inland seismic stations is moderated strongly by the typhoons, as affirmed by the better correlation coefficient. However, even though the seismic energy is high in winter, no significant changes in wave heights are seen due to the sheltering effect of the inland topography, which may prevent strong winds.

We consider the seasonal and geographic effects recognized in our study to provide a valuable guide to model-estimated complexity in wave height prediction based on inland seismometers in an Island country like Taiwan. First, we note that the simulation may be insensitive to the frequency band used to estimate the seismic energy. Also, the seasonal effect indicates that wave height prediction cannot be performed using a single combination of parameters for the year. Therefore, a season-dependent parameter set should be considered. Furthermore, the correlation between seismic energy and wave heights is better under intense weather conditions, but the relationship is not linear. Finally, for a small area like Taiwan, most microseismic energy can be transmitted through the land and should be evaluated carefully.

ACKNOWLEDGMENTS

We thank the Central Weather Bureau of Taiwan for kindly providing significant wave height and broadband seismometer data. C.-L. Lo, W.-N. Wu, and W.-B. Doo are recognized for providing valuable suggestions for the manuscript. All figures were made using the Generic Mapping Tools (Wessel & Smith, 1998). This research was supported by the Central Weather Bureau of Taiwan with grant MOCT-CWB-106-O-04 and by the Ministry of Science and Technology, Taiwan, under contract MOST-105-2116-M-008-010 and MOST-106-3113-M-008-001.

REFERENCES

Ardhuin, F., Stutzmann, E., Schimmel, M., & Mangeney, A. (2011). Ocean wave sources of seismic noise, *Journal of Geophysical Research. Oceans*, *116*(C9). https://doi.org/10.1029/2011JC006952

Bodvarsson, G. M. (1975). *Ocean wave-generated microseisms at the Oregon coast* (M.S. thesis). Oregon State University.

Bromirski, P. D. (2009). Earth vibrations, *Science*, *324*(5930), 1026–1027. https://doi.org/10.1126/science.1171839

Chen, Y. N., Gung, Y., You, S. H., Hung, S. H., Chiao, L. Y., Huang, T. Y., et al. (2011). Characteristics of short period secondary microseisms (SPSM) in Taiwan: The influence of shallow ocean strait on SPSM, *Geophysical Research Letters*, *38*(4). https://doi.org/10.1029/2010GL046290

Ferretti, G., Zunino, A., Scafidi, D., Barani, S., & Spallarossa, D. (2013). On microseisms recorded near the Ligurian coast (Italy) and their relationship with sea wave height. *Geophysical Journal International*, *194*(1), 524–533. https://doi.org/10.1093/gji/ggt114

Ferretti, G., Scafidi, D., Cutroneo, L., Gallino, S., & Capello, M. (2016). Applicability of an empirical law to predict significant sea-wave heights from microseisms along the Western Ligurian Coast (Italy). *Continental Shelf Research*, *122*(1), 36–42. https://doi.org/10.1016/j.csr.2016.03.029

Haubrich, R., Munk, W., & Snodgrass, F. (1963). Comparative spectra of microseisms and swell. *Bulletin of the Seismological Society of America*, *53*(1), 27–37.

Haubrich, R. A., & McCamy, K. (1969), Microseisms: coastal and pelagic sources. *Reviews of Geophysics*, *7*(3), 539–571. https://doi.org/10.1029/RG007i003p00539

Howell, R. K. (1990). *An algorithm for the extraction of ocean wave spectra from narrow beam HF radar backscatter* (PhD thesis). Memorial University of Newfoundland.

Kalra, R., Deo, M., Kumar, R., & Agarwal, V. K. (2005). Artificial neural network to translate offshore satellite wave data to coastal locations. *Ocean Engineering*, *32*(16), 1917–1932. https://doi.org/10.1016/j.oceaneng.2005.01.007

Kedar, S., Longuet-Higgins, M., Webb, F., Graham, N., Clayton, R., & Jones, C. (2008). The origin of deep ocean microseisms in the North Atlantic Ocean. *Proceedings of the Royal Society A: Mathematical, Physical and Engineering Sciences*, *464*(2091), 777–793. https://doi.org/10.1098/rspa.2007.027

Longuet-Higgins, M. S. (1950). A theory of the origin of microseisms. *Philosophical Transactions of the Royal Society A*, *243*(857), 1–35. https://doi.org/10.1098/rsta.1950.0012

Peterson, J. (1993). *Observations and modeling of seismic background noise (open-file report 93-322)*. U.S. Geological Survey. https://doi.org/10.3133/ofr93322

Reul, N., Tenerelli, J., Chapron, B., Vandemark, D., Quilfen, Y., & Kerr, Y. (2012). SMOS satellite L-band radiometer: A new capability for ocean surface remote sensing in hurricanes. *Journal of Geophysical Research: Oceans*, *117*(C2). https://doi.org/10.1029/2011JC007474

Rhie, J., & Romanowicz, B. (2004). Excitation of Earth's continuous free oscillations by atmosphere–ocean–seafloor coupling. *Nature*, *431*(7008), 552–556. https://doi.org/10.1038/nature02942

Schimmel, M., Stutzmann, E., Ardhuin, F., & Gallart, J. (2011). Polarized Earth's ambient microseismic noise. *Geochemistry, Geophysics, Geosystems, 12*(7). https://doi.org/10.1029/2011GC003661

Stutzmann, E., Schimmel, M., Patau, G., & Maggi, A. (2009). Global climate imprint on seismic noise. *Geochemistry, Geophysics, Geosystems, 10*(11). https://doi.org/10.1029/2009GC002619

Sverdrup, H. U., & Munk, W. H. (1947). *Wind, sea, and swell: theory of relations for forecasting (Report 601)*. U.S. Navy Hydrographic Office.

Thompson, E. F., & Vincent, C. (1985). Significant wave height for shallow water design. *Journal of Waterway, Port, Coastal, and Ocean Engineering, 111*(5), 828–842. https://doi.org/10.1061/(ASCE)0733-950X(1985)111:5(828)

Tillotson, K., & Komar, P. D. (1997). The wave climate of the Pacific Northwest (Oregon and Washington): A comparison of data sources. *Journal of Coastal Research, 13*(2), 440–452.

Tindle, C. T., & Murphy, M. J. (1999). Microseisms and ocean wave measurements. *IEEE Journal of Oceanic Engineering, 24*(1). 112–115. https://doi.org/10.1109/48.740159

Traer, J., Gerstoft, P., Bromirski, P. D., & Shearer, P. M. (2012). Microseisms and hum from ocean surface gravity waves. *Journal of Geophysical Research: Solid Earth, 117*(B11), B11307. https://doi.org/10.1029/2012JB009550

4

Listening for Diverse Signals From Emergent and Submarine Volcanoes

Chastity Aiken

ABSTRACT

Underwater volcanoes produce seismic, tsunami, and volcanic hazards. Yet our first clues regarding underwater volcanism are often limited to observations from satellite imagery or regional seismic networks because of infrequent marine surveys and a limited number of seafloor observatories. On regional networks, large explosions and eruptions can be observed as well as volcanic tremor and magnitude >3 volcano-tectonic earthquakes and very-long-period events, albeit with limited resolution. However, at long distances, other types of signals such as long-period events, explosive-type signals, and landslides may go undetected. Thus there is a need to monitor underwater volcanism locally, to increase observations and enhance their quality and, consequently, sharpen our view of underwater volcanic processes. At present, local deployments of hydrophones and seismometers are rare for underwater volcanoes, perhaps because they seldom present a risk to communities. However, in the future, underwater volcano monitoring may shift toward using existing telecommunication cables. In this chapter, underwater volcano signals are discussed: how they are recorded, detected, monitored, and distinguished in underwater data in the context of known subaerial and submarine volcanism. Examples of known and monitored underwater volcanism are presented as well as economical and environment-friendly real-time submarine volcano monitoring options for the future.

4.1. INTRODUCTION

Emergent and submarine volcanic systems, such as central and shield volcanoes, volcanic domes and cones, calderas, and lava fields, constitute the majority of volcanism on Earth. They arise in three primary geological settings: island arcs, intraplate (hotspots), and mid-ocean ridges. Some submarine volcanoes may eventually grow tall enough in these settings to form ocean islands like the Bonin Islands, the Azores, and La Réunion Island, but many do not. Some underwater volcanoes, such as Kick 'em Jenny, have been investigated and monitored for the seismic, volcanic, and tsunami hazards they pose to coastal communities, the maritime industry, and ecosystems (Fig. 4.1). Like subaerial volcanoes, earthquakes of notable magnitude from submarine volcanoes can damage coastal community infrastructure—more so if communities are not built to resist seismic damage, as on Mayotte in 2018 following a volcanic eruption (e.g., Taillefer et al., 2022). The maximum reported magnitude of a volcanic swarm has reached 6 on the Richter scale (McNutt & Roman, 2015), and this was also recently observed at the 2022 submarine Hunga Tonga-Hunga Ha'apai volcanic eruption (Kintner et al., 2022).

Volcanic hazards, such as ballistic ejecta, can also arise from shallow submarine volcanoes (e.g., Fiske et al.,

IFREMER, Centre Bretagne, Plouzané, France

Noisy Oceans: Monitoring Seismic and Acoustic Signals in the Marine Environment, Geophysical Monograph 284, First Edition.
Edited by Gaye Bayrakci and Frauke Klingelhoefer.

DOI:10.1002/9781119750925.ch04

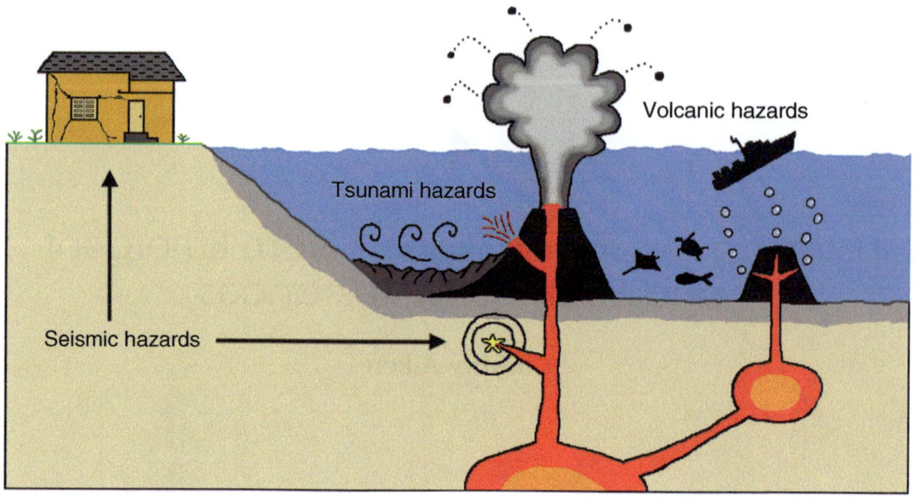

Figure 4.1 Submarine volcano hazards. Submarine volcanoes can present several hazards to coastal communities, the maritime industry, and ecosystems. The hazards are 3 fold: degassing, flank collapse, and earthquakes. Degassing alters the water column's chemistry and could lower water density, whereas flank collapse can lead to tsunamis, plumes, and ballistic ejecta that enter the atmosphere. Earthquakes originate from rising magma and sometimes cause infrastructural damage.

1998). For example, the 2022 Hunga Tonga-Hunga Ha'apai volcano eruption created a plume that infiltrated the mesosphere (Carr et al., 2022). Furthermore, explosive emergent and submarine eruptions and degassing can cause water disturbances and lower water density (e.g., Lindsay et al., 2005). Lower water density could influence vertical ocean currents and cause ships to lose buoyancy and sink, but more certainly, the explosions from a shallow submarine eruption can destroy a ship (e.g., Fiske et al., 1998). In addition, shallow submarine volcanoes can produce chemical imbalances that affect underwater ecosystems and alter atmospheric chemistry, which could be lethal for humans (e.g., Rizzo et al., 2016). Direct observations have shown that explosive underwater eruptions, flank collapse, and mass wasting (landslides) associated with volcanism can produce tsunamis that devastate coastal populations at considerable distances. Submarine volcanoes such as Kick 'em Jenny (1939), Karymskoye Lake (1996), and, more recently, Anak Krakatau (2018) and Hunga Tonga-Hunga Ha'apai (2022) have all produced tsunamis (e.g., Williams et al., 2019; Lipiejko et al., 2021; Moodie, 2022). However, underwater volcanism does not always produce tsunamis. According to numerical models, volcanic-driven tsunamis are a function of several factors, such as eruptive depth, eruptive magnitude, and water-magma interactions (Ulvrova et al., 2014).

Direct evidence of active submarine volcanism includes visual confirmation at the surface (ejecta or steam emission), possibly from satellite imagery (e.g., Tralli et al., 2005; Brandl et al., 2020), but can also arise from detailed bathymetric mapping, measuring the temperature of fluid samples, and dredging lava flows of volcanoes during marine campaigns. However, the oceans are the largest and most inaccessible area on Earth's surface, and locating submarine volcanoes by ship without convincing evidence is exceedingly difficult and costly. While volcanoes expected at plate boundaries and known hotspots have been mapped, identifying which seamounts are active remains a problem for submarine volcano detection and monitoring. This problem stems from the fact that tsunami inundation, breakage of telecommunication cables, and/or seismicity recorded on regional networks are often the first evidence we have of active submarine volcanoes, the latter leading to large errors in earthquake location. For example, seismicity from the Vailulu'u Seamount in the Samoan Archipelago was first recorded in 1973, but the volcano itself was not identified or mapped until 1999 during the AVON 2 and 3 oceanographic campaigns (Konter et al., 2004). A similar situation occurred with the recently discovered volcanic system offshore Mayotte. Seismicity was detected beginning in May 2018 from a single land-based seismometer (Lemoine et al., 2020) and using a global seismic land network (Cesca et al., 2020), but the volcano's edifice and its eruptive activity were not confirmed until May 2019 during the MAYOBS-1 oceanographic mission (Feuillet et al., 2021). Another example is Lo`ihi Seamount, the youngest and smallest Hawaiian volcano, which was first charted as a topographic high in 1940 and later discovered to be seismically active (Garcia, 2015 and references therein). It was not until more earthquake swarms occurred in the 1970s that more dredging expeditions were conducted, during which young lavas were recovered from the Lo`ihi Seamount, confirming the source of the earthquake swarms was of volcanic origin (Garcia et al., 2006).

Emergent and submarine volcanic activity over the last 80 or so years spans a range of depths and tectonic settings, but most of the seismological observations of submarine volcanism are made at regional distances (Tepp & Dziak, 2021). When possible, underwater volcanoes and their eruptive activity have been successfully monitored by Global Positioning System (GPS) networks (e.g., Lemoine et al., 2020) via hydroacoustic arrays (e.g., Fox et al., 1995; 2001; Parnell-Turner et al., 2022), through the deployment of ocean-bottom hydrophones (OBHs) or ocean-bottom seismometers (OBSs) (e.g., Konter et al., 2004; Feuillet et al., 2021), and using regional land-based networks, such as the Polynesian Seismic Network (Schindelé et al., 2002). While regional hydroacoustic networks can detect weaker activity better than land-based seismometers at a similar distance, a regional distance from the source somewhat limits detailed interpretation of the seismic signals. Furthermore, OBS deployments are usually temporary and placed during ongoing eruptive activity after it is first detected by regional seismic networks. On the other hand, some seafloor observatories have been installed to monitor underwater volcanism locally (e.g., Favali et al., 2015; Wilcock et al., 2016), and others are in the process of being constructed (e.g., Aiken et al., 2021; Nomikou et al., 2022). Yet the effort and cost to install seafloor observatories for underwater volcanoes is difficult to justify because most submarine volcanoes are too deep, too remote, or too weak to be considered societal threats. In this regard, regional and global monitoring of submarine volcanoes via seismic and hydroacoustic arrays will likely remain vital due to logistics and costs. In addition, locally deployed instruments will still be useful in real time for the few cases that are threats, as occurred during the 2018 Mayotte seismovolcanic crisis (e.g., Feuillet et al., 2021; Saurel et al., 2022).

Whether regionally or locally detected, underwater volcano signals are indicative of not only seismicity associated with the volcanism itself but also consequences that may arise from the volcanism, such as landslides, magma–water interactions, degassing, etc. (Fig. 4.1). Volcanic signals are often characterized as being low-frequency, hybrid, high-frequency volcano-tectonic (VT), tremor, or explosive (e.g., Lahr et al., 1994; Chouet, 2003; McNutt, 2005; McNutt & Roman, 2015). These characteristic signals are considered to be the "heartbeat" of both emergent/submarine and subaerial volcanoes, offering insight into when eruptions will occur and the hazards they may present, such as flank collapse and tsunamis (e.g., Dondin et al., 2011; Lipiejko et al., 2021) and may even provide an image of their complex reservoir structure (e.g., Foix et al., 2021). In this chapter, Section 4.2 provides a brief introduction to marine instrumentation used to detect and monitor emergent and submarine volcanism. Section 4.3 introduces observed seismic and hydroacoustic signals from emergent and submarine volcanoes and demonstrates how they can be identified in marine recorded data. Examples of these various volcanic signals from emergent and submarine volcanoes in different tectonic settings worldwide are shown, and the future of detecting and monitoring emergent and submarine volcanism is discussed.

4.2. DETECTION AND MONITORING OF SUBMARINE VOLCANISM

Apart from satellite imagery, the detection of submarine eruptions is predominantly a remote "listening" operation using hydrophones and seismometers and consists of analyzing different seismic phases (Fig. 4.2). One example is T-phase arrivals, which are classically the tertiary (T) phase of earthquakes but generally refer to any signal of hydroacoustic path or origin detected seismically (Tepp & Dziak, 2021). T-phase arrivals on regional hydrophones and seismometers (e.g., Bohnenstiehl et al., 2013) often offer the first signals of emergent/submarine volcanic activity. However, emergent/submarine volcano detection and monitoring can be accomplished across various platforms with different time and space scales. These platforms may be fixed, mobile, surface-bound, bottom-mounted, or profiled throughout the water column, and the data is transferred to a ship or the shore by cable, radio, or satellite (Baumgartner et al., 2014) but may also be taken directly from a recovered instrument. For emergent/submarine volcano detection or monitoring, data is typically gathered via hydroacoustic arrays (cabled or not), OBSs, OBHs, or cabled sensors in seafloor observatories (Fig. 4.2).

4.2.1. Hydroacoustic Arrays

Hydroacoustic detection or monitoring of emergent/submarine eruptions is common (Tepp & Dziak, 2021), and volcanism can be monitored at local, regional, or global distances. Long-term hydroacoustic arrays, recording for decades, are typically deployed for purposes other than volcanic monitoring but are sometimes used for it. For example, the Sound Surveillance System (SOSUS) was used to monitor Axial Seamount (Dziak & Fox, 1999), and the Missile Impact Location System (MILS) detected activity in Hawai`i (Johnson et al., 1963). Both networks consist of bottom-mounted hydrophones cabled to onshore stations used by the United States Army. There is also the International

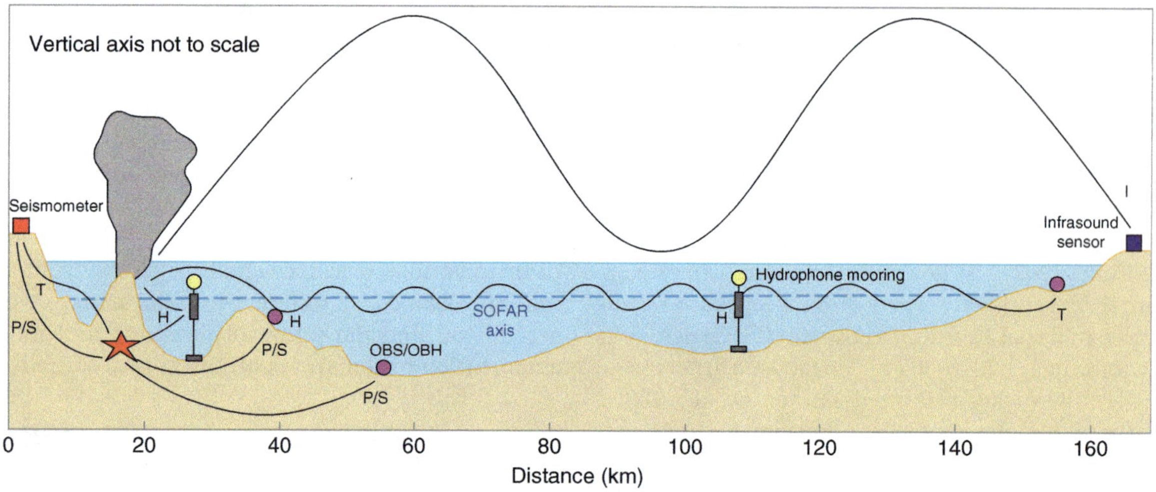

Figure 4.2 Diagram of different recording instruments (OBS = ocean-bottom seismometer, OBH = ocean-bottom hydrophone) and the recorded phases (P, S, T, H, and I) and propagation paths of signals (lines) generated by submarine volcanoes. The star represents earthquakes resulting from volcanism. Adapted from Tepp & Dziak (2021).

Monitoring System (IMS) of the Comprehensive Nuclear-Test-Ban Treaty Organization (CTBTO), which will consist of more than 300 monitoring stations when completed. Eleven of those are/will be hydroacoustic stations for detecting soundwaves from underwater explosions and have already been used to monitor submarine volcanism (e.g., Metz et al., 2016). These arrays offer the detection and long-term monitoring of otherwise unobservable activity associated with emergent and submarine volcanism (e.g., Dziak et al., 2011). Other forms of hydroacoustic detection and monitoring include moored hydrophones, OBHs, and/or the environmentally friendly MERMAIDs (Mobile Earthquake Recorder in Marine Areas by Independent Divers). These instruments are typically deployed and recovered after shorter periods (weeks to a couple of years) during sea campaigns.

Moored hydrophones and OBHs have been used as a targeted approach for monitoring underwater volcanism, where continuous power is otherwise unavailable. Although OBHs are typically used for detecting explosions (e.g., Bazin et al., 2022), gas flux (Chadwick et al., 2008), and activity near or at the surface (e.g., Dziak et al., 2012b), they can also be deployed to detect seismicity such as earthquakes when possible, even if it is more reasonable to use OBSs. Several emergent and submarine volcanoes have been targeted for hydroacoustic monitoring using moored hydrophones and OBHs, such as the NW Rota-1 volcano in the Mariana Arc (Dziak et al., 2012a), West Mata of the Tofua Arc (Dziak et al., 2015), and Mayotte's submarine volcano (Bazin et al., 2022).

MERMAIDS, on the other hand, are floating hydrophones (Simon et al., 2021) that do not require anchors, can be recovered, and do not leave waste on the seafloor. Recently, they have been engineered to land on the ocean bottom (Bonnieux et al., 2021). Being a new technology, MERMAIDs have not yet been intentionally deployed to study submarine volcanism, but a two-year MERMAID experiment in the Indian Ocean yielded evidence of earthquake signals originating from a triple junction of three ocean spreading ridges (Sukhovich et al., 2015). However, because MERMAIDs freely drift, it may be difficult to use them for a targeted, localized, and detailed study such as that of an emergent or submarine volcano.

4.2.2. Seismometer Arrays

OBS/OBH deployments (and recoveries) during sea campaigns are also common for monitoring active underwater volcanism, especially in the geographical region of an active volcano. In general, OBSs and OBHs are deployed simultaneously to detect weak signals (e.g., P and S phases or other types of signals; see Section 4.3) that may be missed by land stations and increase the "seismic resolution" of submarine volcano processes. For example, the East Pacific Rise and Axial Seamount spreading centers have both had local OBS deployments/recoveries, as has the arc volcano Nishinoshima (Tepp & Dziak, 2021).

One notable and recent example of OBS/OBH deployments that have provided new insights into underwater volcanism is the long-term monitoring efforts by REVOSIMA (*RE*seau de surveillance *VO*lcanique-*SI*smologique de *MA*yotte). Mayotte's submarine volcano system was first discovered using global land seismic networks in 2018; later, local OBS deployment and recovery operations were conducted every month for the first few deployments in 2019 and then every six months since 2019, thanks to financing from French public ministries and agencies (Aiken et al., 2021; Saurel et al., 2022). The near-real-time long-term OBS deployments and recoveries offshore Mayotte have proven to be useful for understanding how a new submarine volcano was created at the seafloor (Feuillet et al., 2021) and for conducting local passive tomography that imaged the volcano's plumbing (Foix et al., 2021).

However, the case of Mayotte is exceptional. Continuous OBS deployments and recoveries require funding sources to sustain the costs of ship time, personnel, and OBS consumables (batteries, anchors, etc.), in addition to the fact that a ship must be available to conduct the deployment/recovery operations. Thus, monitoring active submarine volcanism with OBS deployments can be difficult if the volcanism needs to be monitored in real time. But compared to the cost of a seafloor observatory to monitor a submarine volcano (e.g., Section 4.2.3), continual deployment/recovery of OBSs and/or OBHs may prove to be adequate and more cost-effective (e.g., Shinohara et al., 2017).

4.2.3. Cabled Systems

Real-time underwater volcanoes can be monitored with cable monitoring and/or cabled seafloor observatories. Seafloor observatories offer the opportunity to collect seismic and acoustic data as well as geological, chemical, and biological data, which, when considered together, can provide an integral view of a volcano's processes. Seafloor observatories specifically for volcanism are rare, but a few exist. For example, the Hawai`i Undersea Geo-Observatory (HUGO) was installed in the late 1990s to monitor Lo`ihi Seamount (Caplan-Auerbach & Duennebier, 2001), but unfortunately, due to a fault in the fiber-optic cable, it became inoperative after only three months. More recently, seafloor observatories have been created under the Ocean Observatories Initiatives (OOI) off the northwestern coast of the United States (e.g., Favali et al., 2015), monitoring underwater volcanoes such as Axial Seamount (e.g., Wilcock et al., 2016). Another example of a long-term submarine volcano monitoring effort is the EMSO-Azores observatory, which monitored the mid-Atlantic ridge and Lucky Strike submarine volcano for over 10 years (Cannat & Sarradin, 2010). The recent submarine volcanism identified offshore Mayotte (Lemoine et al., 2020; Cesca et al., 2020; Feuillet et al., 2021) has inspired plans to install a seafloor observatory to monitor the active volcanic system in the long term to mitigate risks (Aiken et al., 2021). Similarly, designs have been implemented to install a seafloor observatory to monitor Kolumbo, a submarine volcano in the Aegean Sea whose rim lies just 18 m below sea level (Nomikou et al., 2022). There are few seafloor observatories monitoring underwater volcanoes, likely because many volcanoes do not pose a threat and observatories are expensive to install. For example, a three-station cabled observatory for Mayotte was estimated to cost several million euros.

A less expensive choice for real-time submarine volcano monitoring may already exist in the seafloor: telecommunication cables. In one way, telecommunication cables can be used to attach OBSs and OBHs to create a kind of seafloor observatory. Such was the case for an OBS/OBH installed near active submarine volcanoes on a decommissioned telecommunication cable running between Japan and Guam through the Izu-Bonin forearc (Kasahara et al., 1998). In another way, telecommunication cables have been interrogated using the distributed acoustic sensing (DAS) and Brillouin optical time domain reflectometer (BOTDR) methods to measure variations in the fiber characteristics caused by physical phenomena, such as acoustic signals and seismicity, or to measure strain. Although telecommunication cables have not yet been reported as being used for monitoring emergent/submarine volcano activity, recent advances in cable monitoring indicate their potential (e.g., Lindsey et al., 2019; Williams et al., 2019; Sladen et al., 2019; Lior et al., 2020). For instance, hydroacoustic signals were detected on a fiber-optic submarine cable offshore in Japan. Air guns were used as a source, and DAS methods were applied to interrogate the cable (Matsumoto et al., 2021). In this study, the researchers showed that applying DAS on a fiber-optic submarine cable performs similarly to an OBS in the 4–70 Hz range and that a submarine cable is better coupled to the seafloor than hydrophones, leading to a higher likelihood of detecting volcanic sources. However, it remains to be shown whether such methods and tools can be used for detecting and interpreting those sources.

4.2.4. Limitations in Detecting Submarine Volcanism

Both hydroacoustic and seismic arrays have proven that it is possible to observe and quantify deep-ocean volcanic eruption mechanisms and processes (e.g., Dziak et al., 2015). Although permanent arrays seemingly offer real-time information, one drawback is that they are not always deployed near submarine volcanoes and lie at regional distances. As such, these arrays can be somewhat limited for monitoring submarine volcano eruptions unless they exist near submarine volcanoes.

Therefore, deployment of hydroacoustic and seismic arrays during sea campaigns for recordings of a year or more, targeted at active hazardous submarine volcanoes, can be useful to improve our knowledge of submarine volcano eruptions until long-term options (e.g., real-time monitoring, cabled or not) become viable. However, care should be taken when placing instruments for local monitoring to avoid damage from eruptive volcanism and its consequences (e.g., Chadwick, 2006; Chadwick et al., 2012).

Furthermore, the detection capabilities of submarine telecommunication cables are currently being studied (e.g., Lior et al., 2020). OBSs and OBHs can be attached to telecommunication cables to provide real-time data. However, one drawback to seafloor cables may be the amount of data that DAS or BOTDR interrogation can generate. Interrogating signal properties during data transmission in cables in real time for emergent/submarine volcanism monitoring would require massive data storage capabilities. Despite this, fiber-optic cables may become indispensable for monitoring submarine volcanism in the future, mainly because the cables may already exist in areas where they can be used to monitor underwater volcanism and thus tracking efforts would not require financing for sea campaigns, hydroacoustic or seismometer instrumentation, etc.

4.3. DIVERSE VOLCANO SIGNALS RECORDED UNDERWATER

Volcanoes produce numerous acoustic and seismic waves, which are indirect measurements for interpreting their physical processes and have been observed and studied at many volcanoes around the world (Chouet & Matoza, 2013 and references therein). Most observations of volcanic signals have been made at subaerial volcanoes, as submarine volcanoes are more difficult to monitor continuously and/or locally offshore due to infrequent marine surveys, costs, and availability of power sources (see Section 4.2). However, because subaerial and underwater volcanism share similar physical processes, waveform characteristics of the seismicity are also similar. In addition, seismic signals from underwater volcanoes, like subaerial volcanoes, sometimes share similarities with those found in subduction zones (e.g., Chouet, 1996; Obara, 2002). Thus, more detailed analysis is needed to identify their different sources (i.e., volcanic vs. tectonic origin vs. noise). The following subsections discuss these topics: (1) how to distinguish noise from signal in ocean data and (2) volcanic signals in ocean data and their interpretations. Specifically, volcanic signals will be discussed by category: (i) volcanic earthquakes, which are often classified into three categories (high-frequency, low-frequency, or hybrid-type events); (ii) volcanic tremor; (iii) volcanic explosion-type events; and (iv) volcanic landslides and other types of signals.

4.3.1. Distinguishing Signal from Noise in the Ocean

One difficulty associated with monitoring underwater volcanism, beyond the aforementioned monitoring limitations, is that underwater receivers record signals from various sources—from solid Earth sources (local and distant earthquakes) to a range of sources from the water in which the receivers reside (i.e., oceans and lakes). Often, the variety of transient signals recorded on underwater receivers bear similarities. Thus, the ability to detect or monitor emergent/submarine volcanoes—or any seismically active system from the seafloor—depends on distinguishing local signals from distant signals and other types of "noise" (e.g., Bécel et al., 2011).

The distinction between signals and noise is achieved by examining the spectral characteristics of the phenomena (e.g., harmonics or lack thereof) in the waveforms, as well as their duration, repetitiveness, and phases or lack thereof. Ocean noise in seismic data typically overwhelms signals at low frequencies (<2 Hz) and high frequencies (>10 Hz). It can include long-period gravity waves in the 0.02–0.03 Hz range and currents and turbulence at the seafloor boundary in the 0.03–0.1 Hz range (Babcock et al., 1994; Webb, 1998). The solid Earth produces microseisms, also known as Earth hum, that manifest in the 0.1–5 Hz band and create the highest noise levels. At the same time, storms generate slightly longer-period Rayleigh waves (peak near 0.1 Hz), and wind-driven ocean waves generate even longer-period waves (<0.04 Hz) by comparison (Webb, 1998). It has been suggested that oceanic microseism noise occurs at frequencies below 1.6 Hz (Caplan-Auerbach & Duennebier, 2001), and lake-generated microseism noise can range from 0.5 to 2 Hz, but the spectral peak varies depending on water conditions (e.g., Xu et al., 2017). In addition, some signals above 5 Hz have been attributed to breaking waves; and passing ships can generate signals above 10 Hz (Wenz, 1962), as can certain marine mammals (Pickard & Emery, 1990). Thus it is necessary to examine the widest frequency band possible when distinguishing signals from noise while considering the numerous sources that may contaminate seismic and acoustic data recorded underwater.

4.3.2. High-Frequency Volcanic Signals

High-frequency (HF) events have peak frequencies >5 Hz and are associated with shear failure and slip in a brittle medium. Volcano-tectonic (VT) earthquakes are generally considered HF events (Fig. 4.3a). VT earthquakes are manifestations of brittle failure induced by magma movement, and they typically have high-amplitude P and S phases and broad spectra up to 15 Hz (Lahr et al., 1994). VTs can also display weaker energy

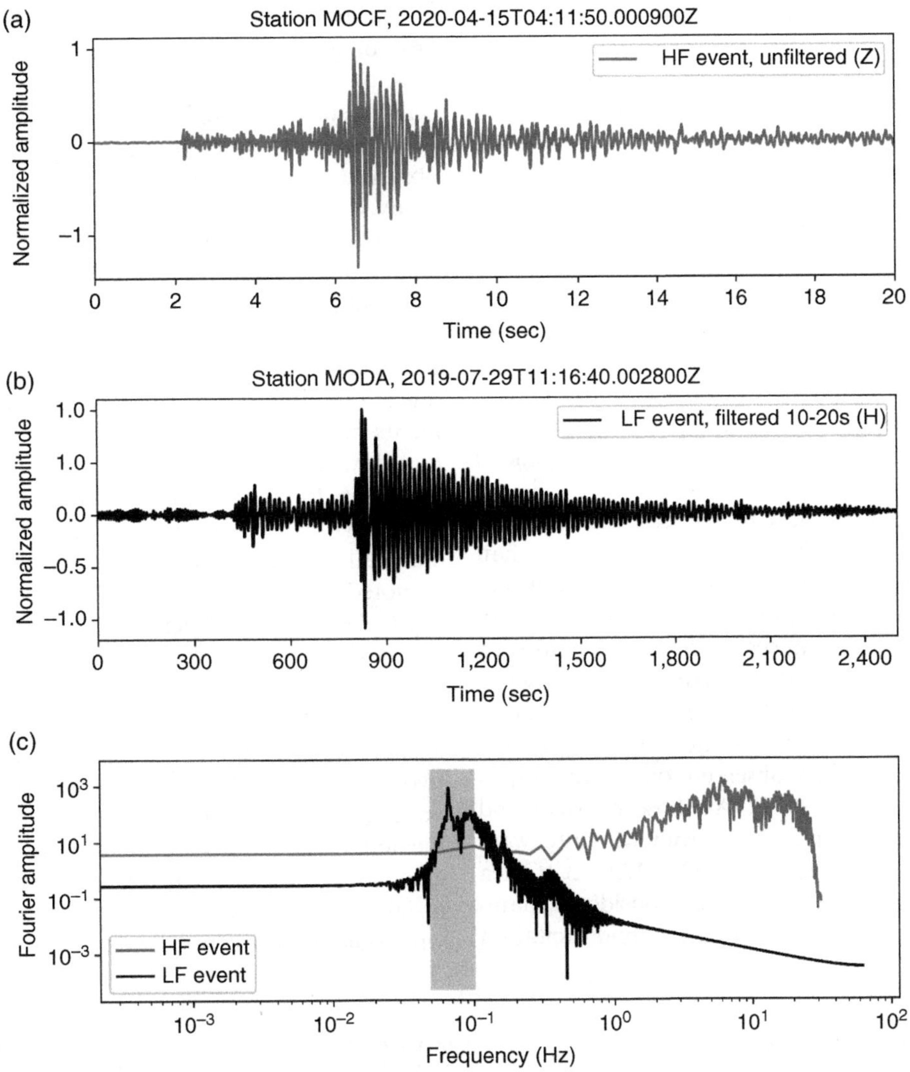

Figure 4.3 Example of high-frequency (HF) and low-frequency (LF) events observed on an ocean-bottom seismometer deployed at Mayotte's submarine volcano system. (a) Time series of a volcano-tectonic earthquake with distinct P- and S-wave arrivals and duration <20 seconds. (b) Time series of a very-long-period event with a long duration (~20 minutes). Adapted from Laurent et al. (2019). (c) Comparison of frequency spectra of the HF and LF events. The LF event in this example is a very long-period event with a dominant period in the 10–20 s range (gray box).

at >15 Hz, but this may depend on how they are recorded (e.g., the sensor, sampling rate, and attenuation). Because VTs often have higher magnitudes than other event types (e.g., volcanic tremor), their high signal-to-noise ratios make them the easiest to detect. As such, they are usually the earliest reported precursor to eruptive activity. VT events can occur days to years in advance of an eruption; their magnitudes are generally lower than 3.0 but may be greater than 5.0, as was the case of a magnitude 5.8 event at the subaerial Kasatochi volcano in August 2008 (McNutt & Roman, 2015) and a magnitude 5.9 at Mayotte's submarine volcano in May 2019 (e.g., Feuillet et al., 2021). The total energy released by VT event swarms is a useful proxy for intruded magma volume regardless of magma type for subaerial volcanoes, volcano islands (e.g., Anatahan), and even seamounts (e.g., Kick 'em Jenny, Lo`ihi) (White & McCausland, 2016). VTs can be extremely useful for forecasting eruptions at long-dormant volcanoes. Unfortunately, volcanoes do not always produce VTs, and when they do, VTs may not be the prevalent type of earthquake produced. Thus, other volcanic signals must be used to infer magma volumes, ongoing eruptive activity, volcanic processes, etc.

4.3.3. Low-Frequency Volcanic Signals

Low-frequency (LF) events have spectral peaks <5 Hz and generally have narrow-band codas and/or poorly defined S phases. Although LF events can have varied signal characteristics from one volcano to the next, they are usually linked to fluid transport and/or degassing in a magmatic system. However, they can also arise from other processes, such as strong path effects and/or shallow sources (e.g., Bean et al., 2014). A hybrid-type event has both LF and HF characteristics (Lahr et al., 1994; Chouet, 1996). Volcanoes generate several types of LF. These include very-long-period (VLP) events, also known as very-low-frequency (VLF) events, and long-period (LP) events (e.g., LP earthquakes, drumbeat earthquakes) (Figs. 4.3, 4.4, and 4.5).

VLP events contain very low frequencies ranging from 0.01 to 0.5 Hz and generally have durations longer than VT events (Fig. 4.3b). Because they exhibit very long wavelengths from tens to hundreds of kilometers, they are very easily observed at long distances. Such was the case for VLP events recorded by the Polynesian Seismic Network at the hotspot submarine volcano Rocard (Talandier et al., 2016) and a VLP event in November 2018 originating offshore Mayotte that was observed on seismographs around the world (Hicks, 2018). At those regional and remote distances, VLPs emanating from Mayotte were found to have surface wave magnitude (M_s) >3 (Cesca et al., 2020). In general, VLP events are considered manifestations of magma transport or resonant fluid excitation in volcanic systems, signifying perturbations in magma flow or gas in conduits. When inverted for their source location, they have imaged cracks, dikes, and sills in volcanic systems, highlighting their complexities (Chouet & Matoza, 2013). Large-magnitude VLP events have been observed and investigated over the last three decades at many volcanoes worldwide (Chouet & Matoza, 2013 and references therein). They are sometimes observed before and during caldera collapses and may be associated with ring faulting (e.g., Cesca et al., 2020 and references therein).

On the other hand, there are fewer observations of LP-type events at submarine or emergent volcanoes, likely due to our ability to detect them without deploying instruments near the volcano. This is because LP events often have low magnitudes (<~2) and may not produce strong T-phases. In addition, LP events look like typical earthquakes (Fig. 4.4) but have a long S-wave coda, which can lead to overestimation of their duration magnitudes, as noted for some subaerial volcanoes (Del Pezzo et al., 2013). One instance where LP events were recorded is offshore Mayotte (Fig. 4.4), where a locally deployed OBS network was installed and recovered several times over a few years (Retailleau et al., 2022). In that study, LP events occurred far less frequently than VTs during a two-year observation period, but in that observation period, roughly 2,000 LP events were detected and located. Despite the few observations we have of LPs from emergent/submarine volcanoes, detecting LPs can offer insight into volcano activity, as LPs have been associated with fluid pressure transients, such as in conduits or a

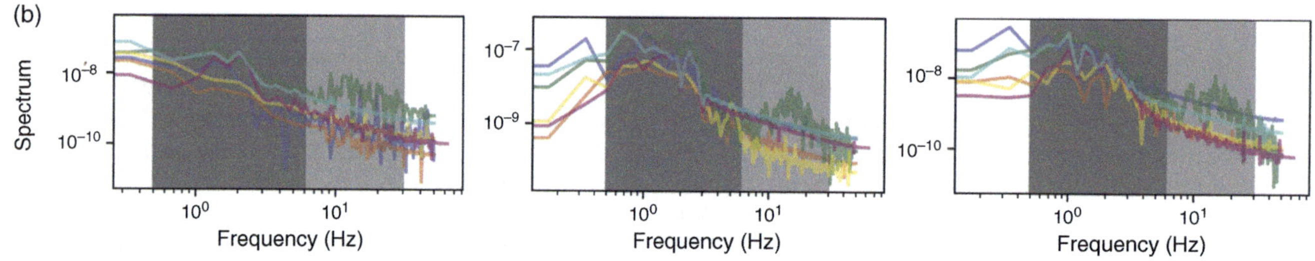

Figure 4.4 Examples of three long-period (LP) earthquake records and spectra recorded offshore Mayotte. The signals (a) and spectra recordings (b) shown here are three-component land stations, but Retailleau et al. (2022) reported that these events were visible on OBS data as phase picks were made on OBS to locate them. Both the P- and S-waves of the LPs appear to be deficient in high frequencies. Retailleau et al. (2022)/ Académie des sciences / CC BY 4.0.

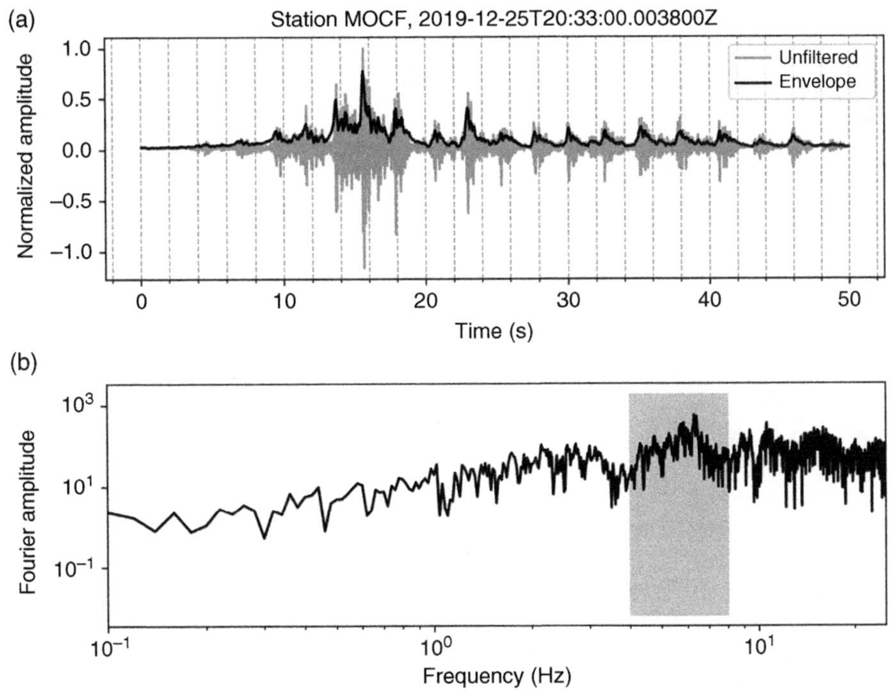

Figure 4.5 Example of a periodic tremor sequence (a.k.a. drumbeat earthquakes) observed from Mayotte's submarine volcanic system. (a) Time series of the drumbeat sequence. In this sequence, the drumbeats occur periodically every two to three seconds. (b) Frequency spectra of the drumbeat sequence have a dominant frequency in the 4–8 Hz range (gray box).

magma chamber (Aki et al., 1977; Chouet, 1988; Neuberg et al., 2000; Chouet & Matoza, 2013).

Finally, a special type of LF is "drumbeat" earthquakes (Fig. 4.5). They are called drumbeat earthquakes because the signals resemble heartbeats due to their periodicity and repetitive nature in time and amplitude over timescales of minutes to days. Volcanoes worldwide have produced long-lasting drumbeat earthquake sequences (e.g., Neuberg et al., 2006; Moran et al., 2008; Bell et al., 2017), but they are rarely observed in submarine and emergent volcanoes. The only known example of drumbeat earthquakes recorded at a submarine volcano is that of Mayotte. Like LP earthquakes (Fig. 4.4), drumbeat earthquakes were observed in OBS data due to long-term local OBS deployments (Fig. 4.5). The physical mechanism driving drumbeat earthquakes is highly debated. However, their occurrence seems linked to magma ascent and pre-eruptive volcanic unrest at very shallow depths (Chouet & Matoza, 2013), suggesting they can be used as precursors for volcanic hazard assessment.

4.3.4. Volcanic Tremor Signals

Volcanic tremor is another type of low-frequency signal (e.g., Fig. 4.6). This is a generalized term for seismic signals characterized as having a sustained amplitude lasting from minutes to days and containing frequencies typically ranging from 1–9 Hz (McNutt, 1992). Besides HF events, volcanic tremor is one of the most commonly observed volcano signals from emergent/submarine volcanoes. Volcanic tremor can be classified as eruptive or noneruptive (magma movement within a system). It can exhibit a variety of durations, amplitudes, and even frequency contents, such that tremor can be narrow-band, broadband, gliding, and/or harmonic (e.g., Girona et al., 2019).

Several mechanisms have been proposed for the occurrence of volcanic tremor, including fluid-elastic resonance (Chouet & Matoza, 2013), frictional processes (Hotovec et al., 2013), and outgassing flux through permeable media when it occurs shallowly (Girona et al., 2019). Gliding tremor (Fig. 4.6a) has been interpreted as the resonance of a magma-gas mixture in a conduit at South Sarigan Seamount. However, it has also been interpreted as a result of changes in magma velocity as its gas content increases and is released (Searcy, 2013 and references therein). Gasses and/or venting can produce harmonic tremor from resonances in the systems, as observed at submarine volcanoes in the volcano island arc south of Japan (Dziak & Fox, 2002) and NW Rota-1 in the Mariana Arc (e.g., Fig. 4.7) (Chadwick et al., 2008). In some cases, tremor has been recognized as the superposition of individual LP events; this has been observed at the emergent

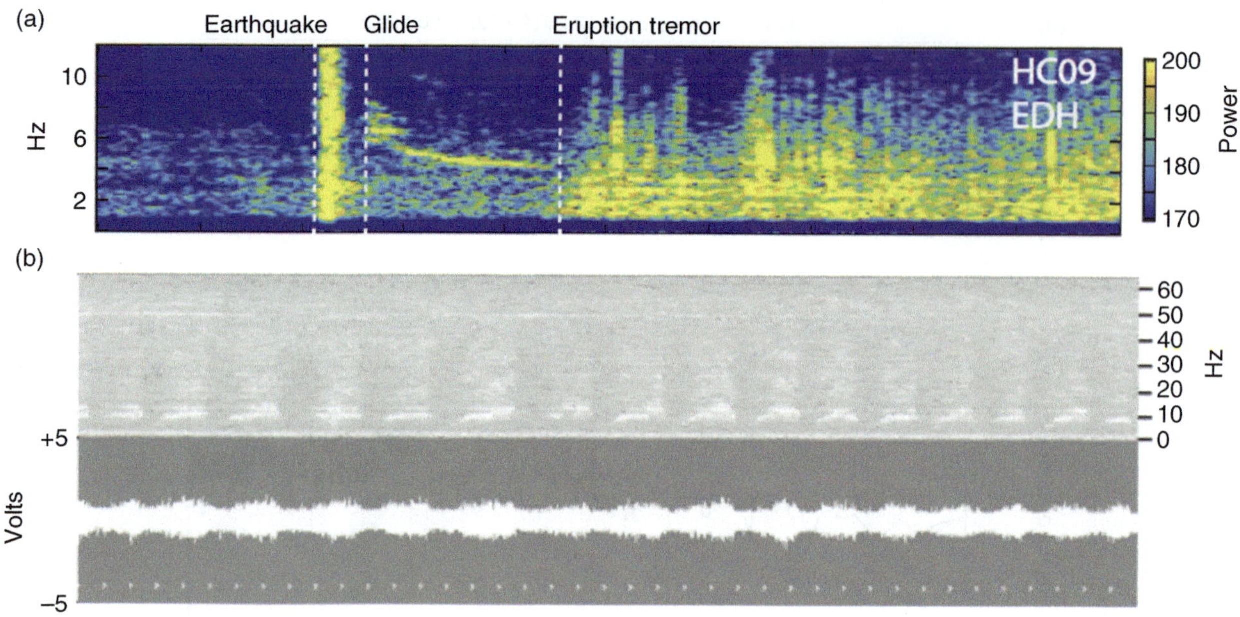

Figure 4.6 Hydroacoustic recordings of volcanic tremor recorded at an emergent and submarine volcano. (a) Power spectrogram of nonharmonic volcanic tremor during an eruption at the emergent Bogoslof volcano. (b) Harmonic tremor originating from the Volcano Islands arc south of Japan. In (b), the spectrogram in grayscale is the top panel, and the hydroacoustic recording is shown in the bottom panel in white. Dziak et al. (2002) / with permission from John Wiley & Sons, Inc.

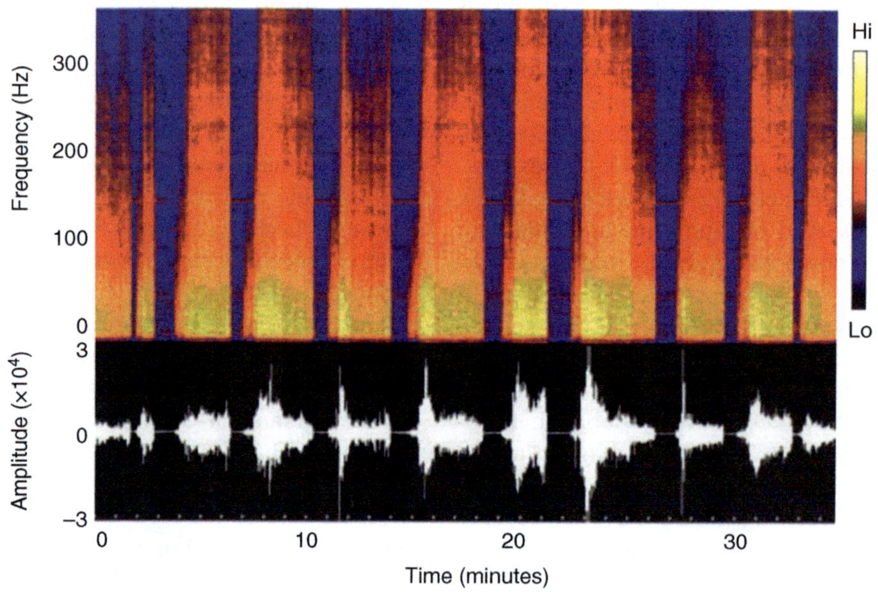

Figure 4.7 Example of large, discrete explosion bursts recorded by a hydrophone at the NW Rota-1 submarine volcano. A time series of explosions is shown on the bottom, and a contrasting spectrogram is shown on top. The color bar corresponds to a scale of low (Lo) to high (Hi) Fourier spectra amplitudes. Chadwick et al. (2008) / with permission from John Wiley & Sons, Inc.

Bogoslof volcano (Tepp et al., 2020) and at subaerial volcanoes (e.g., Hotovec et al., 2013).

4.3.5. Volcanic Explosion-Type Signals

Submarine volcanoes can also emit other signals, such as hydroacoustic signals resembling explosions. In general, volcanic explosions are signals with an impulsive onset ranging in duration and amplitude and can be one-offs or occur hundreds to thousands of times a day, sometimes rhythmically. They can be attributed to different volcanic processes, such as gas-driven explosions, magmatic degassing, gas bubbles bursting, and lava-water interactions during an eruption. Because there are many differing sources of acoustic explosions, they are best understood when confirmed by simultaneous direct video/photo and hydrophone observations made locally (e.g., Chadwick et al., 2008). For example, the NW Rota-1 submarine volcano in the Mariana Arc produced eruptive tremor bursts with sudden, sharp onsets in the acoustic data and a gradual rise in amplitude lasting several minutes, after which the amplitude sharply increased before decaying in discrete steps (Chadwick et al., 2008) (Fig. 4.7). These bursts were associated with explosive volcano activity and ejecta, although not as notable as explosions associated with the South Sarigan and Hunga Tonga-Hunga Ha'apai eruptions (Talandier et al., 2020; Carr et al., 2022). Moreover, sometimes eruptive tremor is less repetitive and discrete and more continuous and associated with climatic explosive events. For example, during both the 2010 South Sarigan Seamount and the 2016–2017 emergent Bogoslof volcano eruptions, explosive events were often followed by coeruptive tremor lasting minutes to tens of minutes (Green et al., 2013; Haney et al., 2020).

As observed by the HUGO seafloor observatory at the Lo`ihi Seamount, Hawai`i, acoustic explosions can sometimes have frequencies that vary from <10 Hz to several kHz and can exhibit short durations (<1 s) and similar time-series patterns: that is, the signals can occur as triplicates due to propagation effects in the water column (Caplan-Auerbach & Duennebier, 2001). It is unclear whether the acoustic explosions observed by HUGO were associated with the Lo`ihi Seamount or lava from the subaerial Kilauea eruption entering the ocean, because they were not locatable. Another example of short-duration acoustic explosions comes from lava–water interactions at the seafloor offshore Mayotte, where short-duration (<10 seconds) impulsive events were observed in hydrophone recordings; their location indicated an association with active lava-flow fields. Their duration implies that they represent energy directly released into the water, such as during a thermal explosion when hot lava is in contact with water (Bazin et al., 2022).

4.3.6. Volcanic Landslide Signals

Landslides, a source of devastating tsunamis for coastal communities, are one consequence of emergent/submarine volcanism (e.g., Dondin et al., 2011) and are sometimes observed on data from underwater instrumentation. In particular, rockfalls are a type of landslide that can alter volcanic domes at emergent and submarine volcanoes (e.g., Watlington et al., 2002; Clouard & Bonneville, 2004). At West Mata volcano in the Lau Basin of the Tonga Arc, for example, landslides are frequent, as evidenced by a series of bathymetric mappings (Caplan-Auerbach et al., 2014). A landslide at this particular volcano was recorded hydroacoustically on 1 January 2010 and offers a very clear instrumentally recorded example (Fig. 4.8). The landslide was recorded on four hydrophones, began with a low-frequency signal (<50 Hz), and exhibited a broadband coda lasting several minutes, similar to landslides recorded on Kilauea volcano (Caplan-Auerbach et al., 2001). Noticeably, hydrophones on the northern and western sides of the volcano recorded a Lloyd's mirror effect, providing evidence of a moving source such as that of a landslide. Furthermore, using the Lloyd's mirror phenomenon and durations of the acoustic signals, Caplan-Auerbach et al. (2014) were able to estimate the average sliding velocities to be between 11 and 25 m/s, a range similar to speeds for other submarine volcanoes.

However, landslides associated with submarine volcanism produce diverse and complex spectral characteristics that can be difficult to interpret without direct observations to verify the source of the signal. For example, Caplan-Auerbach and Duennebier (2001) describe the acoustic recordings (made by HUGO) of a landslide on a submarine flank of Kilauea lasting several hours and occurring 3–10 times per day as low-frequency "rumble" and "roar" signals followed by longer broadband "hiss" signals that were attributed to a more continuous debris flow. Therefore, due to their complexity, submarine volcano landslides are best identified when direct observations can be made, such as video recordings or a series of bathymetric mappings between volcanic events.

4.4. CONCLUSIONS

It is often remote sensing—via satellite imagery, hydroacoustic arrays, or seismic arrays—that gives us our first vision of emergent/submarine volcanism processes, whether the volcanism is known or unknown. In rare cases, local deployments of hydrophones and seismometers, and even video recording equipment, have been used to confirm, detect, and monitor emergent/submarine volcanic activity; even rarer is the use of cabled seafloor observatories and telecommunication cables for this purpose. New technologies (e.g., MERMAIDs) and new

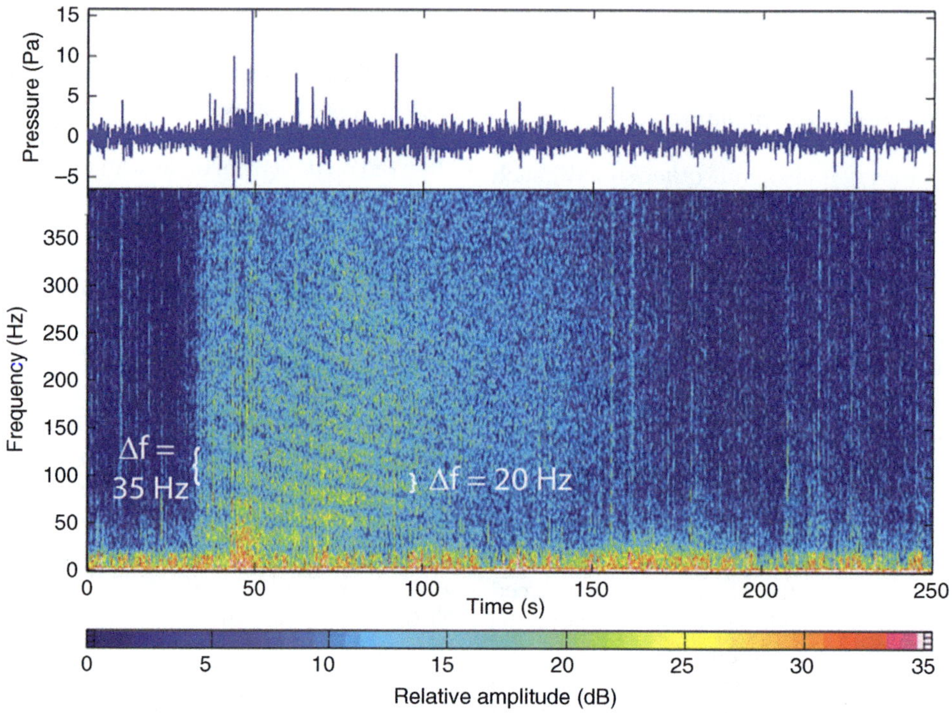

Figure 4.8 Example of a landslide at West Mata volcano. The top panel shows hydrophone pressure recorded on the volcano's north side. The bottom panel is a spectrogram of the landslide signal demonstrating the Lloyd's mirror phenomenon. The relative spacing between spectral minima changes over time (due to time differences between direct and reflected arrivals) and indicates a moving source. Caplan-Auerbach et al. (2014) / American Geophysical Union.

methods (e.g., DAS, BOTDR) applied to telecommunication cables, or attachment of OBSs and OBHs to retired cables, could prove to be useful tools for providing not only the first automatic detection and location of emergent/submarine volcanism but also long-term monitoring without the need for continuous deployments/recoveries of hydrophones and seismometers by sea campaign. However, continuous deployment/recovery of underwater instrumentation may be more cost-effective than installing a cabled observatory. In addition, monitoring by MERMAIDs and interrogation of telecommunications cables using DAS and BOTDR need to be tested in underwater volcano environments to fully understand their detection capabilities. For active underwater volcanoes that pose a risk to communities, these technologies could become the new standard for monitoring underwater volcanoes, but it remains to be proven how reliable they can be through calibration and quantification.

Whatever tools are used, the seismic and hydroacoustic signals that are detected provide insights into the behavior of underwater volcanic systems, from their interior processes (magmatic processes) to their exterior processes (magma-water interactions, landslides, tsunamis, etc.). These signals include well-known event types such as high-frequency volcano-tectonic earthquakes and low-frequency events such as long-period events and very-long period events, as well as volcanic tremor. Although large explosions and eruptions as well as volcanic tremor and magnitude > 3 volcano-tectonic earthquakes and very-long period events can be detected by regional networks (both on land and underwater), other signals require more local underwater monitoring. For example, the lesser-understood and observed impulsive-type signals can vary in duration and source (explosions, lava–water interactions, degassing), and complex and varied landslide signals can cause tsunamis but often need ground truthing for confirmation. Overall, when underwater, the signals that volcanoes can produce are incredibly diverse and complex. While underwater volcanoes produce signals similar to subaerial volcanoes, underwater volcanoes often lack but need localized deployments to be able to observe and understand the full spectrum of signals they can produce.

AVAILABILITY STATEMENT

Most data presented in this chapter was not used or created for this research. Data shown in Figs. 4.3 and 4.5 is available upon request (Mayotte Volcanological and Seismological Monitoring Network (REVOSIMA) et al., 2021). Figure 4.1 was made and Figs. 4.2, 4.6, 4.7, and 4.8 were modified using GIMP 2.1 (open source), and Figs. 4.3 and 4.5 were made using Python, an open source programming language.

ACKNOWLEDGMENTS

I greatly appreciate the constructive feedback of both reviewers. Without their thoughtful and detailed recommendations, this book chapter would not be as complete. I would also like to personally thank the editors, Frauke Klingelhoefer and Gaye Bayrakci, for their invitation to write this book chapter and for their continued support throughout the entire writing process. I would also like to thank Océane Foix for her feedback on a first draft of this review. More information about the REVOSIMA work group for monitoring Mayotte's submarine volcano can be found at https://www.ipgp.fr/fr/revosima/reseau-de-surveillance-volcanologique-sismologique-de-mayotte.

REFERENCES

Aiken, C., Saurel, J.-M., & Foix, O. (2021). Earthquake location and detection modeling for a future seafloor observatory along Mayotte's volcanic ridge. *Journal of Volcanology and Geothermal Research*, *418*, 107322. https://doi.org/10.1016/j.jvolgeores.2021.107322

Aki, K., Fehler, M., & Das, S. (1977). Source mechanism of volcanic tremor: fluid-driven crack models and their application to the 1963 Kilauea eruption. *Journal of Volcanology and Geothermal Research*, *2*(3), 259–287. https://doi.org/10.1016/0377-0273(77)90003-8

Babcock, J. M., Kirkendall, B. A., & Orcutt, J. A. (1994). Relationships between ocean bottom noise and the environment. *Bulletin of the Seismological Society of America*, *84*(6), 1991–2007. https://doi.org/10.1785/BSSA0840061991

Baumgartner, M. F., Stafford, K. M., & Latha, G. (2018). Near real-time underwater passive acoustic monitoring of natural and anthropogenic sounds. In R. Venkatesan et al. (Eds.), *Observing the oceans in real time*. Springer Oceanography. https://doi.org/10.1007/978-3-319-66493-4_10

Bazin, S., Dubost, F., Torterotot, M., Royer, J.-Y., Samaran, F., & Retailleau, L. (2022). *Hydroacoustic observatory of Mayotte volcano: preliminary results*. Paper presented at EMSO Time Series Conference: Observing Ocean Sound, Oct. 2020, Canary Islands, Spain. https://hal.archives-ouvertes.fr/hal-03582697

Bean, C. J., De Barros, L., Lokmer, I., Métaxian, J. P., O'Brien, G. S., & Murphy, S. (2014). Long-period seismicity in the shallow volcanic edifice formed from slow-rupture earthquakes. *Nature Geoscience*, *7*(71–75). https://doi.org/10.1038/ngeo2027

Bécel, A., Diaz, J., Laigle, M., Hirn, A. & the "Thales Was Right" Continuous Recording Group (2011). Searching for unconventional seismic signals on a subduction zone with a submerged forearc: OBS offshore the Lesser Antilles. *Tectonophysics*, *603*, 31–31. https://doi.org/10.1016/j.tecto.2012.10.031

Bell, A. F., Hernandez, S., Gaunt, H. E., Moths, P., Ruiz, M., Sierra, D., & Aguaiza, S. (2017). The rise and fall of periodic 'drumbeat' seismicity at Tungurahua volcano, Ecuador. *Earth Planet Science Letters*, *475*, 58–70. https://doi.org/10.1016/j.epsl.2017.07.030

Bohnenstiehl, D. R., Dziak, R. P., Matsumoto, H., & Lau, T.-K. A. (2013). Underwater acoustic records from the March 2009 eruption of Hunga Ha'apai-Hunga Tonga volcano in the Kingdom of Tonga. *Journal of Volcanology and Geothermal Research*, *249*, 12–24. https://doi.org/10.1016/j.jvolgeores.2012.08.014

Bonnieux, S., Hello, Y., Sigloch, K., Philippe, O., Rocca, F., & Nolet, G. (2021). *Mobile earthquake recording in marine areas by independent divers: now landing on the ocean bottom*. Paper presented at AGU Fall Meeting, New Orleans, LA, United States. https://hal.archives-ouvertes.fr/hal-03555710

Brandl, P. A., Schmid, F., Augustin, N., Grevemeyer, I., Arculus, R. J., Devey, C. W. et al. (2020). The 6-8 Aug 2019 eruption of "Volcano F" in the Tofua Arc, Tonga. *Journal of Volcanology and Geothermal Research*, *390*, 106695. https://doi.org/10.1016/j.jvolgeores.2019.106695

Cannat, M. & Sarradin, P.-M. (2010). MoMARsat: monitoring the mid-Atlantic ridge. https://doi.org/10.18142/130

Caplan-Auerbach, J. & Duennebier, F. (2001). Seismic and acoustic signals detected at Loihi seamount by the Hawaii Undersea Geo-Observatory. *Geochemistry Geophysics Geosystems*, *2*(5). https://doi.org/10.1029/2000GC000113

Caplan-Auerbach, J., Dziak, R. P., Bohnenstiehl, D. R., Chadwick, W. W., Jr., & Lau, T. K. (2014). Hydroacoustic investigation of submarine landslides at West Mata volcano, Lau Basin. *Geophysical Research Letters*, *41*(16), 5927–5934. https://doi.org/10.1002/2014GL060964

Caplan-Auerbach, J., Fox, C. G., & Duennebier, F. K. (2001). Hydroacoustic detection of submarine landslides on Kilauea volcano. *Geophysical Research Letters*, *28*(9), 1811–1813. https://doi.org/10.1029/2000GL012545

Carr, J. L., Horváth, Á., Wu, D. L., & Friberg, M. D. (2022). Stereo plume height and motion retrievals for the record-setting Hunga Tonga-Hunga Ha'apai eruption of 15 January 2022. *Geophysical Research Letters*, *49*(9), 1–7. https://doi.org/10.1029/2022GL098131

Cesca, S., Letort, J., Razafindrakoto, H. N. T., Heimann, S., Rivalta, E., Isken, M. P. et al. (2020). Drainage of a deep magma reservoir near Mayotte inferred from seismicity and deformation. *Nature Geoscience*, *13*, 87–93. https://doi.org/10.1038/s41561-019-0505-5

Chadwick, W. W. Jr. (2006). A submarine volcano is caught in the act. *Science*, *314*(5807), 1887–1888. https://doi.org/10.1126/science.1137082

Chadwick, W. W. Jr., Cashman, K. V., Embley, R. W., Matsumoto, H., Dziak, R. P., de Ronde, C. E. J., et al. (2008). Direct video and hydrophone observations of submarine explosive eruptions at NW Rota-1 volcano, Mariana arc. *Journal of Geophysical Research: Solid Earth*, *113*, B08S10. https://doi.org/10.1029/2007JB005215

Chadwick, W. W. Jr., Dziak, R. P., Hazel, J. H., Embley, R. W., & Matsumoto, H. (2012). Submarine landslide triggered by volcanic eruption recorded by in-situ hydrophone. *Geology*, *40*(1), 51–54. https://doi.org/10.1130/G32495.1

Chouet, B. (1988). Resonance of a fluid-driven crack: radiation properties and implications for the source of long-period events and harmonic tremor. *Journal of Geophysical Research: Solid Earth*, *93*(B5), 4375–4400. https://doi.org/10.1029/JB093iB05p04375

Chouet, B. A. (1996). Long-period volcano seismicity: its source and use in eruption forecasting. *Nature*, *380*(6572), 309–316. https://doi.org/10.1038/380309a0

Chouet, B. (2003). Volcano seismology. *Pure Applied Geophysics*, *160*, 739–788. https://doi.org/10.1007/PL00012556

Chouet, B. A. & Matoza, R. S. (2013). A multi-decadal view of seismic methods for detecting precursors of magma movement and eruption. *Journal of Volcanology and Geothermal Research*, *252*, 108–175. https://doi.org/10.1016/j.jvolgeores.2012.11.013

Clouard, V. & Bonneville, A. (2004). Submarine landslides in French Polynesia. In R. Hekinian, J. L. Cheminée, & P. Stoffers (Eds.), *Oceanic hotspots*. Springer. https://doi.org/10.1007/978-3-642-18782-7_7

Del Pezzo, E., Bianco, F., & Borgna, I. (2013). Magnitude scale for lp events: a quantification scheme for volcanic quakes. *Geophysical Journal International*, *194*(2), 911–919. https://doi.org/10.1093/gji/ggt126

Dondin, F., Lebrun, J.-F., Kelfoun, K., Fournier, N., & Randrianasolo, A. (2011). Sector collapse at Kick 'em Jenny submarine volcano (Lesser Antilles): numerical simulation and landslide behavior. *Bulletin of Volcanology*, *74*, 595–607. https://doi.org/10.1007/s00445-011-0554-0

Dziak, R. P., Bohnenstiehl, D. R., Baker, E. T., Matsumoto, H., Caplan-Auerbach, J., Embley, R. W., et al. (2015). Long-term explosive degassing and debris flow activity at West Mata submarine volcano. *Geophysical Research Letters*, *42*(5), 1480–1487. https://doi.org/10.1002/2014GL062603

Dziak, R. P., Bohnenstiehl, D. R., & Smith, D. K. (2012a). Hydroacoustic monitoring of oceanic spreading centers: past, present, and future. *Oceanography*, *25*(1), 116–127. https://doi.org/10.5670/oceanog.2012.10

Dziak, R. P., & Fox, C. G. (1999). The January 1998 earthquake swarm at Axial volcano, Juan de Fuca Ridge: hydroacoustic evidence of seafloor volcanic activity. *Geophysical Research Letters*, *26*(23), 3429–3432. https://doi.org/10.1029/1999GL002332

Dziak, R. P. & Fox, C. G. (2002). Evidence of harmonic tremor from a submarine volcano detected across the Pacific Ocean basin. *Journal of Geophysical Research*, *107*(B5), 2085. https://doi.org/10.1029/2001JB000177

Dziak, R. P., Haxel, J. H., Bohnenstiehl, D. R., Chadwick, W. W., Jr, Nooner, S. L., Fowler, M. J. et al. (2012b). Seismic precursors and magma ascent before the April 2011 eruption at Axial Seamount. *Nature Geosciences*, *5*, 478–482. https://doi.org/10.1038/ngeo1490

Dziak, R. P., Hammond, S. R., & Fox, C. G. (2011). A 20-year hydroacoustic time series of seismic and volcanic events in the Northeast Pacific Ocean. *Oceanography*, *24*(3), 280–293. https://doi.org/10.5670/oceanog.2011.79

Favali, P., Beranzoli, L., & De Santis, A. (2015). *Seafloor observatories: a new vision of the Earth from the abyss. Springer.* https://doi.org/10.1007/978-3-642-11374-1

Feuillet, N., Jorry, S., Crawford, W. C., Deplus, C., Thinon, I., Jacques, E., et al. (2021). Birth of a large volcanic edifice through lithosphere-scale dyking offshore Mayotte (Indian Ocean). *Nature Geoscience*, *14*, 787–795. https://doi.org/10.1038/s41561-021-00809-x

Fiske, R. S., Cashman, K. V., Shibata, A., & Watanabe, K. (1998). Tephra dispersal from Myojinsho, Japan, during its shallow submarine eruption of 1952–1953. *Bulletin of Volcanology*, *59*, 262–275. https://doi.org/10.1007/s004450050190

Foix, O., Aiken, C., Saurel, J.-M., the MAYOBS / REVOSIMA Seismology Team, Feuillet, N., et al. (2021). Offshore Mayotte volcanic plumbing revealed by local passive tomography. *Journal of Volcanology and Geothermal Research*, *420*, 107395. https://doi.org/10.1016/j.jvolgeores.2021.107395

Fox, C. G., Matsumoto, H., & Lau, T.-K. A. (2001). Monitoring Pacific Ocean seismicity from an autonomous hydrophone array. *Journal of Geophysical Research*, *106*, 4183–4206. https://doi.org/10.1029/2000JB900404

Fox, C. G., Radford, W. E., Dziak, R. P., Lau, T.-K., Matsumoto, H., & Schreiner, A. E. (1995). Acoustic detection of a seafloor spreading episode on the Juan de Fuca Ridge using military hydrophone arrays. *Geophysical Research Letters*, *22*(2), 131–134. https://doi.org/10.1029/94GL02059

Garcia, M. O. (2015). How and why Hawaiian volcanism has become pivotal to our understanding of volcanoes from their source to surface. In R. Carey, V. Cayol, M. Poland, & D. Weis (Eds.), *Hawaiian volcanoes: from source to surface*. AGU Books: Geophysical Monograph Series. https://doi.org/10.1002/9781118872079.ch1

Garcia, M. O., Caplan-Auerbach, J., De Carlo, E. H., Kurz, M. D., & Becker, N. (2006). Geology, geochemistry and earthquake history of Loihi Seamount, Hawaii's youngest volcano. *Geochemistry*, *66*(2), 81–108. https://doi.org/10.1016/j.chemer.2005.09.002

Girona, T., Caudron, C., & Huber, C. (2019). Origin of shallow volcanic tremor: the dynamics of gas pockets trapped beneath thin permeable media. *Journal of Geophysical Research*, *124*, 4831–4861. https://doi.org/10.1029/2019JB017482

Green, D. N., Evers, L. G, Fee, D., Matoza, R. S., Snellen, M., Smets, P. et al. (2013). Hydroacoustic, infrasonic and seismic monitoring of the submarine eruptive activity and sub-aerial plume generation at South Sarigan, May 2010. *Journal of Volcanology and Geothermal Research*, *257*, 31–43. https://doi.org/10.1016/j.jvolgeores.2013.03.006

Haney, M. M., Fee, D., McKee, K. F., Lyons, J. J., Matoza, R. S., Wech, A. G. et al. (2020). Co-eruptive tremor from Bogoslof volcano: seismic wavefield composition at regional distances. *Bulletin of Volcanology*, *82*, 18. https://doi.org/10.1007/s00445-019-1347-0

Hicks, S. (2018). Twitter science: low-frequency seismic rumblings that traveled all around the world yesterday originated close to the tiny island of Mayotte in the Indian Ocean. [Tweet] https://twitter.com/seismo_steve/status/1062042619264581633?s=21

Hotovec, A. J., Prejean, S. G., Vidale, J. E., & Gomberg, J. (2013). Strongly gliding harmonic tremor during the 2009 Eruption of Redoubt Volcano. *Journal of Volcanology and Geothermal Research*, *259*, 89–99. https://doi.org/10.1016/j.jvolgeores.2012.01.001

Johnson, R. H., Northrop, J., & Eppley, R. (1963). Sources of Pacific T phases. *Journal of Geophysical Research*, *68*(14), 4251–4260. https://doi.org/10.1029/JZ068i014p04251

Kasahara, J, Sato, T., Momma, H. & Shirasaki, Y. (1998). A new approach to geophysical real-time measurements on a deep-sea floor using decommissioned submarine cables. *Earth Planets Space*, *50*, 913–925. https://doi.org/10.1016/S0031-9201(98)00090-9

Kintner, J. A., Yeck, W. L., Earle, P. S., Prejean, S., & Pesicek, J. D. (2022). High-precision characterization of seismicity from the 2022 Hunga Tonga-Hunga Ha'apai volcanic eruption. *Seismological Research Letters.* https://doi.org/10.1785/0220220250

Konter, S. R., Staudigel, H., Hart, S. R., Shearer, & P. M. (2004). Seafloor seismic monitoring of an active submarine volcano: local seismicity at Vailulu'u Seamount, *Samoa. Geochemistry Geophysics Geosystems*, *5*(6), Q06007. https://doi.org/10.1029/2004GC000702

Lahr, J. C., Chouet, B. A., Stephens, C. D., Power, J. A., & Page, R. A. (1994). Earthquake classification, location, and error analysis in a volcanic environment: implications for the magmatic system of the 1989–1990 eruptions at Redoubt Volcano, Alaska. *Journal of Volcanology and Geothermal Research*, *62*(1–4), 137–151. https://doi.org/10.1016/0377-0273(94)90031-0

Laurent, A., Satriano, C., & Bernard, P. (2019). *Detection, location and characterization of VLF events during the 2018-2019 seismovolcanic crisis in Mayotte*. Paper presented at Rencontres scientifiques et techniques RESIF 2019, Nov. 2019, Biarritz, France. https://hal.archives-ouvertes.fr/hal-02446050

Lemoine, A., Briole, P., Bertil, D., Roullé, A., Foumelis, M., Thinon, I., et al. (2020). The 2018–2019 seismo-volcanic crisis east of Mayotte, Comoros islands: seismicity and ground deformation markers of an exceptional submarine eruption. *Geophysical Journal International*, *223*, 22–44. https://doi.org/10.1093/gji/ggaa273

Lindsay, J. M., Shepherd, J. B., & Wilson, D. (2005). Volcanic and scientific activity at Kick 'em Jenny submarine volcano 2001–2002: implications for volcanic hazard in the Southern Grenadines, *Lesser Antilles. Natural Hazards*, *34*(1), 1–24. https://doi.org/10.1007/s11069-004-1566-2

Lindsey, N. J., Dawe, T. C., & Ajo-Franklin, J. B. (2019). Illuminating seafloor faults and ocean dynamics with dark fiber distributed acoustic sensing. *Science*, *366*(6469), 1103–1107. https://doi.org/10.1126/science.aay5881

Lior, I., Sladen, A., Rivet, D., Ampuero, J. P., Hello, Y. M., Lamare, P., et al. (2020). On the detection capabilities of underwater DAS. *Journal of Geophysical Research*, *126*(3). https://doi.org/10.1029/2020JB020925

Lipiejko, N., Whittaker, C. N., Lane, E. M., White, J. D. L., & Power, W. L. (2021). Tsunami generation by underwater volcanic explosions: application to the 1952 explosions of Myojinsho Volcano. *Pure Applied Geophysics*, *178*, 4743–4761. https://doi.org/10.1007/s00024-021-02857-1

Mayotte Volcanological and Seismological Monitoring Network (REVOSIMA), et al. (2021). Data collection: http://volobsis.ipgp.fr/revosima/

Mastumoto, H., Araki, E., Kimura, T., Fujie, G., Shiraishi, K., Tonegawa, T., et al.(2021). Detection of hydroacoustic signals on a fiber-optic submarine cable. *Scientific Reports*, *11*, 2797. https://doi.org/10.1038/s41598-021082093-8

McNutt, S. (1992). Volcanic tremor. In L. Brekhovskikh, K. K. Turekian, K. Emery, C. K. Tseng, X. LePichon, J. R. Curray, et al. (Eds.), *Encyclopedia of Earth system science* (pp. 417–425). Academic Press.

McNutt, S. R. (2005). Volcano seismology. *Annual Review Earth and Planetary Sciences*, *32*, 461–491. https://doi.org/10.1146/annurev.earth.33.092203.122459

McNutt, S. R., & Roman, D. C. (2015). Volcano seismicity. In H. Sigurdsson (Ed.), *The encyclopedia of volcanoes* (2nd Ed., pp. 1011–1034). Academic Press. https://doi.org/10.1016/B978-0-12-385938-9.00059-6

Metz, D., Watts, A. B., Grevemeyer, I., Rodgers, M. & Paulatto, M. (2016). Ultra-long-range hydroacoustic observations of submarine volcanic activity at Monowai, Kermadec Arc. *Geophysical Research Letters*, *43*(4), 1529–1536. https://doi.org/10.1002/2015GL067259

Moodie, K. (2022). Tonga eruption likely the world's largest in 30 years—scientist. *Radio New Zealand.* https://www.rnz.co.nz/news/world/459657/tonga-eruption-likely-the-world-s-largest-in-30-years-scientist

Moran, S. C., Malone, S. D., Qamar, A. I., Thelen, W., Wright, A. K., & Caplan-Auerbach, J. (2008). Seismicity associated with the renewed dome-building eruption of Mount St. Helens, 2004–2005. In D. R. Sherrod, W. E. Scott, & P. F. Stauffer (Eds.), *A volcano rekindled: the renewed eruption of Mount St. Helens, 2004–2006 (Ch. 2)*. United States Geological Survey. https://doi.org/10.3133/pp17502

Neuberg, J. Luckett, R., Baptie, B. & Olsen, K. (2000). Models of tremor and low-frequency earthquake swarms on Montserrat. *Journal of Volcanology and Geothermal Research*, *101*(1), 83–104. https://doi.org/10.1016/S0377-0273(00)00169-4

Neuberg, J. W., Tuffen, H., Collier, L., Green, D., Powell, T., & Dingwell, D. (2006). The trigger mechanism of low-frequency earthquakes on Montserrat. *Journal of Volcano and Geothermal Research*, *153*(1 and 2), 37–50. https://doi.org/10.1016/j.jvolgeores.2005.08.008

Nomikou, P., Polymenakou, P. N., Rizzo, A. L., Petersen, S., Hannington, M., Kilias, S. P. et al. (2022). SANTORY: SANTORini's seafloor volcanic observatory. *Frontiers in Marine Science*, *9*, 796376. https://doi.org/10.3389/fmars.2022.796376

Obara, K. (2002). Nonvolcanic deep tremor associated with subduction in southwest Japan. *Science*, *296*(5573), 1679–1681. https://doi.org/10.1126/science.1070378

Parnell-Turner, R., Smith, D. K., & Dziak, R. P. (2022). Hydroacoustic monitoring of seafloor spreading and transform faulting in the Equatorial Atlantic Ocean. *Journal of Geophysical Research*, *127*(7). https://doi.org/10.1029/2022JB024008

Pickard, G. L., & Emery, W. J. (1990). *Descriptive physical oceanography*. Pergamon.

Retailleau, L., Saurel, J.-M., Laport, M., Lavayssière, A., Ferrazzini, V., Zhu, W. et al. (2022). Automatic detection for a comprehensive view of Mayotte seismicity. *Comptes Rendus, Géoscience*, *1–18*. https://doi.org/10.5802/crgeos.133

Rizzo, A. L., Caracausi, A., Chavagnac, V., Nomikou, P., Polymenakou, P. N., Mandalakis, M., et al. (2016). Kolumbo submarine volcano (Greece): an active window into the Aegean subduction system. *Scientific Reports*, *6*, 28013. https://doi.org/10.1038/srep28013

Saurel, J.-M., Jacques, E., Aiken, C., Lemoine, A., Retailleau, L., Lavayssière, A., et al. (2022). Mayotte seismic crisis: building knowledge in near real-time combining land and ocean-bottom seismometers, first results. *Geophysical Journal International*, *228*(2), 1281–1293. https://doi.org/10.1093/gji/ggab392

Schindelé, F., Reymond, D., Hébert, H., & Heinrich, P. (2002). Geophysical hazards in the Marquesas Islands (French Polynesia). *Bureau de Recherches Géologiques et Minières, Report in Géologie de la France*, *2*, 39–52. http://geolfrance.brgm.fr/sites/default/files/upload/documents/gf2-2-2002.pdf (in French)

Searcy, C. (2013). Seismicity associated with the May 2010 eruption of South Sarigan Seamount, *Northern Mariana Islands. Seismological Research Letters*, *84*(6), 1055–1061. https://doi.org/10.1785/0220120168

Shinohara, M., Ichihara, M., Sakai, S. I., Yamada, T., Takeo, M., Sugioka, H., et al. (2017). Continuous seismic monitoring of Nishinoshima volcano, Izu-Ogasawara, by using long-term ocean bottom seismometers. *Earth, Planets and Space*, *69*(1), 1–9. https://doi.org/10.1186/s40623-017-0747-7

Simon, J. D., Simons, F. J., & Irving, J. C. E. (2021). A MERMAID miscellany: Seismoacoustic signals beyond the P wave. *Seismological Research Letters*, *92*(6), 3657–3667. https://doi.org/10.1785/0220210052

Sladen, A., Rivet, D., Ampuero, J. P., De Barros, L., Hello, Y., Calibris, G. & Lamare, P. (2019). Distributed sensing of earthquakes and ocean-solid Earth interactions on seafloor telecom cables. *Nature Communications*, *10*, 5777. https://doi.org/10.1038/s41467-019-13793-z

Sukhovich, A., Bonnieux, S., Hello, Y., Irisson, J.-O., Simons, F. J., & G. Nolet (2015). Seismic monitoring in the oceans by autonomous floats. *Nature Communications*, *6*, 8027. https://doi.org/10.1038/ncomms9027

Taillefer, N., Bertil, D., Negulescu, C., Roullé, A., Iasio, C., Auclair, S. et al. (2022). Seismic damage scenarios for Mayotte: a tool for disaster management. *Comptes Rendus, Géoscience*, 1–17. https:/doi.org/10.5802.crgeos.120

Talandier, J., Hyvernaud, O., Hébert, H., Maury, R. C., & Allgeyer, S. (2020). Seismic and hydroacoustic effects of the May 29, 2010 submarine South Sarigan volcanic explosion: Energy release and interpretation. *Journal of Volcanology*, *394*, 106819. https://doi.org/10.1016/j.jvolgeores.2020.105819

Talandier, J., Hyvernaud, O., & Maury, R. C. (2016). Unusual seismic activity in 2011 and 2013 at the submarine volcano Rocard, Society hot spot (French Polynesia). *Geophysical Research Letters*, *43*(9), 4247–4254. https://doi.org/10.1002/2016GL068342

Tepp, G. & Dziak, R. P. (2021). The seismo-acoustics of submarine volcanic eruptions. *Journal of Geophysical Research*, *126* (4). https://doi.org/10.1029/2020JB020912

Tepp, G., Dziak, R. P., Haney, M. M., Lyons, J. J., Searcy, C., Matsumoto, H., et al. (2020). Seismic and hydroacoustics observations of the 2016-17 Bogoslof eruption. *Bulletin of Volcanology*, *82*(1), 1–21. https://doi.org/10.1007/s00445-019-1344-3

Tralli, D. M., Blom, R. G., Zlotnicki, V., Donnellan, A., & Evans, D. L. (2005). Satellite remote sensing of earthquake, volcano, flood, landslide and coastal inundation hazards. *Journal of Photogrammetry and Remote Sensing*, *59*(4), 185–198. https://doi.org/10.1016/j.isprsjprs.2005.02.002

Ulvrová, M., Paris, R., Kelfoun, K., & Nomikou, P. (2014). Numerical simulations of tsunamis generated by underwater volcanic explosions at Karymskoye lake (Kamchatka, Russia) and Kolumbo volcano (Aegean Sea, Greece). *Natural Hazards and Earth System Sciences*, *14*(2), 401–412. https://doi.org/10.5194/nhess-14-401-2014

Watlington, R. A., Wilson, W. D., Johns, W. E., & Nelson, C. (2002). Updated bathymetry survey of Kick-'em-Jenny submarine volcano. *Marine Geophysical Researches*, *23*, 271–276. https://doi.org/10.1023/A:1023615529336

Webb, S. C. (1998). Broadband seismology and noise under the ocean. *Reviews of Geophysics*, *36*(1), 105–142. https://doi.org/10.1029/97RG02287

Wenz, G. M. (1962). Acoustic ambient noise in the ocean: spectra and sources. *Journal of Acoustical Society of America*, *34* (12). https://doi.org/10.1121/1.1909155

White, R. & McCausland, W. (2016). Volcano-tectonic earthquakes: a new tool for estimating intrusive volumes and forecasting eruptions. *Journal of Volcanology and Geothermal Research*, *309*, 139–155. https://doi.org/10.1016/j.jvolgeores.2015.10.020

Wilcock, W. S. D., Tolstoy, M., Waldhauser, F., Garcia, C., Tan, Y. J., Bohnenstiehl, D. R., et al. (2016). Seismic constraints on caldera dynamics from the 2015 Axial Seamount eruption. *Science*, *354*(6318), 1395–1399. https://doi.org/10.1126/science.aah5563

Williams, E. F., Fernandez-Ruiz, M. R., Magalhaes, R., Vanthillo, R., Zhan, Z., Gonzalez-Herraez, M. & Martins, H. F. (2019). Distributed sensing of microseisms and teleseisms with submarine dark fibers. *Nature Communications*, *10*, 5778. https://doi.org/10.1038/s41467-019-13262-7

Xu, Y., Koper, K. D., & Burlacu, R. (2017). Lakes as a source of short-period (0.5–2s) microseisms. *Journal of Geophysical Research*, *122*, 8241–8256. https://doi.org/10.1002/2017JB014808

5

Seismic and Acoustic Monitoring of Submarine Landslides: Ongoing Challenges, Recent Successes, and Future Opportunities

Michael A. Clare[1], Gwyn Lintern[2], Edward Pope[3], Megan Baker[3], Sean Ruffell[3], Mohammad Zaki Zulkifli[1], Steven Simmons[4], Morelia Urlaub[5], Mohammed Belal[1], and Peter J. Talling[3]

ABSTRACT

Submarine landslides pose a hazard to coastal communities and critical seafloor infrastructure, occurring on all of the world's continental margins, from coastal zones to hadal trenches. Offshore monitoring has been limited by the largely unpredictable occurrence of submarine landslides and the need to cover large regions. Recent subsea monitoring has provided new insights into the preconditioning and run-out of submarine landslides using active geophysical techniques. However, these tools measure a small spatial footprint and are power- and memory-intensive, thus limiting long-duration monitoring. Most landslide events remain unrecorded. In this chapter, we first show how passive acoustic and seismologic techniques can record acoustic emissions and ground motions created by terrestrial landslides. This terrestrial-focused research has catalyzed advances in characterizing submarine landslides using onshore and offshore networks of broadband seismometers, hydrophones, and geophones. We discuss new insights into submarine landslide preconditioning, timing, location, velocity, and down-slope evolution arising from these advances. Finally, we outline challenges, emphasizing the need to calibrate seismic and acoustic signals generated by submarine landslides. Passive seismic and acoustic sensing has a strong potential to enable more complete hazard catalogs to be built and open the door to emerging techniques (such as fiber-optic sensing) to fill key knowledge gaps.

5.1. INTRODUCTION

Submarine landslides can be orders of magnitude larger than those on land, occur on remarkably low-angle (<2 degrees) slopes, and can generate run-out that travels hundreds to thousands of kilometers into the deep sea (Moore et al., 1989; Hampton et al., 1996; Nisbet and Piper, 1998; Piper et al., 1999; Carter et al., 2014). Underwater slope failures can generate tsunamis that inundate coastal communities and sometimes result in major loss of life (e.g., Harbitz et al., 2014; Tappin et al., 2014) and adversely impact critical seafloor infrastructure networks, such as the cables and pipelines on which we rely for global communications, the Internet, and energy supplies (Piper et al., 1999; Carter et al., 2014). To date, most studies of submarine landslides and their run-out have been

[1]*National Oceanography Centre, Southampton, UK*

[2]*Natural Resources Canada, Geological Survey of Canada, Sidney, British Columbia, Canada*

[3]*Department of Earth Sciences, and Department of Geography, University of Durham, Durham, UK*

[4]*Earth and Environment Institute, Hull, UK*

[5]*GEOMAR Helmholtz Centre for Ocean Research, Kiel, Germany*

Noisy Oceans: Monitoring Seismic and Acoustic Signals in the Marine Environment, Geophysical Monograph 284, First Edition.
Edited by Gaye Bayrakci and Frauke Klingelhoefer.

DOI:10.1002/9781119750925.ch05

based on analysis of the deposits that past events left behind. These studies have used combinations of (1) seafloor surveys (e.g., multibeam echosounders and side scan sonars) to image submarine landslides in planform (e.g., Prior et al., 1982; McAdoo et al., 2000; Mountjoy et al., 2009; Casas et al., 2016; Normandeau et al., 2019; Brackenridge et al., 2020); (2) subsurface geophysical surveys to determine the geometry and internal character of submarine landslide deposits (e.g., Gee et al., 2006; Bull et al., 2009; Vardy et al., 2012; Nwoko et al., 2020); (3) intrusive sampling or coring to calibrate geophysical interpretations, provide material for geochronological analysis (age and recurrence determination), and/or determine their source using geochemical and other analytical techniques (e.g., Geist and Parsons, 2010; Dugan, 2012; Urlaub et al., 2013; Vanneste et al., 2014); (4) in situ or laboratory-based testing to understand geotechnical and geomechanical behavior of submarine landslides (e.g., Sultan et al., 2010; Ai et al., 2014; Miramontes et al., 2018); (5) analysis of ancient, exhumed submarine landslides from rock outcrops (e.g., Brookes et al., 2018; Bull et al., 2019; Ogata et al., 2019); and (6) physical and numerical modeling to quantify slope stability, post-failure behavior, and resultant impacts (e.g., to seafloor infrastructure or the tsunami they may initiate) (e.g., Sultan et al., 2010; Harbitz et al., 2014; Tappin et al., 2014; Puzrin et al., 2016).

These techniques provide valuable insights into the location, extent, and nature of past failures, as well as providing the building blocks for generating models that enable inference of the likely behavior of future events; however, they do not capture the behavior of field-scale submarine landslides in action and do not reveal the in situ conditions at the time of failure. Numerical models therefore remain relatively poorly constrained, and many key questions remain unanswered, particularly regarding the initiation, kinematics, and impacts of submarine landslides. For instance, what are the preconditioning factors for failure, and over what time scales are they important? Why are external triggers (e.g., earthquakes and tropical cyclones) more effective in some settings than others? What lag times may be involved following cumulative or sudden perturbations that precede failure, and what controls that delay? Such information is critical to understanding precisely how and when landslides are triggered, developing effective early warning systems, and understanding any climate change feedback (e.g., Maslin et al., 2004; Paull et al., 2007; Harbitz et al., 2014; Brothers et al., 2013; Normandeau et al., 2021). Key questions also remain unanswered on the kinematic behavior of submarine landslides as they propagate. For example, what kinematics are involved during different styles of slope failure, and how does that change as the mass travels down-slope? Why do some slopes fail suddenly, with major and rapid displacements, while others move much more slowly and progressively? Does failure type control the nature and extent of run-out, and if so, how? These and many other questions remain open, largely due to the challenges involved in monitoring failure processes and their precursor conditions in real time in marine settings that often cover very large areas.

Some of the challenges that inhibit monitoring of submarine landslides include the following:

• The *water depths* of many settings where submarine landslides occur place financial and logistical constraints on long-duration and continuous monitoring (e.g., Talling et al., 2013; Clare et al., 2017). Although slope failures can occur in shallow water deltaic or fjord settings (e.g., Prior et al., 1989; Biscara et al., 2012), many occur in hundreds to thousands of meters of water depth and have been documented in the deepest hadal trenches on Earth (Strasser et al., 2013; Kioka et al., 2019).

• The *remote nature* (i.e., location far from shore) of many submarine landslides places constraints on power, communications, and data transfer. This in turn limits the resolution and frequency of measurements that can be made due to a reliance on in situ power supplies and the recovery and redeployment of instruments by expensive vessels (Urlaub & Villinger, 2019).

• The *unpredictable nature of submarine landslides over time*. As many studies have suggested that submarine landslide recurrence may be Poissonian (i.e., approximately random), predicting when a specific seafloor slope may fail is a key challenge (Geist & Parsons, 2010; Urgeles & Camerlenghi et al., 2013; Urlaub et al., 2013). Although relatively small (e.g., $<1000\ m^3$) submarine landslides may be relatively frequent (e.g., one or more per year) in high-sediment-supply settings (e.g., fjord-head deltas), events that pose a greater threat to seafloor infrastructure or are tsunamigenic (e.g., $>>1\ km^3$) tend to recur on much longer timescales (e.g., hundreds or thousands of years). Continuous observations are required to capture these episodic events.

• The *large extent of areas affected by slope instability* means monitoring at point locations may completely miss events that occur elsewhere along the same margin (McAdoo et al., 2000; Urgeles & Camerlenghi et al., 2013; Casas et al., 2016; Collico et al., 2020; Gamboa et al., 2021). Being in precisely the right place at the right time is highly unlikely in all but a few settings where the controls on slope instability are well constrained (e.g., seasonally active, spatially focused sediment supply at steep submarine deltas; Lintern et al., 2016).

• The *powerful nature of a landslide and its run-out* poses a hazard to monitoring infrastructure that is placed in the way of the landslide (e.g., Inman et al., 1976; Khripounoff et al., 2003). Recent examples of field-based monitoring demonstrate how even relatively small flows can damage and displace sensors and associated equipment (e.g., Inman et al., 1976; Clare et al., 2020 and references therein). Large slope failures would likely destroy monitoring arrays in their path.

5.1.1. Recent Advances in Direct Monitoring of Submarine Landslides

Despite these challenges, recent technological advances have enabled the monitoring of several aspects of submarine landslide behavior. Geotechnical monitoring provides information about the preconditioning of submarine landslides through the use of in situ devices that monitor changes in subsurface conditions, such as pore pressure (e.g., Prior et al., 1989; Strout & Tjelta, 2005; Stegmann et al., 2011). Technological advances have triggered recent growth in geophysical monitoring of aspects of submarine landslides including (1) repeated seafloor surveys (at timescales from decades to minutes in some cases) to document elevation changes, evolution of the landslide itself, the effects of landslide run-out on the seascape, and subsequent reworking by other marine processes (e.g., Smith et al., 2007; Biscara et al., 2012; Kelner et al., 2016; Mastbergen et al., 2016; Fujiwara et al., 2017; Chaytor et al., 2020; Heijnen et al., 2020; Guiastrennec-Faugas et al., 2021; Normandeau et al., 2021); (2) time-lapse reflection seismic surveys to monitor changes in subsurface conditions (e.g., Blum et al., 2010; Hunt et al., 2021; Roche et al., 2021; Waage et al., 2021); (3) direct monitoring of turbidity currents (some of which likely initiated from submarine landslides) using moored or vessel-based active acoustic sensors, such as acoustic Doppler current profilers (ADCPs) and multibeam sonars, that enable measurement of flow velocity and estimation of suspended sediment concentrations (e.g., Xu et al., 2010, 2011; Hughes Clarke et al., 2012; Khripounoff et al., 2012; Simmons et al., 2020); and (4) monitoring changes in seafloor movement and elevation via geodetic location of acoustic transponders (e.g., Campbell et al., 2015; Paull et al., 2018; Urlaub et al., 2018; Zhao et al., 2021). Such studies tend to involve either campaign-style surveys (often with years between individual campaigns) or short-duration (i.e., weeks to months long) continuous measurements (as they are limited by power supply and data storage); otherwise, they require a cabled connection for data transfer and external power supply. Therefore, while they provide detailed (i.e., temporally or depth/laterally resolved) measurements, such approaches necessarily cover short periods and limited areas relative to the full extent of continental slopes that may be affected by slope instability, as well as the extent of an individual landslide event (Urgeles & Camerlenghi et al., 2013; Urlaub & Villinger, 2019; Brackenridge et al., 2020; Fan et al., 2020). These studies also primarily focus on slope preconditioning, or the run-out produced by slope failures; hence, significant knowledge gaps remain about the inception of slope failures, their kinematics, and their transition from slope failure to more dilute run-out. These are key but poorly constrained parameters in tsunami modeling, impact assessments for critical seafloor infrastructure, and understanding deep water sediment transport in general.

5.1.2. Aims

There is thus a compelling need for low-power-consumption sensors that can cover large spatial domains and that are capable of monitoring a wide range of environmental conditions (including the timing, location, and behavior of submarine landslides), and for systems that are deployed offshore to generate sufficiently small data volumes that sustained, long-endurance, autonomous monitoring is possible. Here, we show how passive acoustic and seismological monitoring has begun to address these constraints and can contribute to filling key knowledge gaps. First, we explore the application of passive acoustic and seismological monitoring of terrestrial landslides and the relevance of those techniques in the marine realm. Second, we discuss which aspects of submarine landslides may be detected using passive acoustic and seismological monitoring. We explore the kinds of environmental noise or ground motion generated by submarine landslides and what signals we should anticipate recording. Third, we present examples from recent marine studies that have successfully monitored precursor conditions, initiation, movement, and impacts of submarine landslides using passive acoustic and seismological approaches. We conclude with some ongoing technological developments and the future application of other techniques, including those detailed in other chapters of this volume.

5.2. PASSIVE GEOPHYSICAL MONITORING OF TERRESTRIAL LANDSLIDES

Early terrestrial studies successfully identified subsurface mining blasts and earthquakes using seismometers, which stimulated further research into the potential of passive geophysical monitoring to detect onshore slope failures (Galitzin, 1915; Jeffreys, 1923; Antsyferov, 1966; Cadman & Goodman, 1967). The elastic straining of soils and rock, friction due to displacement along a failure plane within the sliding mass, and collision of the landslide mass at its down-slope limit were subsequently recognized as signals that could be recorded by seismologic monitoring techniques (e.g., Cadman and Goodman, 1967). It is now well known that progressive failure and detachment of unstable masses can generate local ground motions that are equivalent to earthquakes; hence, land-based seismological networks are increasingly used to determine the timing and location of slope failures, particularly in remote settings where repeat topographic surveys are rare or nonexistent (e.g., Hibert et al., 2019). Such studies enable the identification of many landslides that would otherwise be missing from historical records, thus significantly improving the completeness of hazard catalogs (Ianucci et al., 2020).

In addition to creating ground motions, terrestrial landslide activity can also generate acoustic emissions, most of which tend to have frequency content similar to, or just below, the spectrum of audible sound (Chichibu et al., 1989; Rouse et al., 1991; Dixon & Spriggs, 2007; Dixon et al., 2015). Acoustic emissions are generated as a slope is subjected to stress or shear and/or the landslide mass starts to move downslope (Dixon et al., 2015); hence, both acoustic emission and seismological monitoring have begun to gain recognition as onshore early warning tools to identify the early stages or precursors to slope failure (e.g., Mainsant et al., 2012; Dixon et al., 2015; Michlmayr et al., 2017; Le Breton et al., 2021). Analysis following the Anak Krakatau volcanic sector collapse in Indonesia (which generated a devastating tsunami in December 2018) revealed that broadband seismic and infrasound monitoring networks detected not only the landslide event but also potential triggering events that preceded the collapse (Walter et al., 2019; Fig. 5.1). Regional tsunami monitoring networks did not detect the resultant surface wave due to its localized point source, as they were designed to detect longer line sources (Ye et al., 2020). The collapse itself was represented by a short-lived (one- to two-minute-long) low-frequency (0.01–0.03 Hz) signal, the first P-wave of which was detected at nine onshore seismometers across the Sumatra and Java region, pinpointing its location at Anak Krakatau (Walter et al., 2019; Fig. 5.1). This event had a moment magnitude (M_W) equivalent to 5.3. A higher-frequency (0.1–4 Hz) seismic event was recorded 115 seconds before the sector collapse, while the collapse signal was followed by five minutes of a continuous tremor-like signal at high frequencies (0.7–4 Hz) attributed to post-failure eruptive activity. Infrasound arrays as far afield as Australia (>1000 km to the southwest) recorded a high-energy impulse that matched the origin times of the short-period seismic signal corresponding to the collapse event (Walter et al., 2019). This example is particularly pertinent not only because it demonstrates the capability of passive seismological and acoustic monitoring to detect terrestrial landslides onshore but also because approximately half of the failed mass was underwater (Hunt et al., 2021).

Onshore seismologic networks have been used to detect terrestrial landslides (e.g., Fig. 5.2), including the Hsiaolin hillslope landslide, which occurred when Typhoon Morakot hit Taiwan in 2009. Broadband seismic data recorded (1) very-long-period seismic signals (20–50 seconds) that were interpreted to be related to the elastic rebound of exposed stratigraphy as the overlying landslide mass was evacuated and moved downslope, and (2) large-amplitude high-frequency signals (1–10 Hz) that were interpreted to result from the impact of the landslide mass at the base of the slope (Lin, 2015). Based on the time between these seismic events and the documented run-out distance of 2,500 m, Lin (2015) estimated a maximum velocity of >80 m/s. A total of 51 other landslides were detected on the same day by the same broadband network (Lin et al., 2010). Perhaps most remarkably, 14 of these were detected offshore. Hence, in instances where broadband seismic networks are sufficiently dense, they can provide a useful tool for hazard monitoring.

Processes that generate ground motion and create collisions within similar gravity-driven mass movements have also been successfully recorded using various passive geophysical techniques, including (1) submerged hydrophones and microphones to measure bedload transport based on noise generated by impacts caused by granular material in rivers and streams (Downs et al., 2016; Geay et al., 2017; Rickenmann, 2017; Geay et al., 2020), (2) bespoke geophone and infrasound arrays that detect self-generated noise and vibrations related to snow avalanches (Suriñach et al., 2005; Bessason et al., 2007; van Herwijnen & Schweizer, 2011; Lacroix et al., 2012; Van Herwijnen et al., 2016), and (3) use of existing seismological stations to detect ice avalanches, rock falls, and glacial outburst floods from sudden ground movements as the mass impacts the ground (Caplan-Auerbach & Hugel, 2007; Deparis et al., 2008; Cook et al., 2018, 2021). The successful demonstration of passive geophysical tools to monitor terrestrial mass movements has opened the door for applying similar techniques in offshore settings to overcome some of the many shortcomings of active source monitoring tools (e.g., particularly about power and spatial limitations). Resolution and signal-to-noise ratio are often orders of magnitude poorer for submerged instruments as a result of (1) poor coupling with the seafloor arising from their crude placement (i.e., where instrument frames are dropped from a ship, as compared to the precise placement of instruments onshore) and (2) the environmental noise of the water column, which is affected by a variety of natural processes and human activities.

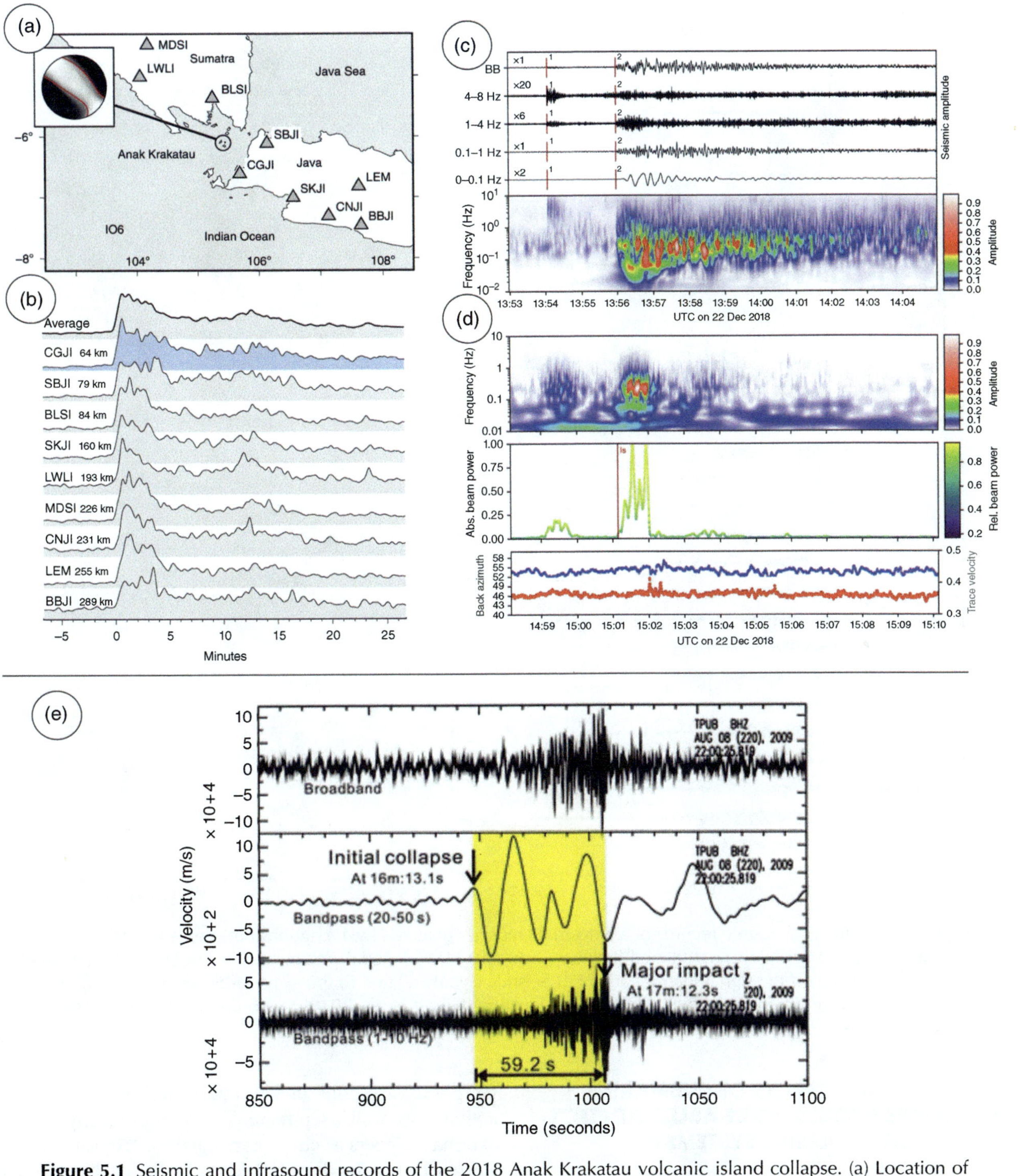

Figure 5.1 Seismic and infrasound records of the 2018 Anak Krakatau volcanic island collapse. (a) Location of regional seismic monitoring network. (b) Vertical component of 0.4–1 Hz seismic records at various stations. (c) Normalized seismic amplitudes at the station nearest to Anak Krakatau recording a high-frequency event before the collapse (top) and a low-frequency signal related to the landslide (bottom). (d) Infrasonic spectrogram records of frequency, beam power, and back azimuth used to locate the event from a seven-element infrasound array. (e) Broadband seismogram (top) and bandpass filtered by 20–50 seconds (middle) and 1–10 Hz (bottom) as recorded before, during, and after the 2009 Hsiaolin onshore landslide in Taiwan. (a–d) Walter et al. (2019) / Springer Nature / CC by 4.0. (e) Lin et al. (2015)/ Springer Nature / CC by 4.0.

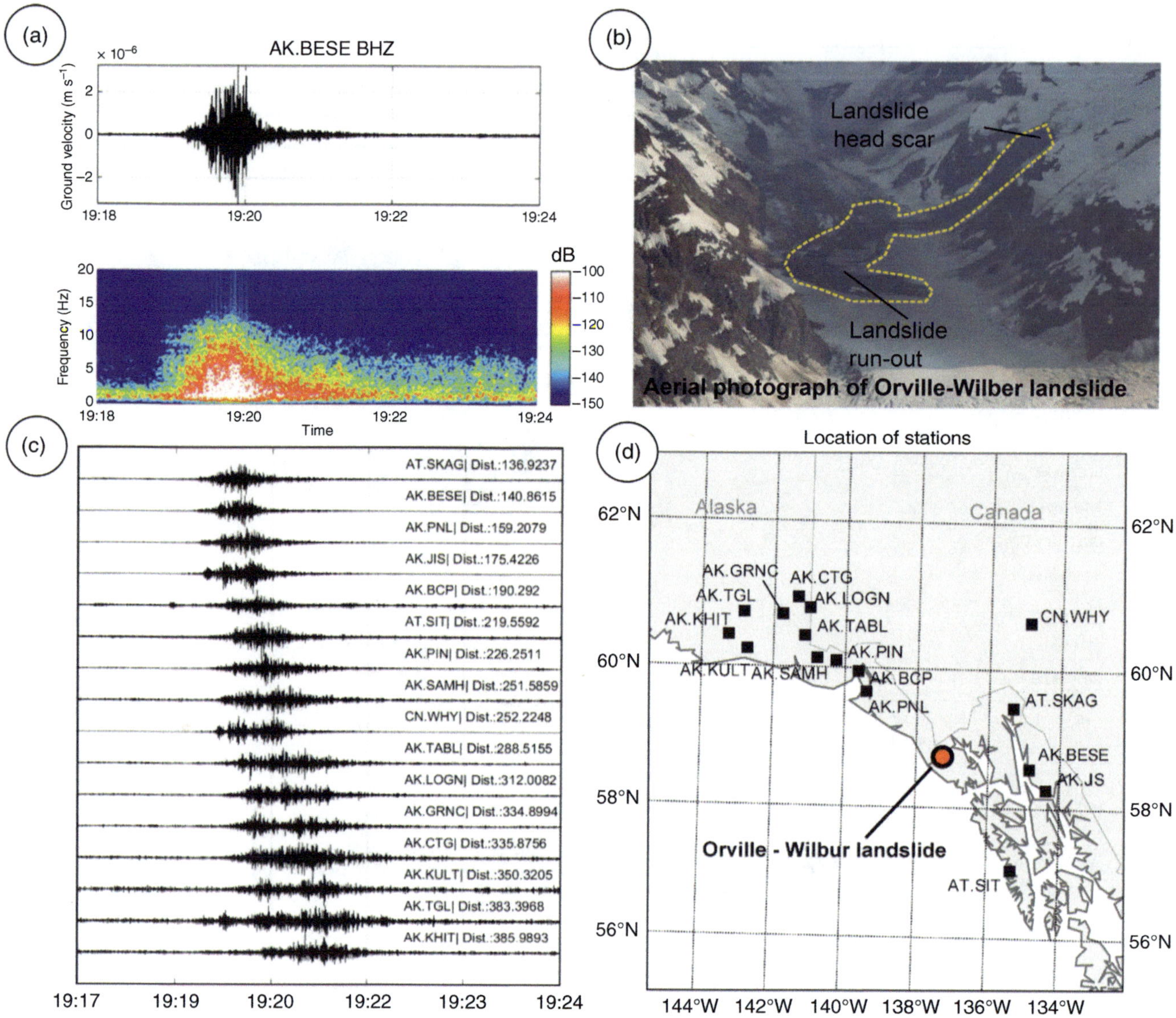

Figure 5.2 Example of seismic records of a terrestrial landslide detected by 17 stations between 136 and 385 km from the event. (a) Seismic signal filtered between 1 and 10 Hz. (b) Aerial photograph of the landslide, which provided validation of the signal interpretation. (c) Seismic signals related to the same landslide at 16 locations, which are shown in (d) in relation to the landslide location. Hibert et al. (2019) / with permission from Oxford University Press.

5.3. WHICH ASPECTS OF SUBMARINE LANDSLIDES SHOULD WE BE ABLE TO DETECT WITH PASSIVE SYSTEMS?

Based on these past studies, a range of information can theoretically be recorded from submarine landslides by passive seismic and acoustic monitoring (Fig. 5.3). We now broadly discuss some of these aspects in the order in which they would occur (i.e., before, during, and following a slope failure).

First, data can be acquired in the build-up to a slope failure, providing information on preconditioning, including temporal variations in volcanic or subsurface fluid flow activity, as well as potentially identifying any plausible external triggers such as earthquakes, tropical cyclones, storms, or rapid sediment delivery by floods. Next, the timing and location of failure can potentially be ascertained by triangulating seismic signals (e.g., the case of the Anak Krakatau event) as a result of ground motions and/or noise generated as shearing occurs along the basal failure surface (and potentially also along its lateral margins), and as that shear propagates through the subsurface (Puzrin et al., 2016). As the landslide mass becomes mobile, ground motions and noise will likely be generated by friction

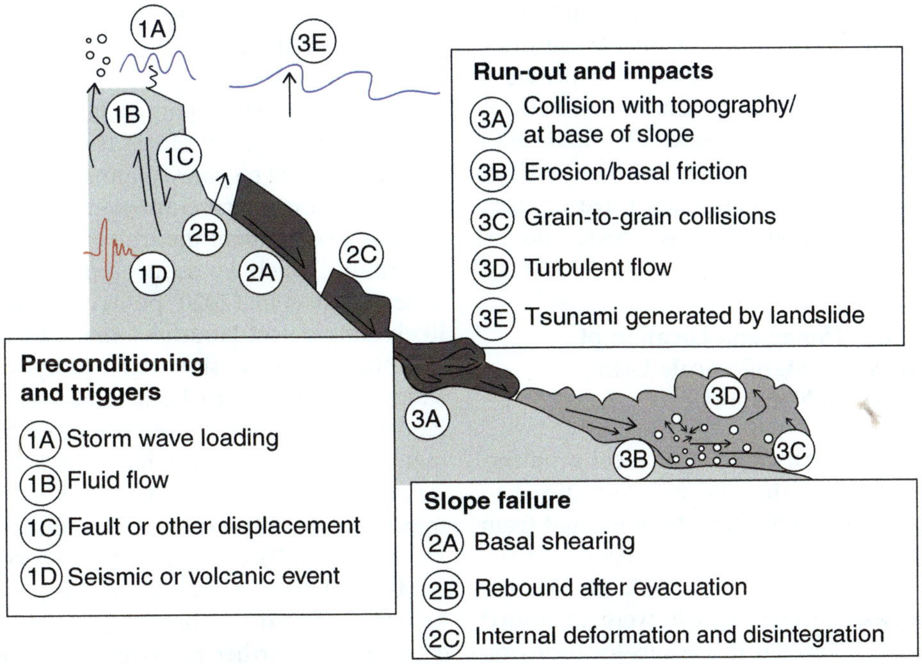

Figure 5.3 Schematic illustration of some aspects of submarine landslides capable of generating seismic and/or acoustic noise as discussed in this chapter, relating to preconditioning and triggering (1A–1D), landslide inception and down-slope displacement (2A–2C), and its subsequent evolution and impacts (3A–3E).

created along the failure plane (Campbell et al., 1995; Brodsky et al., 2003; Gee et al., 2005) due to internal deformation of the moving sediment mass (e.g., Mountjoy et al., 2009; Buss et al., 2019; Sammartini et al., 2021), and if the mass starts to disintegrate as it entrains and mixes with ambient seawater (e.g., Masson et al., 2006). A reduction in vertical effective stress caused by evacuation of overlying sediments or temporary loading and unloading as a result of the transit of an overriding landslide mass can trigger elastic rebound, producing long-period signals (Kanamori & Given, 1982; Kawakatsu, 1989; Lin, 2015) that may occur suddenly, or steadily over prolonged timescales (e.g., >months in onshore examples; Lin, 2015; Leung & Ng, 2016). The displaced mass that moves downslope may travel considerable distances (tens to hundreds of kilometers) over relatively low-angle (<1–2 degrees) slopes. The ground motions and noise generated depend on the changing rheology and kinematics of that moving mass and its interactions with the seafloor and any topography it encounters (e.g., Mountjoy et al., 2018). Friction arising from displacement and erosion at the base of the moving mass should be detectable. However, the nature of the seafloor substrate may enhance or attenuate any signals generated. As the landslide disintegrates, it may switch from a plastically deforming mass of sediment to a dense slurry dominated by grain-to-grain collisions or a more dilute flow characterized by sediment suspended primarily by turbulence (e.g., Talling et al., 2007; Sumner & Paull, 2014; Paull et al., 2018). Each of these modes will likely generate different signals. Where the moving mass generates a sudden physical impact and comes to rest, such as at breaks in slope (ten Brink et al., 2006), following a sudden height drop (Moernaut & De Batist, 2011), or where it meets pronounced topographic barriers (Frey-Martínez et al., 2006), strong, high-frequency signals should be anticipated (Caplan-Auerbach & Hugel, 2007; Deparis et al., 2008). Finally, the indirect effects of the landslide may generate a seismic signal, such as when the displacement of the overlying water mass generates a tsunami (e.g., Yuan et al., 2005).

5.4. RECENT ADVANCES AND OPPORTUNITIES IN PASSIVE MONITORING OF SUBMARINE LANDSLIDES

Motivated by successful terrestrial landslide monitoring and enabled by advances in offshore geophysical monitoring (as highlighted by the other chapters in this volume and Baumgartner et al. 2018), several recent studies have demonstrated that passive monitoring of several aspects of submarine landslides is no longer just a possibility but a tangible reality. Many challenges remain, most notably the fingerprinting of seismic and acoustic signals. The paucity of direct submarine landslide monitoring studies (for reasons discussed earlier in this chapter) means many of these signals remain poorly calibrated; hence, the precise cause and source of signals are often unclear. That being said, recent studies have provided important forward steps in understanding previously unresolved aspects of submarine landslides and offer opportunities to fill other outstanding knowledge gaps. We now highlight studies that illuminate three of these areas: (1) timing and location, (2) kinematics, and (3) run-out.

5.4.1. Determining the Timing and Location of Submarine Landslides at a Margin Scale Using Land-Based Seismological Networks

Using networks of land-based seismological monitoring stations, seismic sources that did not correspond to cataloged earthquakes or storms have been located from radiating surface waves in the Gulf of Mexico and the South China Sea. In the case of the Gulf of Mexico, 85 such nonearthquake seismic signals were identified between 2008 and 2015, located in various locations on the offshore continental slope, and have been attributed to submarine landslide activity (Fan et al., 2020; Fig. 5.4). Individual landslide locations have previously been proposed from analysis of onshore seismological monitoring (e.g., Dewey & Dellinger, 2008; Ten Brink et al., 2008), but this is the first study to identify multiple submarine landslides. The seismic energy recorded is presumed to have arisen from the fast downslope movement of relatively rigid landslide masses (Fan et al., 2020). Given the density of offshore oil and gas infrastructure in the Gulf of Mexico (which is known to be vulnerable to damage by submarine landslides and their run-out), this study provides a much-improved understanding of locations prone to slope failure (Chaytor et al., 2020).

As well as identifying several seismic signals attributed to terrestrial landslides from an onshore broadband seismometer network, Lin (2015) also interpreted arrivals from offshore-generated ground motions in a region offshore SW Taiwan to relate to submarine landslides occurring shortly after the passage of the powerful Typhoon Morakot (8 August 2009). The triangulated location of these seismic sources was clustered around submarine canyon flanks and steep continental slopes, which are known from previous seafloor surveys to be more prone to slope instability (Yeh et al., 2021). Submarine sediment flows, likely triggered by slope failures and related to this event, damaged several telecommunications cables that crossed the Gaoping Canyon (Carter et al., 2014; Pope et al., 2017).

The ability to identify the timing and location of submarine landslides using existing onshore seismological networks (i.e., without the need for deployment of bespoke underwater sensors) offers a potentially game-changing opportunity to fill gaps in hazard catalogs that are required for the robust and resilient design of offshore engineering projects (e.g., oil and gas, renewables, subsea cables) and for developing hazard assessments for coastal communities (e.g., to inform tsunami warnings and guide any related civil response planning). There is still a need to calibrate these interpretations through follow-up seafloor surveys to document the true efficacy of the method and determine which landslide types are reliably detected. Indeed, Fan et al. (2020) point out that this approach will likely only detect landslides that generate sufficient seismic energy; hence, slow (e.g., creep-like) or small landslides are unlikely to be detected. Although submarine landslide catalogs may become more extensive using this method, they will remain incomplete for such events that do not generate discernable seismic noise. Because tsunamigenic landslides tend to be faster and larger, this method provides a promising step forward in submarine landslide detection across very large areas and will only be strengthened as broadband seismic networks are expanded. A further positive is that these passive detection methods provide a low-cost approach to tackle the current biases in submarine landslide studies, given that most of them presently focus on the North Atlantic and offshore from more economically developed countries (Clare et al., 2019).

5.4.2. Quantifying Landslide Kinematics Using Hydrophones

Submarine landslide-generated noise has been used to determine the existence of events in remote oceanic settings that would otherwise likely have gone unnoticed. Networks of hydrophones, including the Hawai`i Undersea Geo-Observatory in the Pacific Ocean, recorded a combination of partly subaerial and fully submarine landslides on the flanks of volcanic islands, which generated acoustic signals that were recorded up to 7,000 km from their source, sometimes exceeding the amplitudes generated by volcanic eruptions by up to an order of magnitude (Caplan-Auerbach et al., 2001; Chadwick et al., 2012; Caplan-Auerbach et al., 2014; Fig. 5.5). The distinctive, extremely low-frequency signals recorded are thought to relate to the collapse of large landslide blocks composed of highly competent volcanic material, as well as from their disintegration as they traveled downslope as an avalanche (Chadwick et al., 2012). In these instances, the triggering event generally relates to enhanced volcanic activity, such as at the West Mata volcano near Tonga and Kilauea volcano, Hawaii (Caplan-Auerbach et al., 2001, 2014). In addition to pinpointing their location and identifying the timing relative to volcanic activity, the multipathing of the propagated sound waves enabled the determination of the landslide transit velocities, which

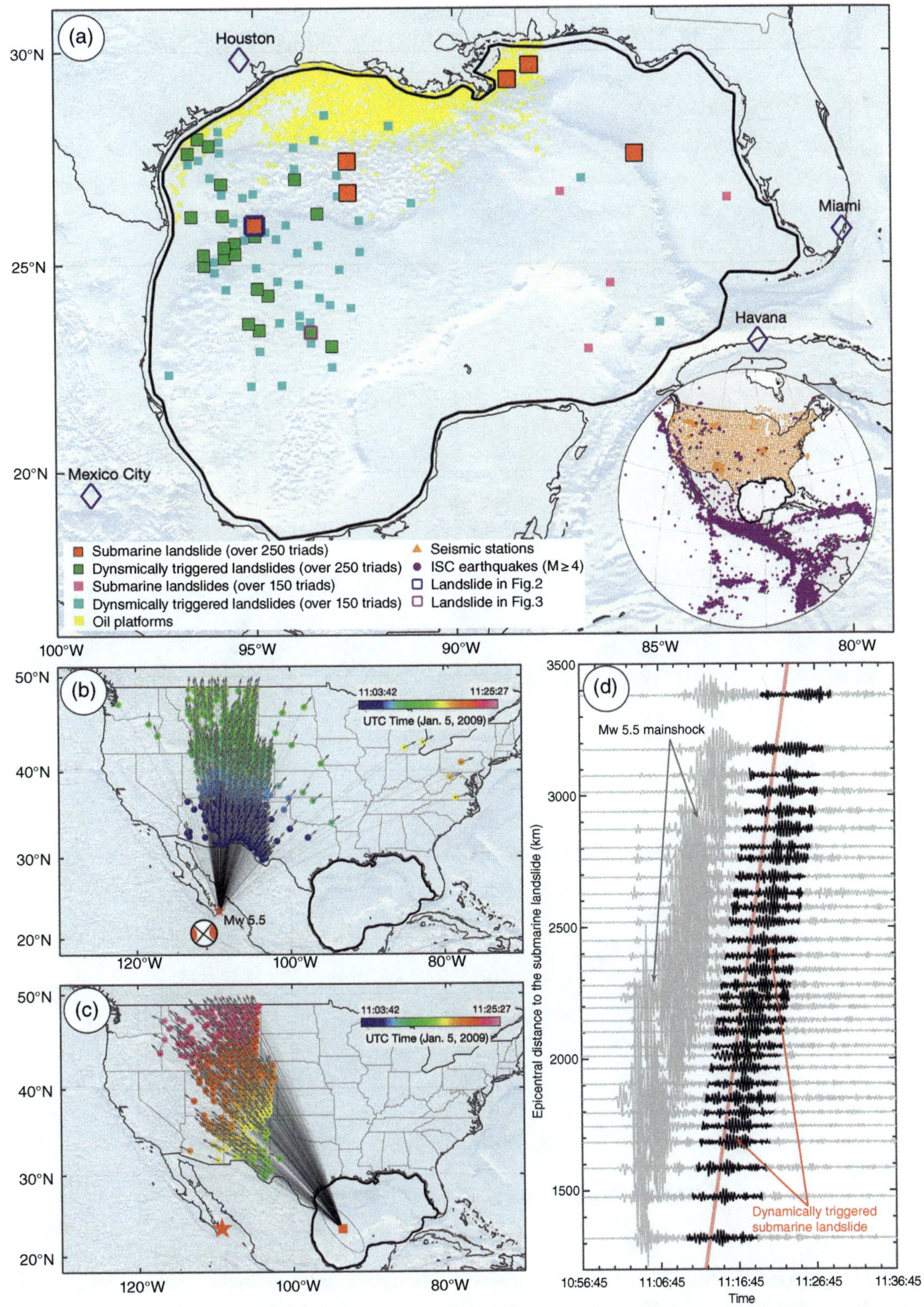

Figure 5.4 Identification of submarine landslides using terrestrial seismometer networks. (a) Locations of seismic signals interpreted to relate to submarine landslides in the Gulf of Mexico using the onshore seismic monitoring network. Inset shows earthquakes in the monitored period (2008–2015) of magnitude 4 or above. (b) Rayleigh wave arrival times (colors of circles) and propagation directions (lines) for an earthquake; (c) a submarine landslide thought to have been triggered by that same earthquake (red star). (d) Bandpass-filtered (20–50 seconds) waveforms for selecting the seismic stations used to locate the submarine landslide. Grey signals show precursor earthquakes, and black shows the signal of the landslide itself. Fan et al. (2020) / with permission from John Wiley & Sons.

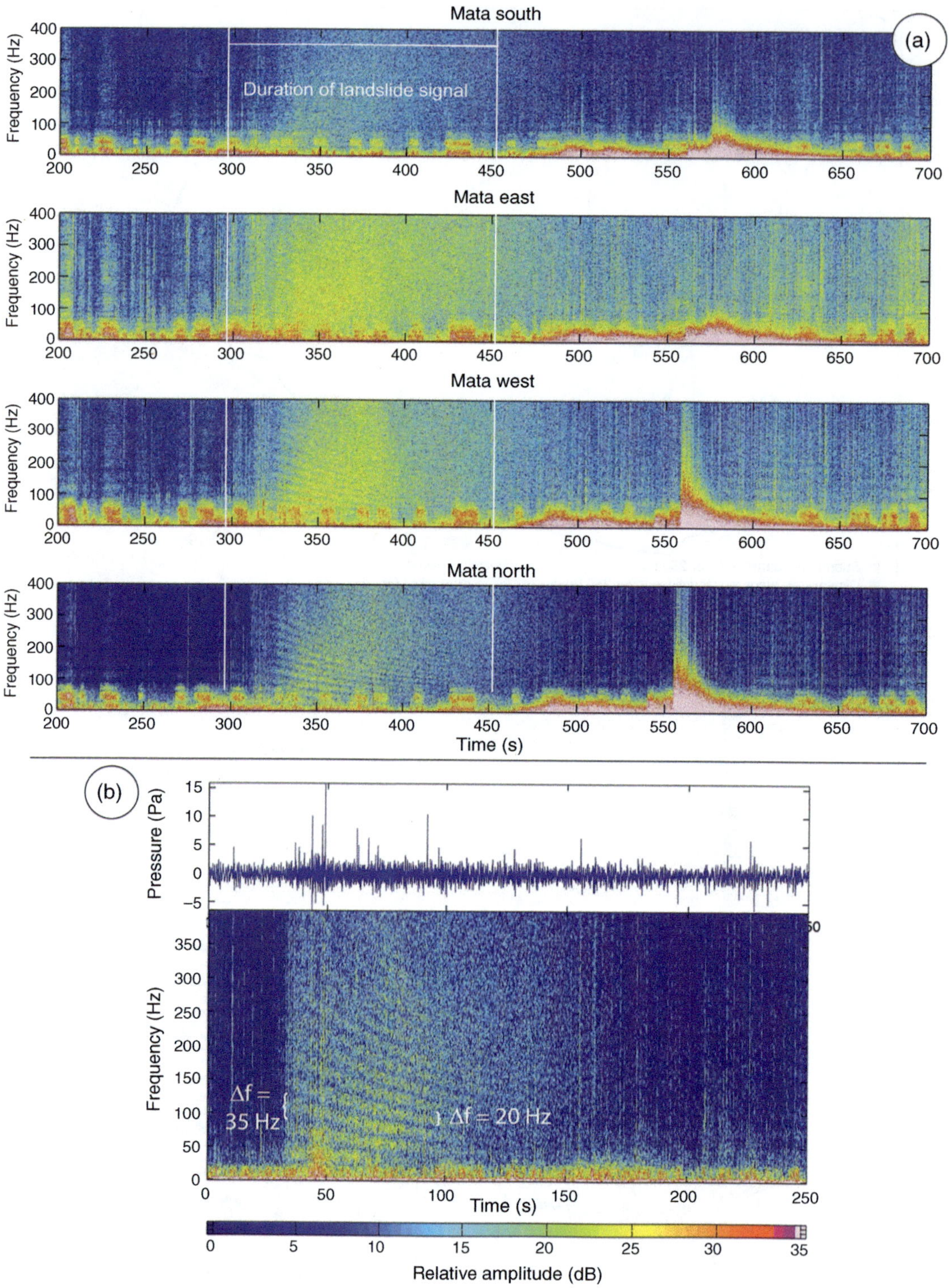

Figure 5.5 Hydrophone spectrograms showing evidence of submarine landslides in the southwest Pacific (2014). (a) The signal between 300 and 450 seconds is interpreted as a submarine landslide due to its broadband spectrum and variable frequency content. It is distinct from degassing explosions (low frequency, <50 Hz) events and regional earthquakes (very low frequency between 470 and 600 seconds). (b) Hydrophone time series and spectrogram for a submarine landslide. Interference bands are noted at 35 Hz at the start of the landslide and 20 Hz near the end. The change in the spacing of frequency bands indicates a moving source and is used to infer the vector velocity of the landslide. Caplan-Auerbach et al. (2014) / with permission from John Wiley & Sons.

are estimated at between 10 and 25 m/s (Caplan-Auerbach et al., 2014).

The location (particularly the water depth) of a landslide source and its initial velocity are key inputs for tsunami modeling (e.g., Løvholt et al., 2015). Typically, these parameters remain poorly or entirely unconstrained, which results in wide uncertainty in any predictive tsunami modeling. Therefore, these and future observations using similar hydrophone networks can improve our understanding of a globally significant hazard (e.g., Tappin et al., 2008; Harbitz et al., 2014; Goff & Terry, 2016). Again, calibration of the hydrophone signals is needed; hence, field studies that make multiparameter measurements of submarine landslide activity will contribute significantly to the fingerprinting of different acoustic responses and enhance confidence in the future acoustic detection and characterization of submarine landslides.

5.4.3. Characterizing Landslide Run-Out to Enhance Hazard Assessments

Submarine landslide run-out is often far more extensive than that on land and may evolve dramatically about its rheology as flows mix with seawater and/or entrain seafloor sediment (Parker, 1982; Zeng et al., 1991; Mohrig & Marr, 2003; Gee et al., 2006; Heerema et al., 2020). These mass flows can travel at high speeds and across large distances, so they can damage sensors placed in their path (e.g., see several examples in Clare et al., 2020). Therefore, using passive, remote-sensing acoustic, and seismic monitoring tools placed out of the direct pathway of such flows can enable undisturbed monitoring and reduce instrument damage or loss. Recent studies have demonstrated how hydrophones may record turbidity currents (some of which were triggered by landslides) in fjord settings that are fed by seasonally active rivers, including on the Fraser and Squamish deltas, both in British Columbia (Lintern et al., 2016; 2019; Hizzett et al., 2018; Hay et al., 2021; Fig. 5.6).

At the Fraser Delta, hydrophones are deployed on a seafloor observatory connected to shore via the VENUS cabled network, enabling real-time data streaming, including measurements by complementary active source geophysical sensors (Lintern & Hill, 2010). Sensors such as ADCPs provide the calibration and confidence that the acoustic signals (which include both high [1–300 kHz] and low [Hz] frequency ranges) can be attributed to turbidity currents (Lintern et al., 2016, 2019). At the

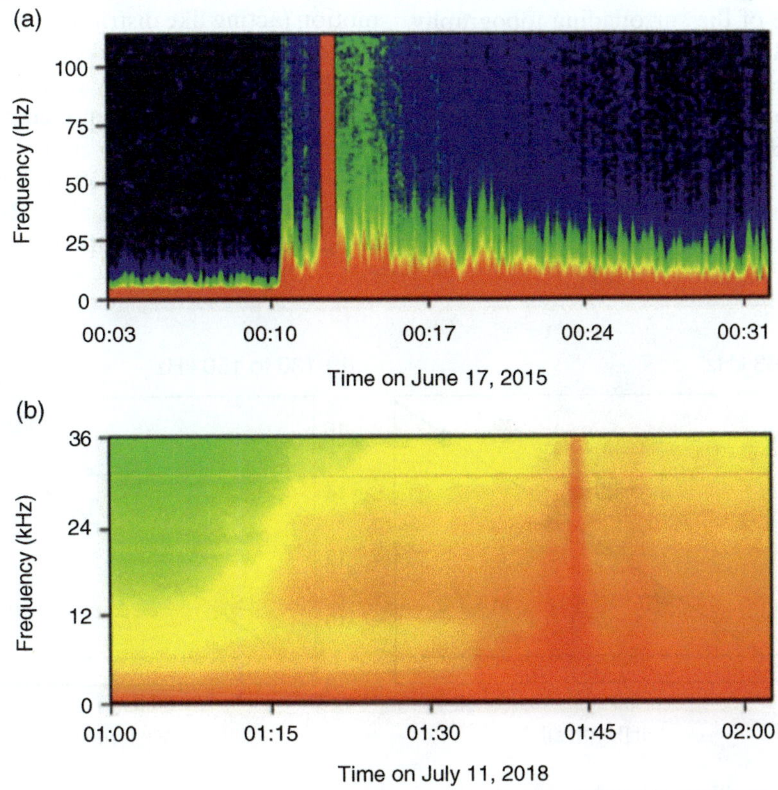

Figure 5.6 Hydrophone spectrograms showing turbidity current events in British Columbia: (a) passing directly through a hydrophone at the Delta Dynamics laboratory at the Fraser Delta; (b) passing below a mooring in the Bute Inlet Channel, British Columbia. Lintern et al. (2019).

Squamish Delta, a hydrophone suspended from a vessel 10 m above an active submarine channel recorded the noise generated from 16 turbidity currents that reached speeds of 2 m/s (Hay et al., 2021). Extensive calibration was provided by a pair of 500 kHz multibeam sonars placed on the same frame as the hydrophone, three echosounders (28, 200, and 70–110 kHz) mounted on the vessel itself, and a 1,200 kHz ADCP suspended from a separate frame, also 10 m above the seafloor (15 m from the hydrophone frame). This novel study showed that where turbidity currents exceeded 1 m/s, they generated noise well above ambient ocean conditions across a spectrum of 10–500 kHz, interpreted to result from highly energetic sand-sized particle collisions. Cross-reference of hydrophone and ADCP-derived velocity measurements revealed that spectral density (at 100 kHz frequencies) is proportional to the speed of the flow front, where a 100-fold increase in spectral density relates to a doubling in frontal speed (Fig. 5.7; Hay et al., 2021). Even more powerful sediment flows (i.e., >2 m/s, and potentially >10 m/s) that occur within larger submarine canyons and channels (e.g., Carter et al., 2014; Azpiroz et al., 2017; Paull et al., 2018), or that travel vast distances across open continental slopes (e.g., Piper et al., 1999), may generate similar acoustic signals and/or generate ground motions (depending on their velocity and interaction with and the nature of the surrounding topography and seafloor substrate) that can be detected with seismometers, but this requires future investigation.

Seafloor cabled-networks, such as the Neutrino Mediterranean Observatory-Submarine Network 1 (NEMO-SN1), have demonstrated that offshore seismic monitoring can also play a future role in detecting the run-out from submarine landslides. Although background environmental noise was often found to complicate conclusive signal identification, high-frequency and long-duration (at least longer than that of earthquakes) signals recorded by an ocean bottom broadband seismometer on the NEMO-SN1 network offshore Mount Etna were attributed to submarine landslide run-out (Sgroi et al., 2014). Therefore, this approach can be readily applied using marine seismometer networks, and similar deployment of more than one seismometer would allow for greater characterization of events, including pinpointing their source. These recordings were particularly high quality because the dedicated installation procedure using a remotely operated vehicle ensured excellent coupling with the seafloor. This is not always the case for ocean bottom seismometer surveys, as these instruments are typically dropped from a surface vessel and land in a relatively uncontrolled manner on the seafloor.

5.4.4. Opportunities Using Distributed Cable-Based Sensing

Spatially resolved distributed sensing along seafloor fiber-optic cables enables broadband seismic and acoustic monitoring across long distances (over tens to thousands of kilometers). Several recent studies have demonstrated temporally and spatially resolved monitoring of ground motion (acting like distributed networks of seismometers) generated by earthquakes, fault displacements, and acoustic noise created by ocean waves and deep-sea bottom currents that resuspend sediment at the seafloor (e.g., Blum et al., 2010; Marra et al., 2018; Ajo-Franklin et al., 2019; Lindsey et al., 2019; Williams et al., 2019; Zhan et al., 2020; Nishimura et al., 2021; Zhan et al., 2021; Wilcock, 2021; Fig. 5.8). Distributed acoustic sensing data

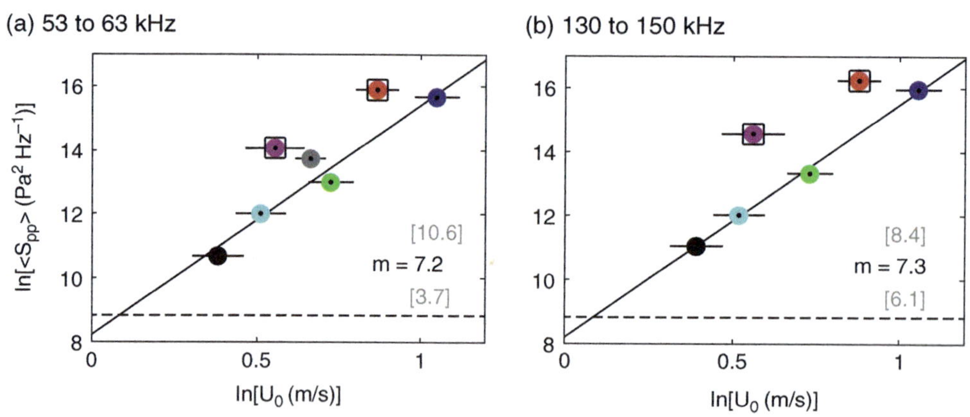

Figure 5.7 Cross-comparison of band-averaged noise spectral densities (S_{pp}) with frontal velocities (U_0) of turbidity currents recorded in the Squamish Delta, British Columbia, for two frequency bands. Colored symbols represent different turbidity current events. The solid black line is a best-fit trend, and the dashed black line shows the detection threshold above ambient noise levels. Hay et al. (2021) / AIP Publishing / CC BY 4.0.

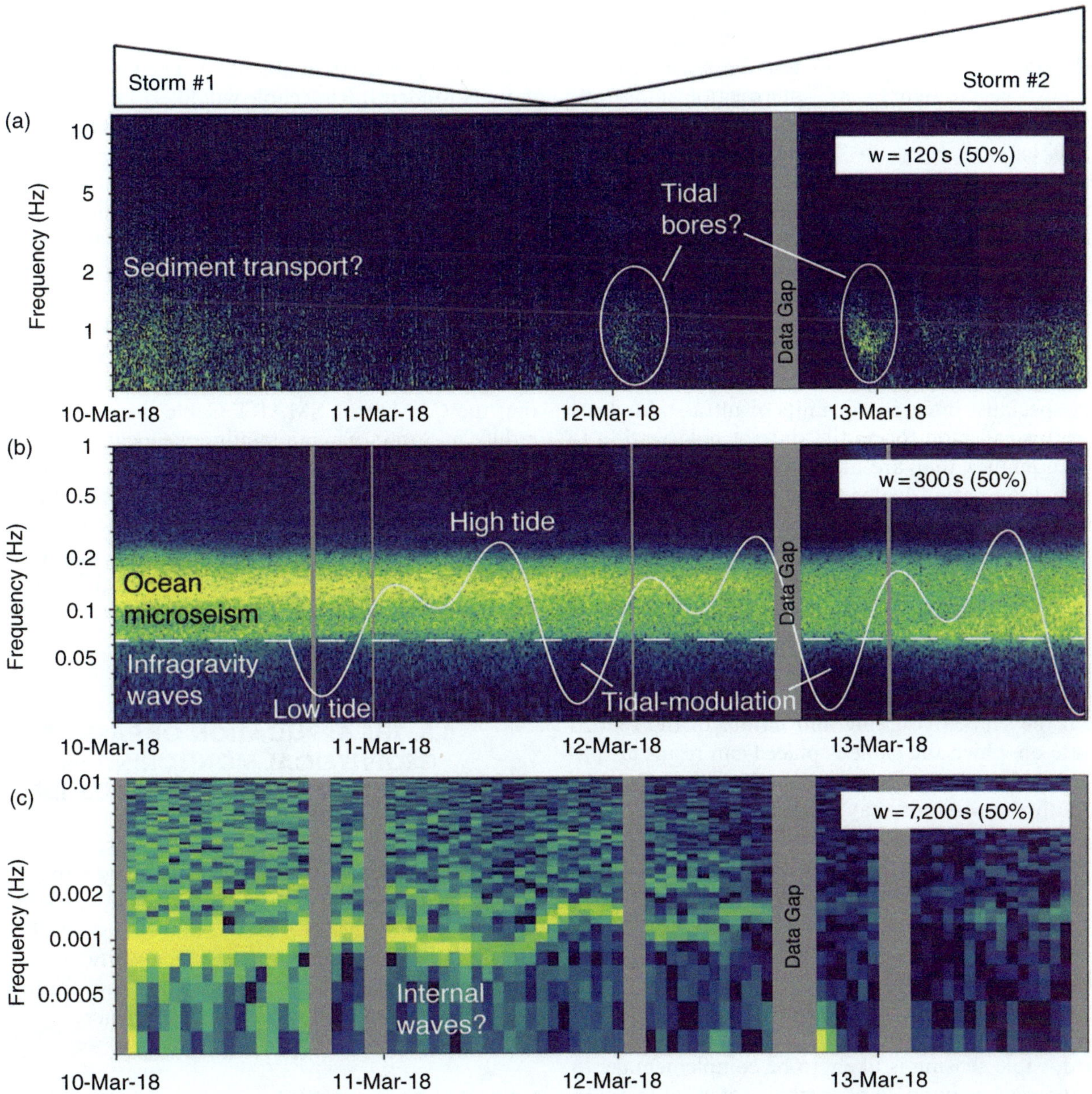

Figure 5.8 Frequency spectrograms from DAS along a seafloor fiber optic cable offshore Monterey Bay, California. These demonstrate how this technique can detect (a) short-lived tidal and near-bed currents, (b) tidal effects due to pressure changes, and (c) internal waves. Therefore, this approach may also be appropriate to measure aspects such as the run-out of submarine landslides (e.g., turbidity currents), displaced water masses in front of a moving mass, or ground motions generated by the landslide itself. Lindsey et al. (2019).

acquired along a fiber-optic cable in the Nankai subduction zone, western Japan, was found to have comparable performance to adjacent broadband ocean bottom seismometers at seismic frequencies of >1 Hz, provided the cable was well-coupled to the seafloor (Ide et al., 2021). The same study concluded that even though such distributed acoustic monitoring records one component of strain, seismic source localization within tens of kilometers of the cable can still be achieved. At lower frequencies (i.e., 0.02–0.05 Hz), the distributed acoustic sensing observations were notably noisier than those from ocean bottom seismometers, thus making detecting such events challenging unless events occur very close to the cable. The low-frequency regime may be critical for regional-scale landslide detection (i.e., Fan et al., 2020); hence, this technique may be most appropriate for detecting relatively local low-frequency or far-field high-frequency signals, but more studies are required that provide similar comparisons to test the sensitivities of different cable configurations, interrogator units, and analytical approaches (Lindsey & Martin, 2021; Lior et al., 2021).

Recent advances have seen the application of sensing to detect changes in the state of polarization over the full length of a 10,000 km telecommunications cable, which enabled recording of moderate to large earthquakes, as well as pressure signals from ocean swells (thus illustrating the potential for real-time tsunami detection over transoceanic distances; Zhan et al., 2021). This distance far exceeds that reached by previous studies and did not require specialist interrogator units or ultra-stable laser sources, instead monitoring the state of polarization to detect anomalies that are attributed to strong seismic waves or long-period water waves (Zhan et al., 2021). It should be noted, however, that these long-range measurements were spatially integrated (i.e., averaged along the entire fiber length); hence, additional independent information is required to locate and further characterize the hazard besides the locally recorded signal timing and magnitude. Zhan et al. (2021) also recognized that site-specific effects can affect the sensitivity of measurements, wherein the seabed coupling and nature of the seabed substrate on which the cable is placed can result in spatially variable signal attenuation. Indeed, coupling the cable with the seafloor may be particularly important for cable-based sensing. When assessing the efficacy of distributed acoustic sensing (DAS) to detect earthquakes, Lioer et al. (2021) found that performance was notably reduced in areas of highly irregular, rocky seafloor, where the cable is locally in free span, raised above the seafloor across sections; and enhanced in areas of low relief, soft seafloor, particularly where cables may become embedded or buried by sedimentation. Going forward, cable sensing is likely to be complementary to other approaches but also has strong potential to fill in major geographic gaps in seismic monitoring, potentially making use of the existing global network of submarine cables and accessing new cables (Ranasinghe et al., 2018; Howe et al., 2019; Mizutani et al., 2020; Mecozzi et al., 2021; Wilcock, 2021).

Although DAS can provide valuable information, such sensing is limited to no more than a few hundred kilometers from shore; hence, additional sensors, which could include hydrophones or geophones, distributed at nodal locations along a cable may help to fill this gap over longer distances. This is the premise of the SMART (Science Monitoring and Reliable Telecommunication) Cables initiative, which proposes attaching instruments, including ground motion, pressure, and temperature sensors, as external nodes on repeaters (typically spaced 60–80 km apart) on new or decommissioned submarine telecommunication cables (Ranasinghe et al., 2018; Howe et al., 2019; Mecozzi et al., 2021). The CAM (Continent-Azores-Madeira) Ring cable will likely be the first operational SMART Cable system (coming online in 2024), connecting mainland Portugal with the Azores. As these nodes are powered and connected by cables, they will potentially provide opportunities for real-time seismo-acoustic monitoring far from shore. This is particularly relevant for the CAM Ring cable, as the strongest earthquakes occur far offshore, and many of the previously mapped submarine landslides are located on the flanks of remote submerged seamounts (Gamboa et al., 2021; Matias et al., 2021). Numerical modeling indicates that the CAM Ring SMART Cable should significantly reduce uncertainty in pinpointing the location and timing of offshore earthquakes and identify events (such as submarine landslides) that are currently undetectable by the existing onshore and coastal monitoring networks, while detection of the passage and direction of tsunamis should provide improved early warning of coastal impacts and thus time to prepare emergency responses (Matias et al., 2021).

5.5. THE APPLICATION OF PASSIVE GEOPHYSICAL MONITORING IN ADVANCING SUBMARINE LANDSLIDE SCIENCE

We now revisit the overarching challenges and science questions outlined in the introductory section, specifically addressing how emerging seismic and acoustic techniques can fill outstanding knowledge and capacity gaps, which uncertainties remain, and a proposed strategy for forward steps to advance submarine landslide science.

5.5.1. Can Passive Seismic and Acoustic Techniques Overcome the Logistical Challenges That Have Previously Hindered the Monitoring of Submarine Landslides?

First we respond to the logistical challenges: the often deep-water and remote location of submarine landslides and their unpredictable and powerful nature. Provided sensors and receivers are protected in appropriate pressure housing, passive seismoacoustic landslide monitoring using hydrophones and geophones can be performed across the full range of water depths in the global ocean, from coastal (e.g., Hay et al., 2021) to thousands of meters of water depth (e.g., Caplan-Auerbach et al., 2014). To date, we are not aware of any landslide monitoring in hadal water depths, but it is certainly logistically possible, and the use of ocean-bottom seismometers and DAS along fiber-optic cables is increasing in many deep ocean settings to record earthquake- and nonearthquake-related noise (e.g., >4,000 m offshore Chile; Batsi et al., 2019; Ide et al., 2021). Given the relatively early stages of many techniques we have described, most previous studies have relied on individual instruments or limited networks of seabed or moored instruments with narrow spatial coverage. Hence, widespread (i.e., basin- to ocean-scale) passive monitoring of submarine landslides remains an open and future challenge. However, onshore seismic monitoring networks are far denser than those offshore. Where margin geometry permits, these can immediately be used to identify submarine landslides that may occur hundreds to thousands of kilometers from shore, where their location is not known a priori (Fan et al., 2020).

The Gulf of Mexico is perhaps an ideal candidate, as its physiographic geometry suits the use of triangulation from an onshore network to pinpoint the offshore location of submarine landslides. Arc-shaped convergent continental margins are also good future candidates for this approach, as are enclosed basins such as the Mediterranean, as they provide similarly well-suited geometries and are often "blind spots" for landslide-tsunami hazards where new insights are required (e.g., Sunda Arc; Goff & Terry, 2016). The geometry of many margins does not suit this approach, however. As recognized by Lin (2015), the deployment of wider and denser networks of onshore and offshore seismometers will be required in such settings for limitations to be overcome. DAS along fiber-optic cables provides opportunities for long-distance monitoring and potentially fills several spatial gaps in an offshore monitoring capacity. However, this technology cannot currently be extended beyond the first repeater and is thus presently limited to distances of less than a few hundred kilometers from shore (Matias et al., 2021). This means many submarine landslides, which occur far offshore, cannot yet be recorded by this technique. Longer-range (i.e., tens of thousands of kilometers) monitoring is possible along cables that connect continents, either as integral measurements with limited spatial resolution (e.g., Marra et al., 2018; Zhan et al., 2021) or at powered SMART sensor nodes along the cable (e.g., Howe et al., 2019).

All the passive monitoring approaches enable the placement of sensing equipment out of the path of submarine landslides and their run-out, which has hindered many previous efforts where sensors, moorings, and platforms have been damaged or lost (e.g., Khripounoff et al. 2003; Clare et al., 2020 and references therein). These new techniques ensure that the timing initiation of a landslide can be recorded and that measurements can be made during the landslide event itself, from which key aspects of landslide behavior can be determined.

5.5.2. What Aspects of Submarine Landslides Can We Assess from Passive Remote Sensing Techniques, and What Needs To Be Resolved?

Landslide Timing

The timing of submarine landslides has been determined from signals detected by submerged hydrophones, geophones, and land-based seismic monitoring systems. Therefore, observational passive monitoring data sets can provide new insights into the timing and frequency of submarine landslides, starting to fill a key knowledge gap. For example, land-based seismic monitoring in the Gulf of Mexico identified 85 previously undetected submarine landslide events (Fan et al., 2020). The ability to identify past submarine landslide signals provides important information to develop more robust hazard assessments and inform resilient routing and siting of offshore infrastructure (Chaytor et al., 2020), while the quick identification of landslide signals has the potential to enhance existing tsunami warning systems (Fan et al., 2020; Matias et al., 2021). An immediate opportunity lies in the back-analysis of legacy data sets to explore whether past landslide signals exist. However, previous studies have ignored them as they focused on different frequency spectral ranges. Such an approach has already shown value in identifying gas- or fluid flow-related processes from ocean-bottom seismometers or ocean currents from DAS data sets (Baksi et al., 2019).

Landslide Triggers and Preconditioning

In addition to determining whether a landslide has occurred, ascertaining the timing of landslide inception allows us to investigate the environmental conditions in the build-up to that event and thus determine likely triggering and/or preconditioning factors. Limitations in the precise determination of submarine landslide timing (e.g., due to uncertainties in radiocarbon dating) have hindered the identification of these links previously, but this is now possible, as highlighted by acoustic and seismic monitoring before the 2018 Anak Krakatau volcanic collapse (Walter et al., 2019) and offshore landslides following earthquakes in Taiwan (Lin, 2005). Future monitoring will likely lead to identifying new or potentially surprising triggers, such as the intriguing link between distant earthquakes and submarine landslides in the Gulf of Mexico (Fan et al., 2020). Passive monitoring technology is now sufficiently mature to test previous hypotheses on landslide triggering and investigate new ones; however, operational early warning systems (i.e., that automatically raise alerts ahead of a potential landslide event) remain a future prospect for all but a few locations, as this requires a robust understanding of the conditions before failure or identification of a landslide as it happens. This will therefore be more feasible at sites where triggers and/or preconditioning are more predictable and closely linked to enhanced sediment supply from rivers (i.e., where landslides are more prone during or following periods of higher river discharge, such as the bedload-dominated Fraser or Squamish Delta; Lintern et al., 2020; Hay et al., 2021) or relate to heightened volcanic activity (e.g., Anak Krakatau, Indonesia; Mount et al.; Urlaub et al., 2018; Walter et al., 2019).

Landslide Location

It is possible to pinpoint the location of submarine landslides through triangulation (to a precision of approximately tens of kilometers; Caplan-Auerbach et al., 2014; Lin, 2015; Fan et al., 2020); however, as already noted, this approach is limited to locations where sensor networks are sufficiently widely and densely distributed. There is a wealth of existing onshore (and, to a limited extent, offshore) broadband seismic data sets in several regions where the same approach could be taken to investigate whether similar landslide-like signals exist, whether they can be located spatially, and whether they correspond to any of the recent direct measurements of slope instability and their run-out or existing maps of landslide vulnerability (e.g., Paul et al., 2018; Urlaub et al., 2018; Obelcz et al., 2020; Gamboa et al., 2021). We suggest that detailed bathymetric surveys should be performed in areas where seismic signals attributed to submarine landslides have been located, to determine which landslides are recorded and test the efficacy of landslide detection. The Gulf of Mexico, the Mediterranean, the Caribbean, and regions offshore southwest Taiwan are prime candidates for this field calibration.

Landslide Kinematics

If a landslide mass is known, it is possible to infer other aspects of submarine landslide behavior from the resultant seismic and acoustic signals. For instance, inversion of these signals can be used to derive force histories. From this volume, velocity, changes in basal friction, run-out distance, and other important aspects of landslide kinematics can be determined, as demonstrated by the successful application of these approaches in terrestrial settings (e.g., Brodsky et al., 2003; Ekström & Stark, 2013; Yamada et al., 2013; Moretti et al., 2015). However, these aspects will likely be more challenging to constrain than landslide timing and location, requiring concerted research and field calibration. Sites where passive and active monitoring are performed concurrently will provide valuable opportunities to calibrate the nature of seismic and/or acoustic signals, such as at the Squamish Delta, Canada, where variations in noise spectra were linked to changes in the concentration and grain size of sediment avalanches, and noise intensity corresponded to the velocity of the flow front of turbidity currents (Hay et al., 2021). Moored arrays of direct active sensors (e.g., ADCPs) have been used to track the passage and evolution of landslide run-out and turbidity currents down submarine canyons over tens of kilometers (e.g., Paull et al., 2018). The addition of co-located hydrophones and geophones to such moored or seafloor arrays will enable the calibration of seismic and acoustic signals as well as determining threshold limits for detection.

Site-Specific Effects

We suggest that different deployment configurations should be trialed to investigate how signals attenuate in various settings and investigate the effects of increasing distance and differing angles of incidence from the source. The effects of seafloor morphology may attenuate the signal generated by submarine landslides and complicate its propagation pathway, which is particularly important for directional instruments such as hydrophones. Although hydrophones have been shown to detect events on relatively open delta slopes (Hay et al., 2021), the extent to which the topographic effects of deeply incised submarine canyons may limit the detection of events by similar sensors placed outside the canyon along narrow fjords or other irregular seafloor terrain remains poorly constrained. It is also important to identify and differentiate landslide-related noise from ambient seismic and acoustic noise that arises from natural background processes and human activities in the ocean. In noisy marine environments, ambient noise may overprint or entirely obscure the records of submarine landslides; hence, understanding the levels, frequency, and time windows of that background noise is important when designing monitoring strategies (Lior et al., 2021).

Societal lockdowns in 2020 and 2021 associated with the COVID-19 pandemic have been identified as periods of significantly reduced seismic noise both on land and at sea (Lecocq et al., 2020; Ryan et al., 2020) and may therefore provide a rare window of reduced background noise to explore local to global data sets. If signals can be sufficiently calibrated and confidently attributed to submarine landslides, historical records can be reanalyzed to extend landslide catalogs in many regions worldwide, adding significant value to already valuable legacy data sets and providing new avenues for research and informing hazard management strategies.

Challenges in Data Transfer

As passive geophysical systems require much less power than active sensors (e.g., ADCPs), they can improve the endurance of offshore monitoring, lengthening monitoring windows in time. Future challenges still lie in data transmission from sites far from shore (i.e., in abyssal to hadal water depths), particularly if an end goal is to support early warning systems. This issue is largely moot for cabled systems supporting high-bandwidth data transfer. However, as already noted, the current reach of DAS limits long-distance (> hundreds of kilometers) monitoring. There is currently no simple solution for ocean-bottom seismometers or hydrophones on seabed frames or moorings, as data file sizes typically exceed that currently permissible by acoustic data transmission.

5.5.3. Suggestions for Future Directions

We now finish by making specific suggestions regarding future activities, listed in a broadly prioritized order.

Focus on Calibrating Hydrophone and Geophone Records at Specific, Well-Known, Active Sites

We suggest that efforts should be focused on specific sites and settings to provide the calibration needed to strengthen confidence in characterizing submarine landslides from passive monitoring data. It is sensible to target active sites where previous monitoring has demonstrated regular landslide activity, where pre- and post-event bathymetric data are available, and where passive instruments (hydrophones, geophones, fiber-optic cables, etc.) can be deployed concurrently with active sensors that record key parameters such as event timing, subsurface pore pressure, elevation changes, and flow velocity and sediment concentration. The Fraser Delta is one such location, connected by seafloor cables to Ocean Networks Canada's VENUS network (which enables real-time data transfer to shore) in an area of high sediment supply with subannual landslide activity. Synchronous deployment of hydrophones and seismometers with current meters, inclinometers, and piezometers (among other sensors) provides an ideal test bed for calibrated interpretation of seismic and acoustic signals attributed to small-scale submarine landslides (Lintern & Hill, 2010; Lintern et al., 2020). Other cable-connected sites, such as the EMSO (European Multidisciplinary Seafloor and Water Column Observatory) Ligure-Nice observatory, provide opportunities to monitor the development of potential precursor conditions, but landslide recurrence is orders of magnitude lower at this site compared to the Fraser Delta; hence, the likelihood of catching a landslide itself is much less likely (Bompais et al., 2019). Similarly, the DONET (Dense Oceanfloor Network system for Earthquakes and Tsunamis) cabled sensor network located offshore southwest Japan provides opportunities to detect submarine landslides; but intriguingly, no evidence for such activity was found within a few hours of two large (>6 M_w) earthquakes, with a study concluding that slopes may not be particularly susceptible to slope failure in that region (Gomberg et al., 2021).

Indeed, most cabled observatories tend to avoid areas of very frequent landslide activity, so it will also be necessary to perform field calibration in other settings using more conventional landers or moorings that are not connected by cables, which requires sea-going research cruises. Sites with high and predictable sediment supply and where active field campaigns are planned will be the preferred candidates for the synchronous deployment of active and passive monitoring equipment, such as major submarine canyons that connect to large rivers or whose head intersects seasonally variable littoral transport cells (e.g., canyons where subannual turbidity currents have previously been recorded: Congo Canyon, West Africa; Gaoping Canyon, Taiwan; Var Canyon, Mediterranean; Capbreton Canyon, Bay of Biscay; Khripounoff et al., 2012; Azpiroz-Zabala et al., 2017; Paull et al., 2018; Zhang et al., 2018). On a larger scale, it will also be important to acquire data at active volcanic sites where onshore and offshore monitoring is ongoing, such as the geodetic monitoring offshore Mount Etna or land-based acoustic emission and GPS monitoring at Anak Krakatau (Urlaub et al., 2018; Walter et al., 2018). These activities should go hand in hand with efforts to determine site-specific effects on attenuation (e.g., due to the topographic, substrate, and other effects), which will benefit from numerical modeling to help identify and design the optimal sensor (e.g., where is the strongest or earliest signal expected?).

Explore and Calibrate Existing Land-Based Seismic Monitoring Data

We suggest that more can and should be made of legacy broadband seismic data sets to explore past catalogs for evidence of submarine landslides. This method has been demonstrated offshore in the United States and Taiwan (Lin, 2015; Fan et al., 2020) and is more widely applicable using data that have already been collected. Analysis of past time-series data should then be complemented with the acquisition of new seafloor survey data to confirm that the spatial locations of landslide signals do indeed correspond with landslide events and determine if a specific type(s) of landslide is more likely to generate a detectable seismic signal. This calibration will be most effective in areas where detailed swath multibeam echo-sounder surveys have already been performed (in particular, high-resolution data sets acquired using autonomous underwater vehicles), as differential elevations between multiple surveys provide a time window for a landslide as well as its location.

Although repeat seafloor surveys are relatively rare, their use is growing, particularly in active submarine canyons (e.g., Smith et al., 2007; Paull et al., 2018), volcanic island flanks (e.g., Le Friant et al., 2010; Caplan-Auerbach et al., 2014; Hunt et al., 2021), and offshore river deltas (e.g., Lintern et al., 2016; Obelcz et al., 2017), including across the Mississippi submarine delta where seafloor infrastructure is threatened by submarine slope failures (Chaytor et al., 2020; Obelcz et al., 2020). The footprint of these surveys tends to be limited; hence, it will most likely be necessary to target specific areas, focusing on acquiring high-resolution bathymetric data that enables the identification of landslide scars and/or deposits. Evidence of landslide impacts on seafloor infrastructure (e.g., cable breaks, pipeline damage) may provide a useful additional point of calibration as this provides constraints on both the timing and location of the event (e.g., Carter et al., 2014).

Analysis of legacy seismic monitoring data sets, as well as new ones, will benefit from the application of emergent machine learning techniques that permit analysis of vast data sets to train algorithms to automatically identify landslide events and potentially identify previously unrecognized environmental patterns that relate to landslide preconditioning (Thirugnanam et al., 2020; Deng et al., 2021).

Calibrate and Test Fiber-Optic Detection of Submarine Landslides

The emergence of fiber-optic sensing to monitor offshore geohazards (in particular, DAS) is an exciting prospect but one that currently remains unproven at field scale for submarine landslide detection. The frequency ranges for detection appear to largely fall within the capabilities of DAS. However, confident identification of low-frequency (<<1 Hz) signals may be challenging, particularly in noisy environments (e.g., due to natural ocean processes and human activities) or where cables are poorly coupled with the seafloor. Synchronous deployment of arrays of ocean-bottom seismometers and hydrophones in active landslide settings is needed to provide confidence in the interpretation of DAS data and assess the distance (both along the cable and adjacent to it) over which measurements can reliably be made of the different signals created by submarine landslides. When installing new pipelines and cables, or when seafloor structures are decommissioned and left in situ, consideration could be given to integrating passive monitoring instrumentation to complement that used for monitoring infrastructure integrity (Wilcock et al., 2021). As SMART cables are installed (planned from 2024), opportunities will arise for accessing distributed sensor packages that may include hydrophones and/or geophones, adding valuable new potential locations for submarine landslide detection (Howe et al., 2019; Matias et al., 2021).

Extending the Global Monitoring Network to Enable Operational Early Warning Systems

In the longer term, there is a need for expanded acoustic and seismic monitoring networks offshore not only for detecting submarine landslides but also for a host of ocean and earth science applications (Wilcock et al., 2021). Future opportunities may exist through collaboration with global monitoring programs such as MERMAIDS (Mobile Earthquake Recording in Marine Areas by Independent Divers), which employs a network of autonomous floats equipped with hydrophones (0.1–50 kHz) that passively drift across the ocean for deployments lasting up to five years and can transmit seismograms in near-real time (Sukhovich et al., 2015; Hello and Nolet, 2020). Data gathered in such a manner may provide useful opportunistic snapshots of certain aspects of landslide activity. However, such monitoring may only yield limited data to answer outstanding science questions, given the mobile nature of the floats and the lack of ground-motion data. Therefore, static seafloor sensors, rather than floating ones, will also be required, but a similar approach (i.e., that focuses on developing low-cost and long-endurance monitoring platforms) is an important future direction.

Whether seismic monitoring networks are extended offshore attached to cables, via distributed fiber-optic sensing, or using stand-alone moorings, seafloor landers, or nodes, the development of more cost-effective sensors that do not require regular servicing and are capable of transmitting data remotely will transform our ability to monitor and ultimately provide early warnings for submarine landslides. It is unlikely that we will fully rely on passive monitoring approaches. However, a future outlook could include automated decision-making by smart networks that determine whether a significant event has occurred, prompting the release of pop-up floats to transmit data or active (high-power-use) systems that are triggered to start recording or switch to a higher sampling rate by passive (lower power-use) sensors.

5.6. CONCLUDING REMARKS

Although there are still many open questions concerning submarine landslides and their hazards, recent growth in marine geophysical monitoring is rapidly providing data to fill many gaps. Passive seismic and acoustic monitoring is in its relative infancy in this field. However, it has already shown significant promise in enabling the detection of submarine landslides over large areas to ascertain their precise timing (and hence determine likely triggers), provide constraints on their transit velocity and run-out distance, and determine their effects (e.g., tsunami generation). We consider that these longer-endurance and often more cost-effective techniques are likely to complement existing approaches rather than replace them entirely, particularly as many aspects of landslides can generate ground motions and/or acoustic noise. Ongoing calibration must be a priority to reliably discern and interpret signals generated by submarine landslides. Greater understanding through modeling will help to identify optimal sensor configurations. Distributed sensing along fiber-optic cables is an intriguing proposition. However, there is a compelling need to understand precisely what a submarine landslide signal looks like and over what spatial scale such signals can be reliably recorded. Studies are immediately required to test the sensitivity of passive landslide detection (using onshore and offshore networks) through concurrent deployment of active sensors. Once that calibration is performed across a breadth of sites and settings, we suggest that passive seismic and acoustic monitoring has the potential to answer a host of key outstanding questions and significantly advance the field of submarine landslide science.

ACKNOWLEDGMENTS

Clare and Talling acknowledge funding from the Natural Environment Research Council (NERC), including "Environmental Risks to Infrastructure: Identifying and Filling the Gaps" (NE/ P005780/1), "New field-scale calibration of turbidity current impact modeling" (NE/ P009190/1), and "Developing a Global Listening Network for Turbidity Currents and Seafloor Processes" (NE/ S009965/1 and NE/S010068/1). The authors also acknowledge discussions with collaborators as part of Talling's NERC International Opportunities Fund grant (NE/ M017540/1) "Coordinating and pump-priming international efforts for direct monitoring of active turbidity currents at global test sites." Clare also acknowledges support from the International Cable Protection Committee. Talling was supported by a NERC and Royal Society Industry Fellowship hosted by the International Cable Protection Committee. Additional funding for Clare was provided by the NERC National Capability Climate Linked Atlantic Sector Science Programme (NE/R015953/1). Pope was supported by a Leverhulme Early Career Fellowship (ECF-2018–267). We thank Bruce Howe for constructive feedback that improved the manuscript.

REFERENCES

Ai, F., Strasser, M., Preu, B., Hanebuth, T. J., Krastel, S., & Kopf, A. (2014). New constraints on oceanographic vs. seismic control on submarine landslide initiation: a geotechnical approach off Uruguay and northern Argentina. *Geo-Marine Letters*, *34*(5), 399–417.

Ajo-Franklin, J. B., Dou, S., Lindsey, N. J., Monga, I., Tracy, C., Robertson, M., et al. (2019). Distributed acoustic sensing using dark fiber for near-surface characterization and broadband seismic event detection. *Scientific Reports*, *9*(1), 1–14.

Antsyferov, M. S. (1966). *Seismo-acoustic methods in mining*. Consultants Bureau.

Azpiroz-Zabala, M., Cartigny, M. J., Talling, P. J., Parsons, D. R., Sumner, E. J., Clare, M. A., et al. (2017). Newly recognized turbidity current structure can explain the prolonged flushing of submarine canyons. *Science Advances*, *3*(10), e1700200.

Batsi, E., Tsang-Hin-Sun, E., Klingelhoefer, F., Bayrakci, G., Chang, E. T., Lin, J. Y., et al. (2019). Nonseismic signals in the ocean: indicators of deep sea and seafloor processes on ocean-bottom seismometer data. *Geochemistry, Geophysics, Geosystems*, *20*(8), 3882–3900.

Baumgartner, M. F., Stafford, K. M., & Latha, G. (2018). Near real-time underwater passive acoustic monitoring of natural and anthropogenic sounds. In *Observing the Oceans in Real Time* (pp. 203–226). Springer.

Bessason, B., Eiríksson, G., Thórarinsson, Ó., Thórarinsson, A., & Einarsson, S. (2007). Automatic detection of avalanches and debris flows by seismic methods. *Journal of Glaciology*, *53* (182), 461–472.

Biscara, L., Hanquiez, V., Leynaud, D., Marieu, V., Mulder, T., Gallissaires, J. M., et al. (2012). Submarine slide initiation and evolution offshore Pointe Odden, gabon-analysis from annual bathymetric data (2004–2009). *Marine Geology*, *299*, 43–50.

Blum, J. A., Chadwell, C. D., Driscoll, N., & Zumberge, M. A. (2010). Assessing slope stability in the Santa Barbara Basin, California, using seafloor geodesy and CHIRP seismic data. *Geophysical Research Letters*, *37*(13).

Bompais, X., Garziglia, S., Blandin, J., & Hello, Y. (2019). EMSO-Ligure Nice, a coastal cabled observatory dedicated to the study of slope stability. In *OCEANS 2019-Marseille* (pp. 1–8). IEEE.

Brackenridge, R. E., Nicholson, U., Sapiie, B., Stow, D. & Tappin, D. R. (2020). Indonesian throughflow as a preconditioning mechanism for submarine landslides in the Makassar Strait. *Geological Society, London, Special Publications*, *500* (1), 195–217.

Brodsky, E. E., Gordeev, E., & Kanamori, H. (2003). Landslide basal friction as measured by seismic waves. *Geophysical Research Letters*, *30*(24).

Brothers, D. S., Luttrell, K. M., & Chaytor, J. D. (2013). Sea-level–induced seismicity and submarine landslide occurrence. *Geology*, *41*(9), 979–982.

Brooks, H. L., Hodgson, D. M., Brunt, R. L., Peakall, J. & Flint, S. S. (2018). Exhumed lateral margins and increasing flow confinement of a submarine landslide complex. *Sedimentology*, *65*(4), 1067–1096.

Bull, S., Arnot, M., Browne, G., Crundwell, M., Nicol, A., & Strachan, L. (2019). Neogene and quaternary mass-transport deposits from the Northern Taranaki Basin (North Island, New Zealand) morphologies, transportation processes, and depositional controls. In K. Ogata, A, Festa, & G. Andrea (Eds.) *Submarine landslides: subaqueous mass transport deposits from outcrops to seismic profiles* (pp. 171–180). Wiley.

Bull, S., Cartwright, J., & Huuse, M. (2009). A review of kinematic indicators from mass-transport complexes using 3D seismic data. *Marine and Petroleum Geology*, *26*(7), 1132–1151.

Buss, C., Friedli, B., & Puzrin, A. M. (2019). Kinematic energy balance approach to submarine landslide evolution. *Canadian Geotechnical Journal*, *56*(9), 1351–1365.

Cadman, J. D., & Goodman, R. E. (1967). Landslide noise. *Science*, *158*(3805), 1182–1184.

Campbell, C. S., Cleary, P. W., & Hopkins, M. (1995). Large-scale landslide simulations: global deformation, velocities and basal friction. *Journal of Geophysical Research: Solid Earth*, *100*(B5), 8267–8283.

Campbell, K. J., Kinnear, S., & Thame, A. (2015). AUV technology for seabed characterization and geohazards assessment. *The Leading Edge*, *34*(2), 170–178.

Caplan-Auerbach, J., Fox, C. G., & Duennebier, F. K. (2001). Hydroacoustic detection of submarine landslides on Kilauea volcano. *Geophysical Research Letters*, *28*(9), 1811–1813

Caplan-Auerbach, J., & Huggel, C. (2007). Precursory seismicity associated with frequent, large ice avalanches on Iliamna volcano, Alaska, USA. *Journal of Glaciology*, *53*(180), 128–140.

Caplan-Auerbach, J., Dziak, R. P., Bohnenstiehl, D. R., Chadwick, W. W., & Lau, T. K. (2014). Hydroacoustic investigation of submarine landslides at West Mata volcano, *Lau Basin. Geophysical Research Letters*, *41*(16), 5927–5934.

Carter, L., Gavey, R., Talling, P. J., & Liu, J. T. (2014). Insights into submarine geohazards from breaks in subsea telecommunication cables. *Oceanography*, *27*(2), 58–67.

Casas, D., Chiocci, F., Casalbore, D., Ercilla, G., & De Urbina, J. O. (2016). Magnitude-frequency distribution of submarine landslides in the Gioia Basin (southern Tyrrhenian Sea). *Geo-Marine Letters*, *36*(6), 405–414.

Chadwick Jr, W. W., Dziak, R. P., Haxel, J. H., Embley, R. W., & Matsumoto, H. (2012). Submarine landslide triggered by volcanic eruption recorded by in situ hydrophone. *Geology*, *40*(1), 51–54.

Chaytor, J. D., Baldwin, W. E., Bentley, S. J., Damour, M., Jones, D., Maloney, J., et al. (2020). Short-and long-term movement of mudflows of the Mississippi River Delta Front and their known and potential impacts on oil and gas infrastructure. *Geological Society, London, Special Publications*, *500*(1), 587–604.

Chichibu, A., Jo, K., Nakamura, M., Goto, T., & Kamata, M. (1989). Acoustic emission characteristics of unstable slopes. *Journal of Acoustic Emission*, *8*(4), 107–112.

Clare, M. A., Vardy, M. E., Cartigny, M. J., Talling, P. J., Himsworth, M. D., Dix, J. K., et al. (2017). Direct monitoring of active geohazards: emerging geophysical tools for deep-water assessments. *Near Surface Geophysics*, *15*(4), 427–444.

Clare, M., Lintern, D. G., Rosenberger, K., Clarke, J. E. H., Paull, C., Gwiazda, R., et al. (2020). Lessons learned from the monitoring of turbidity currents and guidance for future platform designs. *Geological Society, London, Special Publications*, *500*(1), 605–634.

Clare, M., Chaytor, J., Dabson, O., Gamboa, D., Georgiopoulou, A., Eady, H., et al. (2019). A consistent global approach for the morphometric characterization of subaqueous landslides. *Geological Society, London, Special Publications*, *477* (1), 455–477.

Collico, S., Arroyo, M., Urgeles, R., Gràcia, E., Devincenzi, M., & Peréz, N. (2020). Probabilistic mapping of earthquake-induced submarine landslide susceptibility in the South-West Iberian margin. *Marine Geology*, *429*, 106296.

Cook, K. L., Andermann, C., Gimbert, F., Adhikari, B. R. & Hovius, N. (2018). Glacial lake outburst floods as drivers of fluvial erosion in the Himalaya. *Science*, *362*(6410), 53–57.

Cook, K. L., Rekapali, R., Dietze, M., et al. 2021. Detection and potential early warning of catastrophic flow events with regional seismic networks, *Science*, *374*, 87–92.

Deng, L., Smith, A., Dixon, N., & Yuan, H. (2021). Machine learning prediction of landslide deformation behaviour using acoustic emission and rainfall measurements. *Engineering Geology*, *106315*.

Deparis, J., Jongmans, D., Cotton, F., Baillet, L., Thouvenot, F., & Hantz, D. (2008). Analysis of rock-fall and rock-fall avalanche seismograms in the French Alps. *Bulletin of the Seismological Society of America*, *98*(4), 1781–1796.

Dewey, J. W., & Dellinger, J. A. (2008). *Location of the Green Canyon (offshore southern Louisiana) seismic event of February 10, 2006 (open-file report)*. U.S. Geological Survey.

Dixon, N., & Spriggs, M. (2007). Quantification of slope displacement rates using acoustic emission monitoring. *Canadian Geotechnical Journal*, *44*(8), 966–976.

Dixon, N., Spriggs, M. P., Smith, A., Meldrum, P., & Haslam, E. (2015). Quantification of reactivated landslide behaviour using acoustic emission monitoring. *Landslides*, *12*(3), 549–560.

Downs, P. W., Soar, P. J., & Taylor, A. (2016). The anatomy of effective discharge: the dynamics of coarse sediment transport revealed using continuous bedload monitoring in a gravel-bed river during a very wet year. *Earth Surface Processes and Landforms*, *41*(2), 147–161.

Dugan, B. (2012). Petrophysical and consolidation behavior of mass transport deposits from the northern Gulf of Mexico, IODP Expedition 308. *Marine Geology*, *315*, 98–107.

Ekström, G., & Stark, C. P. (2013). Simple scaling of catastrophic landslide dynamics. *Science*, *339*(6126), 1416–1419.

Fan, W., McGuire, J. J. & Shearer, P. M. (2020). Abundant spontaneous and dynamically triggered submarine landslides in the Gulf of Mexico. *Geophysical Research Letters*, *47*(12), e2020GL087213.

Frey-Martínez, J., Cartwright, J., & James, D. (2006). Frontally confined versus frontally emergent submarine landslides: A 3D seismic characterisation. *Marine and Petroleum Geology*, *23*(5), 585–604.

Fujiwara, T., dos Santos Ferreira, C., Bachmann, A. K., Strasser, M., Wefer, G., Sun, T., et al. (2017). Seafloor displacement after the 2011 Tohoku-Oki earthquake in the northern Japan trench examined by repeated bathymetric surveys. *Geophysical Research Letters*, *44*(23), 11–833.

Galitzin, B. (1915). Sur l'angle d'emergence des rayons sismiques. *Bull. (Izv.) Central Seismic Commission Russian Academy of Science*, *7*(2).

Gamboa, D., Omira, R., & Terrinha, P. (2021). A database of submarine landslides offshore West and Southwest Iberia. *Scientific Data*, *8*(1), 1–9.

Geay, T., Belleudy, P., Gervaise, C., Habersack, H., Aigner, J., Kreisler, A., et al. (2017). Passive acoustic monitoring of bed load discharge in a large gravel bed river. *Journal of Geophysical Research: Earth Surface*, *122*(2), 528–545.

Geay, T., Zanker, S., Misset, C., & Recking, A. (2020). Passive acoustic measurement of bedload transport: toward a global calibration curve? *Journal of Geophysical Research: Earth Surface*, *125*(8), e2019JF005242.

Gee, M. J. R., Gawthorpe, R. L., & Friedmann, J. S. (2005). Giant striations at the base of a submarine landslide. *Marine Geology*, *214*(1–3), 287–294.

Gee, M. J. R., Gawthorpe, R. L., & Friedmann, S. J. (2006). Triggering and evolution of a giant submarine landslide, offshore Angola, revealed by 3D seismic stratigraphy and geomorphology. *Journal of Sedimentary Research*, *76*(1), 9–19.

Geist, E. L., & Parsons, T. (2010). Estimating the empirical probability of submarine landslide occurrence. In D. C. Mosher et al. (Eds.), *Submarine mass movements and their consequences* (pp. 377–386). Springer.

Goff, J., & Terry, J. P. (2016). Tsunamigenic slope failures: the Pacific Islands "blind spot"? *Landslides*, *13*(6), 1535–1543.

Gomberg, J., Ariyoshi, K., Hautala, S., & Johnson, H. P. (2021). The finicky nature of earthquake shaking-triggered submarine sediment slope failures and sediment gravity flows. *Journal of Geophysical Research: Solid Earth*, *126*(10), e2021JB022588.

Guiastrennec-Faugas, L., Gillet, H., Peakall, J., Dennielou, B., Gaillot, A., & Jacinto, R. S. (2021). Initiation and evolution of knickpoints and their role in cut-and-fill processes in active submarine channels. *Geology*, *49*(3), 314–319.

Hampton, M. A., Lee, H. J., & Locat, J. (1996). Submarine landslides. *Reviews of Geophysics*, *34*(1), 33–59.

Harbitz, C. B., Løvholt, F., & Bungum, H. (2014). Submarine landslide tsunamis: how extreme and how likely? *Natural Hazards*, *72*(3), 1341–1374.

Hay, A. E., Hatcher, M. G., & Hughes Clarke, J. E. (2021). Underwater noise from submarine turbidity currents. *JASA Express Letters*, *1*(7), 0

Heerema, C. J., Talling, P. J., Cartigny, M. J., Paull, C. K., Bailey, L., Simmons, S. M., et al. (2020). What determines the downstream evolution of turbidity currents? *Earth and Planetary Science Letters*, *532*, 116023.

Heijnen, M. S., Clare, M. A., Cartigny, M. J., Talling, P. J., Hage, S., Lintern, D. G., et al. (2020). Rapidly-migrating and internally-generated knickpoints can control submarine channel evolution. *Nature Communications*, *11*(1), 1–15.

Hello, Y., & Nolet, G. (2020). Floating seismographs (MERMAIDS). *Encyclopedia of Solid Earth Geophysics*, 1–6.

Hibert, C., Michéa, D., Provost, F., Malet, J. P., & Geertsema, M. (2019). Exploration of continuous seismic recordings with a machine learning approach to document 20 yr of landslide activity in Alaska. *Geophysical Journal International*, *219*(2), 1138–1147.

Hizzett, J. L., Hughes Clarke, J. E., Sumner, E. J., Cartigny, M. J. B., Talling, P. J., & Clare, M. A. (2018). Which triggers produce the most erosive, frequent, and longest runout turbidity currents on deltas? *Geophysical Research Letters*, *45*(2), 855–863.

Howe, B. M., Arbic, B. K., Aucan, J., Barnes, C. R., Bayliff, N., Becker, N., et al. (2019). SMART cables for observing the global ocean: science and implementation. *Frontiers in Marine Science*, *6*, 424.

Hughes Clarke, J.E, Brucker, S., Muggah, J., Church, I., Cartwright, D., Kuus, P., et al. (2012). The Squamish ProDelta: monitoring active landslides and turbidity currents. In *Canadian Hydrographic Conference 2012, Proceedings* (p. 15).

Hunt, J. E., Tappin, D. R., Watt, S. F. L., Susilohadi, S., Novellino, A, Ebmeier, S.K, et al. (2021). Submarine landslide megablocks show half of Anak Krakatau island failed on December 22nd, 2018. *Nature Communications*, *12*, 2827.

Iannucci, R., Lenti, L., & Martino, S. (2020). Seismic monitoring system for landslide hazard assessment and risk management at the drainage plant of the Peschiera Springs (Central Italy). *Engineering Geology*, *277*, 105787.

Ide, S., Araki, E. & Matsumoto, H. (2021). Very broadband strain-rate measurements along a submarine fiber-optic cable off Cape Muroto, Nankai subduction zone, Japan. *Earth, Planets and Space*, *73*(1), 1–10.

Inman, D. L., Nordstrom, C. E., & Flick, R. E. (1976). Currents in submarine canyons: An air-sea-land interaction. *Annual Review of Fluid Mechanics*, *8*(1), 275–310.

Jeffreys, H. (1923). The Pamir earthquake of 1911 February 18, in relation to the depths of earthquake foci. *Geophysical Journal International*, *1*, 22–31.

Kanamori, H., & Given, J. W. (1982). Analysis of long-period seismic waves excited by the May 18, 1980, eruption of Mount St. Helens—a terrestrial monopole? *Journal of Geophysical Research: Solid Earth*, *87*(B7), 5422–5432.

Kawakatsu, H. (1989). Centroid single force inversion of seismic waves generated by landslides. *Journal of Geophysical Research: Solid Earth*, *94*(B9), 1 2363–12374.

Kelner, M., Migeon, S., Tric, E., Couboulex, F., Dano, A., Lebourg, T., & Taboada, A. (2016). Frequency and triggering of small-scale submarine landslides on decadal timescales: analysis of 4D bathymetric data from the continental slope offshore Nice (France). *Marine Geology*, *379*, 281–297.

Khripounoff, A., Vangriesheim, A., Babonneau, N., Crassous, P., Dennielou, B., & Savoye, B. (2003). Direct observation of intense turbidity current activity in the Zaire submarine valley at 4000 m water depth. *Marine Geology*, *194*(3–4), 151–158.

Khripounoff, A., Crassous, P., Bue, N. L., Dennielou, B., & Jacinto, R. S. (2012). Different types of sediment gravity flows detected in the Var submarine canyon (northwestern Mediterranean Sea). *Progress in Oceanography*, *106*, 138–153.

Kioka, A., Schwestermann, T., Moernaut, J., Ikehara, K., Kanamatsu, T., McHugh, C. M., et al. (2019). Megathrust earthquake drives drastic organic carbon supply to the hadal trench. *Scientific Reports*, *9*(1), 1–10

Lacroix, P., Grasso, J. R., Roulle, J., Giraud, G., Goetz, D., Morin, S., & Helmstetter, A. (2012). Monitoring of snow avalanches using a seismic array: Location, speed estimation, and relationships to meteorological variables. *Journal of Geophysical Research: Earth Surface*, *117*(F1).

Lecocq, T., Hicks, S. P., Van Noten, K., Van Wijk, K., Koelemeijer, P., De Plaen, R. S., et al. (2020). Global quieting of high-frequency seismic noise due to COVID-19 pandemic lockdown measures. *Science*, *369*(6509), 1338–1343.

Leung, A. K., & Ng, C. W. W. (2016). Field investigation of deformation characteristics and stress mobilisation of a soil slope. *Landslides*, *13*(2), 229–240.

Lin, C. H. (2015). Insight into landslide kinematics from a broadband seismic network. *Earth, Planets and Space*, *67* (1), 8.

Lin, C. H., Kumagai, H., Ando, M., & Shin, T. C. (2010). Detection of landslides and submarine slumps using broadband seismic networks. *Geophysical Research Letters*, *37*(22).

Lindsey, N. J., & Martin, E. R. (2021). Fiber-optic seismology. *Annual Review of Earth and Planetary Sciences*, *49*, 309–336.+

Lindsey, N. J., Dawe, T. C., & Ajo-Franklin, J. B. (2019). Illuminating seafloor faults and ocean dynamics with dark fiber distributed acoustic sensing. *Science*, *366*(6469), 1103–1107.

Lintern, D. G., & Hill, P. R. (2010). An underwater laboratory at the Fraser River delta. *Eos, Transactions American Geophysical Union*, *91*(38), 333–334.

Lintern, D. G., Hill, P. R., & Stacey, C. (2016). Powerful unconfined turbidity current captured by cabled observatory on the Fraser River delta slope, British Columbia, *Canada. Sedimentology*, *63*(5), 1041–1064.

Lintern, D. G., Mosher, D. C. & Scherwath, M. (2019). Advancing from subaqueous mass movement case studies to providing advice and mitigation. *Geological Society, London, Special Publications*, *477*, 1–14. https://doi.org/10.1144/SP 477–2018-190

Lior, I., Sladen, A., Rivet, D., Ampuero, J. P., Hello, Y., Becerril, C., et al. (2021). On the detection capabilities of underwater distributed acoustic sensing. *Journal of Geophysical Research: Solid Earth*, *126*(3), e2020JB020925.

Løvholt, F., Pedersen, G., Harbitz, C. B., Glimsdal, S., & Kim, J. (2015). On the characteristics of landslide tsunamis. *Philosophical Transactions of the Royal Society A: Mathematical, Physical and Engineering Sciences*, *373*(2053), 20140376.

Mainsant, G., Larose, E., Brönnimann, C., Jongmans, D., Michoud, C., & Jaboyedoff, M. (2012). Ambient seismic noise monitoring of a clay landslide: toward failure prediction. *Journal of Geophysical Research: Earth Surface*, *117*(F1).

Marra, G., Clivati, C., Luckett, R., Tampellini, A., Kronjäger, J., Wright, L., et al. (2018). Ultrastable laser interferometry for earthquake detection with terrestrial and submarine cables. *Science*, *361*(6401), 486–490.

Maslin, M., Owen, M., Day, S., & Long, D. (2004). Linking continental-slope failures and climate change: Testing the clathrate gun hypothesis. *Geology*, *32*(1), 53–56.

Masson, D. G., Harbitz, C. B., Wynn, R. B., Pedersen, G., & Løvholt, F. (2006). Submarine landslides: processes, triggers and hazard prediction. *Philosophical Transactions of the Royal Society A: Mathematical, Physical and Engineering Sciences*, *364*(1845), 2009–2039.

Mastbergen, D., van den Ham, G., Cartigny, M., Koelewijn, A., de Kleine, M., Clare, M., et al. (2016). Multiple flow slide experiment in the Westerschelde Estuary, The Netherlands. In D. C. Mosher et al. (Eds.), *Submarine mass movements and their consequences* (pp. 241–249). Springer.

Matias, L. M., Carrilho, F., Sá, V., Omira, R., Niehus, M., Corela, C., et al.(2021). The contribution of submarine optical fiber telecom cables to the monitoring of earthquakes and tsunamis in the NE Atlantic. *Frontiers in Earth Science*, *9*, 611.

McAdoo, B. G., Pratson, L. F., & Orange, D. L. (2000). Submarine landslide geomorphology, US continental slope. *Marine Geology*, *169*(1–2), 103–136.

Mecozzi, A., Cantono, M., Castellanos, J. C., Kamalov, V., Muller, R., & Zhan, Z. (2021). Polarization sensing using submarine optical cables. *Optica*, *8*(6), 788–795.

Michlmayr, G., Chalari, A., Clarke, A., & Or, D. (2017). Fiber-optic high-resolution acoustic emission (AE) monitoring of slope failure. *Landslides*, *14*(3), 1139–1146.

Miramontes, E., Garziglia, S., Sultan, N., Jouet, G., & Cattaneo, A. (2018). Morphological control of slope instability in contourites: a geotechnical approach. *Landslides*, *15*(6), 1085–1095.

Mizutani, A., Yomogida, K., & Tanioka, Y. (2020). Early tsunami detection with near-fault ocean-bottom pressure gauge records based on the comparison with seismic data. *Journal of Geophysical Research: Oceans*, *125*(9), e2020JC016275.

Moernaut, J., & De Batist, M. (2011). Frontal emplacement and mobility of sublacustrine landslides: results from morphometric and seismostratigraphic analysis. *Marine Geology*, *285* (1–4), 29–45.

Mohrig, D., & Marr, J. G. (2003). Constraining the efficiency of turbidity current generation from submarine debris flows and slides using laboratory experiments. *Marine and Petroleum Geology*, *20*(6–8), 883–899.

Moore, J. G., Clague, D. A., Holcomb, R. T., Lipman, P. W., Normark, W. R., & Torresan, M. E. (1989). Prodigious submarine landslides on the Hawaiian Ridge. *Journal of Geophysical Research: Solid Earth*, *94*(B12), 17465–17484.

Moretti, L., Allstadt, K., Mangeney, A., Capdeville, Y., Stutzmann, E., & Bouchut, F. (2015). Numerical modeling of the Mount Meager landslide constrained by its force history derived from seismic data. *Journal of Geophysical Research: Solid Earth*, *120*(4), 2579–2599.

Mountjoy, J. J., McKean, J., Barnes, P. M., & Pettinga, J. R. (2009). Terrestrial-style slow-moving earthflow kinematics in a submarine landslide complex. *Marine geology*, *267* (3–4), 114–127.

Mountjoy, J. J., Howarth, J. D., Orpin, A. R., Barnes, P. M., Bowden, D. A., Rowden, A. A., et al. (2018). Earthquakes drive large-scale submarine canyon development and sediment supply to deep-ocean basins. *Science Advances*, *4*(3), eaar3748.

Mulder, T., & Cochonat, P. (1996). Classification of offshore mass movements. *Journal of Sedimentary Research*, *66*(1), 43–57.

Nisbet, E. G., and Piper, D. J. (1998). Giant submarine landslides. *Nature*, *392*(6674), 329–330.

Nishimura, T., Emoto, K., Nakahara, H., Miura, S., Yamamoto, M., Sugimura, S., et al. (2021). Source location of volcanic earthquakes and subsurface characterization using fiber-optic cable and distributed acoustic sensing system. *Scientific Reports*, *11*(1), 1–12.

Normandeau, A., Campbell, D. C., Piper, D. J., & Jenner, K. A. (2019). Are submarine landslides an underestimated hazard on the western North Atlantic passive margin? *Geology*, *47* (9), 848–852.

Normandeau, A., MacKillop, K., Macquarrie, M., Richards, C., Bourgault, D., Campbell, D. C., et al. (2021). Submarine landslides triggered by iceberg collision with the seafloor. *Nature Geoscience*, *14*(8), 599–605.

Nwoko, J., Kane, I., & Huuse, M. (2020). Megaclasts within mass-transport deposits: their origin, characteristics and effect on substrates and succeeding flows. *Geological Society, London, Special Publications*, *500*(1), 515–530.

Obelcz, J., Xu, K., Georgiou, I. Y., Maloney, J., Bentley, S. J., & Miner, M. D. (2017). Sub-decadal submarine landslides are important drivers of deltaic sediment flux: insights from the Mississippi River Delta Front. *Geology*, *45*(8), 703–706.

Obelcz, J., Wood, W. T., Phrampus, B. J., & Lee, T. R. (2020). Machine learning augmented time-lapse bathymetric surveys: A case study from the Mississippi river delta front. *Geophysical Research Letters*, *47*(10), e2020GL087857.

Ogata, K., Festa, A., & Pini, G. A. (Eds.) (2019). *Submarine landslides: subaqueous mass transport deposits from outcrops to seismic profiles* (Vol. *247*). John Wiley & Sons.

Parker, G. (1982). Conditions for the ignition of catastrophically erosive turbidity currents. *Marine Geology*, *46*(3–4), 307–327.

Paull, C. K., Talling, P. J., Maier, K. L., Parsons, D., Xu, J., Caress, D. W., et al. (2018). Powerful turbidity currents driven by dense basal layers. *Nature Communications*, *9*(1), 1–9.

Paull, C. K., Ussler, W., & Holbrook, W. S. (2007). Assessing methane release from the colossal Storegga submarine landslide. *Geophysical Research Letters*, *34*(4).

Piper, D. J., Cochonat, P., & Morrison, M. L. (1999). The sequence of events around the epicentre of the 1929 Grand Banks earthquake: initiation of debris flows and turbidity current inferred from sidescan sonar. *Sedimentology*, *46*(1), 79–97.

Pope, E. L., Talling, P. J., Carter, L., Clare, M. A. & Hunt, J. E. (2017). Damaging sediment density flows triggered by tropical cyclones. *Earth and Planetary Science Letters*, *458*, 161–169.

Puzrin, A. M., Germanovich, L. N., & Friedli, B. (2016). Shear band propagation analysis of submarine slope stability. *Géotechnique*, *66*(3), 188–201.

Prior, D. B., Bornhold, B. D., Coleman, J. M., & Bryant, W. R. (1982). Morphology of a submarine slide, Kitimat arm, British Columbia. *Geology*, *10*(11), 588–592.

Prior, D. B., Suhayda, J. N., Lu, N. Z., Bornhold, B. D., Keller, G. H., Wiseman, W. J., et al. (1989). Storm wave reactivation of a submarine landslide. *Nature*, *341*(6237), 47–50

Puzrin, A. M., Germanovich, L. N., & Friedli, B. (2016). Shear band propagation analysis of submarine slope stability. *Géotechnique*, *66*(3), 188–201.

Ranasinghe, N., Rowe, C., Syracuse, E., Larmat, C. & Begnaud, M. (2018). Enhanced global seismic resolution using transoceanic SMART cables. *Seismological Research Letters*, *89*(1), 77–85.

Rickenmann, D. (2017). Bedload transport measurements with geophones, hydrophones and underwater microphones (passive acoustic methods). In D. Tsutsumi & J. B. Laronne (Eds.), *Gravel bed rivers: processes and disasters* (pp. 185–208). Wiley & Sons.

Roche, B., Bull, J. M., Marin-Moreno, H., Leighton, T. G., Falcon-Suarez, I. H., Tholen, M., et al. (2021). Time-lapse imaging of CO2 migration within near-surface sediments during a controlled sub-seabed release experiment. *International Journal of Greenhouse Gas Control*, *109*, 103363.

Rouse, C., Styles, P., & Wilson, S. A. (1991). Microseismic emissions from flowslide-type movements in South Wales. *Engineering Geology*, *31*(1), 91–110.

Ryan, J. P., Joseph, J. E., Margolina, T., Peavey Reeves, L., Hatch, L., DeVogelaere, A., et al. (2020). Quieting of low-frequency vessel noise in Monterey Bay National Marine Sanctuary during the COVID-19 pandemic. *The Journal of the Acoustical Society of America*, *148*(4), 2734–2734.

Sammartini, M., Moernaut, J., Kopf, A., Stegmann, S., Fabbri, S. C., Anselmetti, F. S., & Strasser, M. (2021). Propagation of frontally confined subaqueous landslides: insights from combining geophysical, sedimentological, and geotechnical analysis. *Sedimentary Geology*, *105877*.

Sgroi, T., Monna, S., Embriaco, D., Giovanetti, G., Marinaro, G., & Favali, P. (2014). Geohazards in the western Ionian Sea: Insights from non-earthquake signals recorded by the NEMO-SN1 seafloor observatory. *Oceanography*, *27*(2), 154–166.

Simmons, S. M., Azpiroz-Zabala, M., Cartigny, M. J. B., Clare, M. A., Cooper, C., Parsons, D. R., et al. (2020). Novel acoustic method provides first detailed measurements of sediment concentration structure within submarine turbidity currents. *Journal of Geophysical Research: Oceans*, *125*(5), e2019JC015904.

Smith, D. P., Kvitek, R., Iampietro, P. J., & Wong, K. (2007). Twenty-nine months of geomorphic change in upper Monterey Canyon (2002–2005). *Marine Geology*, *236*(1–2), 79–94.

Stegmann, S., Sultan, N., Kopf, A., Apprioual, R., & Pelleau, P. (2011). Hydrogeology and its effect on slope stability along the coastal aquifer of Nice, *France. Marine Geology*, *280*(1), 168–181.

Strasser, M., Kölling, M., Ferreira, C. D. S., Fink, H. G., Fujiwara, T., Henkel, S., et al. (2013). A slump in the trench: tracking the impact of the 2011 Tohoku-Oki earthquake. *Geology*, *41*(8), 935–938.

Strout, J. M., & Tjelta, T. I. (2005). In situ pore pressures: what is their significance and how can they be reliably measured? *Marine and Petroleum Geology*, *22*(1), 275–285.

Sukhovich, A., Bonnieux, S., Hello, Y., Irisson, J. O., Simons, F. J., & Nolet, G. (2015). Seismic monitoring in the oceans by autonomous floats. *Nature Communications*, *6*(1), 1–6.

Sultan, N., Savoye, B., Jouet, G., Leynaud, D., Cochonat, P., Henry, P., et al. (2010). Investigation of a possible submarine landslide at the Var delta front (Nice continental slope, southeast France). *Canadian Geotechnical Journal*, *47*(4), 486–496.

Sumner, E. J., & Paull, C. K. (2014). Swept away by a turbidity current in Mendocino submarine canyon, *California. Geophysical Research Letters*, *41*(21), 7611–7618.

Suriñach, E., Vilajosana, I., Khazaradze, G., Biescas, B., Furdada, G., & Vilaplana, J. M. (2005). Seismic detection and characterization of landslides and other mass movements. *Natural Hazards and Earth System Sciences*, *5*(6), 791–798.

Talling, P. J., Paull, C. K., & Piper, D. J. (2013). How are subaqueous sediment density flows triggered, what is their internal structure and how does it evolve? Direct observations from monitoring of active flows. *Earth-Science Reviews*, *125*, 244–287.

Talling, P. J., Wynn, R. B., Masson, D. G., Frenz, M., Cronin, B. T., Schiebel, R., et al. (2007). Onset of submarine debris flow deposition far from original giant landslide. *Nature*, *450*(7169), 541–544.

Tappin, D. R., Grilli, S. T., Harris, J. C., Geller, R. J., Masterlark, T., Kirby, J. T., et al. (2014). Did a submarine landslide contribute to the 2011 Tohoku tsunami? *Marine Geology*, *357*, 344–361.

Tappin, D. R., Watts, P., & Grilli, S. T. (2008). The Papua New Guinea tsunami of 17 July 1998: anatomy of a catastrophic event. *Natural Hazards and Earth System Sciences*, *8*(2), 243–266.

Ten Brink, U. S., Geist, E. L., & Andrews, B. D. (2006). Size distribution of submarine landslides and its implication to tsunami hazard in Puerto Rico. *Geophysical Research Letters*, *33*(11).

Ten Brink, U., Twichell, D., Geist, E., Chaytor, J., Locat, J., Lee, H., et al. (2008). *Evaluation of tsunami sources with the potential to impact the US Atlantic and Gulf coasts (administrative report to the US Nuclear Regulatory Commission)*. U. S. Geological Survey.

Thirugnanam, H., Ramesh, M. V., & Rangan, V. P. (2020). Enhancing the reliability of landslide early warning systems by machine learning. *Landslides*, *17*(9), 2231–2246.

Urgeles, R., & Camerlenghi, A. (2013). Submarine landslides of the Mediterranean Sea: trigger mechanisms, dynamics, and frequency-magnitude distribution. *Journal of Geophysical Research: Earth Surface*, *118*(4), 2600–2618.

Urlaub, M., Petersen, F., Gross, F., Bonforte, A., Puglisi, G., Guglielmino, F., et al. (2018). Gravitational collapse of Mount Etna's southeastern flank. *Science Advances*, *4*(10), eaat9700.

Urlaub, M., Talling, P. J., & Masson, D. G. (2013). Timing and frequency of large submarine landslides: implications for understanding triggers and future geohazard. *Quaternary Science Reviews*, *72*, 63–82.

Urlaub, M., Talling, P. J., & Clare, M. (2014). Sea-level–induced seismicity and submarine landslide occurrence: comment. *Geology*, *42*(6), e337–e337.

Urlaub, M., & Villinger, H. (2019). Combining in situ monitoring using seabed instruments and numerical modelling to assess the transient stability of underwater slopes. *Geological Society, London, Special Publications*, *477*(1), 511–521.

Vanneste, M., Sultan, N., Garziglia, S., Forsberg, C. F., & L'Heureux, J. S. (2014). Seafloor instabilities and sediment deformation processes: the need for integrated, multi-disciplinary investigations. *Marine Geology*, *352*, 183–214.

Van Herwijnen, A., Heck, M., & Schweizer, J. (2016). Forecasting snow avalanches using avalanche activity data obtained through seismic monitoring. *Cold Regions Science and Technology*, *132*, 68–80.

Van Herwijnen, A., & Schweizer, J. (2011). Seismic sensor array for monitoring an avalanche start zone: design, deployment and preliminary results. *Journal of Glaciology*, *57*(202), 267–276.

Vardy, M. E., L'Heureux, J. S., Vanneste, M., Longva, O., Steiner, A., Forsberg, C. F., et al. (2012). Multidisciplinary investigation of a shallow near-shore landslide, Finneidfjord, Norway. *Near Surface Geophysics*, *10*(4), 267–277.

Waage, M., Singhroha, S., Bünz, S., Planke, S., Waghorn, K. A., & Bellwald, B. (2021). Feasibility of using the P-Cable high-resolution 3D seismic system in detecting and monitoring CO2 leakage. *International Journal of Greenhouse Gas Control*, *106*, 103240.

Walter, T. R., Haghighi, M. H., Schneider, F. M., Coppola, D., Motagh, M., Saul, J., et al. (2019). Complex hazard cascade culminating in the Anak Krakatau sector collapse. *Nature Communications*, *10*(1), 1–11.

Wilcock, W. (2021). Illuminating tremors in the deep. *Science*, *371*(6532), 882–884.

Williams, E. F., Fernández-Ruiz, M. R., Magalhaes, R., Vanthillo, R., Zhan, Z., González-Herráez, M., & Martins, H. F. (2019). Distributed sensing of microseisms and teleseisms with submarine dark fibers. *Nature Communications*, *10*(1), 1–11.

Xu, J. P. (2010). Normalized velocity profiles of field-measured turbidity currents. *Geology*, *38*(6), 563–566.

Xu, J. P. (2011). Measuring currents in submarine canyons: technological and scientific progress in the past 30 years. *Geosphere*, *7*(4), 868–876.

Yamada, M., Kumagai, H., Matsushi, Y., & Matsuzawa, T. (2013). Dynamic landslide processes revealed by broadband seismic records. *Geophysical Research Letters*, *40*(12), 2998–3002.

Ye, L., Kanamori, H., Rivera, L., Lay, T., Zhou, Y., Sianipar, D. & Satake, K. (2020). The 22 December 2018 tsunami from flank collapse of Anak Krakatau volcano during eruption. *Science Advances*, *6*(3), eaaz1377.

Yeh, Y. C., Tsai, C. L., Hsu, S. K., Lin, H. S., Chen, K. T., Cho, Y. Y., & Liang, C. W. (2021). Continental shelf morphology controlled by bottom currents, mud diapirism, and submarine slumping to the east of the Gaoping Canyon, off SW Taiwan. *Geo-Marine Letters*, *41*(1), 1–11.

Yuan, X., Kind, R., and Pedersen, H. A. (2005). Seismic monitoring of the Indian Ocean tsunami. *Geophysical Research Letters*, *32*(15).

Zeng, J., Lowe, D. R., Prior, D. B., Wiseman J. R., W. J., & Bornhold, B. D. (1991). Flow properties of turbidity currents in Bute Inlet, British Columbia. *Sedimentology*, *38*(6), 975–996.

Zhan, Z. (2020). Distributed acoustic sensing turns fiber-optic cables into sensitive seismic antennas. *Seismological Research Letters*, *91*(1), 1–15.

Zhan, Z., Cantono, M., Kamalov, V., Mecozzi, A., Müller, R., Yin, S., & Castellanos, J. C. (2021). Optical polarization–based seismic and water wave sensing on transoceanic cables. *Science*, *371*(6532), 931–936.

Zhang, Y., Liu, Z., Zhao, Y., Colin, C., Zhang, X., Wang, M., et al. (2018). Long-term in situ observations on typhoon-triggered turbidity currents in the deep sea. *Geology*, *46*(8), 675–678.

Zhao, S., Wang, Z., Nie, Z., He, K., & Ding, N. (2021). Investigation on total adjustment of the transducer and seafloor transponder for GNSS/acoustic precise underwater point positioning. *Ocean Engineering*, *221*, 108533.

6

Iceberg Noise

Vera Schlindwein

ABSTRACT

In the 1980s, powerful seismic tremor signals produced by Antarctic icebergs drifting along the shores of Antarctica were first recognized in seismic records in the Pacific Ocean. This chapter reviews studies of iceberg-generated noise and summarizes signal characteristics, their spatial and temporal variations, and source mechanisms. Two classes of iceberg-generated seismic events can be distinguished: bursts with a white spectrum, and tremors with monochromatic or harmonic spectra with gliding frequencies. Bursts occur independently of water depth, with a seasonal peak in summer. They are likely produced by the fracturing of icebergs during their disintegration. Tremors happen independently of season only in the shallow coastal waters of Antarctica. They are produced during stick-slip motion between icebergs or when icebergs scrape over the ground. Both signals are sufficiently powerful to be recorded as seismic phases on land to distances of about 800 km and hydroacoustic phases on hydrophones moored in the Atlantic, Indian, and Pacific Oceans thousands of kilometers away. At frequencies below 40 Hz, iceberg noise notably contributes to the noise budget of the Southern Ocean, especially during austral summer. Noise levels correlate with iceberg volumes such that iceberg noise budgets may be used to monitor temporal variations in iceberg discharge.

6.1. INTRODUCTION

In 2002, Talandier et al. made a curious discovery: the Polynesian Seismic Network had recorded hydroacoustic signals that originated near a large iceberg that calved off Ross Ice Shelf in late 2000 and then drifted in the coastal waters of Antarctica thousands of kilometers away from Polynesia (Fig. 6.1). Müller et al. (2005) found similar signals in records of the Neumayer seismological observatory, Antarctica (Fig. 6.1). Their spectacular harmonic tremor signals sounded like songs when transferred into the frequency range of human hearing. These harmonic tremor signals closely resembled volcanic tremor (e.g., Konstantinou & Schlindwein, 2003; Schlindwein et al., 1995) but were clearly associated with large "singing icebergs" drifting near the observatory.

Since these early observations, a wealth of records has accumulated from local seismic networks on icebergs (MacAyeal et al., 2008), regional seismic networks (e.g., Pirli et al. 2015), regional hydrophone networks (e.g., Dziak et al., 2015) and distant hydrophones of the International Monitoring System (IMS) of the Comprehensive Nuclear-Test-Ban Treaty Organization (CTBTO) in the Pacific, Indian, and Atlantic Oceans (e.g., Matsumoto et al., 2014). It became obvious that in the Southern Ocean, iceberg noise is not a rare curiosity but contributes considerably to the noise budget at

Section Geophysics, Alfred Wegener Institute Helmholtz Centre for Polar and Marine Research, Bremerhaven, Germany; and Faculty of Geosciences, University of Bremen, Bremen, Germany

Noisy Oceans: Monitoring Seismic and Acoustic Signals in the Marine Environment, Geophysical Monograph 284, First Edition.
Edited by Gaye Bayrakci and Frauke Klingelhoefer.

DOI:10.1002/9781119750925.ch06

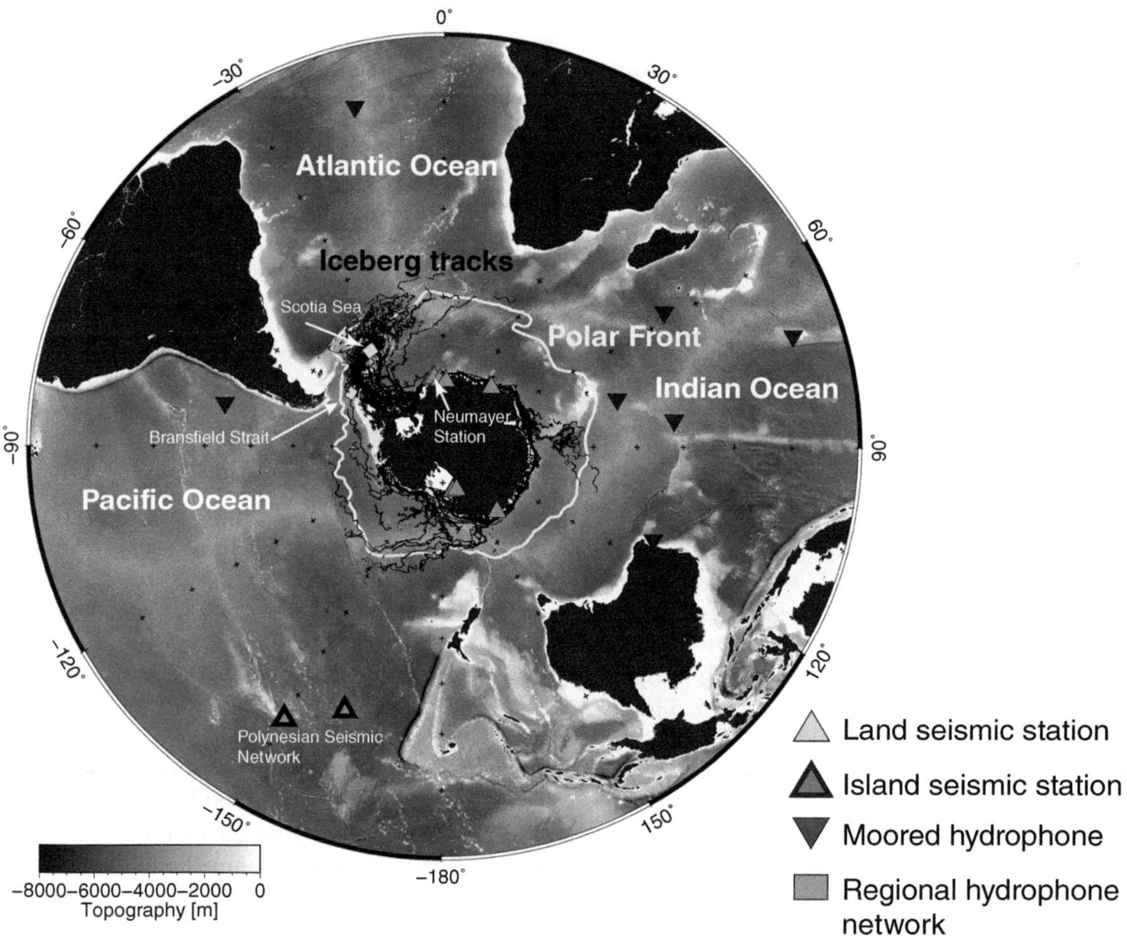

Figure 6.1 Overview of hydrophone and seismic stations with records of iceberg noise as discussed in the text. Irregular, dotted tracks mark iceberg positions from satellite scatterometer data obtained from the database of Budge and Long (2018). The position of the Polar Front separating colder Antarctic waters from warmer waters to the north is from Orsi and Harris (2019). Antarctic ice shelves are highlighted in white. Source: SCAR Antarctic Digital Database. Topography is ETOPO1 (NOAA National Geophysical Data Center, 2009).

frequencies below about 40 Hz (e.g., Matsumoto et al., 2014; Tsang-Hin-Sun et al., 2015), with distinct seasonal (e.g., Royer et al., 2015) variations and in future potentially also decadal variations. As climate change is warming Antarctic waters (Schmidtko et al., 2014) and accelerating the disintegration of ice shelves mainly along the Antarctic Peninsula (Scambos et al., 2000; Rignot et al., 2019), an increased discharge of icebergs can be expected to produce more iceberg noise and affect the natural soundscape of the Southern Ocean.

The purpose of this chapter is to review the available information on iceberg-generated seismic and acoustic noise and give a comprehensive overview of signal characteristics, occurrence patterns, and source mechanisms of iceberg noise as a significant natural but to date underappreciated noise source in the oceans of the Southern Hemisphere.

6.2. WAVEFORMS OF ICEBERG NOISE

The term *iceberg noise* is used here as a generic term to describe various signals emitted by icebergs. The appearance of these signals depends not only on the source process but also on the propagation path and distance. Seismometers installed on an iceberg (MacAyeal et al., 2008) resolve seismic phases of individual stick-slip events as the iceberg moves over ground or collides with another iceberg, whereas distant hydrophones record the extended hydroacoustic wave train of iceberg tremor generated by the same source process (e.g., Royer et al., 2015).

In general, iceberg noise comprises signals with extended duration, much longer than impulsive icequakes caused, for example, by crevassing (Podolskiy & Walter, 2016). Iceberg noise signals last at least a few tens of seconds (Royer et al., 2015), but often the signal duration is

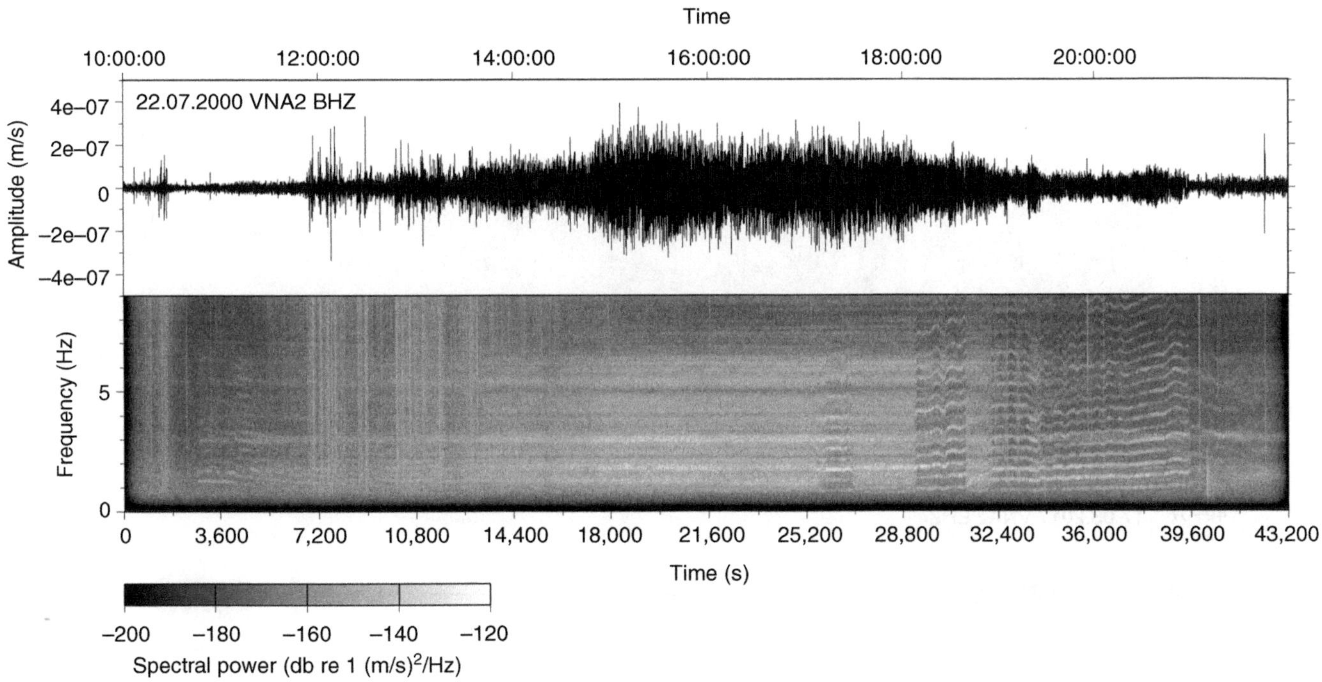

Figure 6.2 Waveform and spectrogram of an emergent iceberg harmonic tremor signal lasting for several hours. The tremor episode is associated with iceberg B09A and was first detected by Müller et al. (2005). The waveform represents the vertical component record of seismic station VNA2 (c.f. Fig. 6.5). An assumed instrument response was removed, and the signal was demeaned and bandpass-filtered between 2 and 12 Hz. The spectrogram was calculated in 10.5 s windows with 95% overlap. Absolute amplitudes and power spectral levels should not be interpreted. For seismogram examples from year 2000 (Figs. 6.2, VNA2, and 6.4, VNA1), documentation of the valid instrument response was not available. The same instrument response as for recent years (Fig. 6.3) was assumed. Absolute values of amplitude and spectral power may therefore not be correct.

in the range of a few minutes to 30 mins in both regional seismic records (Pirli et al., 2015) and distant hydroacoustic records (Chapp et al., 2005; Royer et al., 2015; Talandier et al., 2006). Iceberg harmonic tremor (Fig. 6.2) may persist for many hours (MacAyeal et al., 2008; Müller et al., 2005; Martin et al., 2010).

The onset of iceberg noise signals is typically emergent (Figs. 6.2, 6.3), and individual seismic phases can in most cases not be identified. As a result, classic earthquake location methods that rely on arrival times of P- and S-phases are unsuitable to locate the source of iceberg noise. If the distance to the source is known, estimates of the seismic magnitude of iceberg noise can be made. Müller et al. (2005) could locate an iceberg collision event immediately preceding the harmonic tremor. This collision event had a magnitude of M_l 3.6. Iceberg noise in hydrophone records of Royer et al. (2015), Chapp et al. (2005), and Talandier et al. (2006) reached source levels of 213–245 dB re μPa at 1 m, thus reaching earthquake magnitudes up to m_b 4.2–4.5 according to a b-value based equivalence relation between earthquake magnitudes and hydroacoustic source levels by Royer et al. (2015). Most of the ice events studied by Hanson and Bowman (2006) using hydrophones in the Indian Ocean had magnitude equivalents between m_b 3.25 and 4.

Iceberg noise contributes significantly to the natural soundscape of the Southern Ocean, mainly at frequencies below 50 Hz. Most authors report dominant signal energy between 1 and 40 Hz (Chapp et al., 2005; Martin et al., 2010; Pirli et al., 2015; Talandier et al., 2006) (Figs. 2, 3). Hanson and Bowman (2006) analyzed the spectral bandwidth of about 1800 ice events recorded by hydrophones off southern Australia. The majority of ice events contributed to the frequency band up to 20 Hz and a notable number up to about 40 Hz. While the Nyquist frequency of broadband seismic recording systems limits the observable frequency range to a maximum of typically 50 Hz, hydroacoustic records with higher sampling rates revealed that iceberg noise events may contain even higher frequencies, especially when observed at shorter distances: Royer et al. (2015) report iceberg noise in the frequency band up to 100 Hz at distances of more than 1000 km. At shorter distances of a few hundred kilometers, Dziak et al. (2013) describe a harmonic tremor-type signal containing energy up to 200 Hz and iceberg bursts with frequency content up to 440 Hz.

The characteristic features of iceberg noise are best revealed in spectrograms that show the evolution of the signal's frequency content over time (Fig. 6.2, bottom). Several authors have made attempts to group iceberg noise signals into different categories based on diagnostic features in the spectrograms (Chapp et al., 2005; Dziak et al., 2013; Pirli et al., 2015; Talandier et al., 2002; Talandier et al., 2006; Tanaka et al., 2019). These categories

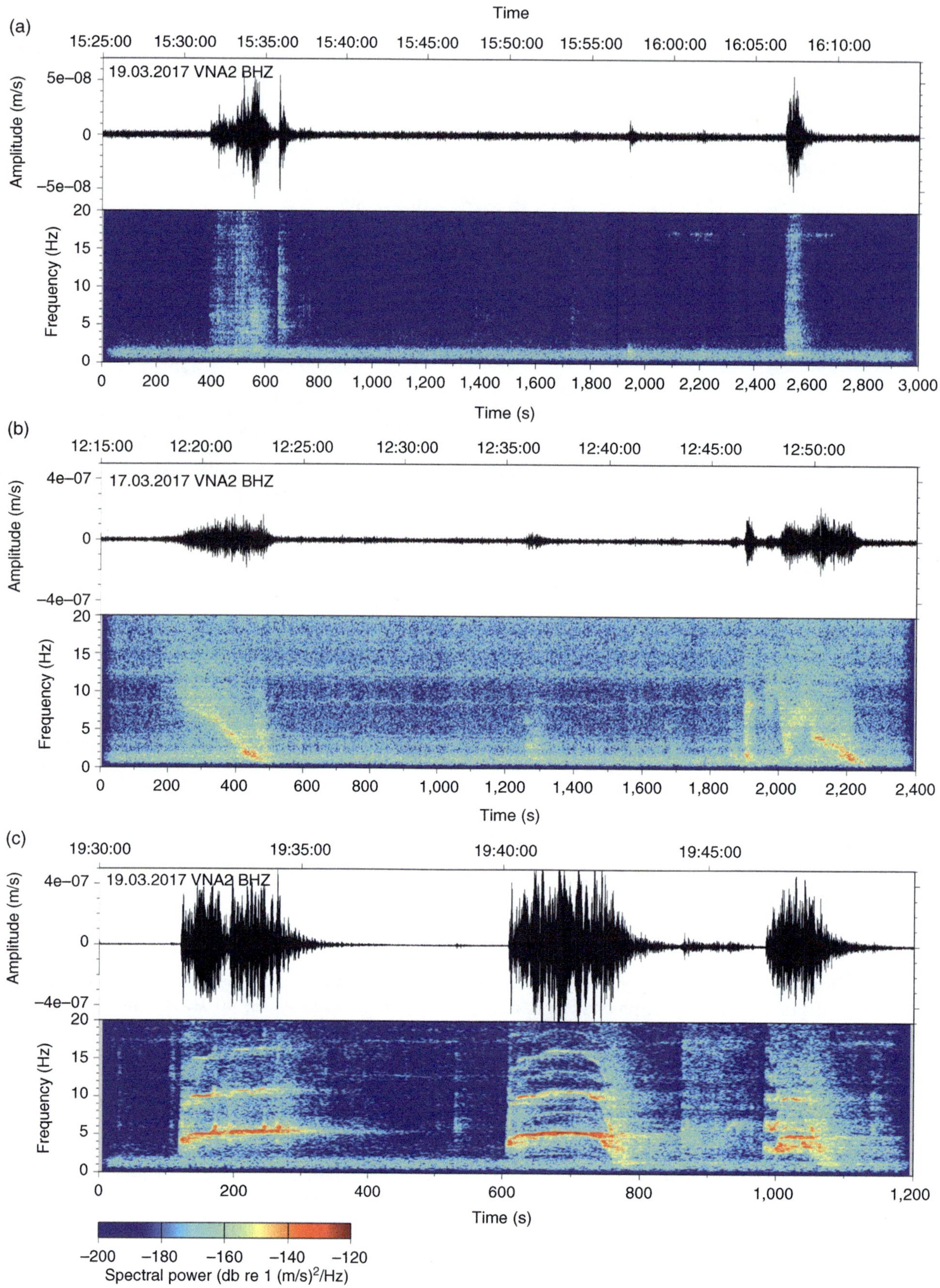

Figure 6.3 Waveforms and spectrograms of (a) iceberg burst signals, (b) iceberg tremor signals, and (c) iceberg harmonic tremor signals. All data was recorded by the vertical component of seismic station VNA2. The known instrument response was removed, the signals demeaned and bandpass-filtered between 2 and 24 Hz, and spectrograms were calculated from 10.5 s windows with 95% overlap.

differ between studies, but two general classes emerge from comparing the signals in these studies. The first class is termed *iceberg bursts* here, based on the terminology of Pirli et al. (2015), who observed "unstructured bursts" as the most frequent iceberg noise signal. Iceberg bursts (Fig. 6.3a) have a broadband frequency content, which clearly distinguishes them from the second class of iceberg noise signal called *iceberg tremor* (Fig. 6.3b,c). Iceberg tremor signals have a monochromatic or polychromatic nature and consist of distinct spectral lines that may smoothly glide over time or experience more rapid jumps in frequency. Many different varieties of iceberg tremor exist, so most authors subdivide this class further. A joint subclass in all studies is *iceberg harmonic tremor* (Figs. 6.2, 6.3c, 6.4), characterized by multiple harmonic overtones of a fundamental frequency gliding over time.

6.2.1. Iceberg Bursts

Iceberg bursts (Fig. 6.3a) are emergent, mostly shorter-duration signals of a few minutes with broad frequency content of up to 40 Hz and more (Dziak et al., 2013) and typically do not show strong amplitude spectral peaks at particular frequencies. Therefore, Pirli et al. (2015) called their spectrograms "unstructured" since they lack narrow spectral lines. Tanaka et al. (2019) observed spindle-shaped iceberg bursts that lasted more than an hour with uniform spectral power between 1 and 8 Hz. In Pirli et al. (2015), iceberg bursts are shorter, giving rise to their name. Likewise, the examples in Fig. 6.3a last only a few minutes. They were recorded in 2017 at seismometer station VNA2 (Fig. 6.5) when iceberg B15AA drifted westward in the coastal current and produced all described classes of iceberg noise (Fig. 6.3a–c). Dziak et al. (2013) detected iceberg bursts during the disintegration of iceberg A53A in open water in the Scotia Sea. Iceberg bursts occur both in the Southern Ocean's deep waters and in Antarctica's shallow coastal waters (Pirli et al., 2015; Talandier et al., 2002; Talandier et al., 2006). Chapp et al. (2005) describe iceberg bursts as forming a sequence of signals termed "cusped tremor." These sequences of iceberg bursts were observed only near outlet glaciers along the coast of East Antarctica.

6.2.2. Iceberg Tremor

Iceberg tremor comprises a large variability of signals. Chapp et al. (2005) therefore call this entire group "variable harmonic tremor," whereas other authors (e.g., Pirli et al., 2015; Tanaka et al., 2019) further subdivide this group based on the number of spectral lines and their temporal stability. Iceberg tremor here comprises all monochromatic signals that do not show the multiple harmonic overtones diagnostic for iceberg harmonic tremor. Iceberg tremor spectrograms consist of only one or two harmonics of a fundamental frequency that may glide by an order of magnitude over time (Fig. 6.3b). Both downward and upward sweeps of this fundamental frequency are observed, as well as more chaotic variations of the spectral content over time (Tanaka et al., 2019). In addition, iceberg tremor includes signal types with one or two spectral lines that are stable over hours or just fluctuate slightly such that the spectral line has the appearance of a snake (Martin et al., 2010; Talandier et al., 2006).

The occurrence of iceberg tremor is consistently confined to shallow waters near the coast of Antarctica. The group of "open water" iceberg noise described in Talandier et al. (2002, 2006) does not include similarly monochromatic signals.

6.2.3. Iceberg Harmonic Tremor

Spectacular examples of iceberg harmonic tremor show up to 30 overtones of a fundamental frequency, which is mostly in the range of 0.5 to a few Hz (Fig. 6.2). Dziak et al. (2013) recorded iceberg harmonic tremor with a fundamental frequency of a few tens of Hertz and harmonic overtones occupying the frequency band beyond 200 Hz. The fundamental frequency and their overtones shift in unison over time, such that icebergs appear to sing (Müller et al., 2005). However, the melody of these songs shows common characteristics. MacAyeal et al. (2008) observed a typical "chevron shape" of the spectrograms in many of their iceberg harmonic tremor examples: signals start with down-sweeping frequencies, followed by a signal pause termed *aseismic eye* (c.f. Fig. 6.6) and then up-sweeping frequencies. The iceberg harmonic tremor in Fig. 6.4 shows a similar chevron shape but without the aseismic eye. However, other shapes of spectrograms are also seen (Figs. 6.2, 6.3c). A further frequently observed characteristic is a sudden switching between phases of polychromatic or chaotic frequency content and phases of a distinctly harmonic signal within the same tremor episode (Fig. 6.2), stemming from the same source location (Martin et al., 2010; Müller et al., 2005). Abrupt switching on and off the harmonic tremor signal is likewise often observed (Chapp et al., 2005; Pirli et al., 2015).

The transition between iceberg tremor and iceberg harmonic tremor is smooth. Figure 6.3c shows three iceberg tremor events, where the first two have several clear harmonics developed such that they classify as iceberg harmonic tremor. In contrast, the harmonic nature of the third signal is less obvious and differs clearly from the well-developed iceberg harmonic tremor, as shown in Figs. 6.2 and 6.4.

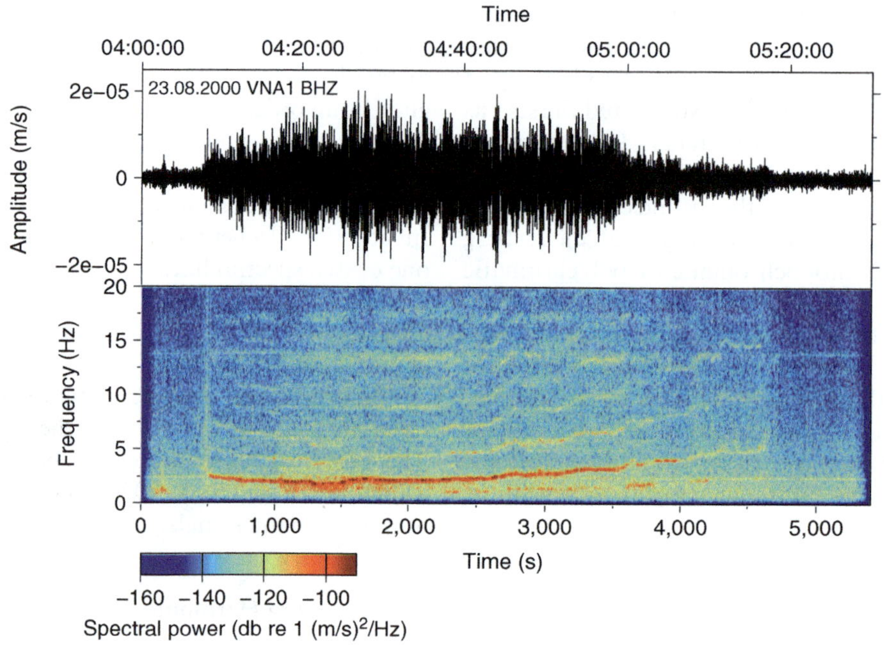

Figure 6.4 Chevron-shaped iceberg harmonic tremor recorded by the vertical component of seismic station VNA1. Fundamental frequencies decrease and subsequently increase again. An assumed instrument response was removed, the signals demeaned and bandpass-filtered between 2 and 24 Hz, and spectrograms were calculated from 10.5 s windows with 95% overlap. Absolute amplitudes and power spectral levels should not be interpreted.

6.3. OBSERVATION AND LOCATION OF ICEBERG NOISE

6.3.1. Hydroacoustic Records at Long Distances

Iceberg noise impacts hydroacoustic data sets over a wide range of distances. Iceberg-generated seismic signals effectively couple into the shallow SOFAR channel of the water column around Antarctica and from there travel as hydroacoustic phases (T-phase) with little amplitude loss across ocean basins. Figure 6.1 gives an overview of moored hydrophone stations that regularly detect iceberg noise. Matsumoto et al. (2014) examined ambient noise levels in the frequency band of iceberg signals in relation to iceberg transport volumes in the South Atlantic, Pacific, and Indian Oceans. Even at distances between 2,800 km and 7,500 km, iceberg noise contributes considerably to the overall ocean noise budget. However, not only is a general increase in noise levels observed, but also individual iceberg noise events are discernible at distances of several thousand kilometers in the Indian Ocean (Chapp et al., 2005; Evers et al., 2013; Royer et al., 2015). In the Pacific Ocean, Talandier et al. (2002, 2006) used onshore seismological stations of the Polynesian Seismic Network to study iceberg noise. Coupling the T-phase of iceberg signals into a seismic phase proved particularly efficient for seismic stations installed near (<1 km) the southern shore of atolls. In these cases, there is an unblocked travel path between the source and the recording site and a short post-conversion distance between the shoreline and the seismometer. This means the seismometer records the same T-phase as a hydrophone, except that the seismometer has a lower cut-off frequency. In these studies, individual iceberg noise events could also be detected as far north as 20°S with propagation distances of about 6,000 km.

Locating the source of iceberg noise events based on distant hydrophone records requires a propagation model for T-phases in the SOFAR channel. Royer et al. (2015) extract sound velocities from the Global Digital Environment Model and average these velocities along great circle paths connecting the source and receiver. Talandier et al. (2006) also use average sound velocities and apply a correction to account for the shallowing of the SOFAR channel south of the Polar Front (Fig. 6.1). To locate the source of iceberg noise, measurements of differential travel times of T-phases at pairs of stations are necessary. Because T-phases are mostly emergent, determining accurate phase arrival times is difficult, and different approaches have been tested. Tanaka et al. (2019) match envelopes of iceberg signals between stations, whereas Royer et al. (2015) use the time of peak spectral power measured at each station. Alternatively, signal cross-correlation between stations works best if the signals contain distinctive features in the waveform, like frequency fluctuations or impulsive components (Talandier et al., 2002).

Because in all of these studies, the source locations are well outside the hydrophone network, source locations are poorly constrained. Talandier et al. (2006) state location errors in epicentral distance as large as 1,000 km when using the small-aperture Polynesian Seismic Network (Fig. 6.1), but as soon as the aperture is increased by including additional distant stations in the Indian Ocean, source locations can be constrained to a few tens of kilometers. Some hydrophone stations consist of three instruments, so-called triads (e.g., Hanson and Bowman, 2006), such that the back-azimuths of incoming signals can be determined. Chapp et al. (2005) achieved location errors <100 km by combining back-azimuth and differential travel-time measurements from just two hydrophone stations in the Indian Ocean spaced at about 30° in latitude and longitude. Errors on the order of 10 km are achievable when distinctive signal characteristics can be correlated between distant hydrophone triads.

Even at these distances, an association of iceberg noise signals with iceberg positions is possible. Larger icebergs (>5 km^2) can be identified and tracked using satellite scatterometer data. Budge and Long (2018) present a recent comprehensive compilation of iceberg positions (c.f. Figs. 6.1, 6.5) that unifies earlier databases. Source locations of iceberg noise signals in Evers et al. (2013) scatter around the trajectories of such tracked icebergs. Chapp et al. (2005) noticed that the back-azimuths of the strongest iceberg signals follow iceberg paths, whereas weaker iceberg noise signals appeared to be produced by smaller, unnamed icebergs that are not included in the iceberg database. Talandier et al. (2002) improved the accuracy of their T-phase locations by including seismic phases of nearby land stations. Their iceberg tremor signals were then located within the 135 km × 40 km perimeter of iceberg B15B as obtained from infrared satellite photographs. In this way, a clear association between icebergs and, at that time, newly observed iceberg tremor signals could be established, and other tremor sources, such as submarine volcanoes, could be ruled out.

6.3.2. Records of Regional Hydroacoustic Networks

With regional networks of moored hydrophones at distances closer to iceberg noise sources, weaker iceberg signals are also detected, and sources can be located with higher accuracy (<5 km) using the same techniques as described previously (Dziak et al., 2010). Dziak et al. (2010, 2015) installed networks of moored hydrophones in the Bransfield Strait and Scotia Sea between 2005 and 2009 primarily to monitor the tectonic activity of the area (Figs. 6.1, 6.5). However, the records contained numerous cryogenic events. At recording distances of 100–500 km, icequakes originating near marine-terminating glaciers are contained in large numbers in the records along with iceberg noise signals (Dziak et al., 2010). Iceberg A53A drifted through the survey area and could be tracked and monitored until its disintegration in the Southern Atlantic Ocean (Dziak et al., 2013).

6.3.3. Seismic Records in Antarctica

Land stations in Antarctica provide high-quality records of all described types of iceberg signals. Observations stem from many seismic stations along the coast of East Antarctica (Eckstaller et al., 2006; Müller et al., 2005; Pirli et al., 2015; Tanaka et al., 2019) as well as from the Ross Sea (MacAyeal et al., 2008; Talandier et al., 2002) (Figs. 6.1, 6.5). In their early study, Talandier et al. (2002) noticed that tremor from icebergs in the Ross Sea was detected by nearby seismic stations but not at more distant seismic stations, such as the South Pole. Müller et al. (2005) and Pirli et al. (2015) recorded iceberg tremor from icebergs drifting along the coast of East Antarctica in a distance range of 200–800 km from stations VNA2 and TROLL (Fig. 6.5). Coastal seismic stations in Lützow-Holm Bay (SYO, Fig. 6.5) provide excellent seismic records of iceberg noise at distances of less than 100 km (Kanao et al., 2017; Tanaka et al., 2019). Three-component observations of seismic land stations allow the determination of back azimuths of incoming arrivals and the monitoring of changes as icebergs pass. Accurate estimates of back-azimuths from frequency-wavenumber analysis of data of the seismological array surrounding station VNA2 consistently point in the direction of iceberg B09A during its 1000 km satellite-tracked travel path from July to October 2000 (Müller et al., 2005). In addition to these associations between directions of incoming tremor signals and satellite-derived iceberg tracks, a few iceberg noise events (<5%) show distinctive P- and S-phase arrivals, notably when icebergs collide with the ground (Müller et al., 2005). In these rare cases, distances to the source can be estimated from travel-time differences, or classic earthquake location techniques can be employed, provided several stations detect the events with sufficient clarity (Pirli et al., 2015).

A breakthrough in understanding iceberg harmonic tremor was achieved by placing broadband seismic stations, or networks of stations, on large tabular icebergs (MacAyeal et al., 2008; Martin et al., 2010). Seismic and hydroacoustic phases of iceberg harmonic tremor were simultaneously recorded by seismic stations a few kilometers from the iceberg collision zone, at a 40 km distance on the same iceberg, and by land stations in the Ross Sea (MacAyeal et al., 2008). In the near-field of the source region, the apparently continuous waveform of iceberg

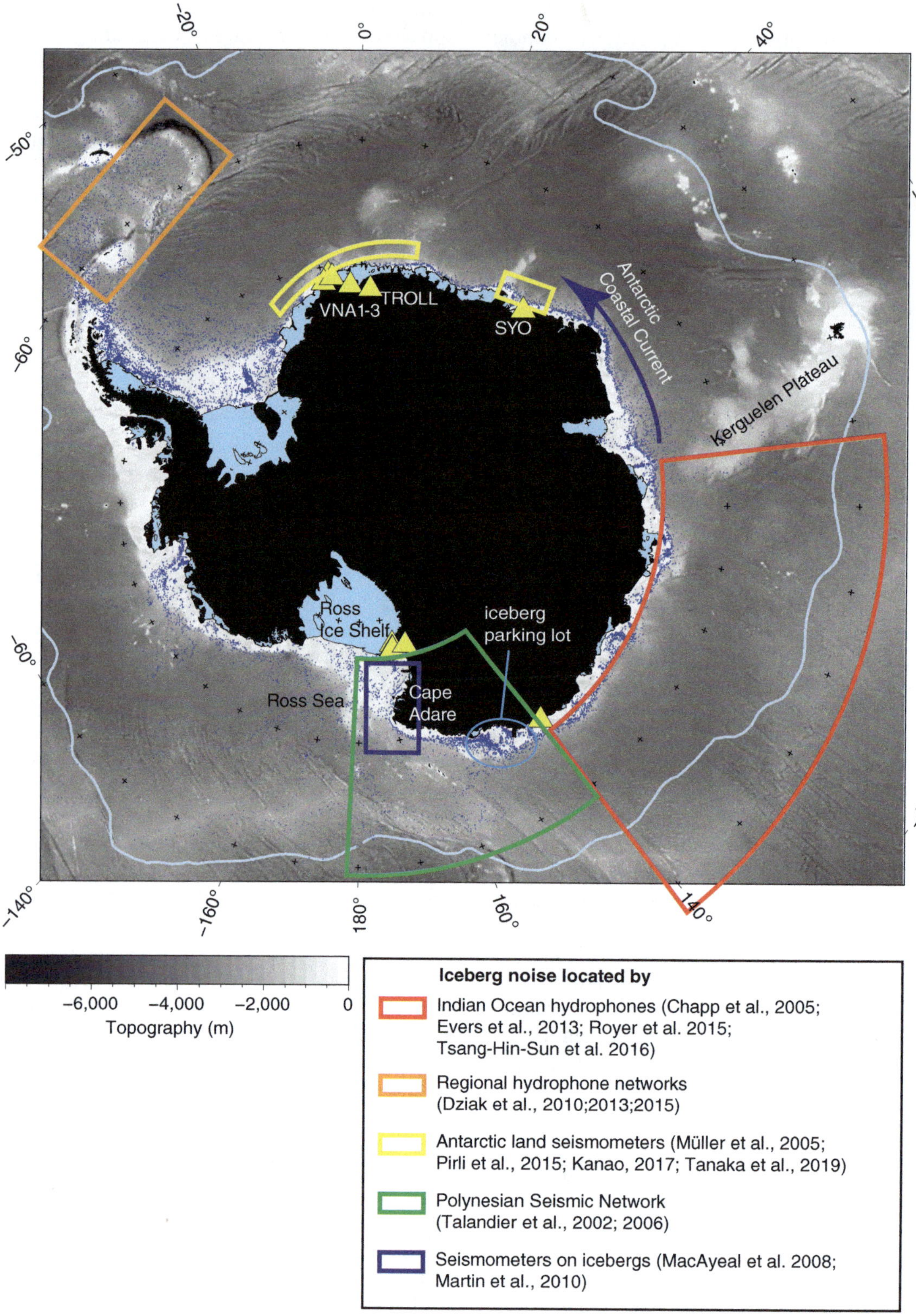

Figure 6.5 Location of iceberg signal occurrence. Areas with iceberg noise generation are outlined and shaded with reference to the respective studies. Data sets as in Fig. 6.1. Only every 10th iceberg position is plotted as a black dot for clarity.

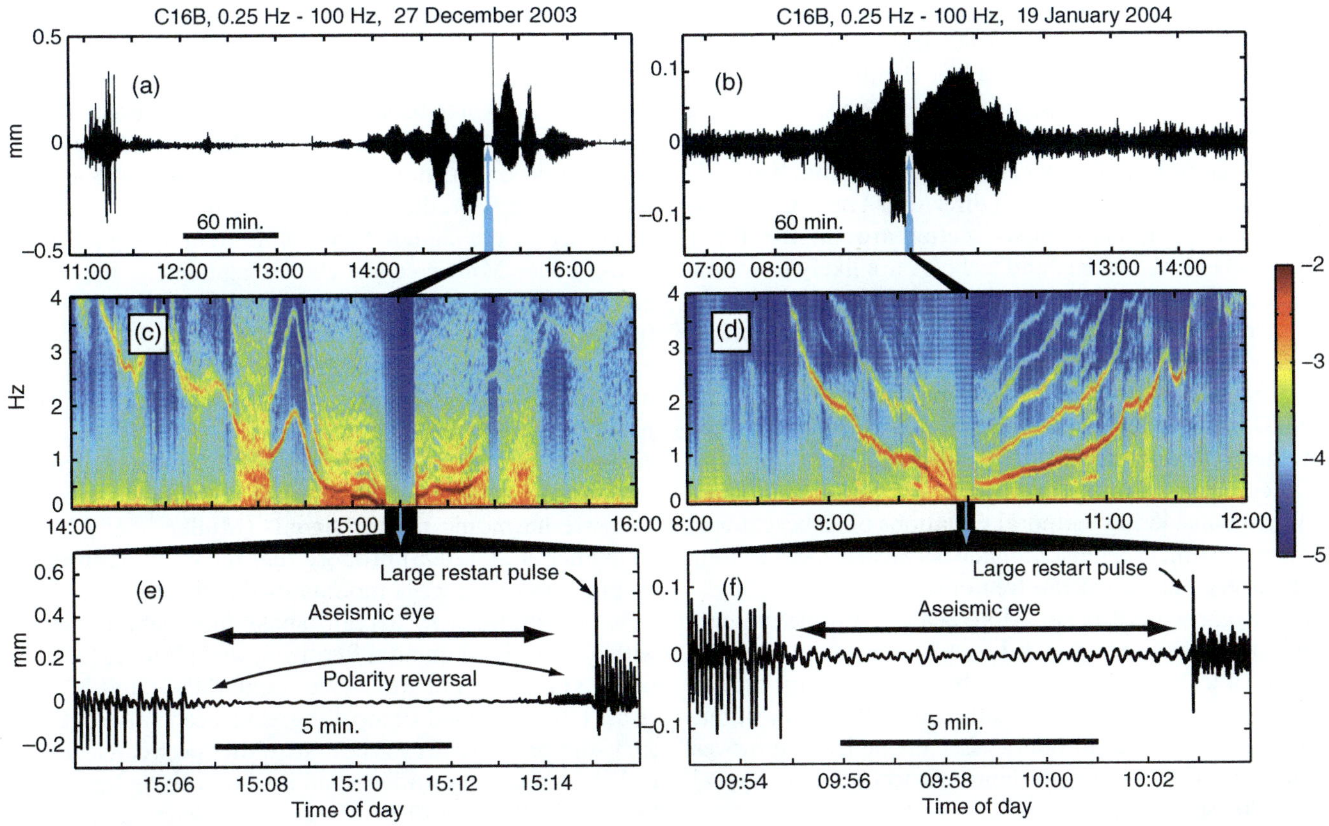

Figure 6.6 Two chevron-patterned iceberg harmonic tremor events. (a, b) Waveforms of the entire tremor episodes. (c, d) Spectrograms of a closeup around the aseismic eye at the "bottom" of the chevron. (e, f) Waveforms immediately before and after the aseismic eye. Tremor resolves into individual repetitive events with a reversal in polarity. Reproduced in black and white from MacAyeal et al. (2008).

harmonic tremor resolved into a series of impulsive repetitive events with distinctive phases (Fig. 6.6). P-phase polarization consistently pointed in the direction of the contact zone between two icebergs, and event first-motion reversed when relative iceberg motion reversed following the "aseismic eye" without relative iceberg motion (Fig. 6.6).

Local observations of iceberg noise thus clearly associate the noise signals with icebergs as the seismogenic source.

6.4. SPATIAL AND TEMPORAL VARIATIONS OF ICEBERG NOISE

Figure 6.5 gives an overview of the source regions of iceberg noise. Iceberg noise can in principle be expected to occur wherever icebergs drift, such that Fig. 6.5 reflects the observational capabilities of the hydrophone and seismometer networks used. Hydrophone observations require an unblocked deep-water travel path, such that, for example, the Indian Ocean hydrophones only record iceberg noise east of the shallow Kerguelen Plateau. Land seismic stations detect iceberg events at distances of about 800 km (Müller et al., 2005) to as far as 2,000 km in exceptional circumstances (Martin et al., 2010, South Pole observations). Despite these observational limitations, spatial patterns in iceberg noise emissions exist. Hanson and Bowman (2006) noticed that event locations of iceberg noise form distinct clusters along the coast of Antarctica. Talandier et al. (2006) call the near-coastal area between 148°E and 155°E the "iceberg parking lot" as many icebergs that calved off the Ross ice shelf became stationary in this area of shallow bathymetry. Ground contact slows larger icebergs, which in turn block the drift path for smaller icebergs such that "parking lots" form, with frequent iceberg collisions and long residence times. From such areas of shallow waters, iceberg noise is preferentially emitted (e.g., Dziak et al., 2013; Müller et al., 2005; Pirli et al., 2015). Martin et al. (2010) directly observed iceberg noise generation when their instrumented iceberg scraped Davey Shoal near Cape Adare (Fig. 6.5).

The described spatial relation between shallow water regions and iceberg noise emissions applies in particular to iceberg tremor and iceberg harmonic tremor (e.g., Dziak et al., 2013; Pirli et al., 2015). Iceberg bursts, in contrast, occur independently of water depth, but they also display some spatial occurrence patterns. Talandier et al. (2002, 2006) noticed that iceberg B19 drifted silently most of the time and only generated iceberg burst signals during sudden changes in drift direction. The source locations of these iceberg bursts scatter around the Polar Front, where waters mix and icebergs are likely to follow irregular drift paths. In addition, when icebergs encounter warmer sea surface temperatures, they disintegrate more quickly, and noise emission increases. Noise-emitting icebergs have been observed to show melt ponds on their surface (Dziak et al., 2013) or be surrounded by many iceberg fragments (Evers et al., 2013) as signs of their ongoing disintegration.

When considering temporal variations of iceberg noise, the same distinction must be made between iceberg tremor and iceberg bursts. Iceberg tremor signals occur throughout the year with little variation. Sea ice, especially land-fast sea ice, is observed to reduce to some extent the generation of iceberg tremor signals but not the occurrence of iceberg harmonic tremor (Tanaka et al., 2019). The latter, however, may occur in tidal cycles as one iceberg moves with tidal currents along a land-fast iceberg (MacAyeal et al., 2008).

Iceberg burst signals, in contrast, show a more prominent seasonal variation (Tanaka et al., 2019). Royer et al. (2015) observed that iceberg signals generated in the open ocean (iceberg bursts) are more numerous during austral summer, whereas iceberg noise signals stemming from the shallow coastal waters of Antarctica (mostly iceberg tremor) show little seasonal variations. In their regional hydrophone networks, Dziak et al. (2015) likewise note increased noise levels during austral summer, which they attribute to greater icequake activity as icebergs and glaciers break due to thermal stress in freeze-thaw cycles. As a result, the contribution of iceberg noise to the general noise budget in the Pacific, Indian, and Atlantic Oceans peaks in austral summer and decreases during austral winter, tightly correlated with seasonal fluctuations in iceberg volumes (Matsumoto et al., 2014; Tsang-Hin-Sun et al., 2015; Leroy et al., 2018). This seasonality of the ocean noise level is observed as far north as 26°S in the Indian Ocean for frequencies up to 30 Hz (Leroy et al., 2018).

6.5. SOURCE MECHANISMS OF ICEBERG NOISE

The spectrogram characteristics of iceberg noise likely reflect an inherent property of the source process. Talandier et al. (2002) observed that spectrograms of iceberg events recorded at distant stations of the Polynesian Seismic Network look remarkably similar to spectrograms of the same event recorded at nearby seismic stations in Antarctica. Therefore, any path effects can be ruled out as explanation for spectral characteristics, and the source mechanism must be capable of producing monochromatic or harmonic spectrograms with gliding frequencies. Early studies of iceberg tremor recognized the similarity of the signals to volcanic tremor (e.g., Chapp et al., 2005; Müller et al., 2005; Talandier et al., 2002) and invoked source mechanisms used to explain volcanic tremor (see Konstantinou & Schlindwein, 2003 for an overview). Common models include the resonance of fluid-filled cracks (Chouet, 1988) or nonlinear fluid flow phenomena (Julian, 2000; Rust et al., 2008). For grounded icebergs, it is conceivable that ocean currents are forced through narrow crevasses in the iceberg and generate harmonic tremor signals (Müller et al., 2005). Talandier et al. (2006) propose that the bobbing, rolling, or pitching of icebergs modulates the eigenfrequency of iceberg vibrations to explain snake-like iceberg tremor with a stable fundamental frequency and periodic fluctuations. The studies of MacAyeal et al. (2008) and Martin et al. (2010) yielded compelling evidence that repetitive stick-slip processes at the contact either between icebergs or between iceberg and ground lead to iceberg harmonic tremor. In the near-field of an iceberg-iceberg collision zone, MacAyeal et al. (2008) resolved individual stick-slip events that merge into harmonic tremor via a Dirac comb process (Aster et al., 2021). The evolution of the fundamental frequency of this tremor matched GPS measurements of iceberg motion: with down-sweeping fundamental frequencies, the iceberg slowed down. No tremor was observed during the aseismic eye as tidal currents turned. When tidal currents picked up again in the opposite direction, a reversed polarity of individual stick-slip events was observed (Fig. 6.6e,f). With increasing iceberg speed, the time interval between the stick-slip events became shorter, and fundamental frequency increased again, producing chevron-shaped tremor signals (Figs. 6.4, 6.6). While this process of collisional motion between a fast and a floating iceberg repeats itself in tidal cycles, iceberg grounding events occur more irregularly. Martin et al. (2010) established a temporal relation between the occurrence of iceberg harmonic tremor and shoaling and subsequent rotation of the iceberg, followed by its breakup. Likewise, examples of less harmonic iceberg tremor, as in Fig. 6.3b, can be explained by the deceleration of icebergs during ground or iceberg contact. Sudden-onset harmonic tremor events (Fig. 6.3c) may result from scraping in a stick-slip manner along the seafloor with a roughly constant drift speed. Stick-slip processes and related harmonic tremor are also reported from the base of ice streams during phases of ice stream advance (Winberry et al., 2013).

Iceberg harmonic tremor often features phases of chaotic, broadband spectra alternating with highly harmonic spectra (Fig. 6.2). A loss of the rhythmic stick-slip behavior due to destructive fracturing at the contact surface may explain the unstructured tremor phases (MacAyeal et al., 2008; Martin et al., 2010).

Because iceberg tremor mostly originates from the shallow, coastal waters of Antarctica, where many icebergs drift in the Antarctic Coastal Current (Fig. 6.5), it appears likely that stick-slip processes play a dominant role in the generation of iceberg tremor. However, special tremor types, such as snake-like tremor with stable fundamental frequencies or structured tremor without harmonics (Fig. 6.3c, last event), suggest that other source mechanisms such as eigenvibrations of cracks may be necessary to explain the full variety of observed waveforms.

In contrast to iceberg tremor, iceberg burst signals are generally attributed to predominantly destructive processes. Dziak et al. (2013) associate the white spectra of iceberg bursts to iceberg disintegration processes such as fracturing, breaking, and calving. Intense icequake activity with high-frequency unstructured broadband sound emission accompanied the fracturing of the Nansen Ice Shelf (Dziak et al., 2019) and occurred seasonally near marine-terminating glaciers in Bransfield Strait (Dziak et al., 2010). Fracturing processes are therefore likely to play a role in generating the broadband spectra of iceberg burst signals. However, it remains difficult to explain how fracturing can produce iceberg burst signals that persist for several hours, as observed, for example, by Tanaka et al. (2019). They suggest that storm-induced swell events act on fast sea ice and icebergs, potentially producing persistent fracturing. In contrast, Pirli et al. (2015) found no temporal or spatial occurrence patterns for iceberg burst signals that could give conclusive hints about their source process. The same iceberg in nearly the same location produced iceberg tremor and iceberg burst signals. Likewise, the described distinction between iceberg signal types is not sharp, and combinations of signal characteristics between burst and tremor signals exist. This suggests that the source processes are more complex. However, the general impression from the existing studies is that the whiter spectra of iceberg burst signals are representative of destructive fracturing processes, whereas the generation of iceberg tremor is best explained with relatively nondestructive stick-slip processes.

6.6. DISCUSSION

Harmonic tremor signals are frequently observed in seismic and hydroacoustic records, and various natural and artificial sources can produce surprisingly similar-looking tremor episodes. Moored recording systems may experience current-induced strumming on cables or recovery ropes, producing extensive harmonic tremor episodes, often with tidal modulation (Essing et al., 2021; Ramakrushana Reddy et al., 2020; Stähler et al., 2018). Volcanoes (e.g., Jellinek & Bercovici, 2011) and hydrothermal systems (e.g., Meier & Schlindwein, 2018) are natural sources of harmonic tremor that may be encountered in the Southern Ocean, especially in tectonically active areas like Bransfield Strait (Dziak et al., 2010). However, observation of volcanic tremor signals is also possible across oceans (Dziak & Fox, 2002). Because the location of tremor sources is difficult due to the emergent nature of the signals, detailed knowledge of the spectrogram characteristics of different source types is important for assigning underlying source processes. Hydrodynamic tremor acting on ocean-bottom seismometers, for example, typically shows turtle-back-shaped tremor signals (Essing et al., 2021) with increasing and decreasing fundamental frequency at the start and end of a tremor episode, respectively, and maximum tremor excitation at a plateau of high fundamental frequencies in between. Iceberg tremor episodes, in contrast, may show a chevron shape and the opposite frequency evolution over time. However, based on waveform or spectrogram characteristics alone, Konstantinou et al. (2019) could not distinguish volcanic tremor episodes from iceberg harmonic tremor at a seismic station on Mt. Erebus volcano in the Ross Sea area. They additionally examined the scaling behavior of the tremor events. Seismic events produced by breakup processes at calving fronts follow a power-law scaling (Kirkham et al., 2017), but the tremor events recorded at Mt. Erebus were better approximated by an exponential scaling law, which led the authors to speculate on a volcanic source rather than iceberg noise in the Ross Sea. However, volcanic tremor on Mt. Erebus has never been confirmed despite many years of observations (R. Aster, personal communication, 2021).

Volcanic, hydrothermal, and artificial tremor generation is confined in time and space to the location and activity of a volcano, a hydrothermal source, or current velocities at an individual mooring location. Icebergs, in contrast, are ubiquitous around Antarctica and frequently produce tremor with considerable acoustic source levels, such that they can be considered the dominant source of tremor signals in the Southern Ocean.

Iceberg tremor signals, particularly harmonic tremor signals, constitute only a small part of iceberg-generated noise (e.g., Pirli et al., 2015). By far the largest number of events are iceberg bursts associated with iceberg disintegration. Noise levels in the Southern Ocean in the frequency band of iceberg noise are dominated by such disintegration events. They follow a distinct seasonality that matches seasonal variations in iceberg volume (Matsumoto et al., 2014; Tsang-Hin-Sun et al., 2015) such that the question arises of whether iceberg noise signals can be used for monitoring iceberg discharge and disintegration. However, a direct causal relation between iceberg discharge or disintegration processes and noise emission is difficult to establish. Iceberg discharge is not necessarily noisy (Dziak et al., 2019). Likewise, noise levels do not scale with iceberg size or keel depth. Large icebergs may drift silently, but conversely, noise may be produced by icebergs that are too small to be tracked (Chapp et al., 2005). Some icebergs continuously produce noise signals, whereas others become noisy only toward the end of their lifetime (Dziak et al., 2013; Evers et al., 2013). Therefore, individual iceberg noise signals do not necessarily reflect the physical state of an iceberg. However, noise budgets integrate over large numbers of iceberg noise signals and thereby may reveal not only seasonal variations but also decadal variations in iceberg disintegration processes, thus offering a possibility to monitor these changes.

Iceberg noise appears to be a particular feature of the Southern Ocean. No comparable contribution to the noise budget of the Arctic Ocean is produced by icebergs discharged from the Greenland ice sheet (Wilcock et al., 2014). Its iceberg production is only about half the production of Antarctica, and Greenland icebergs may not reach the average length of about 750 m of Antarctic icebergs or their average lifetime of about 200 days (Tournadre et al., 2012). Many of the iceberg harmonic tremor signals are associated with large tabular Antarctic icebergs with lengths of tens of kilometers and keel depths of several hundred meters discharged by the large Antarctic shelf ice areas. Marine-terminating glaciers in Greenland, in contrast, tend to produce metastable icebergs that capsize immediately after calving and collide with the calving front (Nettles & Ekström, 2010). This process typically happens when calving occurs close to the grounding line in late summer (Ekström et al., 2006; Nettles & Ekström, 2010). The capsizing and collision process is powerful enough to generate so-called glacial earthquakes (Ekström et al., 2003) with dominant signal periods >30 s and magnitudes around M_s 5. However, it is not connected with a notable noise emission into the ocean comparable to Antarctica's ubiquitous low-frequency iceberg noise. Acoustic receivers deployed 90–150 m below sea ice north of Alaska recorded harmonic tremor signals similar to iceberg harmonic tremor but with fundamental frequencies in the range of kilohertz (Collins et al., 2019). These signals were attributed to colliding sea ice floes above the receivers and are not observable across ocean basins, as is the case for Antarctic iceberg tremor signals.

6.7. CONCLUSION

Icebergs constitute a considerable noise source in the Southern Ocean. Emitted signals show a wide variety, but two broad classes can be distinguished based on spectrogram characteristics: iceberg burst signals have a white spectrum, occur independently of water depth, and are associated with fracturing processes during iceberg disintegration. Iceberg tremor signals have monochromatic or harmonic spectra. They are produced during stick-slip motion between icebergs or when icebergs scrape over ground in the coastal waters of Antarctica. Both noise signals last between minutes and hours and are sufficiently large to be recorded as hydroacoustic phases across the entire Southern Ocean for thousands of kilometers. At frequencies below 40 Hz, iceberg noise represents a notable contribution to the sound scape in the Southern Ocean, particularly during austral summer. Decadal variations of this noise contribution can be expected as iceberg noise levels correlate with iceberg volumes.

ACKNOWLEDGMENTS

Data examples are based on records of the seismological observatory of the Neumayer III station (Wesche et al., 2016). The author gratefully acknowledges the help of A. Eckstaller and C. Müller in pointing out iceberg noise signals. T. Fromm and C. Müller helped to improve the manuscript. The manuscript considerably profited from detailed and thoughtful reviews of R. Aster, M. Pirli, and an anonymous reviewer.

REFERENCES

Aster, R.C., Lipovsky, B. P., Cole, H. M., Bromirski, P. D., Gerstoft, P., Nyblade, A., et al. (2021). Swell-triggered seismicity at the near-front damage zone of the Ross ice shelf. *Seismological Research Letters*, *92*(5), 2768–2792. https://doi.org/10.1785/0220200478

Budge, J. S., & Long, D. G. (2018). A comprehensive database for Antarctic iceberg tracking using scatterometer data. *IEEE Journal of Selected Topics in Applied Earth Observations and Remote Sensing*, *11*(2), 434–442. https://doi.org/10.1109/jstars.2017.2784186

Chapp, E., Bohnenstiehl, D. R., & Tolstoy, M. (2005). Sound-channel observations of ice-generated tremor in the Indian Ocean. *Geochemistry, Geophysics, Geosystems*, *6*(6). https://doi.org/10.1029/2004GC000889

Chouet, B. (1988). Resonance of a fluid-driven crack: Radiation properties and implications for the source of long-period events and harmonic tremor. *Journal of Geophysical Research: Solid Earth*, *93*(B5), 4375–4400. https://doi.org/10.1029/JB093iB05p04375

Collins, M. D., Turgut, A., Menis, R., & Schindall, J. A. (2019). Acoustic recordings and modeling under seasonally varying sea ice. *Scientific Reports*, *9(*8323). https://doi.org/10.1038/s41598-019-44707-0

Dziak, R. P., Bohnenstiehl, D. R., Stafford, K. M., Matsumoto, H., Park, M., Lee, W. S. et al. (2015). Sources and levels of ambient ocean sound near the Antarctic Peninsula. *PLoS ONE*, *10*(4). https://doi.org/10.1371/journal.pone.0123425

Dziak, R. P., Fowler, M. J., Matsumoto, H., Bohnenstiehl, D. R., Park, M., Warren, K., & Lee, W. S. (2013). Life and death sounds of iceberg A53a. *Oceanography*, *26*(2), 10–12. https://doi.org/10.5670/oceanog.2013.20

Dziak, R. P., & Fox, C. G. (2002). Evidence of harmonic tremor from a submarine volcano detected across the Pacific Ocean basin. *Journal of Geophysical Research: Solid Earth*, *107* (B5). https://doi.org/10.1029/2001jb000177

Dziak, R. P., Lee, W. S., Haxel, J. H., Matsumoto, H., Tepp, G., Lau, T. K., et al. (2019). Hydroacoustic, meteorologic and seismic observations of the 2016 Nansen ice shelf calving event and iceberg formation. *Frontiers in Earth Science*, *7*. https://doi.org/10.3389/feart.2019.00183

Dziak, R. P., Park, M., Lee, W. S., Matsumoto, H., Bohnenstiehl, D. R., & Haxel, J. H. (2010). Tectonomagmatic activity and ice dynamics in the Bransfield Strait back-arc basin, Antarctica. *Journal of Geophysical Research*, *115*(B1). https://doi.org/10.1029/2009jb006295

Eckstaller, A., Müller, C., Ceranna, L., & Hartmann, G. (2006). The Geophysics Observatory at Neumayer Stations (GvN and NM-II) Antarctica. *Polarforschung*, *76*(1–2), 3–24.

Ekström, G., Nettles, M., & Abers, G. A. (2003). Glacial earthquakes. *Science*, *302*(5645), 622–624. https://doi.org/10.1126/science.1088057

Ekström, G., Nettles, M., & Tsai, V. C. (2006). Seasonality and increasing frequency of Greenland glacial earthquakes. *Science*, *311*(5768), 1756–1758. https://doi.org/10.1126/science.1122112

Essing, D., Schlindwein, V., Schmidt-Aursch, M. C., Hadziioannou, C., & Stähler, S. C. (2021). Characteristics of current-induced harmonic tremor signals in ocean bottom seismometer records. *Seismological Research Letters*, https://doi.org/10.1785/0220200397.

Evers, L. G., Green, D. N., Young, N. W., & Snellen, M. (2013). Remote hydroacoustic sensing of large icebergs in the southern Indian Ocean: implications for iceberg monitoring. *Geophysical Research Letters*, *40*(17), 4694–4699. https://doi.org/10.1002/grl.50914

Hanson, J.A. & Bowman, J.R., (2006). Methods for monitoring hydroacoustic events using direct and reflected T waves in the Indian Ocean. *Journal of Geophysical Research*, *111*, B02305, https://doi.org/10.1029/2004JB003609.

Jellinek, A. M., & Bercovici, D. (2011). Seismic tremors and magma wagging during explosive volcanism. *Nature*, *470* (7335), 522–526. https://doi.org/10.1038/nature09828

Julian, B. R. (2000). Period doubling and other nonlinear phenomena in volcanic earthquakes and tremor. *Journal of Volcanology and Geothermal Research*, *101*(1–2), 19–26. https://doi.org/10.1016/s0377-0273(00)00165-7

Kanao, M., Murayama, T., Yamamoto, M.-Y., & Ishihara, Y. (2017). Seismic tremors and their relation to cryosphere dynamics in April 2015 around the Lützow-Holm Bay, East Antarctica. *International Journal of Geosciences*, *08*(08), 1025–1047. https://doi.org/10.4236/ijg.2017.88058

Kirkham, J. D., Rosser, N. J., Wainwright, J., Jones, E. C. V., Dunning, S. A., Lane, V. S., et al. (2017). Drift-dependent changes in iceberg size-frequency distributions. *Scientific Reports*, *7*. https://doi.org/10.1038/s41598-017-14863-2

Konstantinou, K. I., Ardiani, M. A., & Sudibyo, M. R. P. (2019). Scaling behavior and source mechanism of tremor recorded at Erebus volcano, Ross Island, Antarctica. *Physics of the Earth and Planetary Interiors*, *290*, 99–106. https://doi.org/10.1016/j.pepi.2019.03.010

Konstantinou, K. I., & Schlindwein, V. (2003). Nature, wavefield properties and source mechanism of volcanic tremor: a review. *Journal of Volcanology and Geothermal Research*, *119*(1–4), 161–187.

Leroy, E.C., Royer, J.-Y., Bonnel, J. & Samaran, F. (2018). Long-term and seasonal changes of large whale call frequency in the southern Indian Ocean, *Journal of Geophysical Research*, *123*, 8568–8580, https://doi.org/10.1029/2018JC014352R

MacAyeal, D. R., Okal, E. A., Aster, R. C., & Bassis, J. N. (2008). Seismic and hydroacoustic tremor generated by colliding Icebergs. *Journal of Geophysical Research: Earth Surface*, *113*(3). https://doi.org/10.1029/2008JF001005

Martin, S., Drucker, R., Aster, R., Davey, F., Okal, E., Scambos, T., & MacAyeal, D. (2010). Kinematic and seismic analysis of giant tabular Iceberg breakup at Cape Adare, Antarctica. *Journal of Geophysical Research: Solid Earth*, *115*(6). https://doi.org/10.1029/2009JB006700

Matsumoto, H., Bohnenstiehl, D. R., Tournadre, J., Dziak, R. P., Haxel, J. H., Lau, T. K. A., et al. (2014). Antarctic icebergs: a significant natural ocean sound source in the Southern Hemisphere. *Geochemistry Geophysics Geosystems*, *15*(8), 3448–3458. https://doi.org/10.1002/2014gc005454

Meier, M., & Schlindwein, V. (2018). First in situ seismic record of spreading events at the ultraslow spreading southwest Indian Ridge. *Geophysical Research Letters*, *45*(19), 10,360–310,368. https://doi.org/10.1029/2018gl079928

Müller, C., Schlindwein, V., Eckstaller, A., & Miller, H. (2005). Singing icebergs. *Science*, *310*(5752), 1299–1299. https://doi.org/Doi 10.1126/Science.1117145

Nettles, M., & Ekström, G. (2010). Glacial earthquakes in Greenland and Antarctica. *Annual Review of Earth and Planetary Sciences*, *38*(1), 467–491. https://doi.org/10.1146/annurev-earth-040809-152414

NOAA National Geophysical Data Center. (2009). *ETOPO1 1 Arc-Minute Global Relief Model.*

Orsi, A. H., & Harris, U. (2019). *Fronts of the Antarctic Circumpolar Current - GIS data, Ver. 1*. Retrieved from https://data.aad.gov.au/metadata/records/antarctic_circumpolar_current_fronts

Pirli, M., Matsuoka, K., Schweitzer, J., & Moholdt, G. (2015). Seismic signals from large, tabular icebergs drifting along the Dronning Maud Land coast, Antarctica, and their significance for iceberg monitoring. *Journal of Glaciology*, *61*(227), 481–492. https://doi.org/10.3189/2015JoG14J210

Podolskiy, E. A., & Walter, F. (2016). Cryoseismology. *Reviews of Geophysics*, *54*(4), 708–758. https://doi.org/10.1002/2016rg000526

Ramakrushana Reddy, T., Dewangan, P., Arya, L., Singha, P., & Kamesh Raju, K. A. (2020). Tidal triggering of the harmonic noise in ocean-bottom seismometers. *Seismological Research Letters*, *91*(2A), 803–813. https://doi.org/10.1785/0220190080

Rignot, E., Mouginot, J., Scheuchl, B., van den Broeke, M., van Wessem, M. J., & Morlighem, M. (2019). Four decades of Antarctic ice sheet mass balance from 1979–2017. *Proceedings of the National Academy of Sciences USA*, *116*, 1095–1103. https://doi.org/10.1073/pnas.1812883116

Royer, J. Y., Chateau, R., Dziak, R. P., & Bohnenstiehl, D. R. (2015). Seafloor seismicity, Antarctic ice-sounds, cetacean vocalizations and long-term ambient sound in the Indian Ocean basin. *Geophysical Journal International*, *202*(2), 748–762. https://doi.org/10.1093/gji/ggv178

Rust, A. C., Balmforth, N. J., & Mandre, S. (2008). The feasibility of generating low-frequency volcano seismicity by flow through a deformable channel. *Geological Society, London, Special Publications*, *307*, 45–56. https://doi.org/10.1144/SP307.4

Scambos, T. A., Hulbe, C., Fahnestock, M., & Bohlander, J. (2000). The link between climate warming and break-up of ice shelves in the Antarctic Peninsula. *Journal of Glaciology*, *46*(154).

Schlindwein, V., Wassermann, J., & Scherbaum, F. (1995). Spectral-analysis of harmonic tremor signals at Mt-Semeru Volcano, Indonesia. *Geophysical Research Letters*, *22*(13), 1685–1688.

Schmidtko, S., Heywood, K. J., Thompson, A. F., & Aoki, S. (2014). Multidecadal warming of Antarctic waters. *Science*, *346*(6214), 1227–1231. https://doi.org/10.1126/science.1256117

Stähler, S. C., Schmidt-Aursch, M. C., Hein, G., & Mars, R. (2018). A self-noise model for the German DEPAS OBS pool. *Seismological Research Letters*, *89*(5), 1838–1845. https://doi.org/10.1785/0220180056

Talandier, J., Hyvernaud, O., Okal, E., & Piserchia, P.-F. (2002). Long-range detection of hydroacoustic signals from large icebergs in the Ross Sea, Antarctica. *Earth and Planetary Science Letters*, *203*, 519–534.

Talandier, J., Hyvernaud, O., Reymond, D., & Okal, E. A. (2006). Hydroacoustic signals generated by parked and drifting icebergs in the Southern Indian and Pacific Oceans. *Geophysical Journal International*, *165*(3), 817–834. https://doi.org/10.1111/j.1365-246X.2006.02911.x

Tanaka, Y., Hiramatsu, Y., Ishihara, Y., & Kanao, M. (2019). Characteristics of non-tectonic tremors around the Lützow-Holm Bay, East Antarctica, during 2013–2015. *Polar Science*, *19*, 77–85. https://doi.org/10.1016/j.polar.2018.11.010

Tournadre, J., Girard-Ardhuin, F., & Legrésy, B. (2012). Antarctic icebergs distributions, 2002–2010. *Journal of Geophysical Research: Oceans*, *117*(C05004). https://doi.org/10.1029/2011JC007441

Tsang-Hin-Sun, E., Royer, J.-Y. & Leroy, E., (2015). Low-frequency sound level in the Southern Indian Ocean, *Journal of the Acoustical Society of America*, *138*, 1–8, https://doi.org/10.1121/1.4936855

Tsang-Hin-Sun, E., Royer, J.-Y. & Perrot, J., (2016). Seismicity and active accretion processes at the ultraslow-spreading Southwest and intermediate-spreading Southeast Indian ridges from hydroacoustic data, *Geophysical Journal International*, *206*, 1232–1245, https://doi.org/10.1093/gji/ggw201

Wesche, C., Weller, R., König-Langlo, G., Fromm, T., Eckstaller, A., Nixdorf, U., & Kohlberg, E. (2016). Neumayer III and Kohnen Station in Antarctica operated by the Alfred Wegener Institute. *Journal of Large-Scale Research Facilities*. https://doi.org/10.17815/jlsrf-2-152

Wilcock, W. S., Stafford, K. M., Andrew, R. K., & Odom, R. I. (2014). Sounds in the ocean at 1–100 Hz. *Ann Rev Mar Sci*, *6*, 117–140. https://doi.org/10.1146/annurev-marine-121211-172423

Winberry, J. P., Anandakrishnan, S., Wiens, D. A., & Alley, R. B. (2013). Nucleation and seismic tremor associated with the glacial earthquakes of Whillans Ice Stream, Antarctica. *Geophysical Research Letters*, *40*(2), 312–315. https://doi.org/10.1002/grl.50130

7

The Sound of Hydrothermal Vents

Brendan Smith and David Barclay

ABSTRACT

The rapid flow of geothermally heated seawater through a complex subterranean hydrothermal system and the turbulent expulsion of the fluid into cold surrounding seawater at a vent are possible sources of underwater noise. Since their discovery in 1977, ambient sound-level anomalies have been documented near vent fields, spurring interest in the use of passive acoustics for monitoring of hydrothermal vent activity. This chapter summarizes the extant acoustic measurements of hydrothermal vents, including location, date, frequency range, and estimated source level. The acoustic source mechanisms anticipated at hydrothermal vents are explained theoretically, and examples of expected source levels for each mechanism are provided for realistic black smoker vent properties. Common confounding ambient sound sources are identified to assist in source discrimination. Finally, considerations for future acoustic measurements are provided, including discussion of the effects of flow noise and sensor proximity to the vent. Despite a relatively small number of measurements and inconclusive results to date with respect to monitoring of vent activity with passive acoustics, the theoretical source levels suggest this approach is feasible. Recommendations toward increasing the success of passive acoustic monitoring, such as the use of coherent processing of hydrophone arrays to aid in source identification, are provided.

7.1. INTRODUCTION

Hydrothermal vents are seafloor structures where the discharge of geothermally heated seawater occurs. They are found near seafloor spreading centers and hotspots and expel vent fluids rich in dissolved minerals that can rapidly precipitate upon cooling to form chimneys, stacks jutting up from the seabed, and dense and dynamic black or white plumes. Since the discovery of low-temperature hydrothermal fields in the Galapagos in 1977 (Corliss et al., 1979) and high-temperature black smoker vents at the East Pacific Rise in 1978 (Spiess et al., 1980), they have been of scientific interest to understand their thermal, chemical, and biological contributions to the world oceans.

Shortly after the discovery of hydrothermal vents, reports began emerging of acoustic anomalies measured by hydrophones in the vicinity of vent fields (Riedesel et al., 1982; Bibee & Jacobsen, 1986). This spurred research into possible sound-generation mechanisms by hydrothermal vents, as well as additional field studies to measure their acoustic signature (Little et al., 1990). Researchers' conclusions on the success of these studies have varied, largely due to the difficulty in separating vent signals from ambient ocean background noise.

A primary motivation for passive acoustic monitoring of hydrothermal vents has been its potential as a remote monitoring technique (Bibee & Jacobson, 1986; Little et al., 1990; Crone et al., 2006; Chadwick et al., 2008). Hot and caustic hydrothermal fluid can quickly degrade

Department of Oceanography, Dalhousie University, Halifax, Nova Scotia, Canada

Noisy Oceans: Monitoring Seismic and Acoustic Signals in the Marine Environment, Geophysical Monograph 284, First Edition.
Edited by Gaye Bayrakci and Frauke Klingelhoefer.

DOI:10.1002/9781119750925.ch07

instruments placed directly into the flow, complicating long-term direct measurements (Crone et al., 2006). Additionally, the ability of acoustic signals to propagate in the ocean with low loss may allow larger areas of the seafloor to be passively surveyed for vent detection more efficiently compared with chemical plume-detection methods (e.g., Baker et al., 2016).

At more explosive volcanic sites, passive acoustic methods have provided data regarding explosion strength and frequency, which augmented direct visual observations (Chadwick et al., 2008). The dependence of vent-related acoustic source mechanisms on parameters such as flow speed, sound speed (a proxy for temperature and fluid composition), chimney length, and cavity size should allow passive acoustics to be used to monitor many characteristics of hydrothermal vents. This long-term time series data would contribute valuable information towards geophysical and oceanographic studies of hydrothermal vent activity. Although monitoring of hydrothermal vents using passive acoustics has not yet been achieved, hydrophone arrays can better constrain source mechanisms, increasing the likelihood of successful remote monitoring by passive acoustic means.

In addition to monitoring hydrothermal vent activity, it has been proposed that understanding hydrothermal vent soundscapes may be important in a biological context (Crone et al., 2006; Lin et al., 2019; Dziak et al., 2020; Chen et al., 2021). It has been shown that some ocean species may use acoustic cues to inform their choice of settlement location (Stanley et al., 2010; Lillis et al., 2013; Eggleston et al., 2016), and hydrothermal vent populations may be colonized by species originating up to 300 km away (Mullineaux et al., 2010). Thus, sound production by hydrothermal vents may be utilized as an acoustic cue to aid settlement by some organisms (Crone et al., 2006; Chen et al., 2021).

The suggestion of vent soundscapes as critical biological settlement cues raises additional concerns regarding deep-sea mining (Lin et al., 2019; Chen et al., 2021). Industrial activities such as drilling, piling, and seismic exploration produce high noise levels that travel long distances (Hildebrand, 2009). Industrial operations extracting materials near hydrothermal vent sites will significantly increase local ambient noise levels. If vent soundscapes are important cues for biological settlement, then it is possible that deep-sea mining activities at hydrothermal vent sites may disrupt this behavior. In addition to the acoustic implications of deep-sea mining, these operations can potentially disrupt hydrothermal vent ecosystems in many ways (Van Dover et al., 2018). Benchmarking of hydrothermal vent soundscapes and continued ambient noise monitoring during mining activities are critical components of an effective environmental impact mitigation strategy (Lin et al., 2019; Chen et al., 2021).

It has been suggested that hydrothermal vent sound generation may be valuable in the search for extraterrestrial life (Dziak et al., 2020). Hydrothermal vents have been proposed as a catalyst for life on Earth (Martin et al., 2008). For this reason, hydrothermal vents on other planets or moons, such as Titan, Enceladus, and Europa, may be ideal targets for the search for life outside Earth (Dziak et al., 2020). Passive acoustic methodologies have properties that make them attractive for space exploration, such as robustness, remote sensing capability, and low cost. Hydrothermal vents may be efficiently detected from large distances on other ocean worlds by passive acoustic methods, informing subsequent sampling by direct methods to test for chemical and biological composition.

In this chapter, the theories of source generation mechanisms related to vent processes are presented and classified by their radiation efficiency. Analytical models are used to compute, tabulate, and compare nominal values of their sound source pressure level. This is followed by a comprehensive survey and comparison of relevant acoustic measurements, and power spectral density measurements are discussed in three classifications: very low frequency (<10 Hz), narrowband (containing strong tonal components at discrete frequencies), and broadband (acoustic energy distributed relatively evenly across the spectrum). The measurements and source mechanisms discussed here are predominantly below 1 kHz, as the measurements to date have identified acoustic anomalies associated with hydrothermal vents below this frequency. A discussion of potentially confounding noise sources in the ocean is presented. Finally, the challenges of making acoustic measurements at hydrothermal vent sites and analyzing and interpreting recordings of vent noise are discussed, with some recommendations for designing future field expeditions.

7.2. THEORY OF SOUND PRODUCTION BY HYDROTHERMAL VENTS

Sound production by hydrothermal vents is due to the interaction of turbulent flow with the vent structure and subsurface cavities, the hot turbulent jet discharging into cold ambient seawater, and the phase transitions of seawater driven by rapidly changing temperature and pressure (Little et al., 1990; Crone et al., 2006). The type of acoustic source mechanisms present differ between high-temperature black smokers and cooler, shallower, boiling or bubbling vent fields. At black smoker vents, unsteady turbulent flow provides a hydrodynamic noise source via interaction with the vent structure. At shallower boiling or bubbling vent sites, the formation, ringing, and collapse of bubbles can generate significant noise levels.

A key factor in the detectability of a given source mechanism is the radiation efficiency of the source, defined as the ratio of acoustic power output to the total power due to both acoustic and hydrodynamic fluid motions (Ross, 1976). Acoustic sources can be categorized by their radiation characteristics as monopoles, dipoles, or quadrupoles, with monopoles being the most efficient radiators and quadrupoles the least efficient. Monopole source radiation efficiency is also the least dependent on Mach number, M, the ratio of vent fluid flow velocity to the sound speed in the medium, whereas the radiation efficiency of quadrupole sources is dependent on M^5. As vent outflows are subsonic (e.g., $M < 1$), monopole source mechanisms tend to dominate the sound field when present. Narrowband signals generated by dipole sources may also be evident in vent acoustic spectra. Quadrupole sources are rarely significant in ocean acoustics (Ross, 1976), although amplification due to flow inhomogeneity may increase the radiation efficiency in particular cases (Morfey, 1973). All three of these source radiation types may be present in hydrothermal vent systems (Little et al., 1990).

In this section, the acoustic source mechanisms most likely to be present in hydrothermal vent systems are categorized according to their radiation pattern. The anticipated acoustic spectral characteristics (e.g., narrowband or broadband) expected to be evident in hydrophone measurements and the acoustic source location of each mechanism are given to assist with the interpretation of acoustic field measurements and identification of the most probable source mechanisms. An illustration of the type and location of the dominant hydrothermal vent acoustic source mechanisms is shown in Fig. 7.1.

The measurement distance of the hydrophone with respect to the hydrothermal vent source affects the amplitude and spectrum of the measured acoustic signal. Measurements taken in the *near field*, approximately within one acoustic wavelength, exhibit different characteristics than those taken in the *far field*, greater than one wavelength from the source. For reference, low-frequency hydrothermal vent acoustic signals between 10 and 100 Hz have wavelengths ranging from 150 to 15 m, respectively. These distances estimate the near- to far-field transition range, which is frequency dependent. Further discussion on this topic and its effect on measured acoustic signals is included in section 7.7.5.3. A table of nomenclature at the end of this chapter defines all variables used.

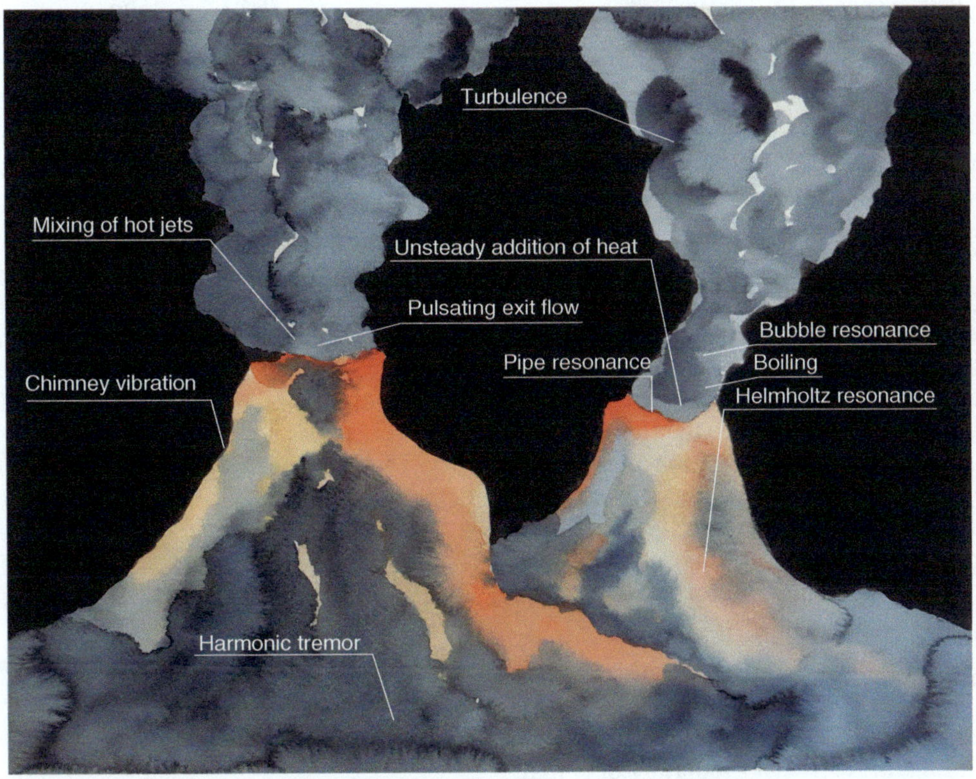

Figure 7.1 Acoustic sources at a hydrothermal vent field. Artwork by Dr. Katja Fennel.

7.2.1. Radiation Efficiency

A key factor in the detectability of an acoustic source is the radiation efficiency of the sound-generating mechanism. The radiation efficiency of a multipole depends on pole order and the size of the acoustic source relative to the wavelength. At low frequencies, where the wavelength is usually much larger than the source radius ($ka \ll 1$), the radiation efficiency is given approximately by the ratio of the specific radiation resistance σ_r and reactance σ_x (Ross 1976):

$$\sigma_r = \frac{(ka)^{2m+2}}{(2m+1)(m+1)^2(2m-1)^2} \qquad \sigma_x = \frac{ka}{(2m+1)(m+1)} \tag{7.1}$$

$$\eta_{rad} = \frac{\sigma_r}{\sigma_x} = \frac{(ka)^{2m+1}}{(m+1)(2m-1)^2} \tag{7.2}$$

where $m = 0$ for monopoles, $m = 1$ for dipoles, $m = 2$ for quadrupoles, a is the acoustic source radius, λ is the wavelength, and k is the wavenumber.

In the case of hydrodynamic noise sources where a flow speed U exists, the expression for radiation efficiency can be written in terms of angular frequency ω and Mach number M (Ross, 1976) or the ratio of flow speed to the speed of sound in the fluid, using the relationship

$$ka = \frac{\omega a}{c_w} = \left(\frac{\omega a}{U}\right)\left(\frac{U}{c_w}\right) = \left(\frac{\omega a}{U}\right)M. \tag{7.3}$$

For hydrodynamic sources, the acoustic radiation efficiency's dependence on Mach number is found by combining equations 7.2 and 7.3 to give

$$\eta_{rad} = \frac{1}{(m+1)(2m-1)^2}\left(\frac{\omega a}{U}\right)^{2m+1} M^{2m+1}. \tag{7.4}$$

For reference, typical jet exit velocities from hydrothermal vents range between approximately 0.1 and 2 m/s (Germanovich et al., 2015). Assuming a nominal sound speed in the surrounding seawater of 1,500 m/s, this gives a maximum hydrothermal vent jet Mach number of $M \approx 0.001$. The dependence of dipole and quadrupole radiation efficiency on the Mach number to the power of the pole order coefficient contributes to the dominance of monopole sound source mechanisms at hydrothermal vents.

7.2.2. Monopole

Monopole acoustic sources are caused by fluctuations in volume and are the most efficient radiators of sound. Monopole acoustic pressure amplitude decays as $1/r$ from the source, with no difference between near-field and far-field decay. Several sources of monopole sound radiation are possible at hydrothermal vents and are discussed here.

Pulsating Exit Flow

Vigorous turbulent discharge of hydrothermal fluid can lead to pulsations in exit flow, generating acoustic radiation. This can be approximated as discharge from an open-ended pipe in an infinite baffle (Little et al., 1990). At low frequencies, the root-mean-square (rms) radiated sound pressure from this source is given by (Ross, 1976)

$$p_a(r) \approx \frac{3\pi a}{16r}\overline{p_i} \tag{7.5}$$

where a is the vent orifice radius, r is the distance from the source, and $\overline{p_i}$ is the rms pressure fluctuation within the pipe.

The dynamic pressure from a hydrothermal jet is given approximately by $\frac{1}{2}\rho_h U^2$ (Kundu & Cohen, 2002). If the rms pressure fluctuation is assumed to be a fraction b_{rms} of the dynamic pressure (Little et al., 1990), then

$$\overline{p_i} = \frac{b_{rms}\rho_h U^2}{2}, \tag{7.6}$$

which yields the following expression for the pressure radiated by the pulsating exit flow:

$$p_a(r) \approx \frac{3\pi a b_{rms}\rho_h U^2}{32r}, \tag{7.7}$$

where ρ_h is the density of the hydrothermal fluid.

Spectrum: Broadband, with a spectral peak at the dominant period of exit flow fluctuation
Locus: Vent orifice

Helmholtz Resonance

Hydrothermal systems contain cavities within the vent and chimney structure, which may be excited by unsteady flow (Crone et al., 2006). Resonance within structural cavities in the seafloor or hydrothermal chimneys is analogous to Helmholtz resonators in air (Ross 1976). Helmholtz resonators consist of a spring, given by the compressibility of the relatively large fluid reservoir, and a mass, given by the fluid in the narrow chimney neck, as shown in Fig. 7.2. The

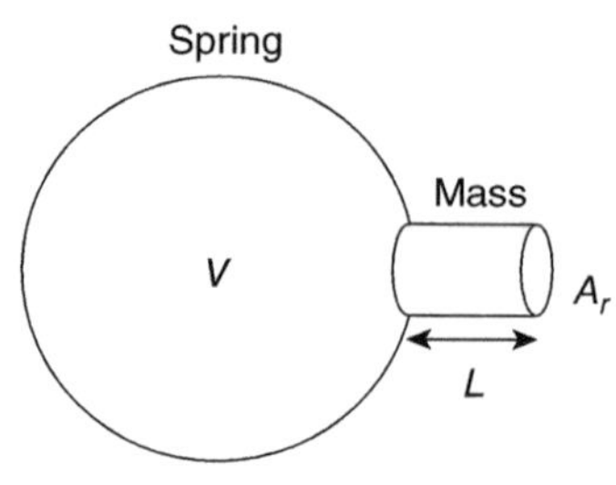

Figure 7.2 Helmholtz resonator diagram.

relative composition of hydrothermal fluid to standard seawater, as well as the temperature and chemistry of the fluid, ultimately determines the compressibility of the fluid reservoir and the mass in the chimney neck. The fluid mass in the neck oscillates against the spring reservoir at the junction between the narrow chimney neck and the much larger fluid reservoir.

Typically, cavity resonance in water is only supported through spring action provided by the walls of an enclosure due to the incompressibility of water (Ross, 1976). However, the high-temperature conditions of hydrothermal fluid can increase its compressibility relative to standard seawater (Bischoff & Rosenbauer, 1985). The frequency of subsurface cavity resonance in hydrothermal systems can be estimated using the expression for the natural frequency of a Helmholtz resonator (Crone et al., 2006):

$$f = \frac{c_h}{2\pi}\sqrt{\frac{A_r}{V\left(L + \frac{1}{2}\sqrt{\pi A_r}\right)}}, \quad (7.8)$$

where c_h is sound speed in the hydrothermal fluid, A_r is the area of the cavity opening, V is the cavity volume, and L is the length of the cavity opening neck.

The pressure amplitude from a Helmholtz resonator depends on the proximity of the unsteady turbulent flow frequency to the natural resonance frequency of the system.

Spectrum: Narrowband at the resonance frequency of the system
Locus: Resonator orifice

Harmonic Tremor

Harmonic tremor is a narrowband, low-frequency seismic and acoustic signal often observed at submarine and subaerial volcanic sites (Dziak & Fox 1992; Dziak et al., 2008). This signal can be generated by many possible periodic physical mechanisms, such as resonance of magma and/or hydrothermal fluid in cracks, cavities, and conduits in the seafloor (Chouet, 1996), the repetitive stick-slip behavior of colliding icebergs (MacAyeal et al., 2008), and "magma-wagging" caused by oscillations of magma columns in volcanoes (Jellinek & Bercovici, 2011). Harmonic tremor has been observed at hydrothermal fields using ocean-bottom hydrophones (Dziak et al., 2008). Due to the low-frequency nature of typical harmonic tremor signals (~1–10 Hz), the scale of associated resonant conduits and cavities has been estimated to be 500 m or greater (Dziak et al., 2008). This implies that harmonic tremor originates not within individual vent chimneys but rather in a larger area beneath a hydrothermal field. Alternatively, physical source mechanisms that do not require such large resonators may be responsible for harmonic tremor observed at hydrothermally active sites.

Harmonic tremor has been detected as far as 14,000 km from volcanic sources due to minimal attenuation during propagation at low frequencies, particularly in the ocean sound channel (Dziak & Fox, 1992). It is not clear if tremor produced by hydrothermal circulation would be detectable at similarly large distances. However, due to the long-range detection potential, harmonic tremor may provide an efficient means of detecting possible hydrothermal vent sites. It should be noted that this signal can also occur without the presence of hydrothermal vents due to other sources, as mentioned earlier.

Spectrum: Narrowband, low-frequency with higher harmonics
Locus: Cracks, cavities, and conduits in the seafloor (for hydrothermal sources)

Pipe Resonance

Hydrothermal vent chimneys can be approximated as cylindrical pipes. Resonance due to standing waves along the length of the chimney can be modeled as pipe resonance (Crone et al., 2006), like that found in many musical wind instruments. The frequency of this source is dependent on the sound speed in the fluid, the length of the chimney, and whether the pipe (chimney) can be considered open on both ends or open on just one end.

The resonance frequency and higher harmonics for pipes open at both ends or closed at one end are given by Fletcher and Rossing (1991) for an

open pipe:

(half-wave resonator) $$f = \frac{mc_h}{2H}, \quad (7.9)$$

and a pipe closed at one end:

(quarter-wave resonator) $$f = \frac{(2m-1)c_h}{4H}, \quad (7.10)$$

where m is mode number, c_h is sound speed in the hydrothermal fluid, and H is the chimney height.

It should be noted that the calculated resonance frequencies are strongly dependent on the assumed sound speed in the hydrothermal fluid. As with Helmholtz resonance, the pressure amplitude for pipe resonance depends on the similarity between the excitation frequency due to unsteady flow and the natural frequency of the pipe.

Spectrum: Narrowband at pipe fundamental frequency with higher-order harmonics
Locus: Vent or other pipe structure orifice

Bubble Resonance

In hydrothermal fields with visible gas bubbles, bubble resonance can be a significant source of monopole sound generation due to the volume oscillation of the individual bubbles upon creation or when excited by the pressure fluctuations in turbulent flow. The fundamental frequency radiated by each bubble is inversely proportional to its size and is given by (Minnaert, 1933)

$$f = \frac{1}{2\pi a}\sqrt{\frac{3p_w\gamma}{\rho_w}} \tag{7.11}$$

where a is bubble (acoustic source) radius, p_w is ambient pressure, γ is the ratio of gas specific heat at constant pressure and volume, and ρ_w is ambient water density. Additional harmonic and nonlinear resonance modes are observed to radiate efficiently (Longuet-Higgins, 1989, 1990).

The release of subsurface gases can sometimes occur in bubble streams with relatively constant bubble radius (Longo et al., 2021). In these cases, the bubble resonance signals are confined to discrete frequencies. Due to the dependence of bubble resonance frequency on ambient pressure, tidal fluctuations may be evident in acoustic measurements of bubbly hydrothermal fields, as frequency fluctuations correlate to changes in static pressure (Longo et al., 2021).

Spectrum: Narrowband at individual bubble breathing modes. Broadband signals, including low-frequency (tens of Hz) components, are created by cascading bubble fragmentation in turbulent flows (Deane, 2010).
Locus: Site of bubble formation until bubble ringing decays (near vent exit).

Boiling

The presence of boiling at hydrothermal vent sites is strongly dependent on depth. Most deep-sea hydrothermal vent sites are unlikely to exhibit boiling due to large hydrostatic pressures; however, at shallower sites, boiling may be a significant source of sound (Little et al., 1990). Boiling has been observed at shallow hydrothermal vent sites (Hannington et al., 2001), and subsurface boiling has been observed seismoacoustically (Vandemeulebrouck, 2010). Boiling results in bubble growth, condensation, and collapse of many different-sized bubbles dependent on local temperature differences between the ambient fluid and the heat source, as well as the presence of nucleation sites for bubble formation (Tang et al., 2018). Bubble resonance (as discussed in the previous section), as well as bubble growth, fractionation, and collapse, contribute to the sound generated by boiling. The diversity of bubble sizes results in a broadband acoustic spectrum (Aoki, 1970; Tang et al., 2018), in contrast to monochromatic signals observed in subsurface bubble streams discussed in the previous section.

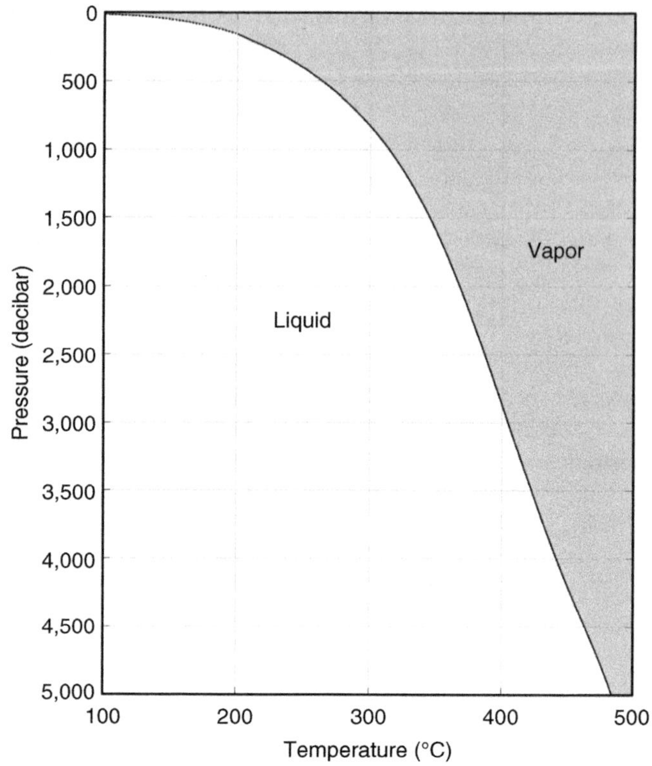

Figure 7.3 Two-phase boundary of seawater for <200 °C (dashed) (Wagner & Pruß, 2002; Sharqawy et al., 2010) and >200 °C (solid) (Bischoff & Rosenbauer, 1984).

The empirically derived two-phase boundary for seawater at temperatures between 100–500 °C as a function of pressure is shown in Fig. 7.3. The data below 200 °C is derived from the vapor-pressure equation for ordinary water given by Wagner and Pruß (2002), modified for the boiling point elevation due to the salinity of seawater (Sharqawy et al., 2010). The data above 200 °C is given by the polynomial from Bischoff and Rosenbauer (1984) for the two-phase boundary of seawater. For example, at 2,200 m depth, the boiling point of seawater is ~375 °C, while observed vent temperatures at the Main Endeavour Vent Field were 355–360 °C (Crone et al., 2006). However, from the curve, it is apparent that shallower hydrothermal sites may indeed boil. When boiling is present, such as at vents in the Okinawa trough (Chen et al., 2021), monopole radiation due to bubble formation and collapse may be a significant noise source.

Spectrum: Broadband (Aoki, 1970; Tang et al., 2018)
Locus: Site of bubble formation and collapse (near vent exit)

Cavitation

Cavitation is rapid bubble formation and subsequent collapse caused by local reductions in pressure, which can result if fluid is forced to accelerate relative to a fluid–fluid or fluid–solid interface (Plesset, 1949). Although the sound source for boiling and cavitation is similar, bubble formation in cavitation is due to a reduction in pressure rather than an increase in temperature as in boiling. Cavitation can produce high levels of noise due to monopole radiation from the generated bubbles (Ross, 1976). It has been suggested that cavitation could occur at some vent sites where cavitation nuclei are available, potentially due to the remnants of boiling, and where constriction of flow within the vent chimney causes a reduction in pressure below the vapor pressure as the velocity increases (White, 2000). The subsequent increase in pressure after discharge beyond the constriction could then lead to the implosion of the vapor cavities. Due to the requirement for pressure to drop below the vapor pressure, this effect is more likely to be observed at shallow vent sites with lower vapor pressure (Little et al., 1990). The associated acoustic spectrum is broadband, with a peak frequency associated with the most prevalent bubble decay time (Ross, 1976).

Spectrum: Broadband, with a peak frequency at the dominant bubble decay time
Locus: Site of bubble formation and collapse

Unsteady Addition of Heat

Along with acceleration in the fluid and unsteady boundaries, Ffowcs Williams (1981) demonstrated that the unsteady addition of heat can be a significant source of monopole sound. The sound power W_{ac} generated by this source is dependent on the heat power Q, the angular frequency ω, and the ratio of specific heats at constant pressure to constant volume and is given by

$$W_{ac} = \frac{(\gamma - 1)^2}{4\pi} \frac{\omega^2}{c_w^2} \frac{Q^2}{\rho_w c_w^3}. \tag{7.12}$$

At typical ocean temperatures, the ratio of specific heats γ is approximately unity, and in most cases, the heat input power is low, resulting in very weak acoustic power from this source. However, at hydrothermal vents, the ratio of specific heats can differ strongly from unity. For example, for pure water at 336 °C and 2,179 m depth, as at the Dante vent in the Main Endeavour Vent Field (Germanovich et al., 2015), the ratio of specific heats is ~2.1 (Lemmon et al., 2010).

Assuming spherical spreading, the sound power can be converted to a sound pressure at a distance r using

$$p(r) = \sqrt{\frac{W_{ac}\rho_w c_w}{4\pi r^2}} \tag{7.13}$$

Spectrum: Broadband
Locus: Vent orifice

7.2.3. Dipole

In hydrothermal vents, dipole acoustic sources are caused by oscillating forces acting on rigid surfaces, resulting in flexural waves in the structure. Following monopoles, dipoles are the second most efficient radiators of sound with acoustic pressure amplitude that decays as $1/r^2$ in the near field ($r < \lambda$) and as $1/r$ in the far field ($r > \lambda$).

Chimney Vibrations

Vibration of the hydrothermal sulfide chimneys can act as a dipole acoustic source. The maximum acoustic pressure amplitude for a dipole source is given by (Ross, 1976)

$$p_a(r) = \frac{2\pi i f F}{4\pi r c_w}, \tag{7.14}$$

where F is the magnitude of the fluctuating force.

The magnitude of the fluctuating force F was approximated by Little et al. (1990) as a fraction b_{rms} of the dynamic pressure due to fluid flow acting on the inside wall of the vent chimney of area

$$A_c = \pi d_i H, \tag{7.15}$$

where d_i is the inner diameter of the chimney.

The fluctuating force due to fluid flow is then given by

$$F \approx b_{rms} \frac{\rho_h U^2}{2} A_c. \tag{7.16}$$

When the diameter of the hydrothermal discharge pathway is much smaller than the outer chimney diameter, the resonance frequency f for a hydrothermal vent sulfide chimney can be approximated by that of a uniform circular beam (Little et al., 1990). The resonance frequency for a beam that is clamped at one end and free at the other is given by (Ross, 1976)

$$f = \frac{(2m-1)^2 \pi d_o c_b}{8\sqrt{8}H^2\sqrt{1 + \frac{\rho_w}{\rho_b}}} \tag{7.17}$$

where m is the mode number, d_o is the outer beam diameter, c_b is the speed of sound in the beam (chimney), ρ_w is the ambient mean fluid density (water), and ρ_b is the beam (chimney) density.

The fluctuating force magnitude F and resonance frequency f can then be substituted into equation 7.14 to estimate the acoustic pressure amplitude for a chimney vibration source. The magnitude of this source ultimately depends on the similarity between the turbulent forcing frequency and the resonance frequency of the chimney.

Sulfide chimney structures commonly taper from a wider base to a narrow peak, and the complex internal structure influences damping and stiffness, ultimately affecting the source level and resonance frequency of chimney vibrations.

Spectrum: Narrowband, fundamental frequency, and higher-order harmonics
Locus: Sulfide chimney structure

Flow Inhomogeneity Amplification

Hot jets with low Mach numbers have been shown to produce more noise than cold jets of equivalent velocity (Ffowcs Williams, 1977). Morfey (1973) showed that inhomogeneity due to nonuniform density of turbulent flow can lead to the amplification of quadrupole jet noise in low-Mach-number jets, resulting in more efficient dipole radiation. In hydrothermal vents, this is relevant due to the low Mach number ($M \approx 0.001$) and the drastic differences in temperature and composition between hot hydrothermal fluid (often 300 °C+) and cold seawater in the deep ocean (Little et al., 1990). The amplification factor is given as a ratio of the magnitudes of the dipole pressure field p_d to the quadrupole pressure field p_q (Morfey, 1973):

$$\frac{p_d}{p_q} = \frac{c_w \rho_w \Delta V}{U} \tag{7.18}$$

where ΔV is the specific volume difference between the jet and the ambient fluid.

Spectrum: Broadband
Locus: Jet mixing region

Mixing of Hot Jets

The mixing of hot hydrothermal jets and their entrainment of surrounding cold seawater can lead to dipole sound radiation (Ffowcs Williams, 1981). The magnitude of this sound is dependent on the jet velocity U, the length scale of the jet source l, the source frequency ω, and the strength of the density inhomogeneity $\Delta\rho$ between the hydrothermal fluid and surrounding seawater. The sound power of this source is given by (Ffowcs Williams, 1981)

$$W_{ac} = \frac{\omega^4 l^4}{c_w^4} \frac{(\Delta\rho)^2}{\rho_w^2} \frac{U^2}{c_w^2} \rho_w c_w^3 l^2. \tag{7.19}$$

This sound power can be converted to sound pressure at a specified distance using equation 7.13. The pure jet region of a hydrothermal vent plume is generally <0.5 m (Crone et al., 2008), and this may be used as an approximate length scale to estimate the sound power of this source.

Spectrum: Broadband
Locus: Jet mixing region

7.2.4. Quadrupole

Quadrupole acoustic sources are caused by free turbulence and are the least efficient radiators of sound. Their radiation efficiency is also the most dependent on Mach number, and the acoustic pressure amplitude decays as $1/r^3$ in the near field ($r < \lambda$) and as $1/r$ in the far field ($r > \lambda$).

Low-Mach-Number Turbulent Jets

Turbulent jets generate sound from point quadrupole sources distributed in the turbulent flow (Lighthill, 1954). The strong dependence of quadrupole radiation on Mach number results in much more efficient sound radiation from high-velocity jets. However, subsonic jets have been shown to produce detectable noise (Lush, 1971). Subsonic jets produce a broadband acoustic spectrum with a peak frequency determined by the velocity of the jet and the diameter of the orifice (Lighthill, 1952, 1954; Lush, 1971). The approximate pressure magnitude for a turbulent jet in the near and far fields can be approximated by the following expressions (Little et al., 1990):

$$p_a = b_{rms} \rho_h U^2 \quad \text{for } r < D,$$

$$p_a(r) = b_{rms} \rho_h U^2 \left(\frac{d}{r}\right)^3 \quad \text{for } r < \lambda, \text{ and} \tag{7.20}$$

$$p_a(r) = b_{rms} \rho_h U^2 \left(\frac{d}{\lambda}\right)^2 \frac{d}{r} \quad \text{for } r > \lambda,$$

where d is the vent orifice diameter.

The peak frequency of the broadband spectrum is between $f_p \approx 0.2U/d$ (Fahy & Walker, 1998) and $f_p \approx 0.8U/d$ (Lush, 1971). The spectrum rolls off at 6–9 dB per octave below this peak frequency and 2–3 dB per octave above the peak frequency (Lush, 1971; Little et al., 1990).

Spectrum: Far-field spectrum is broadband, with a peak frequency at f_p. The near-field spectrum depends on proximity to the two source mechanisms mentioned below.
Locus: Higher frequencies ($>f_p$) produced by mixing region near the jet orifice.
Lower frequencies ($<f_p$) produced by fully developed turbulence downstream.

7.2.5. Estimated Source Sound Pressure Levels

Source sound pressure levels (at 1 m) for many of the generation mechanisms described have been estimated based on the nominal vent parameter values shown in Table 7.1. Many values used here represent a vent at the Grotto vent cluster at the Main Endeavour Vent Field, including measurements by Germanovich et al. (2015)

Table 7.1 Vent parameter values for sound pressure level estimation.

Vent parameter	Value
Depth, z	[a]2,188 m
Fluid velocity, U	[a]1.34 m/s
Temperature, T	[a]320 °C
Hydrothermal fluid density, ρ_h	[a]650 kg/m^3
Seawater density, ρ_w	1,000 kg/m^3
Orifice diameter, d	[a]0.079 m
Chimney length, L	[b]10 m
Ratio of rms pressure oscillation to dynamic pressure, b_{rms}	[c]0.1
Sound speed, c_w (ambient seawater)	[c]1,500 m/s
Sound speed, c_h (hydrothermal fluid)	[c]1,500 m/s
Sulfide chimney sound speed (solid), c_b	[c]3,000 m/s
Sulfide chimney density, ρ_b	[c]4,000 kg/m^3
Sulfide chimney outer diameter, d_o	[c]1 m
Length scale of hot jet mixing source, l	[d]0.25 m
Frequency for unsteady addition of heat and mixing of hot jets, f	10 Hz
Heat output, Q	[a]8,400 W
Ratio of specific heat at constant pressure and volume for vent fluid, γ	1.9

[a] (Germanovich et al., 2015).
[b] (Lelièvre et al., 2017).
[c] (Little et al., 1990).
[d] (Crone et al., 2008).

and Lelièvre et al. (2017). The sound speed in hydrothermal fluid may differ substantially from the nominal value used here, as discussed in Section 7.7.5.2. The value of the ratio of the rms pressure oscillation in the vent fluid jet to the total dynamic pressure, b_{rms}, is also not well constrained. Here it is assumed to be 10% of the total dynamic pressure, following the approach by Little et al. (1990).

These nominal values are used with the equations in the prior sections to estimate acoustic pressure in Pa and sound pressure level in dB re 1 μPa at a 1 m distance from the source, shown in Table 7.2, for comparison with noise measurements near hydrothermal vents presented in Section 7.7.3. The sound pressure level is calculated as $20\log_{10}\left(\frac{p_a}{p_{ref}}\right)$, where $p_{ref} = 1 \times 10^{-6}$ Pa. Extrapolations to other measurement distances can be made using geometrical spreading or other more sophisticated sound propagation models.

7.2.6. Estimated Source Spectra

The measured acoustic spectra associated with the source mechanisms discussed above will heavily depend on mean flow velocity, structure size and stiffness, and properties of the hydrothermal fluid (Little et al., 1990). Table 7.3 summarizes the expected spectral characteristics

Table 7.2 Estimated hydrothermal vent acoustic source levels based on nominal vent parameters.

Source type	Radiation pattern	Estimated acoustic pressure at 1 m (Pa)	Estimated sound pressure level at 1 m (dB re 1μPa)
Pulsating exit flow	Monopole	1.36	123
Unsteady addition of heat	Monopole	3.6	131
Chimney vibrations	Dipole	0.19	105
Mixing of hot jets	Dipole	5.4	135
Turbulent jet, near field	Quadrupole	5.8×10^{-3}	75
Turbulent jet, far field	Quadrupole	4.7×10^{-7}	<0
Turbulent jet with flow inhomogeneity amplification, far field	Dipole	1.8×10^{-4}	45

Table 7.3 Spectral characteristics of hydrothermal vent acoustic source mechanisms.

		Spectrum		
Radiation pattern	Source mechanism	Very low frequency (<10 Hz)	Narrowband	Broadband
Monopole	Pulsating exit flow			✓
	Helmholtz resonance	✓	✓	
	Harmonic tremor	✓	✓	
	Pipe resonance	✓	✓	
	Bubble resonance		✓	
	Boiling			✓
	Cavitation			✓
	Unsteady addition of heat			✓
Dipole	Chimney vibrations	✓	✓	
	Turbulence with flow inhomogeneity amplification			✓
	Mixing of hot jets			✓
Quadrupole	Low-Mach-number turbulent jet			✓

of each source type categorized into very low frequency (below 10 Hz), narrowband, and broadband signals. These spectral characteristics will be used to group field measurements of hydrothermal vent noise in the following section.

7.3. SURVEY OF ACOUSTIC MEASUREMENTS

Underwater acoustic measurements of hydrothermal vents have been documented since their discovery, with the first acoustic anomalies reported near a vent field during seismic studies on the East Pacific Rise in 1980 (Riedesel et al., 1982). Similar reports have continued through today, although in most, there remains uncertainty regarding the attribution of acoustic signatures to or a sound-generating mechanism associated with hydrothermal vent activity. A primary difficulty in identifying acoustic signals from hydrothermal vents is distinguishing them from other sources of ocean ambient noise (Little et al., 1990). For this reason, it is useful to present extant hydrothermal vent noise measurements in the context of the limits of prevailing ocean noise (Wenz, 1962). Additional low-frequency reference spectra of noise processes in the deep ocean, where many hydrothermal vent mechanisms are most prominent, are also shown. These include the Holu spectrum due to saturation of short-wavelength wind-driven surface waves (McCreery et al., 1993), a low-frequency flow noise spectrum due to turbulent flow over the hydrophone body following an $f^{-5/3}$ spectral slope (Basset et al., 2014), and the spectrum due to microseisms (Urick, 1986).

Power spectral densities of all known hydrothermal vent acoustic measurements are presented together with these reference ocean ambient noise levels in Fig. 7.4. New measurements from the Main Endeavour Vent Field (MEF), including a six-month average from 2019 utilizing Ocean Networks Canada's Neptune cabled observatory and a recent short-term deployment with the Barclay Noise Lab's Deep Acoustic Lander (DAL) hydrophone array in 2021, are reported.

The variability in sound level and spectrum highlights the differences in measurement conditions and the challenges in recording at hydrothermal vent environments.

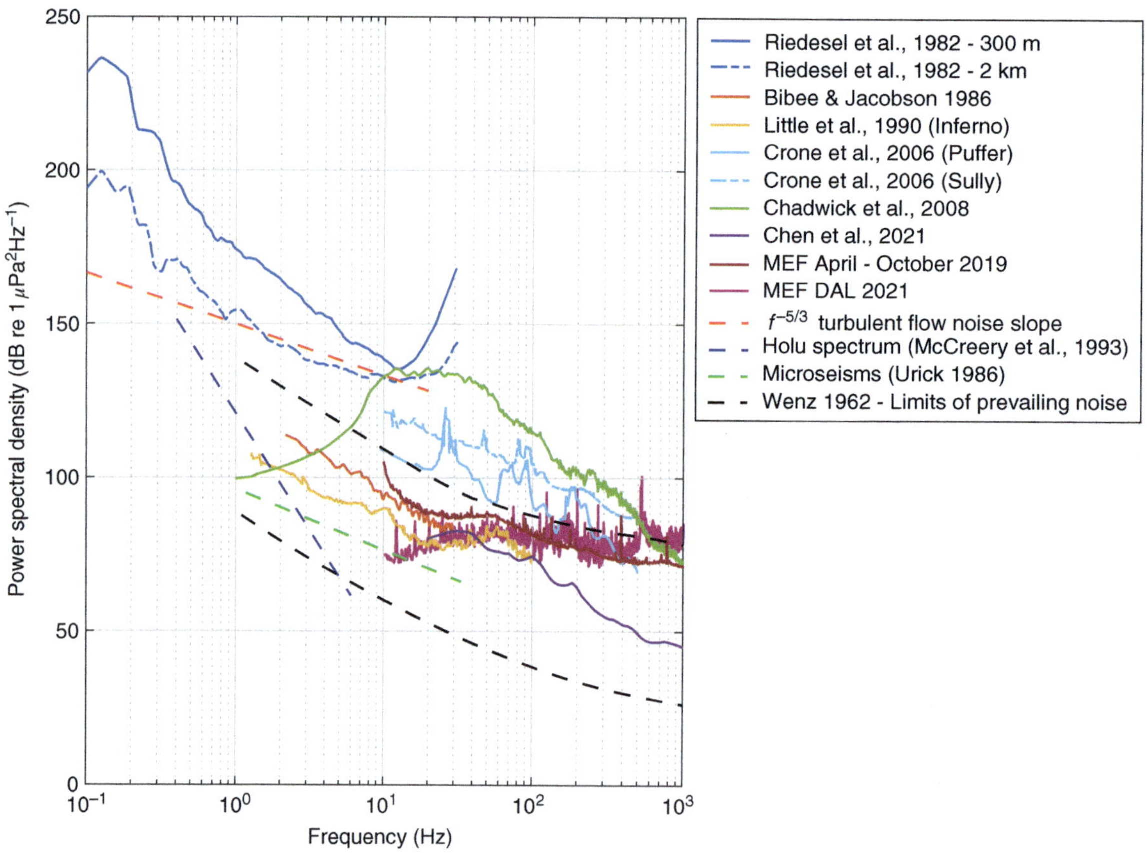

Figure 7.4 Hydrothermal vent noise measurements (solid lines) with reference spectra (dashed) for the Wenz limits of natural processes and distant ship noise, surface and shore breaking wind waves, microseisms, and a reference slope for turbulent flow noise. The median is taken from measurements of Chen et al. (2021).

Although some variation can be attributed to site-specific acoustic source mechanisms, the high sound levels reported by Riedesel et al. (1982) may be influenced by flow noise over the hydrophone body. Many measurements exhibit narrowband and broadband peaks below 100 Hz, which are expected to be characteristic of hydrothermal vents.

Acoustic signals attributed to hydrothermal vents can be grouped into three categories based on their frequency range and bandwidth (see Table 7.3):

- *Very low frequency* (<10 *Hz*): Harmonic tremor and low-frequency pressure pulses with periods ranging from several seconds to minutes
- *Narrowband:* Resonant sources such as chimney structural vibration, bubble ringing, Helmholtz cavity resonance, and standing wave resonance
- *Broadband:* Mixing, boiling and cavitation, pulsating exit flow, and subsonic jet noise, potentially amplified by flow inhomogeneity due to heterogenous vent fluid composition and temperature differences between hot vent fluid and cold seawater

The extant literature regarding observations of hydrothermal vent noise will be organized in accordance with these three categories. Potential acoustic source mechanisms introduced in Section 7.7.2 will be associated with the observations where possible.

7.3.1. Very Low Frequency (<10 Hz)

Studies have documented anomalous low-frequency signals below 10 Hz detected by hydrophones near hydrothermally active sites. These range from extremely low-frequency pressure fluctuations, with periods on the order of minutes (Kasahara et al., 1995; Kasahara & Sato, 2001; Fujioka & Mitsuzawa, 2011), to broadband increases in low-frequency energy between 0.1 and 10 Hz near black smoker hydrothermal vent fields (Riedesel et al., 1982).

The first documented evidence of noise associated with hydrothermal vents was reported by Riedesel et al. (1982) at 21°N on the East Pacific Rise. Hydrophones deployed as part of a seismic study at varying distances from black smoker hydrothermal vent fields were found to have dramatically different automated gain control settings upon retrieval. The hydrophone 300 m from a hydrothermal vent had adjusted to have 16–64 times less gain than the hydrophone 2 km away, translating to a decrease in acoustic noise of 12–18 dB over the 1,700 m that separated the two receivers. Spectral analysis indicated this difference in level was driven by a broadband source with the bulk of the radiated energy between 0.1 and 10 Hz, shown in Fig. 7.4. Reported power levels are very high relative to other observations and ocean noise processes, exceeding 200 dB re 1μPa^2/Hz for the measurements 300 m from the vent field. Seismometers at the same locations did not record corresponding differences in ground motion, suggesting that the signal originated within the water column rather than the seabed and may be associated with hydrothermal vent discharge. The high noise levels and spectral slope reported from these measurements indicate possible contamination with flow noise, although received levels at 10 Hz are equal to those made in a tidal channel with a 2.5 m/s mean flow speed (Auvinen, 2019), an unlikely scenario at this deep ocean site. If these signals are caused by hydrothermal vents, the most likely source mechanisms are pulsating exit flow, the unsteady addition of heat, or the mixing of hot hydrothermal jets, all monopole sources, due to the high noise levels and broadband increase in low-frequency acoustic energy.

An additional unusual signal from the study by Riedesel et al. (1982) was discussed by Little et al. (1990). Large-amplitude, low-frequency signals at ~0.16 Hz (~6 second period) were observed on the hydrophone 300 m from the vents but not on the hydrophone 2 km away. Although similar signals have been observed from microseisms due to nonlinear surface wave interaction (Kedar et al., 2008), the presence of this signal on only one of the hydrophones suggests it is due to a source near the vent field (Little et al., 1990). Similar long-period signals were measured at other vent sites, including the Southern Mariana Trough, Okinawa Trough, and Trans-Atlantic Geotraverse (TAG) hydrothermal field, where low-frequency "pressure pulses" were measured on hydrophones near the vent fields (Fujioka & Mitsuzawa, 2001; Kasahara & Sato, 2001). The periods of these signals were on the order of minutes. The authors attribute these pressure pulses to upwelling of hydrothermal fluid. The intensity of the pressure pulses was correlated with tidal fluctuations in areal strain, which was hypothesized to increase seawater input to the subsurface hydrothermal system. The theoretical mechanisms for this observation have not been discussed in this chapter. However, if this hypothesis is correct, then the same phenomenon may be responsible for the low-frequency signals observed by Riedesel et al. (1982).

Harmonic tremor has also been documented near hydrothermally active sites (Dziak et al., 2008). This is well-known in the context of seismic signals measured near subaerial and submarine volcanoes and is defined as a "continuous seismic signal with regular or irregular sine wave appearance and characterized by low frequencies (0.5–5 Hz). Harmonic tremor has very uniform appearance, whereas spasmodic tremor is pulsating and consists of higher frequencies with a more irregular appearance" (McNutt & Roman, 2015). Dziak et al. (2008) detected harmonic tremor signals on a hydrophone near Brothers volcano 350 km northwest of New Zealand. The tremor had a fundamental frequency of ~3 Hz, which varied by 2–3 Hz, an effect known as *gliding*. Harmonics spaced at 3 Hz intervals were apparent up to ~60 Hz. This was attributed to resonant seismic waves originating in a fluid-filled crack or conduit, where pressure fluctuations associated with unsteady flow provided excitation of the resonant system. The authors analyzed the hydrophone signals for tidal variability but did not find evidence that the tides influenced this signal. However, other authors have documented seismic tremor detected on ocean bottom seismometers, which were strongly modulated by the tides, with the source of the tremor postulated to be due to hydrothermal circulation (Meier & Schlindwein, 2018).

7.3.2. Narrowband

Narrowband sources are expected at hydrothermal sites, particularly at black smoker vents where the tubular chimney structures can lead to standing wave resonance and chimney vibration (Little et al., 1990; Crone et al., 2006). Cavities below the seafloor and within the hydrothermal vent chimneys can also lead to Helmholtz-type resonance with the system's spring provided either by flexion of the cavity walls or compressibility of the hydrothermal fluid (Ross, 1976; Crone et al., 2006).

Riedesel et al. (1982) reported unusual monochromatic narrowband signals at ~20 Hz at the East Pacific Rise with durations of 15–20 seconds. Analysis of seafloor particle motions suggested that this source originated in the seabed, not the water column. The authors suggested subsurface cavity resonance as a potential source, which could be excited by the movement of hydrothermal fluid below the seafloor.

Narrowband and broadband signals from 15–30 Hz were observed near black smoker hydrothermal vents between 10° and 12° N on the East Pacific Rise (Little et al., 1990). Power level measurements within several meters of the vents were up to 10 times louder than measurements >1 km from the vents.

Narrowband signals were also detected by hydrophones within 0.3 m of black smoker hydrothermal vents "Puffer" and "Sully" at the Main Endeavour Vent Field (Crone et al., 2006). These signals, which range in frequency from 10–250 Hz and are 10–20 dB above ambient noise levels, were attributed to either Helmholtz or quarter-wave resonance. The most prominent narrowband signals were observed at Puffer with frequencies of ~27, ~67, and ~92 Hz. Tonal signals at Sully were similar, with peaks at ~48, ~80, and ~91 Hz, as shown in Fig. 7.4. Based on typical vent dimensions and an assumed sound speed of 450 m/s, the authors predicted the existence of resonance frequencies at 120 Hz (Helmholtz) and 113 Hz (quarter-wave), and the approximate alignment with those observed suggests that cavity and standing wave resonance generated the narrowband signals at this site.

Further examples of narrowband signals were reported at the Suiyo Seamount near Japan (Chen et al., 2021). Tonal signals were reported at 30–50, 70–80, and 100–110 Hz, with power spectral densities 6–10 dB above ambient noise levels. These signals were evident on a hydrophone immediately adjacent to a visibly bubbling vent orifice but not at hydrophone positions more distant from the vent. A photograph showing the hydrophone deployed near the vent shows distortion and refraction of light due to hot vent fluid, but bubbles are not visible. If gas bubbles were present at this site, these tonal signals could be due to monopole radiation from bubble volume fluctuations. In the absence of bubbles, monopole Helmholtz resonance from subsurface cavities is the most likely source of narrowband signals. Dipole sources due to chimney vibrations are unlikely, given the apparent absence of sulfide chimney structures at this site, but subsurface conduits could support standing wave resonance.

Bubbles are strong sources of monopole narrowband sound radiation when present. Shallower hydrothermal vent sites, or sites with significant gas release, may release streams of bubbles in addition to hot hydrothermal fluid. These bubbles can "ring" at frequencies dependent on their size, composition, and ambient pressure (Minnaert, 1933). Evidence of bubble resonance at hydrothermal sites was observed by Longo et al. (2021) at the Panarea shallow hydrothermal field in the Mediterranean Sea from 100 to 2,000 Hz. Tidal variations in bubble resonance frequency were observed due to the corresponding changes in ambient pressure. Detrended power spectral densities over a 1.5-year period showed an increase in sound level due to bubbling, which was correlated with an increase in volcanic activity at the nearby subaerial Stromboli volcano. This correlation suggests a subsurface link between the magma reservoirs at Panarea and Stromboli, allowing passive acoustic monitoring of hydrothermal vents to offer a volcanic detection system in this case.

7.3.3. Broadband

Pulsating exit flow, boiling, cavitation, mixing, and subsonic jet noise are sources of broadband sound production at hydrothermal vents. Although the character of this noise has a wider bandwidth than the narrowband sources discussed previously, some sources exhibit a peak frequency. For example, jet noise typically exhibits a peak frequency dependent on the jet velocity and orifice diameter (Little et al., 1990; Fahy & Walker, 1998), and the peak frequency of cavitation depends on the dominant bubble decay time (Ross, 1976). Broadband noise level anomalies near hydrothermal vent fields have been reported in multiple studies.

An increase in broadband low-frequency noise levels from 2 to 30 Hz was observed within 400 m of a low-temperature vent field in the caldera of Axial Seamount on the Juan de Fuca Ridge (Bibee & Jacobson, 1986). Power spectral densities in this frequency range were increased up to 25 dB above ambient noise levels measured at other sites in the vicinity. The same study did not observe increased noise levels at a hydrophone near a black smoker hydrothermal vent in the caldera. Low-temperature vent fields are unlikely to produce detectable noise through the source mechanisms described above, so it is possible that this signal was caused by turbulent boundary layer pressure fluctuations due to flow over the hydrophone body, known as *flow noise*, or another high-temperature black smoker vent in the area.

Additional acoustic measurements at Ashes Vent Field, Axial Seamount, indicated possible jet noise amplified due to flow inhomogeneities, with a peak frequency of 40 Hz (Little et al., 1990). The measured power of this signal was approximately 1×10^{-4} Pa^2/Hz, which was deemed too low to be produced by monopole or dipole source mechanisms (Little et al., 1990). However, the power level also exceeds the maximum quadrupole radiation noise level expected for a cold jet, leading to the conclusion that the temperature contrast between the hot hydrothermal fluid and cold ambient seawater was leading to flow inhomogeneity amplification (Little et al., 1990; Morfey, 1973).

The measurements by Crone et al. (2006) within 0.3 m of vent orifices at the Main Endeavour Vent Field show evidence of broadband acoustic sources 10–30 dB above ambient background noise levels from 1–500 Hz. Based on the observed pressure levels, these are attributed to pulsating exit flow and turbulent jet noise amplified by fluid heterogeneity, which aligns with the observations by Little et al. at Axial Seamount (1990). The proximity of the hydrophones to the vent orifice and the similarity of spectral slope to an $f^{-5/3}$ flow-noise slope (see Fig. 7.4) suggests that at frequencies below ~20 Hz, the measurements are dominated by flow noise.

Explosive volcanic and hydrothermal acoustic signals were measured at NW Rota-1 volcano on the Mariana arc (Chadwick et al., 2008). Broadband acoustic signals correlated with video evidence of explosive bursts were measured up to 50 dB above background noise levels. The frequency range of the explosion signals was between 10 and 1,500 Hz, with most energy concentrated between 10 and 50 Hz. Despite being broadband, the explosions typically exhibited a spectral peak near 12 Hz, which correlated visually to the plume oscillation frequency and visible formation/condensation of steam at the seawater-magma interface. The high noise levels and formation of steam suggest that monopole boiling noise sources due to bubble formation and collapse were occurring during the eruption. Due to the vigorous and explosive nature of this event, these signals represent an upper limit for sound levels expected at less explosive, steady, vigorous black smoker hydrothermal vent sites.

Broadband pulses below 120 Hz were measured at the Suiyo Seamount in August, 2019 (Chen et al., 2021). The pulses were approximately 30 dB above ambient background noise levels. These appeared in bursts with many pulses spaced closely together, followed by hours without pulse activity. The source of these pulses is unclear, although they may be due to explosive or seismic sources.

7.3.4. Tidal Variability

Multiple studies have shown tidal variability of sound levels measured near hydrothermal vents (Kasahara et al., 1995; Kasahara & Sato, 2001; Fujioka & Mitsuzawa, 2001; Crone et al., 2006; Longo et al., 2021). This acoustic variability aligns with observations of flow rate and temperature variability from hydrothermal vents, which have been correlated with tidal hydrostatic pressure changes (Barreyre & Sohn, 2016; Xu & Di Iorio, 2012; Larson et al., 2007; Crone & Wilcock, 2005; Crone et al., 2006). Many of the acoustic source mechanisms at hydrothermal vents are sensitive to both flow rate and temperature (Little et al., 1990). For example, the sound produced by turbulent jets is dependent on the jet velocity, the excitation of resonant sources is dependent on the proximity of the turbulent forcing frequency to the resonance frequency, and the resonance frequency of Helmholtz and quarter-wavelength resonators is dependent on the speed of sound in the hydrothermal fluid, which varies with temperature, pressure, and ratio of liquid/gas composition (Greenspan & Tschiegg, 1957, 1959; Bischoff & Rosenbauer, 1985; Kumagi & Chouet, 2000). For these reasons, tidal variability of sound production by hydrothermal vents is anticipated to be evident in multiday hydrothermal vent acoustic measurements.

7.3.5. Summary of Acoustic Measurements

Table 7.4 summarizes all extant hydrothermal vent acoustic observations, including the presence or absence of tidal modulation, frequency range of the observed anomalous signals, and an estimated source level where enough information was available to back-calculate to 1 m using assumed spherical spreading loss.

Review of these measurements suggests that hydrothermal vents are predominantly low-frequency acoustic sources. Most field studies have indicated anomalies below 500 Hz, with many studies showing very low-frequency fluctuations below 10 Hz. The estimated source levels suggest that long-range detection of hydrothermal vent signals may be feasible depending on the propagation environment and ambient noise conditions. The low-frequency nature of hydrothermal vent noise production

Table 7.4 Summary of hydrothermal vent acoustic measurements and their properties.

Source	Location	Measurement date	Description	Tidal modulation	Frequency [Hz]	Measurement distance [m][a]	Sound pressure level (dB re 1μPa)[b]	Estimated source level (dB re 1μPa at 1 m)[c]
Bibee and Jacobsen, 1986	Axial Seamount	July 1985	Low-frequency noise levels near low-temperature vent field elevated up to 25 dB relative to ambient		2–30	400	115	167
Chadwick et al., 2008	NW Rota-1 volcano, Mariana Arc	April 2006	Explosive eruptions measured by hydrophone; broadband signals with peak frequency ~12 Hz		1–1,500	60	150	185
Chen et al., 2021	Suiyo Seamount	August 2019	Increase in broadband low-frequency noise levels (<100 Hz) near vent		20–100	1	98	98
Crone et al., 2006	Main Endeavour Vent Field, Sully	September 2004	Low-frequency narrowband and broadband signals detected within 0.3 m of vent orifice	✓	10–500	0.3	132	122
	Main Endeavour Vent Field, Puffer	September 2005				0.3	126	116
Dziak et al., 2008	Brothers volcano	October 2004–May 2005	Harmonic tremor measured on hydrophone near volcanic and hydrothermal site; fundamental frequency near 3 Hz with higher harmonics		2–70	–	–	–
Kasahara et al., 1995	Iheya Knolls, Okinawa Trough	October 1994	Large-amplitude "pressure pulses" with >1 minute period when hydrophone is near hydrothermal vent	✓	<0.02	1	–	–
Fujioka and Mitsuzawa, 2001	TAG	August 1994	Large-amplitude "pressure pulses" with >1 minute period when hydrophone is near hydrothermal vent	✓	<0.02	–	–	–
Kasahara and Sato, 2001	South Mariana Trough; Okinawa Trough; TAG	October 1993, October 1994, August 1994	Large-amplitude "pressure pulses" with >1 minute period when hydrophone is near hydrothermal vent	✓	<0.02	1–10	–	–
Little et al., 1990	Axial Seamount	September 1987	Hydrothermal vent jet noise with fluid heterogeneity amplification, peak near 40 Hz		~40	3	107	117
	East Pacific Rise 10°–12° N	December 1985	Narrowband and broadband increases in low-frequency noise levels near vents		15–30	<10	–	–
Riedesel et al., 1982	East Pacific Rise	July–August 1980	Increased low-frequency noise near vent		0.1–10	300	225	275

[a] Measurement distances are approximate based on information provided in each source.
[b] Sound pressure levels calculated by integrating power spectral density graphs published by each source.
[c] Source sound pressure levels at 1 m calculated assuming spherical spreading.

is particularly favorable for long-range detection due to low attenuation during propagation.

7.4. OTHER SOURCES OF AMBIENT NOISE

A difficulty in attributing received acoustic signals to hydrothermal vents is distinguishing them from confounding low-frequency ocean ambient noise sources. Low-frequency signals can travel great distances with little attenuation in the ocean, which results in many other possible signal sources in the same band as hydrothermal vent noise. The sources most likely to interfere with hydrothermal vent acoustic measurements are briefly discussed here. Many of these noise sources are addressed in greater detail in other chapters of this book.

7.4.1. Microseisms

Microseisms in the ocean are low-frequency pressure disturbances between 0.05 and 1 Hz due to nonlinear interference between oppositely traveling surface waves, with the most typical periods between 5 and 7 seconds (Kedar et al., 2008). Depending on the study area and local and basin-scale sea states, these events can persist for hours (Little et al., 1990). Below 10 Hz, microseisms are the dominant source of ambient noise in the ocean (Urick, 1986). Microseisms should be considered when evaluating large-amplitude, low-frequency (<10 Hz) signals recorded near hydrothermal vent fields.

7.4.2. Local and Teleseismic Events

Earthquakes originating near hydrothermal vent fields or from more distant sites have been documented at hydrothermal sites (Riedesel et al., 1982; Dziak et al., 2008). These include seismic P-waves and acoustic T-waves, which are detected on hydrophones as low-frequency (<100 Hz), impulsive signals that can persist from seconds to minutes in duration and be coupled into the deep ocean sound channel through scattering from internal and gravity waves (Godin, 2021).

7.4.3. Biological Sources

Many cetacean species are known to vocalize at low frequencies. For example, fin whales emit high-intensity, low-frequency pulses at ~20 Hz (Weirathmueller et al., 2013). During migrations, these calls can persist for many days at a time, and over long periods, annual cycles become evident due to migration patterns (Curtis et al., 1999). Due to long-distance low-frequency propagation with little attenuation, distant calls may blend to form an apparent continuous signal near 20 Hz depending on time averaging and window length settings during signal processing routines. Due to changes in fin whale behavior throughout the day, these signals may also exhibit a diurnal fluctuation (Širović et al., 2013). The narrowband nature of this signal, combined with the potential for diurnal variability, can lead to confusion with resonant hydrothermal vent signals.

Other cetacean species, such as blue and humpback whales, are also known to vocalize at low frequencies (Curtis et al., 1999). Species endemic to vent environments may also generate sounds specific to local vent sites, although there is currently no direct evidence to support this (Chen et al., 2021). If present, sounds produced by local vent organisms could provide additional signals to aid vent detection. For the purposes of geophysical vent monitoring, when evaluating anomalous signals measured near vent fields, biological sources should be considered before attributing a signal to hydrothermal vent activity.

7.4.4. Anthropogenic Sources

Anthropogenic sound sources, including commercial shipping and seismic exploration, dominate the ocean ambient noise spectrum in the frequency range from 10 to 500 Hz (Urick, 1986; Hildebrand, 2009), where most hydrothermal vent sound sources are expected. Increases in commercial shipping traffic have only exacerbated this contribution to the ambient soundscape, with estimates of up to 12 dB increases in shipping noise over the past few decades (Hildebrand, 2009). Sound sources from shipping include propeller and other propulsion system sources (Urick, 1986). Cavitation caused by propeller motion can generate broadband monopole radiation from bubble formation and collapse. Reciprocating motions such as propeller rotation and internal engine and machinery components generate tonal signals.

The steady rotation of machinery involved in ship propulsion often causes the measured tonal signals to be stable over time. Changes in engine speed are often evident as smooth transitions in the fundamental frequency and harmonics of tonal signals on a spectrogram. Natural tonal sources such as hydrothermal vent resonance or harmonic tremor are unlikely to exhibit similar frequency stability due to the turbulent forcing of these source mechanisms.

Seismic surveys, often for offshore oil and gas exploration, commonly utilize air gun arrays. These devices generate high-intensity, broadband, impulsive signals that can repeat for hours or days. Due to high source levels, low-frequency energy from air guns can be detected at large distances and observed at hydrophones near the Main Endeavour Hydrothermal Vent Field.

Seismic studies performed for industrial applications often follow distinct patterns with clearly defined regular blast periods, which is unusual for a natural source of impulsive sound measured near vent fields. These differences allow vent signals to be distinguished from anthropogenic sources.

7.5. MEASUREMENT AND ANALYSIS CONSIDERATIONS

There are factors unique to acoustic measurements of hydrothermal vents that should be considered when planning field measurements or conducting data analysis. The most salient are flow noise, sound speed in hydrothermal fluid, and differences between near and far field conditions, which are discussed here.

7.5.1. Flow Noise and Coupled Vibration

Flow noise is caused by turbulent boundary layer pressure fluctuations due to the presence of the hydrophone body within a flow. These pressure fluctuations are recorded in the same manner as propagating sound waves by the hydrophone; however, flow noise is a local effect and does not propagate, and thus is sometimes known as *pseudosound*. Placement of hydrophones near (within 1–2 m) of a hydrothermal vent plume increases the probability of flow over the hydrophone, resulting in flow noise. This can be due to shifting of the vent plume by local currents, which could engulf the hydrophone, diffuse flow from secondary openings in the vent system (Bemis, 2012), or cause entrainment and convective flow of seawater by the hot turbulent vent (Kundu & Cohen, 2002). Flow noise can sometimes saturate the dynamic range of the recorder, causing the digital time series to be clipped and masking data over the entire bandwidth of the recorder. Other times, it can be more insidious, increasing low-frequency noise levels without saturating the system.

Flow noise is most apparent at frequencies below 20 Hz. Due to the turbulent pressure fluctuations that cause flow noise, it can often be recognized by its spectral slope of $f^{-5/3}$ following Kolmogorov's turbulent length scale theory (Basset et al., 2014). Given the likelihood of hydrothermal vent acoustic signals being generated below 20 Hz, care should be taken to minimize the effects of flow noise by ensuring the hydrophone is placed beyond the direct influence of the turbulent plume motions or entrainment.

Flow-induced signals resembling harmonic tremor have been documented on ocean-bottom seismometers (Essing et al., 2021). In this case, the source of the tremor-like signal was suggested to be due to ocean bottom currents causing periodic vortex shedding over protruding elements of the instrumentation, such as antennas, ropes, or other structures. Similar signals may be transferred to hydrophones as sound or coupled vibrations in the recording platform. For this reason, when conducting studies of possible harmonic tremor measured by hydrophones near hydrothermal vents, protruding surfaces on the recording package should be minimized. Measurements of bottom current velocity should be compared to harmonic tremor records to inspect the possibility of a flow-induced harmonic tremor signal.

Seafloor vibrations may also be a source of unwanted pseudosound on recording systems placed on or near vent structures. Ceramic or piezoelectric hydrophones are particularly susceptible to mechanical noise that may propagate along cabling or instrument housings. For the case of a typical oceanographic mooring, local flow can cause strum on the mooring line, which is then coupled to the sensor. For bottom-mounted instrument packages, efficient coupling of the hydrophone to the mounting hardware and then directly to the vent chimney may allow mechanical noise generated by vent vibrations to enter the system. Applying the principles of vibration isolation in hydrophone mounting and mooring design can reduce the impact of coupled mechanical noise on acoustic data.

7.5.2. Sound Speed in Hydrothermal Fluid

Estimates of sound speed in hydrothermal fluid have ranged from 450 m/s (Crone et al., 2006) to 2,500 m/s (Dziak et al., 2008). The sound speed in hydrothermal fluid strongly influences the frequency of resonant sources such as Helmholtz and standing wave resonators. Most widely used sound speed models (e.g., Del Grosso, 1974), including the Equation of State of Seawater (McDougall et al., 2013), do not extend to the pressures and temperatures expected in hydrothermal fluids. Some specialized models have been extended to these conditions. Bischoff and Rosenbauer (1985) determined the two-phase boundary for seawater at high temperatures and pressures, while Kumagai and Chouet (2000) provide a model of compressional sound speed in vent fluid, which depends heavily on the gas-volume fraction. Neither model incorporates the high density of dissolved solids present in discharged fluids. Due to the strong relationship between resonant sound sources and assumed hydrothermal fluid sound speed, field measurements of hydrothermal fluid sound speed would further constrain vent parameters inferred from passive acoustic monitoring.

7.5.3. Near Field vs. Far Field

The acoustic near field, for a source with a poorly known size, can be approximated as the region within one wavelength of the sound source, while the far field is the region beyond this point. This section discusses differences that may be found by measuring in either the near or far field of a hydrothermal vent.

The far field represents the region with propagating pressure disturbances: that is, sound waves. This is the acoustic field that would be sampled in the remote detection of hydrothermal vents, which is particularly relevant since the utility of underwater acoustic methods often lies in the ability to detect sound at a great distance from the source. As a result, the characterization of hydrothermal vent sound fields is most usefully achieved with measurements in the far field or at least one wavelength from the vent based on the frequency range of interest. For reference, underwater low-frequency signals between 10 and 100 Hz have wavelengths ranging from 150 to 15 m, respectively.

Measurements taken within one wavelength, or the near field, will be detected similarly to far-field pressure fluctuations by a hydrophone. However, the near field is a region that contains not only propagating pressure disturbances but also fluid motions and pressure fluctuations that do not propagate (Lighthill, 1963; Ffowcs Williams, 1969; Ross, 1976). For this reason, the near-field region often has substantially different acoustic characteristics than the far field. A key difference is that the sound pressure level decay with distance varies between the near and far field for dipole and quadrupole sources. Sound pressure decays as $1/r^2$ in the near field for dipoles and $1/r^3$ for quadrupoles, while in the far field, sound pressure decays as $1/r$ for both, as shown in Fig. 7.5.

The acoustic pressure magnitude also fluctuates more rapidly in the near field than the far field due to interference determined by the specific source geometry and radiation mechanism (Kinsler et al., 2000). The key parameter in this case is the ratio of the distance r to the nominal acoustic source dimension a. For example, Fig. 7.6 illustrates the difference between the near-field and far-field pressure magnitude for the axial response of a

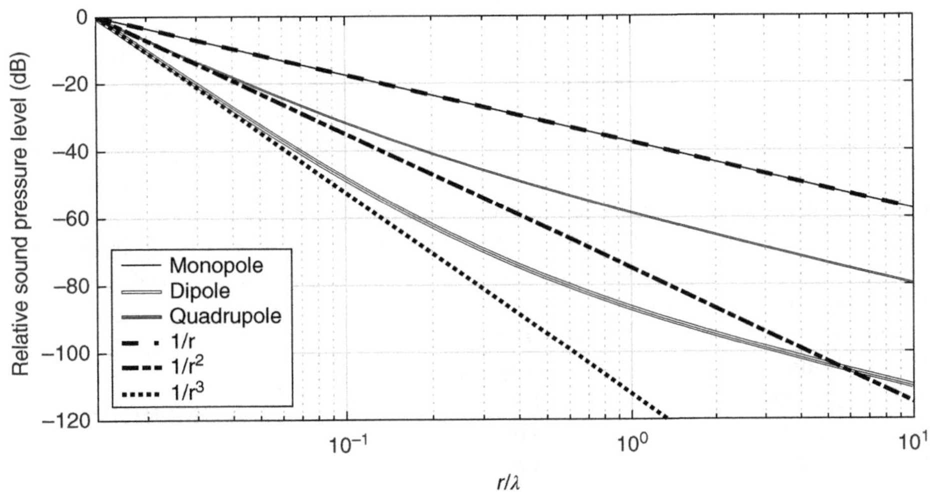

Figure 7.5 Sound pressure level decay with distance for monopole (light gray solid), dipole (gray solid), and quadruple (dark gray solid) sources, with power-law reference lines (dashed).

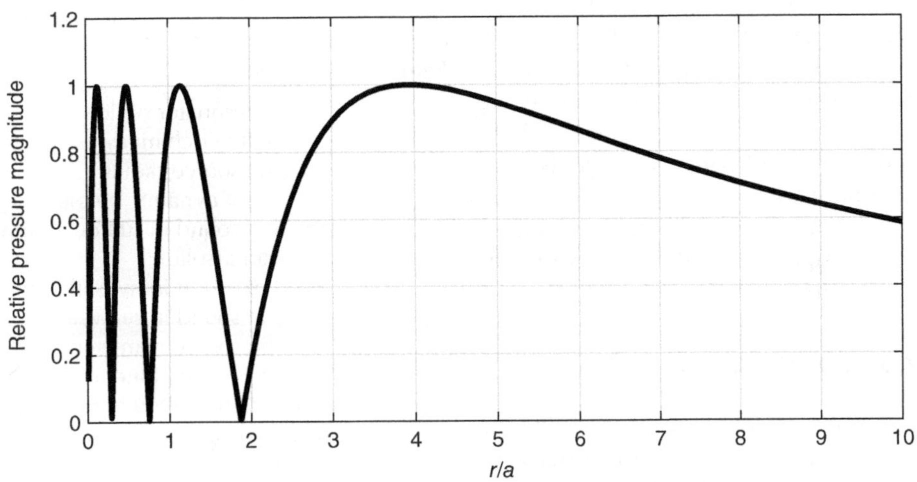

Figure 7.6 Relative pressure magnitude in near and far fields for a plane circular piston source (Kinsler et al., 2000).

plane circular piston (Kinsler et al., 2000). Measurements with $r/a < 3$ show large changes in pressure with distance. Although the exact response will vary based on the source mechanism, this is another possible source of complication with near-field acoustic measurements.

Many studies to date have focused on near-field measurements of hydrothermal vent noise. These measurement positions were likely chosen to give the highest likelihood of detecting sound from the hydrothermal vent above the ambient ocean background noise. However, it should be noted that this sound field may not be representative of that received in the far field, which may be more useful in long-range detection applications. Near-field hydrophone positions may be acceptable for purposes of hydrothermal vent monitoring, allowing sensor placement in closer proximity to the vent orifice. However, these should be chosen to be beyond the region of flow noise interference, and the near-field effects presented here should be considered. Future studies detailing the near-field and far-field acoustic signatures of hydrothermal vents would assist detection and monitoring efforts.

7.5.4. Hydrophone Array Measurements

Because many of the acoustic source mechanisms are expected to be generated local to the hydrothermal vent structure, the use of hydrophone arrays can assist with separating vent signals from other sources of ocean ambient noise using coherent processing. In the simplest case, hydrothermal vent noise should originate below the array, while most confounding sources will originate at the sea surface (e.g., surface waves, rain, shipping) or horizontally (e.g., distant shipping and biological sources). A short vertical array capable of differentiating between vertical arrival angle can thus be paired with source mechanism theories to identify hydrothermal vent signals. Closer to the vent, a hydrophone array can help separate the source mechanisms due to their differing locus with respect to the vent structure (Crone et al., 2006). Identifying incoherent processes like flow noise, perfectly coherent processes such as electronic noise or mechanical vibrations, and true acoustic signals can be done using an array as short as two elements (Auvinen, 2019). Because the separation of hydrothermal vent signals from other sources of ocean ambient noise has been an issue in previous studies, hydrophone arrays of suitable geometry to study low-frequency signals (<500 Hz) are recommended for future hydrothermal vent noise surveys.

7.6. CONCLUSION

The varied acoustic source mechanisms responsible for hydrothermal vent sound production have been discussed, and expected source levels and spectral characteristics identified where possible. The known field studies, including acoustic observations of hydrothermal vents, have been reviewed to establish references for future comparison of additional vent acoustic measurements. Uncertainty remains in almost all extant measurements regarding the source of the measured signals, where flow noise or other more typical ocean ambient noise sources provide confounding possibilities. The spectra reported from existing field studies suggest hydrothermal vents produce primarily low-frequency noise (below ~500 Hz), and source levels are likely sufficient to enable both monitoring and long-range detection.

It should be noted that the disparate fields of geophysics and underwater acoustics present a challenge in uncovering acoustic measurements of hydrothermal vents in the literature due to differences in terminology and topic of focus. To the best of our knowledge, the studies surveyed here constitute all existing field measurements of sound production by hydrothermal vents.

The potential utility of passive hydrothermal vent acoustic monitoring and detection, ranging from scientific applications in geophysics and oceanography, deep-sea conservation, and even extraterrestrial exploration, lend support for further investigation. Laboratory experiments investigating acoustic source mechanisms and additional field measurements utilizing hydrophone arrays will help constrain the mechanisms of sound generation by hydrothermal vents.

NOMENCLATURE

All variables and their respective definitions and units used in this chapter are identified in Table 7.5.

Table 7.5 Variables, definitions, and units used in the chapter.

	Roman Letters	
Symbol	Definition	Units
A_r	Area of resonator orifice	m^2
A_c	Area of inner chimney walls	m^2
a	Acoustic source radius	m
b_{rms}	Fraction of dynamic pressure	–
c_b	Speed of sound in sulfide chimney structure (solid)	m/s
c_h	Speed of sound in hydrothermal fluid	m/s
c_w	Speed of sound in seawater	m/s
d	Acoustic source diameter	m
d_i	Chimney (beam) inner diameter	m
d_o	Chimney (beam) outer diameter	m
F	Magnitude of fluctuating force	N
f	Frequency	Hz
f_p	Peak source frequency	Hz

Table 7.5 (*Continued*)

Symbol	Definition	Units
	Roman Letters	
H	Height of vent chimney	m
k	Acoustic wavenumber	m^{-1}
L	Length of resonator neck	m
l	Length scale of hot jet mixing sound source	m
M	Mach number	–
m	Mode number	–
p_a	Acoustic pressure	Pa
p_d	Dipole pressure field	Pa
$\overline{p_i}$	rms pressure oscillation	Pa
p_q	Quadrupole pressure field	Pa
p_{ref}	Underwater acoustic reference pressure	Pa
p_w	Ambient pressure	Pa
Q	Heat power	W
r	Distance from source	m
T	Vent fluid temperature	°C
U	Flow velocity	m/s
V	Helmholtz cavity volume	m^3
ΔV	Specific volume difference between jet and ambient fluid	m^3
W_{ac}	Acoustic power	W
z	Depth	m
	Greek Letters	
Symbol	**Definition**	**Units**
γ	Ratio of specific heat at constant pressure and volume	–
η_{rad}	Radiation efficiency	–
λ	Wavelength	m
ρ_b	Density of sulfide chimney structure (solid)	kg/m^3
ρ_h	Density of hydrothermal fluid	kg/m^3
ρ_w	Density of seawater	kg/m^3
$\Delta\rho$	Density difference between hydrothermal fluid and ambient seawater	kg/m^3
σ_r	Specific radiation resistance	–
σ_x	Specific radiation reactance	–
ω	Angular frequency	rad/s

REFERENCES

Aoki, T., & Welty, J. R. (1970). Frequency distribution of boiling-generated sound. *Journal of Heat Transfer*, *92*(3), 542–544. https://doi.org/10.1115/1.3449713

Auvinen, M. F., Barclay, D. R., & Coffin, M. E. (2019). Performance of a passive acoustic linear array in a tidal channel. *IEEE Journal of Oceanic Engineering*, *45*(4), 1564–1573. https://doi.org/10.1109/JOE.2019.2944444

Baker, E. T., Resing, J. A., Haymon, R. M., Tunnicliffe, V., Lavelle, J. W., Martinez, F., et al. (2016). How many vent fields? New estimates of vent field populations on ocean ridges from precise mapping of hydrothermal discharge locations. *Earth and Planetary Science Letters*, *449*, 186–196. https://doi.org/10.1016/j.epsl.2016.05.031

Barreyre, T., & Sohn, R. A. (2016). Poroelastic response of mid-ocean ridge hydrothermal systems to ocean tidal loading: Implications for shallow permeability structure. *Geophysical Research Letters*, *43*, 1660–1668, doi:10.1002/2015GL066479

Bassett, C., Thomson, J., Dahl, P. H., & Polagye, B. (2014). Flow-noise and turbulence in two tidal channels. *The Journal of the Acoustical Society of America*, *135*(4), 1764–1774. https://doi.org/10.1121/1.4867360

Bemis, K., Lowell, R. P., & Farough, A. (2012). Diffuse flow: on and around hydrothermal vents at mid-ocean ridges. *Oceanography*, *25*(1), 182–191. https://doi.org/10.5670/oceanog.2012.16

Bibee, L. D., & Jacobson, R. S. (1986). Acoustic noise measurements on Axial Seamount, Juan De Fuca Ridge. *Geophysical Research Letters*, *13*(9), 957–960. https://doi.org/10.1029/GL013i009p00957

Bischoff, J. L., & Rosenbauer, R. J. (1984). The critical point and two-phase boundary of seawater, 200–500°C. *Earth and Planetary Science Letters*, *68*(1), 172–180. https://doi.org/10.1016/0012-821X(84)90149-3

Bischoff, J. L., & Rosenbauer, R., J. (1985). An empirical equation of state for hydrothermal seawater (3.2 percent NaCl). *American Journal of Science*, *285*, 725–763.

Chadwick, W. W., Cashman, K. V., Embley, R. W., Matsumoto, H., Dziak, R. P., de Ronde, C. E. J., et al. (2008). Direct video and hydrophone observations of submarine explosive eruptions at NW Rota-1 volcano, Mariana arc. *Journal of Geophysical Research: Solid Earth*, *113*(B8). https://doi.org/10.1029/2007JB005215

Chen, C., Lin, T., Watanabe, H. K., Akamatsu, T., & Kawagucci, S. (2021). Baseline soundscapes of deep-sea habitats reveal heterogeneity among ecosystems and sensitivity to anthropogenic impacts. *Limnology and Oceanography*, *66*(10), 3714–3727. https://doi.org/10.1002/lno.11911

Chouet, B. A. (1996). Long-period volcano seismicity: its source and use in eruption forecasting. *Nature*, *380*(6572), 309–316. https://doi.org/10.1038/380309a0

Corliss, J. B., Dymond, J., Gordon, L. I., Edmond, J. M., von Herzen, R. P., Ballard, R. D., et al. (1979). Submarine thermal springs on the Galápagos Rift. *Science*, *203*(4385), 1073–1083. https://doi.org/10.1126/science.203.4385.1073

Crone, T. J., & Wilcock, W. S. D. (2005). Modeling the effects of tidal loading on mid-ocean ridge hydrothermal systems. *Geochemistry, Geophysics, Geosystems*, *6*, Q07001, doi:10.1029/2004GC00905

Crone, T. J., Wilcock, W. S. D., Barclay, A. H., & Parsons, J. D. (2006). The sound generated by mid-ocean ridge black smoker hydrothermal vents. *PLoS ONE*, *1*(1), e133. https://doi.org/10.1371/journal.pone.0000133

Crone, T. J., McDuff, R. E., & Wilcock, W. S. D. (2008). Optical plume velocimetry: a new flow measurement technique for use in seafloor hydrothermal systems. *Experiments in Fluids*, *45*(5), 899–915. https://doi.org/10.1007/s00348-008-0508-2

Curtis, K. R., Howe, B. M., & Mercer, J. A. (1999). Low-frequency ambient sound in the North Pacific: long time series observations. *The Journal of the Acoustical Society of America*, *106*(6), 3189–3200. https://doi.org/10.1121/1.428173

Deane, G. B., & Stokes, M. D. (2010). Model calculations of the underwater noise of breaking waves and comparison with experiment. *The Journal of the Acoustical Society of America, 127*(6), 3394–3410. https://doi.org/10.1121/1.3419774

Del Grosso, V. A. (1974). New equation for the speed of sound in natural waters (with comparisons to other equations). *The Journal of the Acoustical Society of America, 56*(4), 1084–1091. https://doi.org/10.1121/1.1903388

Dziak, R. P., & Fox, C. G. (2002). Evidence of harmonic tremor from a submarine volcano detected across the Pacific Ocean basin. *Journal of Geophysical Research, 107*(B5), 2085. https://doi.org/10.1029/2001JB000177

Dziak, R. P., Haxel, J. H., Matsumoto, H., Lau, T. K., Merle, S. G., de Ronde, C. E. J., et al. (2008). Observations of regional seismicity and local harmonic tremor at Brothers volcano, south Kermadec arc, using an ocean bottom hydrophone array. *Journal of Geophysical Research: Solid Earth, 113* (B8). https://doi.org/10.1029/2007JB005533

Dziak, R., Banfield, D., Lorenz, R., Matsumoto, H., Klinck, H., Dissly, R., et al. (2020). Deep ocean passive acoustic technologies for exploration of ocean and surface sea worlds in the outer solar system. *Oceanography, 33*(2). https://doi.org/10.5670/oceanog.2020.221

Eggleston, D. B., Lillis, A., & Bohnenstiehl, D. R. (2016). Soundscapes and larval settlement: larval bivalve responses to habitat-associated underwater sounds. In A. N. Popper & A. Hawkins (Eds.), *The Effects of Noise on Aquatic Life II* (Vol. *875*, pp. 255–263). Springer. https://doi.org/10.1007/978-1-4939-2981-8_30

Essing, D., Schlindwein, V., Schmidt-Aursch, M. C., Hadziioannou, C., & Stähler, S. C. (2021). Characteristics of current-induced harmonic tremor signals in ocean-bottom seismometer records. *Seismological Research Letters, 92*(5), 3100–3112. https://doi.org/10.1785/0220200397

Fahy, F., & Walker, J. (1998). *Fundamentals of Noise and Vibration*. Spon Press.

Fletcher, N. H., & Rossing, T. D. (1991). *The Physics of Musical Instruments. Springer.* https://doi.org/10.1007/978-1-4612-2980-3

Ffowcs Williams, J. E. (1969). Hydrodynamic noise. *Annual Review of Fluid Mechanics, 1*(1), 197–222. https://doi.org/10.1146/annurev.fl.01.010169.001213

Ffowcs Williams, J. E. (1977). Aeroacoustics. *Annual Review of Fluid Mechanics, 9*(1), 447–468. https://doi.org/10.1146/annurev.fl.09.010177.002311

Ffowcs Williams, J. E. (1981). Sound Sources in Aerodynamics—Fact and Fiction. Paper presented at 19th Aerospace Sciences Meeting, St. Louis, MO, U.S.A. https://doi.org/10.2514/6.1981-56

Fujioka, K., & Mitsuzawa, K. (2001). Tide-related variability of hydrothermal activity at the TAG hydrothermal mound, Mid-Atlantic Ridge and the East Pacific Rise. *Journal of the Geodetic Society of Japan, 47*(1), 434–440.

Germanovich, L. N., Hurt, R. S., Smith, J. E., Genc, G., & Lowell, R. P. (2015). Measuring fluid flow and heat output in seafloor hydrothermal environments. *Journal of Geophysical Research: Solid Earth, 120*(12), 8031–8055. https://doi.org/10.1002/2015JB012245

Godin, O. A. (2021). Contributions of gravity waves in the ocean to T-phase excitation by earthquakes. *The Journal of the Acoustical Society of America, 150*(5), 3999–4017. https://doi.org/10.1121/10.0007283

Greenspan, M., & Tschiegg, C. E. (1957). Speed of sound in water by a direct method. *Journal of Research of the National Bureau of Standards, 59*(4), 249. https://doi.org/10.6028/jres.059.028

Greenspan, M., & Tschiegg, C. E. (1959). Tables of the speed of sound in water. *The Journal of the Acoustical Society of America, 31*(1), 75–76. https://doi.org/10.1121/1.1907614

Hannington, M., Herzig, P., Stoffers, P., Scholten, J., Botz, R., Garbe-Schönberg, D., et al. (2001). First observations of high-temperature submarine hydrothermal vents and massive anhydrite deposits off the north coast of Iceland. *Marine Geology, 177*(3–4), 199–220. https://doi.org/10.1016/S0025-3227(01)00172-4

Hildebrand, J. (2009). Anthropogenic and natural sources of ambient noise in the ocean. *Marine Ecology Progress Series, 395*, 5–20. https://doi.org/10.3354/meps08353

Jellinek, A. M., & Bercovici, D. (2011). Seismic tremors and magma wagging during explosive volcanism. *Nature, 470* (7335), 522–525. https://doi.org/10.1038/nature09828

Kasahara, J., Sato, T., & Koresawa, S. (1995). Evidence for periodic hydrothermal activity at the hydrothermal vent in the Iheya Knolls, 1n the Okinawa Trough, Japan, measured by an OBS/ H with the assistance of the submersible "Shinkai 2000." *Journal of Deep Sea Research, 11*, 167–188.

Kasahara, J., & Sato, T. (2001). Tidal effects on volcanic earthquakes and deep-sea hydrothermal activity revealed by ocean bottom seismometer measurements. *Journal of the Geodetic Society of Japan, 47*(1), 424–433.

Kedar, S., Longuet-Higgins, M., Webb, F., Graham, N., Clayton, R., & Jones, C. (2008). The origin of deep ocean microseisms in the North Atlantic Ocean. *Proceedings of the Royal Society A: Mathematical, Physical and Engineering Sciences, 464*(2091), 777–793. https://doi.org/10.1098/rspa.2007.0277

Kinsler, L. E., Frey, A. R., Coppens, A. B., & Sanders, J. V. (2000). *Fundamentals of acoustics* (4th Ed.). John Wiley & Sons.

Kumagai, H., & Chouet, B. A. (2000). Acoustic properties of a crack containing magmatic or hydrothermal fluids. *Journal of Geophysical Research: Solid Earth, 105*(B11), 25493–25512. https://doi.org/10.1029/2000JB900273

Kundu, P. K., & Cohen, I. M. (2002). *Fluid mechanics* (2nd Ed.). Elsevier Science (USA).

Larson, B. I., Olson, E. J., Lilley, M.D. (2007). In situ measurement of dissolved chloride in high temperature hydrothermal fluids. *Geochimica et Cosmochimica Acta, 71*, 2510–2523, doi:10.1016/j.gca.2007.02.013

Lemmon, E. W., Bell, I. H., Huber, M. L., and McLinden, M. O. (2010). Thermophysical properties of fluid systems. *NIST Chemistry WebBook, NIST Standard Reference Database Number 69*. National Institute of Standards and Technology. https://doi.org/10.18434/T4D303

Lelièvre, Y., Legendre, P., Matabos, M., Mihály, S., Lee, R. W., Sarradin, P.-M., et al. (2017). Astronomical and atmospheric impacts on deep-sea hydrothermal vent invertebrates.

Proceedings of the Royal Society B: Biological Sciences, *284*(1852), 20162123. https://doi.org/10.1098/rspb.2016.2123

Lighthill, M. J. (1952). On sound generated aerodynamically I. General theory. *Proceedings of the Royal Society of London. Series A. Mathematical and Physical Sciences*, *211*(1107), 564–587, doi:10.1098/rspa.1952.0060

Lighthill, M. J. (1954). On sound generated aerodynamically II. Turbulence as a source of sound. *Proceedings of the Royal Society of London. Series A. Mathematical and Physical Sciences*, *222*(1148), 1–32. https://doi.org/10.1098/rspa.1954.0049

Lighthill, M. J. (1963). Jet noise. *AIAA Journal*, *1*(7), 1507–1517. https://doi.org/10.2514/3.1848

Lillis, A., Eggleston, D. B., & Bohnenstiehl, D. R. (2013). Oyster larvae settle in response to habitat-associated underwater sounds. *PLoS ONE*, *8*(10), e79337. https://doi.org/10.1371/journal.pone.0079337

Lin, T.-H., Chen, C., Watanabe, H. K., Kawagucci, S., Yamamoto, H., & Akamatsu, T. (2019). Using soundscapes to assess deep-sea benthic ecosystems. *Trends in Ecology & Evolution*, *34*(12), 1066–1069. https://doi.org/10.1016/j.tree.2019.09.006

Little, S. A. (1988). *Fluid Flow and Sound Generation at Hydrothermal Vents* (PhD thesis). Massachusetts Institute of Technology and Woods Hole Oceanographic Institution.

Little, S. A., Stolzenbach, K. D., & Purdy, G. M. (1990). The sound field near hydrothermal vents on Axial Seamount, Juan de Fuca Ridge. *Journal of Geophysical Research*, *95*(B8), 12927. https://doi.org/10.1029/JB095iB08p12927

Longo, M., Lazzaro, G., Caruso, C. G., Corbo, A., Sciré Scappuzzo, S., Italiano, F., et al. (2021). Hydro-acoustic signals from the Panarea shallow hydrothermal field: new inferences of a direct link with Stromboli. *Geological Society, London, Special Publications*, *SP519*, 2020–2184. https://doi.org/10.1144/SP519-2020-184

Longuet-Higgins, M. S. (1989). Monopole emission of sound by asymmetric bubble oscillations. Part 1. Normal modes. *Journal of Fluid Mechanics*, *201*, 525–541. https://doi.org/10.1017/S0022112089001035

Longuet-Higgins, M. S. (1990). Bubble noise spectra. *The Journal of the Acoustical Society of America*, *87*(2), 652–661. https://doi.org/10.1121/1.398934

Lush, P. A. (1971). Measurements of subsonic jet noise and comparison with theory. *Journal of Fluid Mechanics*, *46*(3), 477–500. https://doi.org/10.1017/S002211207100065X

MacAyeal, D. R., Okal, E. A., Aster, R. C., & Bassis, J. N. (2008). Seismic and hydroacoustic tremor generated by colliding icebergs. *Journal of Geophysical Research*, *113*(F3), F03011. https://doi.org/10.1029/2008JF001005

Martin, W., Baross, J., Kelley, D., & Russell, M. J. (2008). Hydrothermal vents and the origin of life. *Nature Reviews Microbiology*, *6*(11), 805–814. https://doi.org/10.1038/nrmicro1991

McCreery, C. S., Duennebier, F. K., & Sutton, G. H. (1993). Correlation of deep ocean noise (0.4–30 Hz) with wind, and the Holu Spectrum—a worldwide constant. *The Journal of the Acoustical Society of America*, *93*(5), 2639–2648. https://doi.org/10.1121/1.405838

McDougall, T. J., Feistel, R., & Pawlowicz, R. (2013). Thermodynamics of seawater. In G. Siedler et al. (Eds.), *International geophysics* (Vol. *103*, pp. 141–158). Elsevier. https://doi.org/10.1016/B978-0-12-391851-2.00006-4

McNutt, S. R., & Roman, D. C. (2015). Volcanic seismicity. In H. Sigurdsson (Ed.), *The encyclopedia of volcanoes* (pp. 1011–1034). Elsevier. https://doi.org/10.1016/B978-0-12-385938-9.00059-6

Meier, M., & Schlindwein, V. (2018). First in situ seismic record of spreading events at the ultraslow spreading southwest Indian ridge. *Geophysical Research Letters*, *45*(19), 10,360–10,368. https://doi.org/10.1029/2018GL079928

Minnaert, M. (1933). On musical air-bubbles and the sounds of running water. *The London, Edinburgh, and Dublin Philosophical Magazine and Journal of Science*, *16*(104), 235–248. https://doi.org/10.1080/14786443309462277

Morfey, C. L. (1973). Amplification of aerodynamic noise by convected flow inhomogeneities. *Journal of Sound and Vibration*, *31*(4), 391–397.

Mullineaux, L. S., Adams, D. K., Mills, S. W., & Beaulieu, S. E. (2010). Larvae from afar colonize deep-sea hydrothermal vents after a catastrophic eruption. *Proceedings of the National Academy of Sciences*, *107*(17), 7829–7834. https://doi.org/10.1073/pnas.0913187107

Plesset, M. S. (1949). The dynamics of cavitation bubbles. *Journal of Applied Mechanics (1949)*, *16*(3): 277–282. https://doi.org/10.1115/1.4009975

Riedesel, M., Orcutt, J. A., MacDonald, K. C., & McClain, J. S. (1982). Microearthquakes in the black smoker hydrothermal field, East Pacific Rise at 21°N. *Journal of Geophysical Research: Solid Earth*, *87*(B13), 10613–10623. https://doi.org/10.1029/JB087iB13p10613

Ross, D. (1976). *Mechanics of underwater noise*. Pergamon Press.

Sharqawy, M. H., Lienhard, J. H., & Zubair, S. M. (2010). Thermophysical properties of seawater: a review of existing correlations and data. *Desalination and Water Treatment*, *16*(1–3), 354–380. https://doi.org/10.5004/dwt.2010.1079

Širović, A., Williams, L. N., Kerosky, S. M., Wiggins, S. M., & Hildebrand, J. A. (2013). Temporal separation of two fin whale call types across the eastern North Pacific. *Marine Biology*, *160*(1), 47–57. https://doi.org/10.1007/s00227-012-2061-z

Spiess, F. N., Macdonald, K. C., Atwater, T., Ballard, R., Carranza, A., Cordoba, D., et al. (1980). East Pacific Rise: hot springs and geophysical experiments. *Science*, *207*(4438), 1421–1433. https://doi.org/10.1126/science.207.4438.1421

Stanley, J. A., Radford, C. A., & Jeffs, A. G. (2010). Induction of settlement in crab megalopae by ambient underwater reef sound. *Behavioral Ecology*, *21*(1), 113–120. https://doi.org/10.1093/beheco/arp159

Tang, J., Xie, G., Bao, J., Mo, Z., Liu, H., & Du, M. (2018). Experimental study of sound emission in subcooled pool boiling on a small heating surface. *Chemical Engineering Science*, *188*, 179–191. https://doi.org/10.1016/j.ces.2018.05.002

Urick, R. (1986). *Ambient noise in the sea*. Peninsula.

Van Dover, C. L., Arnaud-Haond, S., Gianni, M., Helmreich, S., Huber, J. A., Jaeckel, A. L., et al. (2018). Scientific rationale and international obligations for protection of active hydrothermal vent ecosystems from deep-sea mining. *Marine Policy*, *90*, 20–28. https://doi.org/10.1016/j.marpol.2018.01.020

Vandemeulebrouck, J., Roux, P., Gouédard, P., Legaz, A., Revil, A., Hurst, A. W., et al. (2010). Application of acoustic noise and self-potential localization techniques to a buried hydrothermal vent (Waimangu Old Geyser site, New Zealand). *Geophysical Journal International, 180*(2), 883–890. https://doi.org/10.1111/j.1365-246X.2009.04454.x

Wagner, W., & Pruß, A. (2002). The IAPWS Formulation 1995 for the thermodynamic properties of ordinary water substance for general and scientific use. *Journal of Physical and Chemical Reference Data, 31*(2), 387–535. https://doi.org/10.1063/1.1461829

Wenz, G. M. (1962). Acoustic ambient noise in the ocean: spectra and sources. *The Journal of the Acoustical Society of America, 34*(12), 1936–1956. https://doi.org/10.1121/1.1909155

Weirathmueller, M. J., Wilcock, W. S. D., & Soule, D. C. (2013). Source levels of fin whale 20 Hz pulses measured in the Northeast Pacific Ocean. *The Journal of the Acoustical Society of America, 133*(2), 741–749. https://doi.org/10.1121/1.4773277

White, S. N. (2000). *An Investigation into the characteristics and sources of light emission at deep-sea hydrothermal vents* (PhD thesis). Massachusetts Institute of Technology and Woods Hole Oceanographic Institution. https://doi.org/10.1575/1912/4053

Xu, G., and Di Iorio, D. (2012). Deep sea hydrothermal plumes and their interaction with oscillatory flows. *Geochemistry, Geophysics, Geosystems, 13*, Q0AJ01, doi:10.1029/2012GC004188

8

Atypical Signals: Characteristics and Sources of Short-Duration Events

Jean Baptiste Tary

ABSTRACT

Signals recorded on ocean-bottom seismometers deployed on the seafloor are of multiple origins. Some of them are very well known (e.g, [micro]seismicity), and some are mostly unknown due to the difficulty of determining their origin. One of these seismic signals, generally dismissed as noise, are short-duration events (SDEs). Their characteristics are relatively vague, usually different from those of microseismicity, and generally have these characteristics: (1) short duration (<5 s), (2) single-wave arrival, (3) frequency contents with one or few dominant frequencies between 3 and 30 Hz, and (4) recorded by one instrument at a time. Main interpretations include a biological origin, including marine flora/fauna interacting with the instrument, microseismic events, the resonance of fluid-filled conduits, and gas migration and expulsion at the seafloor. Recently, several studies explored possible correlations between SDE signals and spatiotemporal characteristics to better understand their origin. Together with controlled experiments, these studies indicate that the different interpretations may not exclude each other but rather correspond to different SDE groups. These studies also highlight the relevance of conducting long-term multidisciplinary surveys, controlled experiments and tailored modeling taking into account seafloor-specific conditions. Understanding SDEs could then contribute significantly to quantifying seafloor-related biogeological processes.

8.1. INTRODUCTION

The seafloor is a largely unknown environment due to the difficulty of carrying out large-scale, long-term, detailed surveys. Today we know the surface of Mars and the Moon better than we know most of the seafloor. This is mostly due to technical challenges arising from the high pressures, low temperatures, and dark conditions at the bottom of the oceans. Nonetheless, many important phenomena involve the seafloor, including major subduction processes, tsunamis, gas hydrate formation, fluid exchanges, biogeological processes, seafloor generation at mid-ocean ridges, basin sedimentation, and more. Studying these phenomena occurring at the seafloor constitutes an important challenge to better understand the processes at work on the ocean bottom.

Ocean-bottom seismometers (OBSs) are autonomous instruments, generally including geophones or broadband seismometers that record vibrations in three orthogonal directions and a hydrophone that records pressure variations above the seafloor. These instruments are primarily deployed for seismic studies and seismicity monitoring. Major differences exist in knowledge levels about the various signals detected by OBSs, including their characteristics and sources, coming from the water

Department of Geosciences, University of the Andes, Bogotá, Colombia

Noisy Oceans: Monitoring Seismic and Acoustic Signals in the Marine Environment, Geophysical Monograph 284, First Edition.
Edited by Gaye Bayrakci and Frauke Klingelhoefer.

DOI:10.1002/9781119750925.ch08

column, seafloor, and deeper substratum. Some marine seismic signals are well-studied because of their recognized importance for seismic hazards (e.g., [micro]seismicity), resource exploration (e.g., marine seismic surveys, geochemical sampling, coring), or ecological information (e.g., whale vocalizations). Other signals correspond to well-known noises such as instrumental noises, vibrations due to ship traffic, instrument–seafloor coupling, and oceanic currents (e.g., Webb, 1998).

Other signals of unknown origins, such as short-duration events (SDEs), are generally discarded as noise. These signals are thus not systematically reported (e.g., Ostrovsky, 1989), which limits our understanding of their generation process and their properties. In the case of SDEs, their signal properties are not completely identified; nor is the meaning of these properties regarding the influence of source, path, and instrumental effects. Several different origins have been proposed based on these signal properties, different correlations, and experiments, such as "fish bumps" (Buskirk et al., 1981), the resonance of fluid-filled conduits (Díaz et al., 2007), and gas migration and expulsion at the seabed (Tary et al., 2012). Due to the difficulty of distinguishing between these origins based on signal properties alone, different source mechanisms may explain the SDEs observed in various studies/contexts.

This chapter begins with a review of the signal properties of SDEs and compares them with those of microearthquakes. Next, an inventory is provided of the various contexts around the world where SDEs have been reported, based on the literature to date. This is followed by a section presenting a series of key studies and experiments illustrating some of the necessary approaches to improve our understanding of SDEs (i.e., multidisciplinary instruments and approaches, long-term deployments, in situ and laboratory experiments). Then a discussion is provided on the main interpretations of SDEs, emphasizing the three primary conceptual models proposed as their source mechanism. Finally, the chapter concludes with the primary characteristics of SDEs and proposes future directions to further the study of SDEs. Improving our understanding of these signals can, depending on their origin(s), provide important information about other marine processes such as marine flora and fauna, gas emissions, near-surface deformation, natural hazards, and resources.

8.2. SIGNAL CHARACTERISTICS

The primary signal characteristics distinguishing SDEs from microearthquakes are given in Table 8.1. These characteristics correspond to their frequency content (i.e., frequency band and localization), their duration, their amplitude, and which seismic phases are recorded. The features of microearthquakes are well known because of their importance for many studies, although those generally focus on far-field signals, whereas the features of SDEs are more difficult to define as there are always exceptional cases. This arises due to the large variety of signals previously discarded as noise and now potentially labeled SDEs without distinctions about their possible sources. The primary characteristics of SDEs are then relatively vague and correspond to the following (see Fig. 8.1 for some examples): (1) short duration (generally less than 4–5 s); (2) one or a few dominant frequencies between ~3 and ~30 Hz that are not overtones of a fundamental mode; (3) generally consist of only one seismic phase; (4) emergent or impulsive with amplitudes well above the background noise level; (5) on the same instrument, recorded by geophones and seismometers but generally not picked up by hydrophones; (6) seen at only one instrument at a time, pointing to a local source.

The only indisputable characteristic of SDEs is their short duration (Fig. 8.1). Their frequency content is generally between 3 and 30 Hz, but higher frequencies up to the instrument Nyquist frequency of some deployments at

Table 8.1 General characteristics of SDEs and microearthquakes ($M_w < 3$).

	SDEs	Microearthquakes
Frequency band (Hz)	>3 to 30 Hz	>~0.1 Hz to corner frequency
Frequency content	Monochromatic or few dominant frequencies	Broadband
Seismic phases	Only one for most events	P- and S-waves
Presence on sensors	Mostly on geophones/seismometers, hardly on hydrophones	On geophones/seismometers and hydrophones
Duration	<4–5 s	5–10 s depending on magnitude and distance event-station
Amplitudes	From noise level to instrument saturation	From noise level to instrument saturation
Local vs. regional	At only one station for most events: local	At one station or more: local and regional

Exceptions have been observed for most of these characteristics (see text).

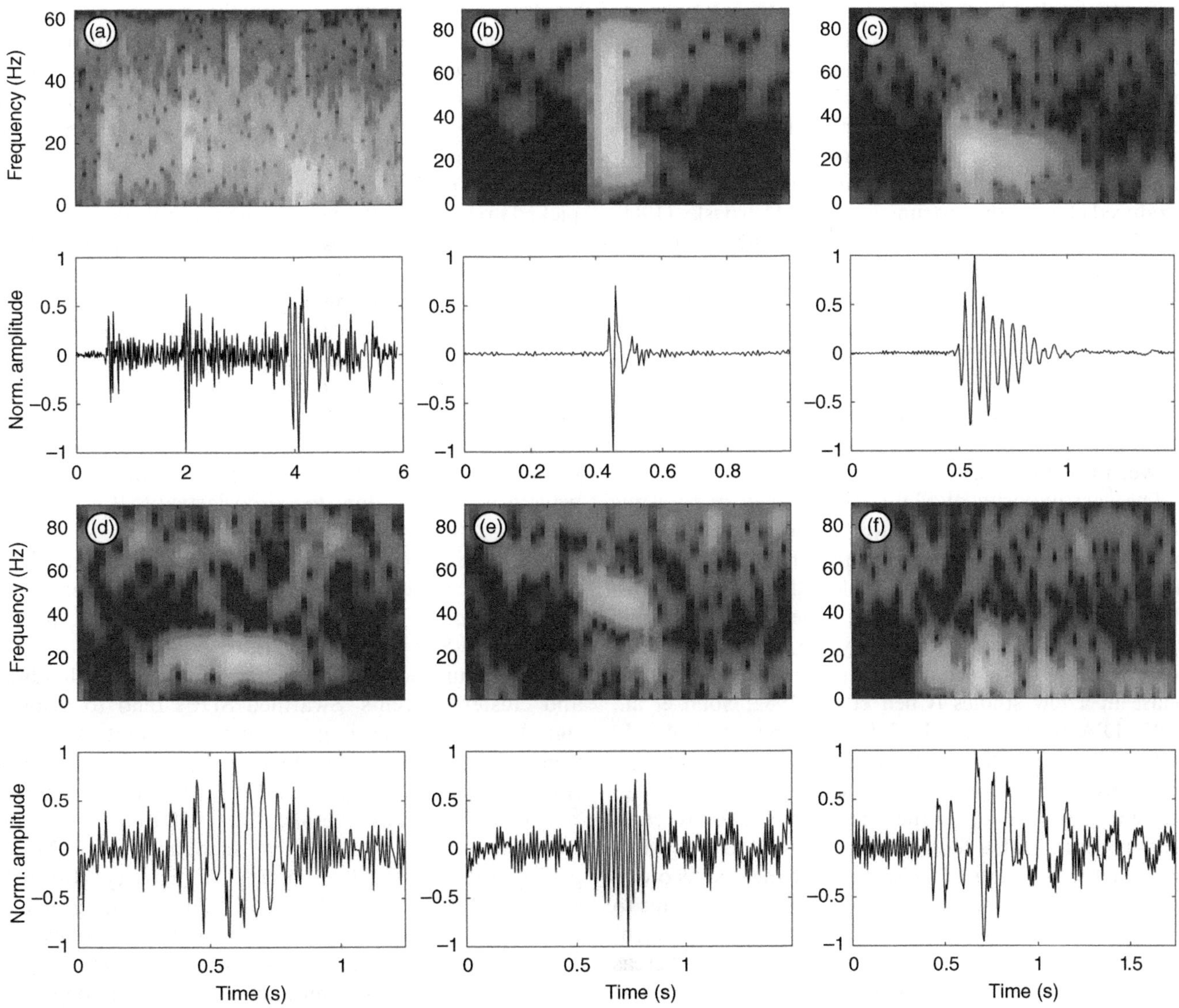

Figure 8.1 Examples of a microearthquake (a) and SDEs (b–f) with their spectrograms and waveforms (vertical component of a short-period geophone). The microearthquake and SDEs are from the Sea of Marmara, Türkiye. SDEs in (b) and (c) are examples of impulsive events, while those in (d) and (e) are examples of emergent events. The last SDE in (f) potentially presents two arrivals. The SDE waveforms were band-pass filtered between 3 and 90 Hz.

50 Hz have been observed (Sgroi et al., 2014; Giovanetti et al., 2016) (Fig. 8.1). The lower limit at ~3 Hz may be due to the lower-frequency limit of short-period geophones (natural frequency of 4.5 Hz) generally used for marine seismic surveys, although no lower frequencies have been observed when broadband seismometers were employed (e.g., Bowman & Wilcock, 2014; Embriaco et al., 2014; Giudicepetro et al., 2019; Hilmo & Wilcock, 2020). By analogy with the frequency content of seismic events scaling with the inverse of their magnitude or resonance frequencies scaling with the inverse of object dimensions, the high-frequency character of SDEs may reflect the local, small size of their source.

In terms of waveforms, SDEs are often impulsive, but a large part of these signals exhibits emergent waveforms as well (Fig. 8.1). Many SDEs also show a regular amplitude decrease following an impulsive onset (e.g., Buskirk et al., 1981; Díaz et al., 2007; D'Alessandro et al., 2009; Hilmo & Wilcock, 2020). Generally, SDEs are recorded by geophones and seismometers but not by hydrophones, even in cases where the signal amplitude is much higher than the amplitude of microearthquakes recorded by both geophones/seismometers and hydrophones. It has been suggested that only strong SDEs can be recorded by hydrophones (Tary et al., 2012; Sgroi et al., 2014; Ugalde et al., 2019), although there isn't a clear correlation yet between amplitude and the ability of hydrophones to record these events. Furthermore, many instrument designs exist for OBSs (e.g., different sensors and sensor positions), complicating the comparison between surveys (e.g., Tary et al., 2012). Despite their high signal-to-noise ratios, SDEs are generally not recorded by more than one instrument at a time, which indicates that these events are produced close to the instruments. Only Ostrovsky (1989) and Tary et al. (2012) report signals with waveforms similar to SDEs recorded at more than one instrument only a few meters apart. In the case of Tary et al. (2012), two instruments located 25 m apart did not record any SDEs simultaneously despite some having amplitudes higher than those of microearthquakes observed at multiple instruments. The possibility of simultaneously recording these signals fades quickly with increasing distance between instruments.

The vast majority of SDEs only display one seismic phase, which may be the case for several reasons: (1) only one phase is recorded well by the instrument (some phases may be too low-amplitudes or below the noise level), (2) the different phases arrive at the same time, or (3) only one seismic phase is generated by the source. However, some SDEs were observed with more than one seismic phase in a few studies (Chen et al., 1982; Sohn et al., 1995; D'Alessandro et al., 2009; Bowman & Wilcock, 2014) (Fig. 8.1). SDE waveforms are generally monochromatic, with consistent signal characteristics along the signal (e.g., Fig. 8.1c–e). In the case of D'Alessandro et al. (2009), some SDEs show a clear change in frequency between the two parts of the signal, with the first recorded by the hydrophone while the second is not. This seems to indicate that the first, high-frequency part with frequencies between 5 and 20 Hz corresponds to P-waves, whereas the second, monochromatic part with a frequency peak at ~8 Hz corresponds to a resonating part coming from either the source or the instrument–seafloor system. In many studies, particle motions of SDEs are mostly rectilinear, meaning particle motions follow one direction, which also corresponds to a body wave and are more likely P-waves (e.g., Embriaco et al., 2014; Franek et al., 2017). On the other hand, P-waves are generally picked up by hydrophones, which may indicate that SDEs consist mostly of S-waves or that P-waves are propagating at a small incidence angle from the seafloor, leading to little of their energy being transmitted to the water column (Stein & Wysession, 2003; Tary, 2011). Some SDEs show elliptical particle motions (e.g., Sgroi et al., 2014), which point to surface waves propagating along the seafloor, such as Scholte waves (Tary et al., 2012); complex interactions between body waves and the seafloor; near-field waves (Vavryčuk, 1992); or anelastic effects on the waveforms (e.g., due to viscoelasticity [Červený & Pšenčík, 2005]).

Considering the general spatiotemporal distribution of SDEs, these signals have been classified into two categories: background and swarmed signals (Tsang-Hin-Sun et al., 2019; Batsi et al., 2019), similar to the occurrence of microearthquakes as individual and clustered events. Swarmed SDEs tend to display signal characteristics (e.g., frequency content, polarization) that are more consistent than background SDEs (Fig. 8.2) (e.g., Tary et al., 2012; Tsang-Hin-Sun et al., 2019). Some studies indicate that SDE distributions sometimes show periodicities, such as the 12–24 hr periodicity attributed to biological activity (Buskirk et al., 1981) or tides (Hsu et al., 2013; Chang et al., 2016), and significant correlations with other phenomena such as instrument depth (e.g., Buskirk et al., 1981), gas emissions (Embriaco et al., 2014; Bayrakci

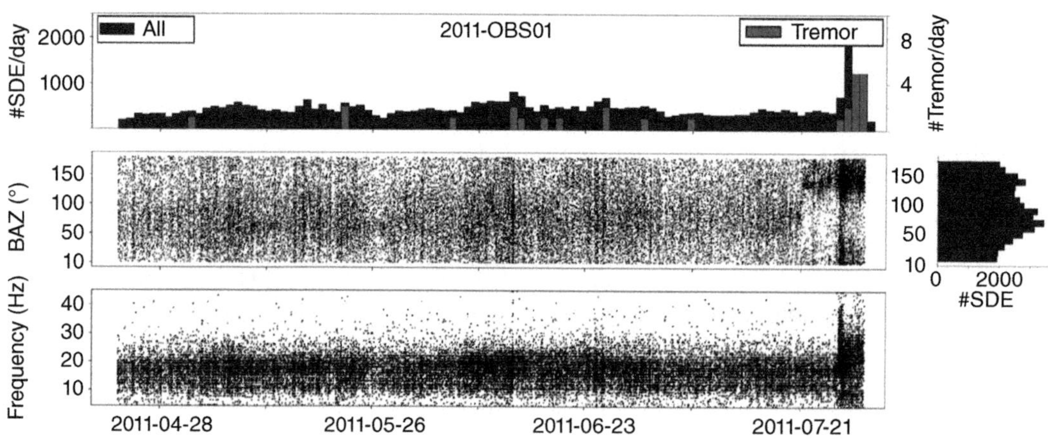

Figure 8.2 Daily number of SDEs for OBS-01 of the 2011 deployment in the Sea of Marmara (Géli et al., 2018) showing background (black) and tremor (swarm-like, gray) SDEs. Back azimuths and the dominant frequency of each SDE are shown in the middle and bottom panels, respectively. Adapted from Tsang-Hin-Sun et al. (2019).

et al., 2014), earthquakes (Tsang-Hin-Sun et al., 2019), and tides (Hsu et al., 2013; Chang et al., 2016). These periodicities and correlations are difficult to observe during surveys but may be attributed to different sources or settings influencing the generation of SDEs. To distinguish between different impacts and sources, it is imperative to minimize or control noise sources arising from instruments, both in terms of noise level (background) and transient noises (e.g., pieces hitting the instrument, electronic noises), and carry out multidisciplinary surveys.

8.3. WORLDWIDE DISTRIBUTION OF SDES

Since the first reference to SDEs in the literature by Buskirk et al. (1981), this kind of signal has been observed in many parts of the world. However, an exhaustive inventory of these signals is difficult because they are usually discarded as noise and hence not systematically reported. This is mentioned, for example, by Ostrovsky (1989), who notes that SDEs have been observed during all experiments conducted by the P. P. Shirshov Institute of Oceanology since 1969 but not reported.

Regarding the spatial distribution of SDEs in the literature, SDEs were described in a variety of contexts worldwide:

- In shallow (e.g., Buskirk et al., 1981; Bowman & Wilcock, 2014; Embriaco et al., 2014), intermediate (e.g., Brocher & Iwatake, 1982; Kovachev et al., 1997; Franek et al., 2014; Batsi et al., 2019), and deep waters (e.g., Buskirk et al., 1981; Prothero & Reid, 1982; Ostrovsky, 1989; Frohlich et al., 1990)
- On continental shelves (e.g., Franek et al., 2017; McDonald et al., 2006)
- At active (e.g., Buskirk et al., 1981; Chang et al., 2016; Hilmo & Wilcock, 2020) and passive (e.g., Díaz et al., 2007; Batsi et al., 2019) margins
- At mid-ocean ridges (e.g., Prothero & Reid, 1982; Sohn et al., 1995)
- Close to hydrothermal vents and submarine volcanoes (e.g., D'Alessandro et al., 2009; Bowman & Wilcock, 2014; Sgroi et al., 2014; Giovanetti et al., 2016)
- Close to mud volcanism and mud diapirs (e.g., Hsu et al., 2013; Franek et al., 2014)
- In areas prone to gas hydrates and seafloor gas releases (e.g., Franek et al., 2014; Franek et al., 2017)
- On the accretionary wedge (e.g., Chen et al., 1982; Frohlich et al., 1990; Batsi et al., 2019)
- In tectonically active basins (e.g., Tary et al., 2012; Tsang-Hin-Sun et al., 2019)
- In onshore settings (e.g., Albarello et al., 2012; Giudicepietro et al., 2019)

These different geodynamic contexts correspond to many different sites such as the Marsili submarine volcano in the Tyrrhenian Sea (D'Alessandro et al., 2009), the Sea of Marmara (Tary et al., 2012), the Galicia passive margin (Díaz et al., 2007), the flooded caldera of Deception Island volcano offshore Antarctica (Bowman & Wilcock, 2014), the corner of East Pacific Rise and Rivera fracture zones (Prothero & Reid, 1982), and the Håkon Mosby mud volcano in the SW Barents Sea continental slope (Franek et al., 2014). The diversity of contexts and sites likely means that more than one source exists for all SDE-like signals observed and that different environmental factors are important to explain the presence and dynamic of SDE activity.

8.4. OBSERVATIONS AND STUDIES ADVANCING SDE UNDERSTANDING

8.4.1. Observations from Different Types of Ocean Bottom Instruments

One of the earlier interpretations of SDEs observed on OBSs attributed these signals to mechanical and/or electronic noises produced by the instruments. Instruments are indeed one of the main sources of noise for OBSs. Potential causes of noise include loose parts (e.g., cables) hitting the instrument and electronic noises due to the hard disk powering up/down. Some of these noises can be ruled out due to their specific signals and/or periodicities, but special care has to be given to the signal-processing steps applied to the time series. Band-pass filtering, for example, as routinely done for event detection, can transform step-like signals into SDE-like signals (Fig. 8.3).

Many types of OBSs and ocean-bottom observatories are developed by institutions and companies worldwide. All of these instruments have very different designs in terms of frame, size, sensor disposition, and sensor type (i.e., hydrophones, short-period geophones, broadband seismometers) (e.g., Tary et al., 2012; Embriaco et al., 2014; Hilmo & Wilcock, 2020). Although some noises may come from instruments, instrument self-noise is characterized before many seismic surveys, showing that, in general, SDE-like signals are not generated directly by the instrument.

Ostrovsky (1989) observes that SDE-like signals are more abundant just after the instrument lands on the seafloor, after which the signal number progressively settles to a background rate. Tary et al. (2012) observe something similar with a very strong event occurring just after the instrument touches bottom (i.e., with an amplitude more than 50 times the average amplitude of other SDEs). This means that, in these cases, the instruments disturbed their immediate surroundings, leading to the generation of SDEs. Potential causes are then in the superficial

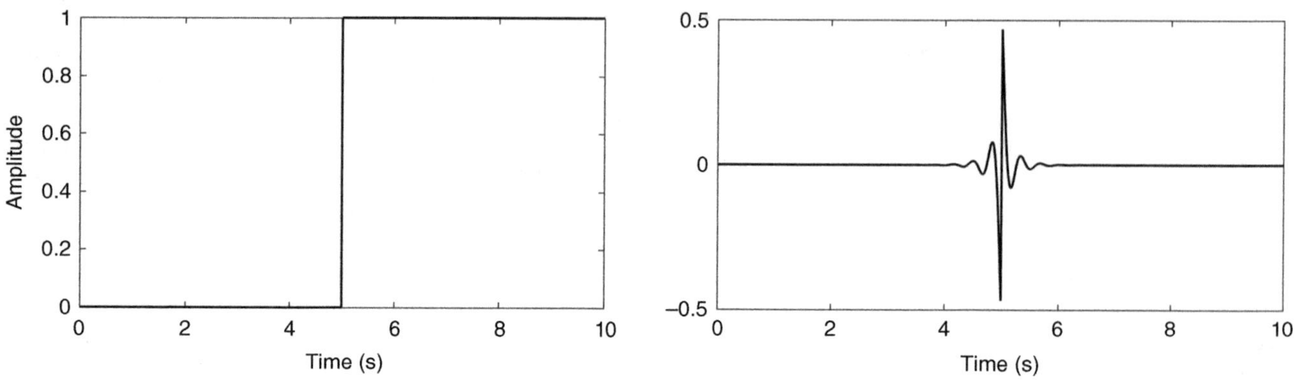

Figure 8.3 Example of band-pass filtering of a step function (left) to obtain an SDE-like signal (200 Hz sampling frequency, filtering between 3 and 90 Hz).

sediments or the water column. Because two instruments located more than a few tens of meters apart do not record SDEs simultaneously, these events are produced in the environment of the instrument's vicinity.

8.4.2. Continuous Long-Term, Multidisciplinary Monitoring of Gas Emissions

Multidisciplinary approaches involving seismological, acoustical, oceanographical, and geochemical observations at a given site are essential to distinguish between possible causes of SDEs (e.g., Embriaco et al., 2014; Monna et al., 2014). The study of Embriaco et al. (2014) is particularly interesting in that regard, as they analyze a time series of a seafloor observatory (SN-4 seafloor observatory, INGV) in shallow waters (~166 m) of the Izmit Gulf at the eastern end of the Sea of Marmara (SoM), Türkiye. The SN-4 seafloor observatory is a multiparametric benthic observatory including a broadband seismometer, a current meter, methane and oxygen sensors, a conductivity, temperature and depth (CTD) meter, and a turbidity meter. With these sensors, the observatory focused primarily on gas emission monitoring.

Considerable evidence of fluid manifestations was observed at the SoM seafloor, such as acoustic flares in the water column (Géli et al., 2008; Bayrakci et al., 2014; Dupré et al., 2015), reduced sediment patches, bacterial mats and authigenic carbonate precipitates (Zitter et al., 2008), gas within sediments (Tary, 2011), and gas hydrates (Bourry et al., 2009). In the SoM, most fluid manifestations are located close to the fault trace of the North Anatolian Fault and on basin edges (Henry et al., 2018). The seafloor observatory was located in the Izmit Gulf close to the North Anatolian fault trace, where gas emissions were observed after the M_w 7.4 Izmit earthquake on 17 August 1999 (Kuşçu et al., 2005). This site is then of great interest to study potential links between fluid emissions and seismicity.

Apart from regional low-magnitude events ($M_L < 3.6$), the seismometer recorded numerous SDEs with standard properties: durations less than 1 s, frequency content with two main peaks around 10 and 15 Hz, and only one seismic phase visible. Approximately 5 to 10 SDEs per day were inventoried with particle motions polarized primarily in the NE–SW direction. More SDE events were inventoried at the beginning of the experiment than at the end (Fig. 8.4), and no 24-hour periodicity in the event temporal distribution was observed.

Simultaneously, the methane sensor detected methane peak events corresponding to abrupt increases in seawater methane concentration that lasted from minutes to an hour. The detection of methane peak events could be related to the sea-bottom current at the time, with a weak current in the direction of the instrument being necessary to carry the methane from the emission site to the instrument and a strong current scattering the methane in a larger area. Regarding the SDEs, more methane peak events were observed at the beginning of the experiment than at the end as well (Fig. 8.4). According to Embriaco et al. (2014), the SDEs and methane peak events have similar temporal distributions. They propose a conceptual model in which periodic methane discharges from local emission sites at the seafloor are simultaneously detected by the methane sensor and the seismometer.

8.4.3. Correlation with Acoustic Monitoring of Gas Emissions

In 2009 and 2011, an acoustic gas bubble detector called BOB (Bubble Observatory) from IFREMER was deployed in the SoM (Bayrakci et al., 2014), where gas emissions were previously observed at the seafloor using ship-borne acoustic methods (Géli et al., 2008) and CHIRP data (Tary, 2011). The BOB instrument is an echosounder, laid down on the seafloor, that insonifies angular sectors of 7° for approximately one hour before

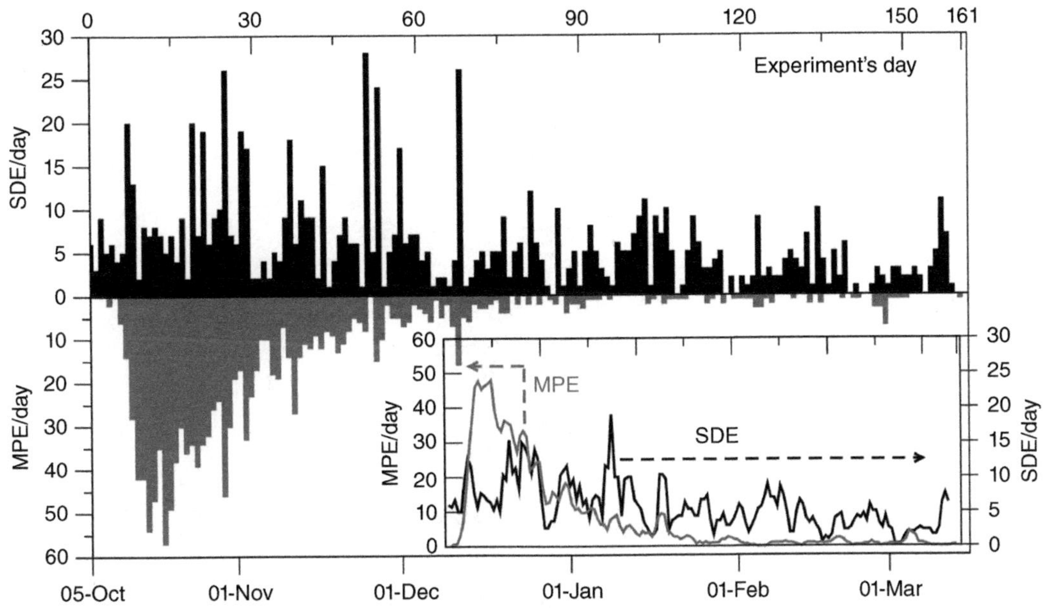

Figure 8.4 Daily number of SDEs (top) and MPEs (bottom) detected by the SN-4 multidisciplinary observatory deployed in the Izmit Gulf, SoM. The inset shows a smoothed version of the histograms (3 days average). Embriaco et al. (2014) / Oxford University Press.

moving to the adjacent sector and in this way insonifies ~170° around the instrument in about one day. Different objects and gas emission sites are detected based on variations of mean volume back-scattering strength.

In 2009, BOB was deployed in the Çınarcık basin together with an OBS located 150 m away. Due to the BOB split-beam transducer frequency of 120 kHz limiting the maximum range to ~100 m, the distance between instruments is unfortunately too large to monitor gas emissions close to the OBS. In 2011, BOB was deployed on the Central High at ~330 m depth, together with an OBS and the KATERINA radon meter (HCMR, Greece) at distances of ~50 m and ~80 m from BOB, respectively (Fig. 8.5). The Central High is a transpressive ridge situated between two basins (i.e., the Çınarcık and Central basins) where intense gas flares (Dupré et al., 2015) and gas in sediments (Bayrakci et al., 2014) were observed. The echograms and their integration per sector show that gas emissions display temporal variations. These may be attributed to variations in gas supply within the sediments or variations in sea-bottom currents (Bayrakci et al., 2014).

During the 2011 deployment, two stable and one temporary sources were observed (GES5 and GES6 in Fig. 8.5). Moreover, the backscatter character associated with the OBS seemed to indicate that this instrument was located directly on a gas source (Bayrakci et al., 2014). This OBS also recorded an average of ~500 SDEs per day during the entire experiment (15 April to 31 July 2011), with peaks at 800 SDEs per day (Fig. 8.6). This is at the high end of the average daily number of SDEs detected at the other OBSs of the network (Tsang-Hin-Sun et al., 2019) and more than 10 times the daily number of SDEs reported from the Tekirdağ basin (Tary et al., 2012) and the Izmit Gulf (Embriaco et al., 2014). The presence of gas emissions suggested by BOB in the immediate vicinity of the OBS and the high number of SDEs detected at this instrument indicate that in this case, these signals may be related to gas emissions at the seafloor.

8.4.4. Correlation with Earthquakes

Using the same 2011 OBS deployment as described in the previous section (10 OBSs in total) and another deployment of 10 OBSs in the western part of the SoM from IFREMER in 2014, Tsang-Hin-Sun et al. (2019) studied the spatiotemporal distribution of SDEs and their potential correlation with local seismicity. Two main types of SDE activity are observed: individual background SDEs and swarm-like SDEs called *tremors* by analogy with volcanic tremors. These tremors contain a series of SDEs with similar properties (i.e., frequency content, back azimuth, corresponding to the angle from north of the direction from the seismic station to the seismic source), generally different from background SDEs showing inconsistent properties, and tremor durations from minutes to an hour at a time. Sites with tremors further exhibit high SDE activity, which may imply a large supply of the SDE source (e.g., high gas content) needed to produce these tremors.

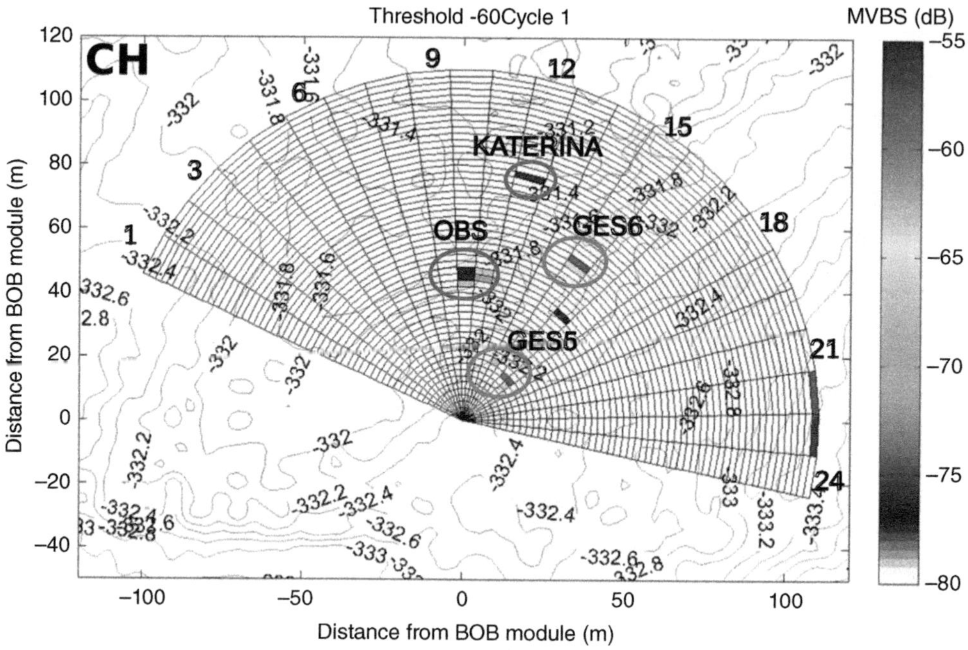

Figure 8.5 Polar diagram showing the echo-integration of acoustic data (1 cycle of ~1 day) of the BOB deployed on the Central High, SoM. Positions of an OBS deployed nearby, the KATERINA radonmeter, and gas sources (GES5 and GES6) are also indicated. MVBS: mean volume back-scattering strength. Bayrakci et al. (2014) / Springer Nature / CC BY 4.0.

Interestingly, SDE normalized temporal distributions observed in the Central Basin in 2011 correlate well with OBSs inside this basin (cross-correlation coefficients between 0.7 and 0.8) and poorly correlate with OBSs outside this basin. This could indicate that in this basin, the source of SDEs at the different sites respond to basin-wide stimuli with response times of a few days to weeks, such as pre- or post-seismic stress variations, changes in deep or near-surface gas supplies, or variations in formation or fault zone permeabilities. This type of correlation is not observed for OBSs deployed in the other parts of the western SoM (Central High, Western High, Tekirdağ Basin).

For the OBS located on the northeastern border of the Tekirdağ Basin, a sudden increase in SDE number was detected approximately 2 hours after a M_w 5 earthquake that occurred on 25 July 2011 ~10 km away, below the Western High (Géli et al., 2018; Tary et al., 2019) (Fig. 8.6), where intense gas emissions in the water column potentially linked to the presence of a buried mud volcano were described by Dupré et al. (2015). A very interesting—yet unexplained—observation is also presented by Tsang-Hin-Sun et al. (2019), leaving open the question of the potential significance of SDE occurrence before earthquakes: back azimuths calculated for SDEs occurring just before and during this sequence show a direction readjustment to 150°, which may correspond to the activation of the same preexisting source located close to the instrument.

Moreover, a significant correlation between temporal distributions of SDEs and cumulative seismic energy was also observed during the 2011 deployment (Fig. 8.6), although this correlation is neither systematic nor observed at all instruments for all events (Tsang-Hin-Sun et al., 2019). Such a correlation implies that some SDEs may be directly or indirectly triggered dynamically by earthquakes. As such, a correlation is not observed in other areas of the SoM or elsewhere (e.g., Embriaco et al., 2014; Franek et al., 2017). Some specific conditions may be required for SDE triggering, such as a large gas supply, links between the fault system and fluid sources, high-permeability pathways channeling fluids, and high local seismic activity in terms of numbers and magnitude.

8.4.5. Correlation with Tides

Two studies carried out around Taiwan investigated the relationship between seismic signals such as SDEs, gas emissions, and tides. In one case, eight short-period OBSs were deployed offshore southwest Taiwan (Hsu et al., 2013), and in the other case, four short-period OBSs were deployed offshore southeast Taiwan (Chang et al., 2016). In southwest Taiwan, the OBS deployment occurred close to fluid manifestations such as active mud volcanoes and pockmarks on the seafloor. Different seismic signals were detected there, including SDEs (called "individual impulse tremors" by Hsu et al. [2013]), long-duration

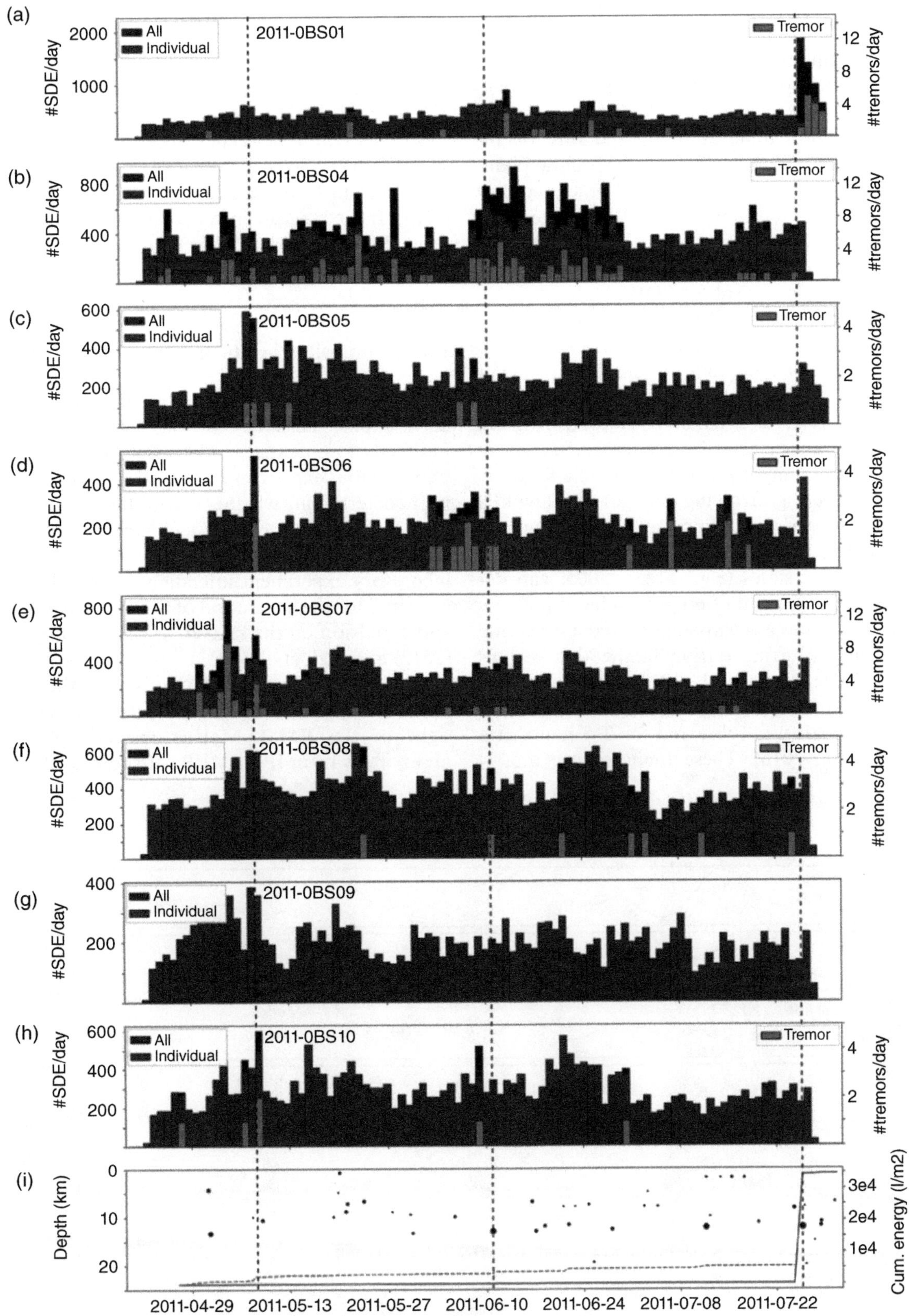

Figure 8.6 (a–h) Daily number of SDEs for all (black), background (dark gray), and tremor (swarm-like, gray) events detected by the OBS deployed in the SoM in 2011 (Géli et al., 2018). (i) Cumulative energy of regional seismicity (black dots) and cumulative energy times 100 of earthquakes until 25 July 2011, recorded by the instruments, are indicated by the solid and dotted lines, respectively. The black vertical lines show the timings of the largest earthquakes. Tsang-Hin-Sun et al. (2019) / Oxford University Press.

SDE tremors, earthquakes, and very-low-frequency earthquakes. The analysis of individual SDE shows three different parts, including a high-frequency wavetrain first, followed by a hybrid phase, and finishing with a monochromatic coda with a frequency at ~7 Hz and a progressive decrease in amplitude. Particle motions of the different wavetrains are primarily in the horizontal plane, consistent with P-waves or surface waves generated by a source close to the seafloor. Hsu et al. (2013) suggest that in most cases, for this kind of signal, only the monochromatic coda is well-recorded, which would explain the lack of more than one wave arrival for SDEs.

The temporal distribution of SDEs shows some correlation with ocean tides (Fig. 8.7), with increases in SDE activity potentially corresponding to semidiurnal and/or diurnal tide periods (Hsu et al., 2013). In addition, larger tides seem to lead to larger signal amplitudes. During this deployment, tides range primarily between ±0.5 m, corresponding to changes in hydrostatic pressure of a few kPa. Seismic event triggering involving changes of a few kPa have also been proposed for microseismicity, generally involving underlying factors such as high pore pressures or near-failure stress states (e.g., Freed, 2005; van der Elst & Brodsky, 2010). In addition, tide-induced pressure variations can influence gas emissions with the triggering of gas emissions occurring during falling tide periods (Römer et al., 2016; Sultan et al., 2020).

In southeast Taiwan, long-duration tremors involving short-duration signals were observed for 2–5 hours at a time (Chang et al., 2016). These tremors have mostly monochromatic frequency content at approximately 4 Hz, which is close to the geophone's natural frequency. The temporal distribution of these signals also follows the semidiurnal periodic pattern of ocean tides. In this case, the monochromatic frequency content close to the geophone's natural frequency suggests that these tremors are generated by the interaction between tide currents and the instrument–seafloor system (Chang et al., 2016). Indeed, ocean currents can generate tremors at different frequencies depending on the resonance frequencies of some parts of the ocean-bottom instruments (Essing et al., 2021).

8.4.6. Controlled in situ and Laboratory Experiments

Most studies attributing SDEs to different sources use correlations between various parameters, pointing indirectly to potential sources. Batsi et al. (2019) present two essential experiments: (1) an in situ experiment aiming at determining whether marine life can generate SDEs, as proposed for many OBS deployments around the world (e.g., Buskirk et al., 1981; Olofsson, 2010); and (2) a laboratory experiment intending to provide a "proof of concept" for the generation of SDEs due to gas migration and expulsion on the seabed, as proposed by Tary et al. (2012) and Hsu et al. (2013).

The in situ experiment consists of the deployment of two collocated OBSs during 48 days in the line of sight of the EMSO-Molène multidisciplinary cabled observatory from IFREMER, offshore Brittany, France.

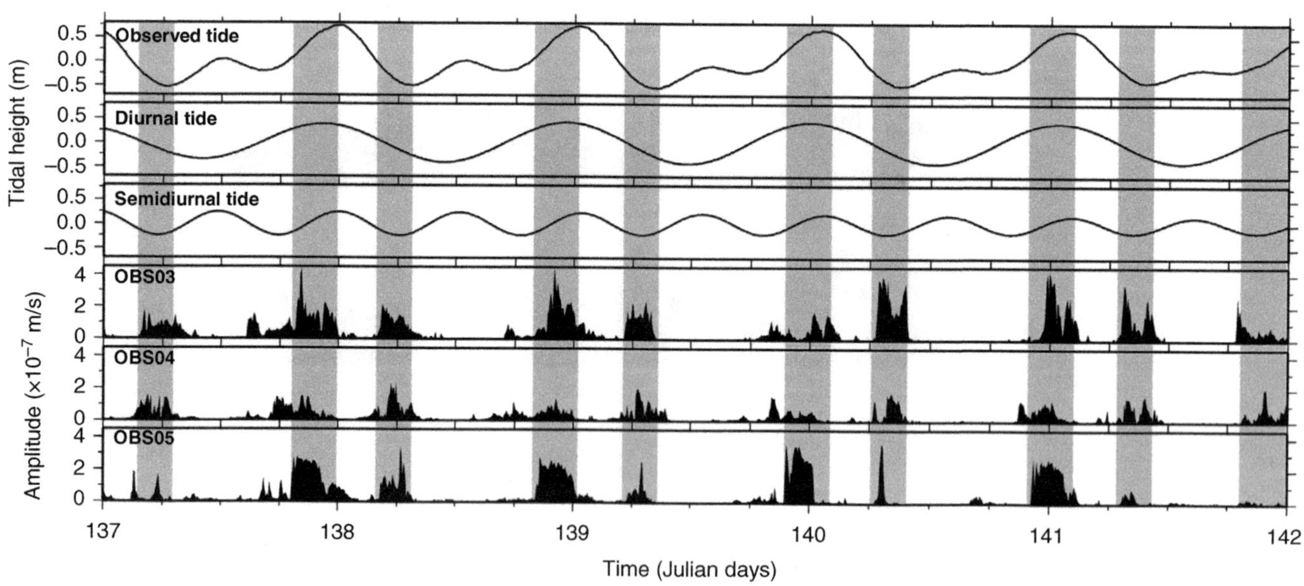

Figure 8.7 Tidal height and tremor amplitude for three OBSs deployed offshore southwest Taiwan. Vertical bands correspond to periods of higher-amplitude tremors. Adapted from Hsu et al. (2013).

This observatory and the two OBSs are located at shallow depths between 12 and 18 m, on a coarse sand layer over a granitic substratum, and in a very dynamic environment with a large seaweed field, active marine fauna, ocean tides with large amplitudes up to 8 m, and strong currents. The observatory includes a video camera taking high-resolution images every 2–3 s during daylight hours, which can monitor activity around the two OBSs. These OBSs included a hydrophone mounted on the frame and released their three-component (3-C) geophone on the seafloor after touching bottom.

Marine fauna around the OBSs includes fish, mollusks, and crustaceans. When interactions between fauna and the instruments occurred, the signals produced on the geophones were generally different from SDEs. Nonetheless, the temporal distribution of SDEs at the OBSs displays a diurnal periodicity and potentially a semidiurnal periodicity. Considering that the back azimuth of these SDEs stays relatively stable during the experiment and the coarse sand does not contain a substantial amount of gas, these periodicities are, in this case, not related to cycles of marine life activities but could be associated with ocean tide currents (Batsi et al., 2019).

The laboratory experiment of Batsi et al. (2019) consists of a 60 × 30 × 30 cm water tank filled with 12 cm of water and 12.5 cm of sand, on which is placed the same case containing a 3-C geophone used for the in situ experiment and in the field (Fig. 8.8). An outlet to inject air or water, connected to a pressure gauge, is located at a distance of 38 cm from the geophone and can inject these fluids into either the sand or water layer. Using air or water injected in the sediments, signals very similar to SDEs are recorded by the geophone when these fluids are migrating through the sediments (Fig. 8.8). Based on the observed waveforms, it is not possible to distinguish signals generated by air or water injections into the sediment layer. The signals generated by fluid injection display higher frequencies (i.e., between 70 and 200 Hz) than "natural" SDEs. This difference in frequency content may result from the smaller scale of the sources in this laboratory experiment compared with real-case fluid migrations on the seafloor. Injecting air directly into the water did not lead to clear observable signals on the geophone, indicating that small-scale air movements in water, in contrast to fluid motion through the upper sediment cover, do not correspond to the source of SDEs. As a result, the laboratory experiment of Batsi et al. (2019) represents an essential proof of concept for producing SDEs by fluid migration in sediments close to the seafloor.

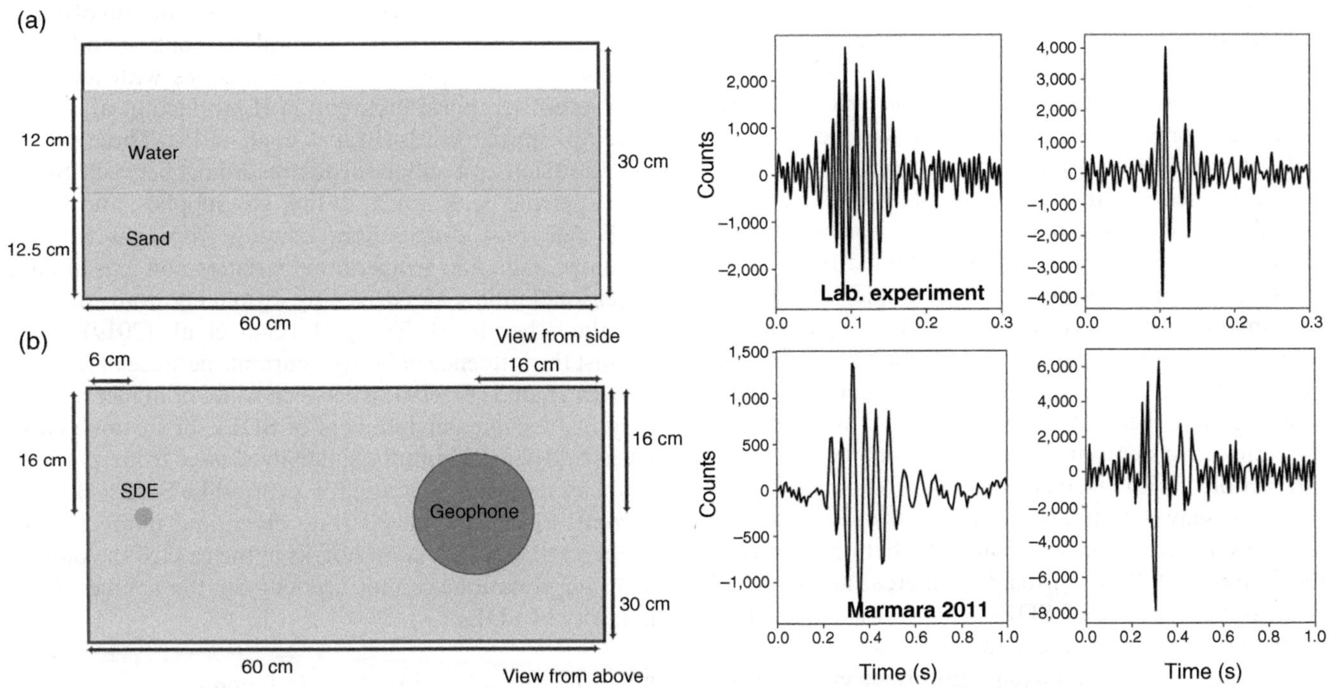

Figure 8.8 (a) Setup of the laboratory experiment where an outlet injecting air or water is 38 cm from a geophone. The fluids can be injected in either the sand or the water. This setup is designed to reproduce fluid movements that may correspond to the source of SDEs. (b) SDEs observed during the laboratory experiment and recorded by OBS-01 of the 2011 SoM deployment (geophone vertical component in all cases). A band-pass filter between 70 and 200 Hz was applied to the geophone data of the laboratory experiment. Adapted from Batsi et al. (2019).

8.5. DISCUSSION OF SDE POTENTIAL SOURCES

8.5.1. Biological Origin

One of the first and main interpretations of the source of SDEs is the biological origin (Buskirk et al., 1981; Olofsson, 2010). This interpretation is based on three primary pieces of indirect evidence: 24-hour and/or 12-hour periodicities potentially corresponding to biological cycles, the recovery of marine flora and fauna attached to the instruments, and a correlation between the number of SDEs and instrument depth corresponding to marine fauna and flora vertical distribution (i.e., more SDEs observed in shallower water). In theory, the variety of SDE waveforms would correspond to different interactions between marine life and the instruments (e.g., different ways of touching or hitting the instruments).

Although some SDEs can be the result of marine flora and fauna touching the instrument, bumping either the geophones or the instrument frame, other indirect evidence discussed below indicates that the majority of SDEs do not have a biological origin involving species in the water column. Diurnal and semidiurnal periodicities in SDE temporal distributions are seldom observed at instruments/surveys and can also be explained by the periodic action of ocean tides and currents on the instruments (Hsu et al., 2013; Chang et al., 2016). A similar point can be made for the correlation between instrument depth and SDE number. This correlation is rarely observed and could be explained, for example, by the influence of tidal hydrostatic pressure variations on gas supplies in superficial sediments (e.g., Römer et al., 2016).

During their in situ experiment, Batsi et al. (2019) did not observe a strong interest in the instruments from marine fauna. Moreover, noises generated by organisms (crustaceans and octopuses, for example) touching the instruments show waveforms different from those of SDEs (e.g., longer signals, higher-frequency content, "disorganized" waveforms). This indicates that SDEs generally are not generated by marine fauna, although different ecosystems may have different behaviors and interactions with instruments on the seafloor. The increased seaweed attached to the instrument also seems to have shielded it from external perturbations, decreasing the number of SDEs instead of increasing it. Finally, swarm-like episodes of SDEs with consistent back azimuth and frequency content are not consistent with a biological origin for all SDEs (e.g., Embriaco et al., 2014). It cannot be excluded that some SDEs may correspond to interactions of the instrument with marine flora/fauna, but these cannot explain the majority of SDEs observed. Still, it is important to note that all scenarios described so far concern marine life on or above the seafloor (e.g., fish bumps). Bioturbation (i.e., the reworking of marine near-surface sediments by animals) has not been investigated yet.

8.5.2. Action of Ocean/Sea Currents

Several studies discuss the impact of bottom currents on OBS data. These currents are one of the primary sources of background noise for OBSs (Webb, 1998), and their influence has to be considered when unknown signals like SDEs are observed in OBS data. Bottom currents can generate noise for various reasons, such as moving the OBS case/frame, vibrations of instrument parts (e.g., Essing et al., 2021), or transport of particles hitting the instrument. The susceptibility to current-induced noises depends on the OBS design (Duennebier et al., 1981; Hilmo & Wilcock, 2020). In Hilmo and Wilcock (2020), on the Cascadia continental slope, seismic noise in the 5–12 Hz band corresponding to the frequency band of local microseisms is not predicted by the bottom currents in the ocean circulation model. In this region, noise in the 5–12 Hz band could be related to SDE occurrence due to local sources.

In some instances, the generation of SDEs is attributed directly to bottom currents, often associated with tides, such as in studies focusing on the eastern Canary Islands (Ugalde et al., 2019) and southeast Taiwan (Chang et al., 2016). In other instances, bottom currents are involved in correlations between SDE temporal distribution and other time series. Examples include correlations with methane peak event temporal distribution (Embriaco et al., 2014) or temperature records (Franek et al., 2017). These studies use the (lack of) simultaneous recording between SDEs and methane peak events to link gas supplies, and hence SDE temporal distribution, coming from gas hydrate decomposition and temperature variations on tens of days to seasonal time scales driven by regional bottom currents. On the other hand, Tsang-Hin-Sun et al. (2019) argue against the influence of bottom currents because of the lack of clear spatial correlation between zones of higher bottom currents and greater numbers of SDEs, or in some cases when SDE back azimuths suddenly change from scattered to focused in one direction for swarm-like SDEs. Bottom currents can hence have strong direct and indirect influences on SDE production but, as in the case of the biological interpretation, cannot account for the source of the majority of SDEs.

8.5.3. Fluids in Near-Surface Sediments

Conceptual Source Model with the Resonance of Fluid-Filled Cracks

Many SDEs exhibit frequency content characterized by one or several frequencies (Díaz et al., 2007;

D'Alessandro et al., 2009; Franek et al., 2017; Tsang-Hin-Sun et al., 2019) (Fig. 8.1), similar to long-period seismicity observed at volcanoes (Chouet, 1996). This similarity led to the proposition that SDEs could result from a sudden pressure change followed by the resonance of a cavity or conduit most likely located in superficial sediments. The source of pressure transients could correspond to small, local changes in the stress field (Díaz et al., 2007), sudden overpressures/underpressures due to gas migration (Tary et al., 2012), or any phenomena that influence these processes (e.g., ocean tides, temperature variations). The resonance of fluid-filled cracks is the most widely used model to explain long-period seismicity, although other models were proposed, such as nonlinear fluid flow (Julian, 1994; Hellweg, 2000; Tary et al., 2014b) or channel wall deformations (Rust et al., 2008).

The resonance of a fluid-filled crack corresponds to the sudden perturbation of its internal pressure followed by the interaction of an acoustic wave within the crack and the crack walls, which generates an interface wave at the walls called *crack wave* (e.g., Maeda & Kumagai, 2017) or *Krauklis wave* (Korneev, 2011). As the Krauklis wave is an interface wave, its amplitude decreases quickly away from the crack walls, and the wavefield observed at seismological stations corresponds to its diffraction at the crack tips (Frehner & Schmalholz, 2010). The model assumes that either the fluid is inviscid or the fracture is large enough to neglect viscosity effects. The viscosities of the main fluids present on the seafloor (e.g., gas, water, oil) are generally low enough to be neglected even for thin fractures ($\leq 10^{-2}$ m) at seismic frequencies (i.e., thick-fracture regime of Korneev [2008]). The phase velocity of the Krauklis wave is particularly low, which allows even low-frequency resonances in the range of 0.1–10 Hz to correspond to smaller-size resonators on the order of tens to hundreds of meters. Another interesting feature of the rectangular fluid-filled crack model is the presence of higher modes that are not integer multiples of the fundamental frequency (e.g., Maeda & Kumagai, 2017). To characterize the resonator, the frequency and quality factor can be extracted for each frequency peak. These measures depend on the excitation characteristics, the crack geometry, and macro properties of the fluid and surrounding medium.

In the case of SDEs, only peak frequencies are generally reported, although quality factors contain interesting information about the oscillator damping, which depends on the fluid properties (Kumagai & Chouet, 2000). The resonance frequency of a thin rectangular crack filled with inviscid fluid is given by (Maeda & Kumagai, 2017)

$$f_m^L = \frac{(m-1)a}{2L\left(1+2\epsilon_m^L C\right)^{1/2}}, \qquad (8.1)$$

for the longitudinal mode, where L is the crack length ($L > W$ the crack width), m the mode (for wavelengths $\frac{2L}{m}$, $m \geq 2$), C the crack stiffness ($C = \left(\frac{b}{\mu}\right)/\left(\frac{L}{d}\right)$), a the fluid sound velocity, b the fluid bulk modulus, d the crack aperture ($d \ll W$), μ the rigidity modulus of the surrounding medium, and ϵ_m^L is given by

$$\epsilon_m^L = \begin{cases} (2/9m)(mW/4\gamma L)^{1/2} & \text{for } (W/L < 4\gamma/m) \\ (1/3m)(1-4\gamma L/3mW) & \text{for } (4\gamma/m < W/L < 1) \end{cases}, \qquad (8.2)$$

where γ quantifies spatial crack edge effects ($\gamma = 0.22$). For transverse modes of resonance, the resonance frequency is given by (Maeda & Kumagai, 2017)

$$f_m^W = \frac{(m-1)a}{2W\left(1+2\epsilon_m^W C\right)^{1/2}}, \qquad (8.3)$$

where ϵ_m^W is expressed as

$$\epsilon_m^W = (W/L)(1/3m)(1-4\gamma W/3mL). \qquad (8.4)$$

Then I evaluate the possibility of generating frequency peaks with a fluid-filled crack close to the seafloor using the mechanical properties of superficial unconsolidated sediments listed in Table 8.2 and either methane or water as the infilling fluid. Resonance frequencies for the fundamental mode of both longitudinal and transverse modes are presented in Fig. 8.9 for a range of crack lengths. Following these calculations for a crack filled with methane, a range of resonance frequencies f_m^L between 10 and 30 Hz (fundamental mode $m = 2$) would correspond to crack

Table 8.2 Properties of fluids and unconsolidated sediments for calculating resonance frequencies given by equations 8.1 and 8.3.

Properties	Values
Water density	1,000 kg/m^3
Sediment density	1,500 kg/m^3
Methane gas density	90 kg/m^3
Sound speed of water	1,500 m/s
Sound speed of methane gas	420 m/s
Bulk modulus of water	2.1 GPa
Bulk modulus of methane gas	~10 MPa
Sediment rigidity modulus	60 MPa ($\mu = \beta^2 \rho_s$)
Sediment shear wave velocity β	200 m/s
Crack width	0.1 m
Crack aperture	0.01 m

A water depth of 1,000 m, corresponding to a hydrostatic pressure of about 10 MPa and a temperature of 0°C, is assumed for some of the values presented.

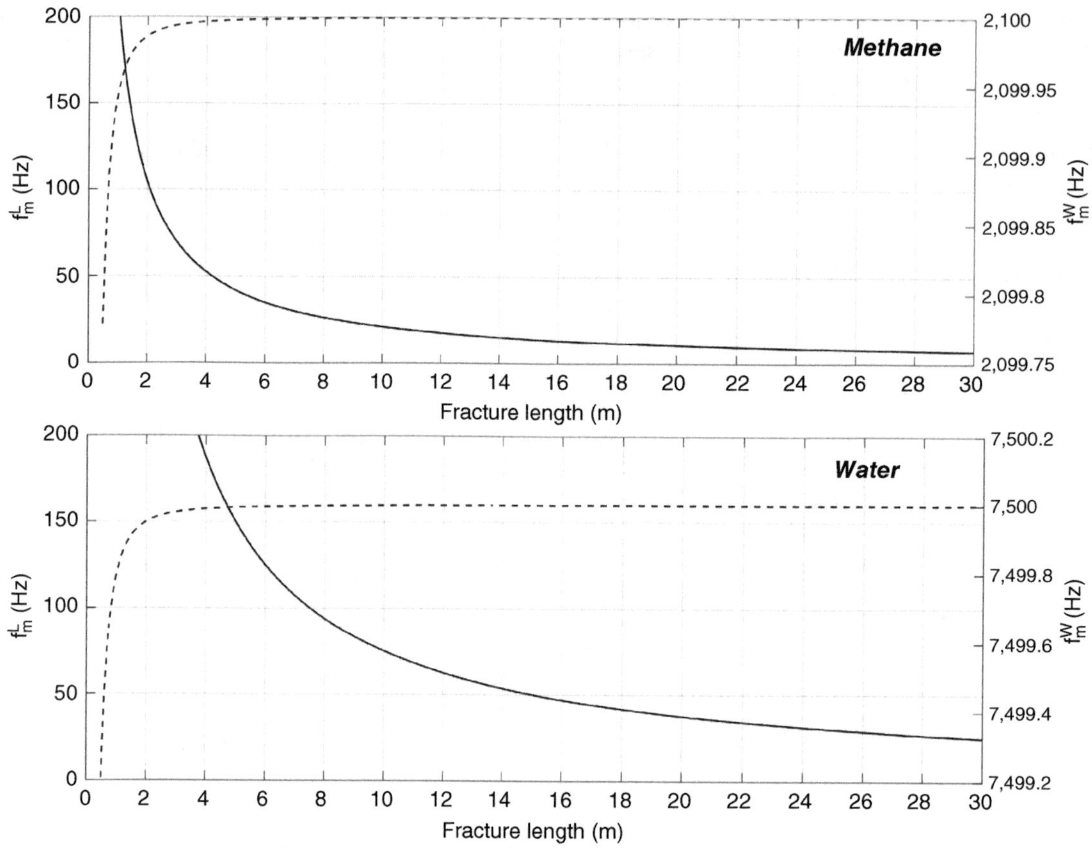

Figure 8.9 Resonance frequencies corresponding to the fundamental mode of the longitudinal (f_m^L, solid line) and transverse (f_m^W, dashed line) modes of a rectangular fluid-filled crack (filled with methane at top and water at bottom), calculated using equations 8.1 and 8.3 and material properties corresponding to an unconsolidated sediment (see Table 8.2).

lengths between 21 and 7 m. In the case of a water-filled crack, the same range of resonance frequencies f_m^L (fundamental mode) would correspond to crack lengths greater than 25 m. The resonance frequencies for the transversal mode show values much larger than the resonance peaks observed.

Some fluid and fracture properties influence the Krauklis wave velocity during fracture resonance. An increase in fluid viscosity leads to increased wave attenuation (e.g., Liang et al., 2017) and decreased phase velocity (Cao et al., 2021a). Low porosities, defined in Korneev (2011) as the ratio between the thicknesses of fluid and elastic layers, decrease the phase velocity or prevent the appearance of Krauklis waves. A recent experimental study focused on the potential effects of fracture properties on Krauklis wave propagation. In this study, Cao et al. (2021b) observed that increases in fracture roughness and fracture stiffness lead to a decrease and an increase in phase velocity, respectively. A decrease/increase in Krauklis wave velocity results in lower/higher resonance frequencies and larger/smaller resonator sizes.

Conceptual Source Model of Gas Migration in Sediments

The presence of fluid manifestations on the seabed is extremely common. Depending on the geological context, these fluid manifestations can take diverse expressions such as black smokers close to magmatic zones; hydrocarbons seeping from deeper reservoirs; gas hydrates, where pressure, temperature, and methane concentration conditions are met; brackish or freshwater seeps; gas flares in the water column; bacterial mats; pockmarks; and mud volcanoes. In addition to the ubiquitous character of fluid manifestations, the primary characteristics of SDEs (see Section 8.2 and Table 8.1) often point to a source in the superficial part of the seafloor (i.e., <20–30 m), usually consisting of a layer of unconsolidated sediments. With gas in the SDE source, some of the gas properties can help explain other observations, such as source responses to hydrostatic pressure variations (i.e., gas solubility changes due to different instrument depths or ocean tides) (e.g., Sultan et al., 2011; Tary et al., 2012; Römer et al., 2016), and temperature variations in the case of gas hydrates (Franek et al., 2017).

The specific signal characteristics of SDEs are similar to those produced by the opening and resonance of a fluid-filled cavity (e.g., Ferrazzini et al., 1990; Vidal et al., 2006). This led Tary et al. (2012) to suggest that SDEs may be generated by gas migration and expulsion at the seafloor (Fig. 8.10). Gas migrates through unconsolidated sediments by fracturing the sediment or using preexisting fractures (Boudreau et al., 2005). These fractures are relatively unstable, and fluids must repeatedly produce new conduits to reach the seafloor. Due to the lack of sediment cohesion close to the seafloor, the resistance to vertical fluid migration is low. The movements generated by the rising gas pockets are then primarily horizontal movements of the conduit walls (Fig. 8.10), consistent with many SDEs showing particle motions that are mostly in the horizontal plane (Tary et al., 2012; Bowman & Wilcock, 2014). These movements can be generated either within the sediment during gas migration or at the seabed during gas expulsion.

The sidewall horizontal movements are then modeled by Tary et al. (2012) using a horizontal force located at different depths (2 and 5 m depth) within very attenuating sediments (see Tary et al. [2012] for more details on its properties). This modeling shows that low-amplitude body waves and larger-amplitude surface waves are generated (Fig. 8.11). These surface waves, called *Scholte waves*, are attenuated quickly in both the sediments and the water column, consistent with some SDE characteristics (i.e., signals recorded at a single instrument at a time, difficulty of hydrophones to record SDEs). However, many SDEs exhibit either linear particle motions or larger signal amplitudes on the vertical component and cannot be explained by the results of this modeling.

8.5.4. Low-Magnitude Seismicity

In some active regions, either magmatically, tectonically, or both, SDEs are sometimes attributed to microseismic events of very low magnitudes (Sohn et al., 1995; Sgroi et al., 2014). These SDEs generally have the same signal characteristics as those listed in Table 8.1. Seismic events corresponding to small ruptures, including source types such as double-couple, tensile opening and closing, or CLVD (compensated vector linear dipole), are characterized by broader spectra than sources involving the resonance of fluid-filled cavities. In many instances, the frequency content of SDEs displays such a broad spectrum around a dominant frequency; however, applying the fast Fourier transform to a short-duration signal results in spectral smearing, which can widen frequency peaks "artificially" (Tary et al., 2014a). On the other hand, a peaked spectrum doesn't guarantee that the source is of the fluid-filled cavity type due to the many potential sources of resonance for a seismic signal (i.e., the problem of seafloor-geophone coupling, instrument resonance, and path effects).

In general, SDEs exhibit one-phase arrival, except for a fraction of them that present two-phase arrivals spaced by about a second or less (e.g., Chen et al., 1982; Prothero & Reid, 1982; Hsu et al., 2013; Sgroi et al., 2014; Bowman &

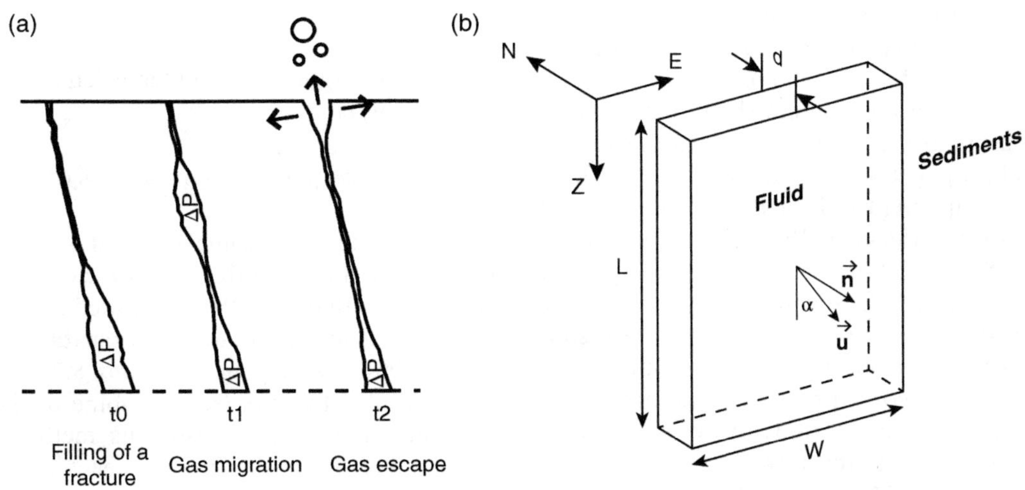

Figure 8.10 (a) Conceptual model of the generation of SDEs due to gas migration and expulsion at the seafloor. SDEs are generated by any sudden wall movements due to cavity opening/closing during fluid migration and pore pressure adjustments. The resonance of these cavities can follow pore pressure changes. Tary et al. (2012) / Oxford University Press. (b) Fluid-filled crack model used to model the SDEs as tensile opening/closing (L: crack length, W: crack width, d: crack aperture). The slip vector $\vec{u}$ makes an angle α with the crack plane. Vector $\vec{n}$ is the normal to the crack plane.

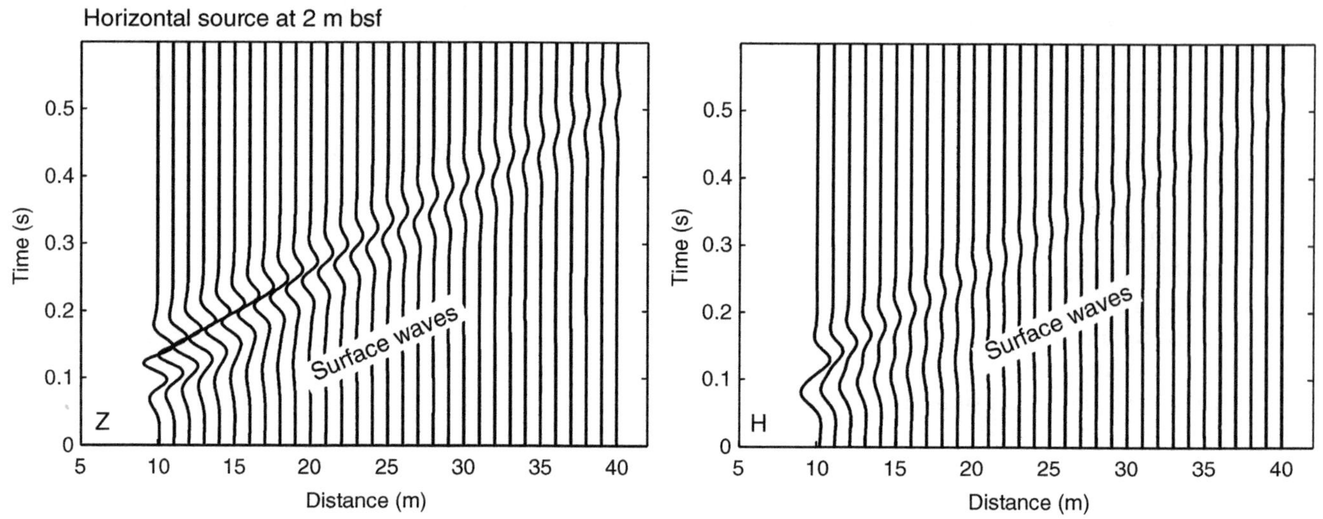

Figure 8.11 Synthetic seismograms calculated using the SKB code (Dietrich, 1988) for a horizontal force source at 2 m bsf (below seafloor) and receivers at 1 cm within the unconsolidated sediments and at distances of 10 to 40 m from the source. The surface waves correspond to Scholte waves. H: horizontal component. Z: vertical component. Tary et al. (2012) / Oxford University Press.

Wilcock, 2014), and polarizations either rectilinear or elliptical. Having only one-phase arrival with an elliptical polarization doesn't preclude these SDEs from being small seismic ruptures. Wavefields generated by seismic sources in the near field (i.e., up to a few P-wave wavelengths) display one-phase arrival and a strong elliptical polarization for a wide range of source back azimuths (Vavryčuk, 1992). However, the ellipticity proportion in the near-field arrivals decreases for sources involving significant volumetric components (e.g., tensile cracks) (Lokmer & Bean, 2010). Elliptical polarizations for body waves can also arise from their propagation in an anelastic medium (Červený & Pšenčík, 2005). Theoretically, P-waves generated by seismic sources should be recorded by pressure sensors in the water column. In this case, a near-horizontal propagation, subparallel to the seafloor, could explain the apparent lack of P-wave transmission to the water column (Stein & Wysession, 2003; Tary, 2011). Many SDEs exhibit near-horizontal particle motions (Tary et al., 2012; Bowman & Wilcock, 2014; Sgroi et al., 2014). This suggests that the potential seismic source of SDEs would be located (1) close to the instrument so as to not be recorded by more than one instrument, and (2) close to the seafloor to avoid near-field and P-wave transmission to the seawater.

The seismic moment M_0 associated with wall movements due to pressure changes within the fracture can be estimated using (Lipovsky and Dunham, 2016)

$$M_0 \sim \mu^* L^2 \Delta w, \tag{8.5}$$

where μ^* is $\mu^* = \mu/(1 - \nu_s)$, ν_s the Poisson ratio of the solid medium, L the fracture length, and Δw the wall movements. Assuming the material has a linear, elastic behavior, the pressure change due to the wall movements can be expressed as

$$\Delta p \sim \frac{\mu^* \Delta w}{L}. \tag{8.6}$$

Using equations 8.5 and 8.6, the seismic moment can be expressed as a function of the pressure change inside the fracture:

$$M_0 \sim \Delta p\, L^3. \tag{8.7}$$

Moment magnitudes are then calculated using (Lee & Stewart, 1981)

$$M_w = \frac{2}{3} \log_{10} M_0 - 6.03. \tag{8.8}$$

To estimate the seismic moment associated with an SDE due to a sudden fluid release close to the seafloor, let's consider that an SDE is generated by a rapid pressure drop from a lithostatic to a hydrostatic level in a fluid-filled fracture. Applying equations 8.7 and 8.8, a 1 m methane-filled fracture (reusing some of the parameters listed in Table 8.2), releasing its methane due to the overpressure inside the fracture reaching the lithostatic pressure, corresponds to a moment magnitude of about –3.3. Such moment magnitudes correspond to events called *picoseismicity* (M_w <–3) and are generally below the detection capabilities of seismic networks (Kwiatek & Ben-Zion, 2016). It may be possible for sensors to detect events of moment magnitudes in the range –5 to –3 if the sensor is less than a few tens of meters away.

In the case of geological contexts such as hydrothermal fields and mid-ocean ridges, this model provides a possible mechanism to generate picoseismicity close to the seafloor due to small changes in temperatures, for example (Sohn et al., 1995), although another necessary condition is the presence of suitable active tectonic structures close to the instrument. In other contexts where the overpressure would arise from gas migration in unconsolidated sediments, the source of SDEs would need to be within a few meters to a few tens of meters from the source. Pressure build-up within sediments due to some material preventing fluid migration (e.g., impermeable or gas hydrate layers) can also lead to natural hydraulic fracturing in superficial sediments and generate picoseismic events (Blouet et al., 2017). On the other hand, a seismic source located in unconsolidated sediments may not involve fracturing but could correspond to the tensile opening and/or closing of a conduit.

8.5.5. Source Modeling

To model SDE waveforms that may correspond to the opening/closing of a conduit, the model configuration used is a vertical crack in a half-space consisting of a very attenuating medium corresponding to unconsolidated sediments (i.e., the same parameters listed in Table 8.2, a P-wave velocity of 1550 m/s, and quality factors for P- and S-waves of 10). The vertical crack is aligned in the east–west direction and located in an isotropic, homogenous, elastic medium close to the free surface at 2 m depth (Figs 8.10 and 8.12). The receivers are laid out at distances between 10 and 50 m from the source in two lines and one circular array to examine the resulting wave types and waveform characteristics.

The Green's functions are calculated using the frequency-wavenumber integration method of Bouchon (1981) implemented by Zhu and Rivera (2002). The Green's functions are sampled at 0.0005 s. The resulting Green's functions are then convolved with a triangular source-time function, taking into account the source rise time. The rise time of a circular tensile crack, taken here as the time needed for a complete opening or closing of the crack, is calculated as (Walter & Brune, 1993)

$$\tau_R \approx \frac{2r}{0.8\,\beta}, \tag{8.9}$$

where r is the crack radius and β the S-wave velocity. Assuming that cracks open/close completely, cracks at depths between 0.5 and 2 m are associated with rise times between 0.003 and 0.013 s. The tensile mechanism of the event source is defined by its moment tensor. The moment tensor for a vertical tensile source with an opening/closing motion along the north–south direction can be expressed as (Vavryčuk, 2001)

$$\boldsymbol{M} = u\begin{bmatrix} \lambda\sin\alpha - \mu(1-\sin\alpha) & 0 & 0 \\ 0 & \lambda\sin\alpha + \mu(1+\sin\alpha) & 0 \\ 0 & 0 & \lambda\sin\alpha \end{bmatrix}, \tag{8.10}$$

where u is the slip and α is the angle between the crack plane and the slip vector ($\alpha = 90°$ for a pure tensile opening, and $\alpha = -90°$ for a pure closing motion; Fig. 8.10). In this modeling, the seismic moment is defined to correspond to a $M_w \sim -3$ event, and the components of $\boldsymbol{M}$ are normalized by its maximum component ($\lambda + 2\mu$).

The synthetic seismograms obtained for an opening crack with rise times of 0.003 and 0.013 s are presented in Figs. 8.13 and 8.14. These seismograms show one or two wave arrivals, depending on the source–receiver distance, corresponding to the P- and S-wave. No surface

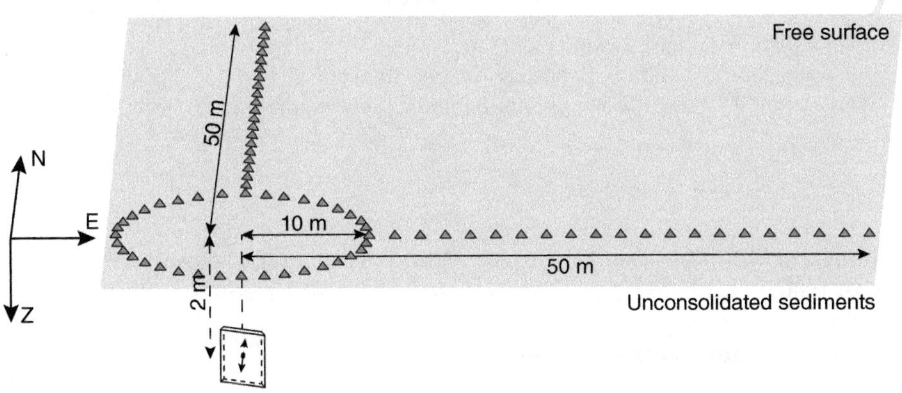

Figure 8.12 Modeling setup for calculating synthetic seismograms using a vertical opening/closing crack at 2 m depth, a very attenuating medium corresponding to unconsolidated sediments, and different receiver arrangements. Receivers are spaced by 2 m along the lines, and every 10° at a 10 m distance for the receivers laid out in a circle.

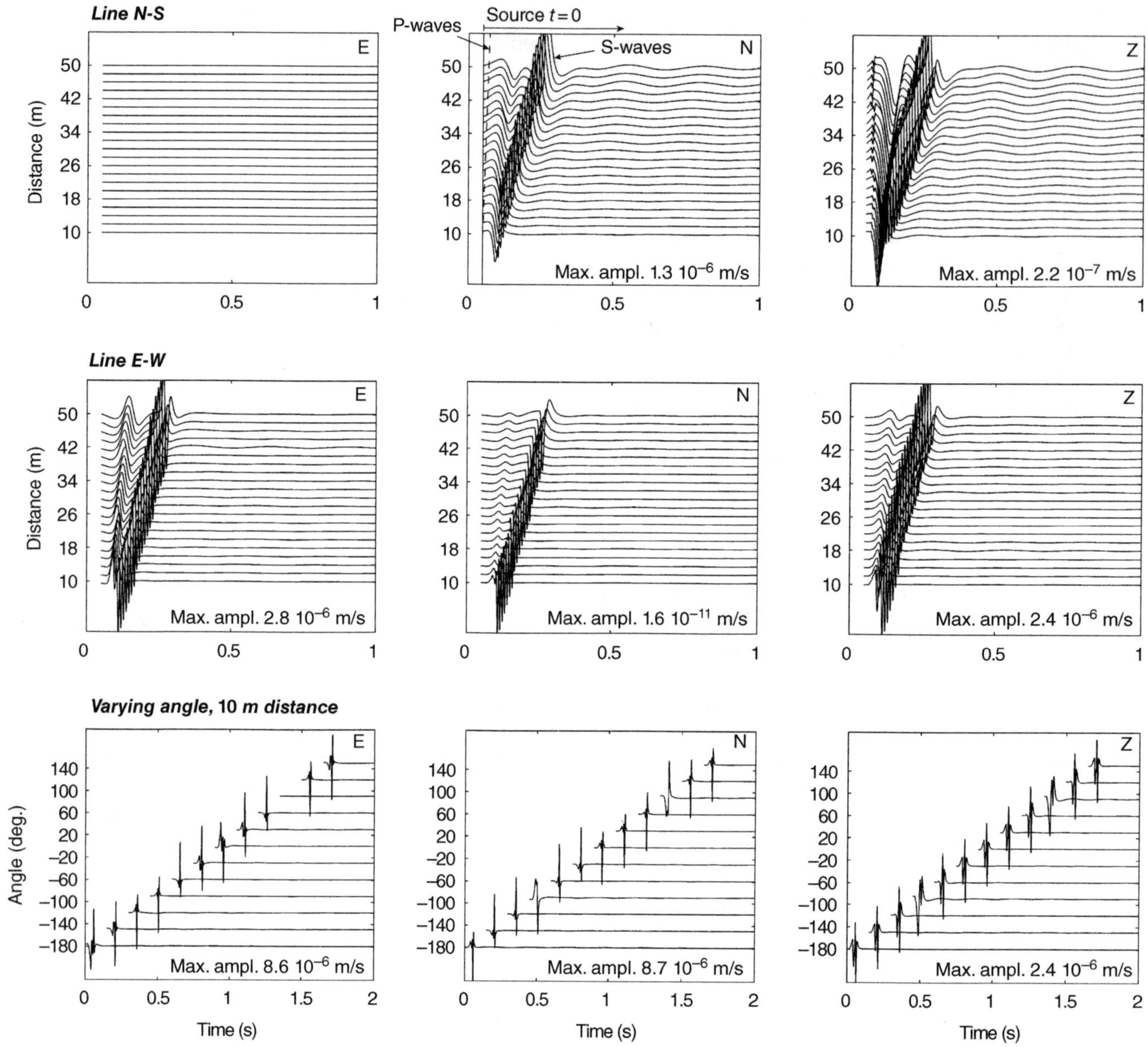

Figure 8.13 Normalized synthetic seismograms calculated using the setup presented in Fig. 8.12, an opening crack at 2 m depth, and a rise time of 0.003 s, band-pass filtered between 5 and 150 Hz (E: east-west component, N: north-south component, Z: vertical component). Synthetic seismograms for varying angles are time-shifted for display only.

waves are observed in this configuration. Using a closing crack with a rise time of 0.003 s leads to similar signals (Fig. 8.15). Decreasing the rise time, corresponding to smaller sources according to equation 8.9, results in higher-frequency waveforms. Hence, variations in signal frequency for SDEs and the observation of one or two wave arrivals could be related to different source dimensions, source depths, and distances between sources and receivers. Longer-duration SDE signals could then be composed of a series of time-delayed opening and closing conduits or cracks close to the seafloor, potentially followed by their resonance for SDEs exhibiting peaked frequency contents.

8.6. CONCLUSION

SDE signals distinguish themselves from microseismicity by a few specific properties: their short durations, peaked frequency content, lack of secondary arrivals,

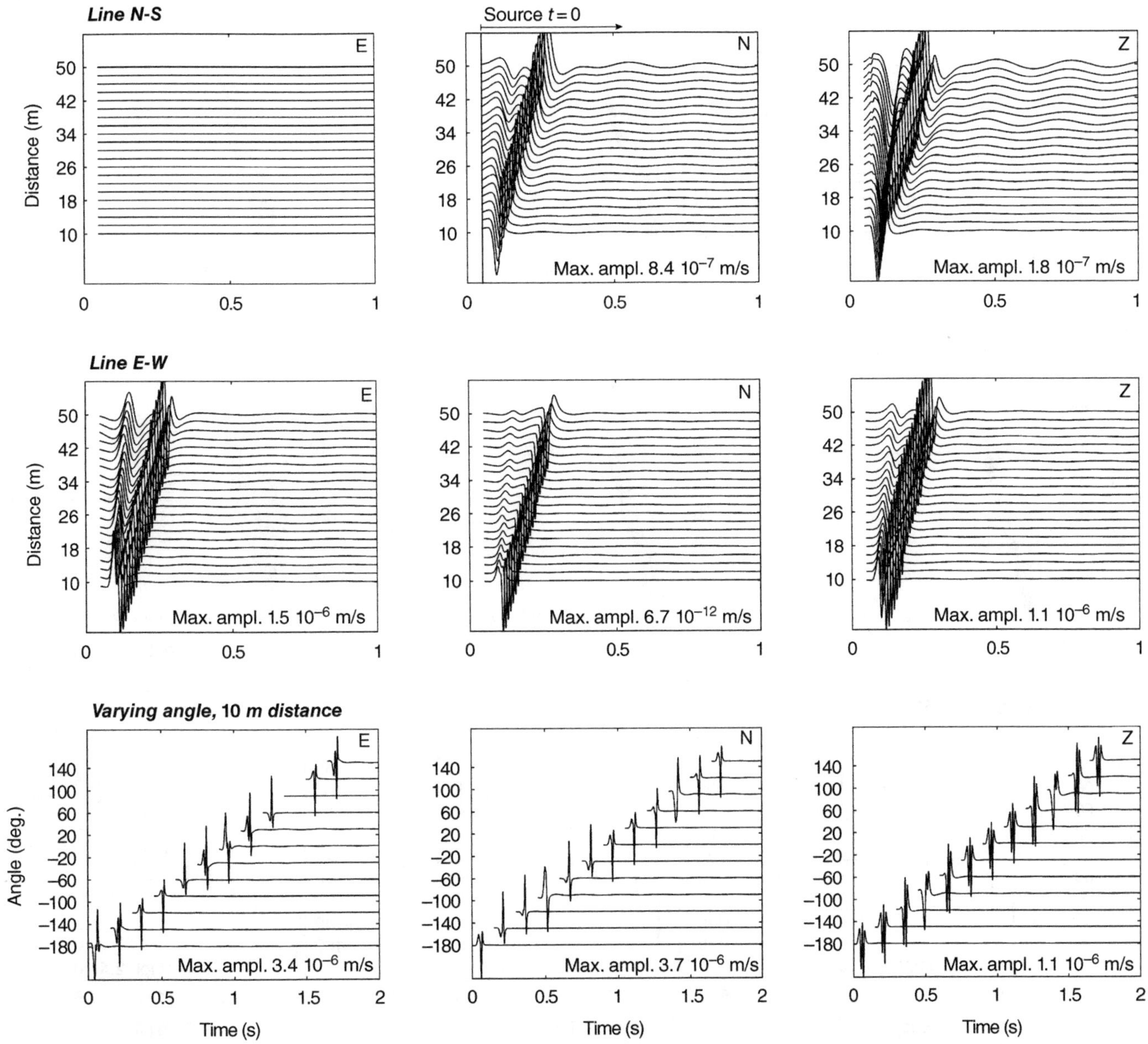

Figure 8.14 Normalized synthetic seismograms calculated using the setup presented in Fig. 8.12, an opening crack at 2 m depth and a rise time of 0.013 s and band-pass filtered between 5 and 150 Hz (E: east-west component, N: north-south component, Z: vertical component). Synthetic seismograms for varying angles are time-shifted for display only.

(at times) large amplitudes, and in particular their recording on one instrument at a time. These properties point to a local source of small dimensions relatively close to the seafloor. SDEs are detected in various geological contexts worldwide, although their actual spatial distribution remains undetermined as they are often not reported. The temporal distribution of SDEs sometimes reveals diurnal and semidiurnal periodicities that were first used as an indication of biological origin but may be related to ocean tides, as is sometimes the case for microseismicity as well (e.g., Scholz et al., 2019). As for microseismicity, SDEs occur both as individual events (background events) and in swarms of tens to thousands of events. These swarmed events exhibit very stable properties (back azimuth and frequency content) that are generally different from background events.

Several interpretations provide explanations for most of these observations. However, accounting for all observations may not be relevant as some SDEs may have different origins. The first primary interpretation of SDEs was the biological origin, which, in light of recent evidence and in situ experiments, cannot explain the majority of

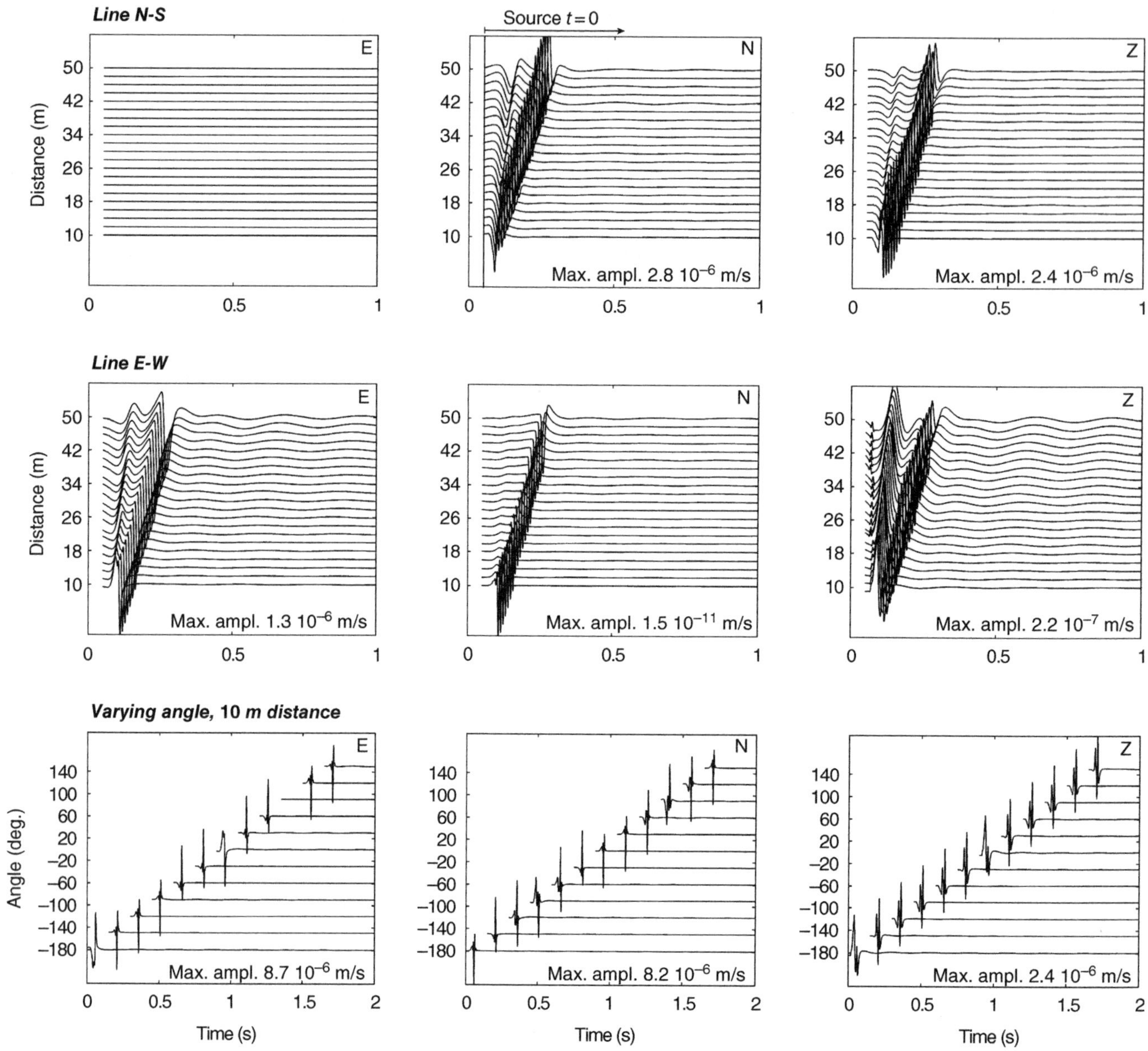

Figure 8.15 Normalized synthetic seismograms calculated using the setup presented in Fig. 8.12, a closing crack at 2 m depth and a rise time of 0.003 s and band-pass filtered between 5 and 150 Hz (E: east-west component, N: north-south component, Z: vertical component). Synthetic seismograms for varying angles are time-shifted for display only.

SDEs. Other interpretations involve the presence of gas, which can be essential to account for some of the spatio-temporal patterns observed (e.g., correlation between instrument depth and number of SDEs). For instance, methane-filled fractures of lengths between 7 and 21 m could produce a resonance frequency corresponding to those of SDEs. On the other hand, picoseismicity close to the seafloor or small opening/closing crack events corresponding to fluid overpressures and draining of cracks with relatively small lengths of a few meters or less could also produce SDEs.

Several essential studies and experiments have been carried out to better understand SDE generation processes (e.g., multidisciplinary deployments, laboratory, and in situ experiments). However, more studies and modeling are required to propose one or more reliable source mechanisms for SDEs including, for instance, laboratory experiments with different kinds of sediments, in situ experiments in other contexts, integrated modeling including viscoelastic material deformation and wave propagation (e.g., due to gas migration and expulsion in unconsolidated sediments), more long-term multidisciplinary deployments searching for a correlation between SDE dynamics and other phenomena (e.g., geochemical variations, fluid flow and current meters, seismicity), and improved SDE signal characterization and classification to determine whether more than one mechanism/interpretation accounts for the different SDE families observed. Consequently, SDE signals are promising events that could provide interesting complementary information about some largely unknown processes occurring on the seafloor.

ACKNOWLEDGMENTS

I thank all crew members and technical and scientific staff who participated in the many marine surveys reported in this study. I also thank Louis Géli and four anonymous reviewers for their valuable comments.

REFERENCES

Albarello, D., Palo, M., & Martinelli, G. (2012). Monitoring methane emission of mud volcanoes by seismic tremor measurements: a pilot study. *Natural Hazards and Earth System Sciences*, *12*(12), 3617–3629.

Batsi, E., Tsang-Hin-Sun, E., Klingelhoefer, F., Bayrakci, G., Chang, E. T., Lin, J. Y., et al. (2019). Nonseismic signals in the ocean: indicators of deep sea and seafloor processes on ocean-bottom seismometer data. *Geochemistry, Geophysics, Geosystems*, *20*(8), 3882–3900.

Bayrakci, G., Scalabrin, C., Dupré, S., Leblond, I., Tary, J. B., Lanteri, N., et al. (2014). Acoustic monitoring of gas emissions from the seafloor. Part II: a case study from the Sea of Marmara. *Marine Geophysical Research*, *35*(3), 211–229.

Blouet, J. P., Imbert, P., & Foubert, A. (2017). Mechanisms of biogenic gas migration revealed by seep carbonate paragenesis, Panoche Hills, *California. AAPG Bulletin*, *101*(8), 1309–1340.

Bouchon, M. (1981). A simple method to calculate Green's functions for elastic layered media. *Bulletin of the Seismological Society of America*, *71*(4), 959–971.

Boudreau, B. P., Algar, C., Johnson, B. D., Croudace, I., Reed, A., Furukawa, Y., et al. (2005). Bubble growth and rise in soft sediments. *Geology*, *33*(6), 517–520.

Bourry, C., Chazallon, B., Charlou, J. L., Donval, J. P., Ruffine, L., Henry, P., et al. (2009). Free gas and gas hydrates from the Sea of Marmara, Turkey: Chemical and structural characterization. *Chemical Geology*, *264*(1–4), 197–206.

Bowman, D. C., & Wilcock, W. S. (2014). Unusual signals recorded by ocean bottom seismometers in the flooded caldera of Deception Island volcano: volcanic gases or biological activity?*Antarctic Science*, *26*(3), 267.

Brocher, T. M., & Iwatake, B. T. (1982). Sources of low-frequency ambient seafloor noise on a continental shelf. *Bulletin of the Seismological Society of America*, *72*(4), 1129–1142.

Buskirk, R. E., Frohlich, C., Latham, G. V., Chen, A. T., & Lawton, J. (1981). Evidence that biological activity affects ocean bottom seismograph recordings. *Marine Geophysical Researches*, *5*(2), 189–205.

Cao, H., Medici, E., & Askari, R. (2021a). Physical modeling of fluid-filled fractures using the dynamic photoelasticity technique. *Geophysics*, *86*(1), T33–T43.

Cao, H., Nakagawa, S., & Askari, R. (2021b). Laboratory measurements of the impact of fracture and fluid properties on the propagation of Krauklis waves. *Journal of Geophysical Research: Solid Earth*, e2020JB021593.

Červený, V., & Pšenčík, I. (2005). Plane waves in viscoelastic anisotropic media—II. Numerical examples. *Geophysical Journal International*, *161*(1), 213–229.

Chang, E. T., Chao, B. F., Chen, G. Y., & Liau, J. M. (2016). Internal tides recorded at ocean bottom off the coast of Southeast Taiwan. *Journal of Geophysical Research: Oceans*, *121*(5), 3381–3394.

Chen, A. T., Frohlich, C., & Latham, G. V. (1982). Seismicity of the forearc marginal wedge (accretionary prism). *Journal of Geophysical Research: Solid Earth*, *87*(B5), 3679–3690.

Chouet, B. A. (1996). Long-period volcano seismicity: its source and use in eruption forecasting. *Nature*, *380*(6572), 309–316.

D'Alessandro, A., D'Anna, G., Luzio, D., & Mangano, G. (2009). The INGV's new OBS/H: analysis of the signals recorded at the Marsili submarine volcano. *Journal of Volcanology and Geothermal Research*, *183*(1–2), 17–29.

Díaz, J., Gallart, J., & Gaspà, O. (2007). Atypical seismic signals at the Galicia Margin, North Atlantic Ocean, related to the resonance of subsurface fluid-filled cracks. *Tectonophysics*, *433*(1–4), 1–13.

Dietrich, M. (1988). Modeling of marine seismic profiles in the t-x and τ-p domains. *Geophysics*, *53*(4), 453–465.

Duennebier, F. K., Blackinton, G., & Sutton, G. H. (1981). Current-generated noise recorded on ocean bottom seismometers. *Marine Geophysical Researches*, *5*(1), 109–115.

Dupré, S., Scalabrin, C., Grall, C., Augustin, J. M., Henry, P., Şengör, A. C., et al. (2015). Tectonic and sedimentary controls on widespread gas emissions in the Sea of Marmara: results from systematic, shipborne multibeam echo sounder water column imaging. *Journal of Geophysical Research: Solid Earth*, *120*(5), 2891–2912.

Embriaco, D., Marinaro, G., Frugoni, F., Monna, S., Etiope, G., Gasperini, L., et al. (2014). Monitoring of gas and seismic energy release by multiparametric benthic observatory along the North Anatolian Fault in the Sea of Marmara (NW Turkey). *Geophysical Journal International*, *196*(2), 850–866.

Essing, D., Schlindwein, V., Schmidt-Aursch, M. C., Hadziioannou, C., & Stähler, S. C. (2021). Characteristics of current-induced harmonic tremor signals in ocean-bottom seismometer records. *Seismological Research Letters*, *92*(5), 3100–3112.

Ferrazzini, V., Chouet, B., Fehler, M., & Aki, K. (1990). Quantitative analysis of long-period events recorded during hydrofracture experiments at Fenton Hill, New Mexico. *Journal of Geophysical Research: Solid Earth*, *95*(B13), 21871–21884.

Franek, P., Mienert, J., Buenz, S., & Géli, L. (2014). Character of seismic motion at a location of a gas hydrate-bearing mud volcano on the SW Barents Sea margin. *Journal of Geophysical Research: Solid Earth*, *119*(8), 6159–6177.

Franek, P., Plaza-Faverola, A., Mienert, J., Buenz, S., Ferré, B., & Hubbard, A. (2017). Microseismicity linked to gas migration and leakage on the Western Svalbard Shelf. *Geochemistry, Geophysics, Geosystems*, *18*(12), 4623–4645.

Freed, A. M. (2005). Earthquake triggering by static, dynamic, and postseismic stress transfer. *Annual Review of Earth Planetary Science*, *33*, 335–367.

Frehner, M., & Schmalholz, S. M. (2010). Finite-element simulations of Stoneley guided-wave reflection and scattering at the tips of fluid-filled fractures. *Geophysics*, *75*(2), T23–T36.

Frohlich, C., Louat, R., & Nakamura, Y. (1990). Earthquake activity in the southern Vanuatu arc recorded by the Texas digital OBS. *Marine Geophysical Researches*, *12*(4), 253–267.

Géli, L., Henry, P., Grall, C., Tary, J. B., Lomax, A., Batsi, E., et al. (2018). Gas and seismicity within the Istanbul seismic gap. *Scientific Reports*, *8*(1), 1–11.

Géli, L., Henry, P., Zitter, T., Dupré, S., Tryon, M., Çağatay, M. N., et al. (2008). Gas emissions and active tectonics within the submerged section of the North Anatolian Fault zone in the Sea of Marmara. *Earth and Planetary Science Letters*, *274*(1–2), 34–39.

Giovanetti, G., Monna, S., Lo Bue, N., Embriaco, D., Frugoni, F., Marinaro, G., et al. (2016). Observing volcanoes from the seafloor in the central Mediterranean area. *Remote Sensing*, *8* (4), 298.

Giudicepietro, F., Chiodini, G., Caliro, S., De Cesare, W., Esposito, A. M., Galluzzo, D., et al. (2019). Insight into Campi Flegrei caldera unrest through seismic tremor measurements at Pisciarelli fumarolic field. *Geochemistry, Geophysics, Geosystems*, *20*(11), 5544–5555.

Hellweg, M. (2000). Physical models for the source of Lascar's harmonic tremor. *Journal of Volcanology and Geothermal Research*, *101*(1–2), 183–198.

Henry, P., Grall, C., Kende, J., Viseur, S., Özeren, M. S., Şengör, A. M. C., et al. (2018). A statistical approach to relationships between fluid emissions and faults: the Sea of Marmara case. *Deep Sea Research Part II: Topical Studies in Oceanography*, *153*, 131–143.

Hilmo, R., & Wilcock, W. S. (2020). Physical sources of high-frequency seismic noise on Cascadia initiative ocean bottom seismometers. *Geochemistry, Geophysics, Geosystems*, *21* (10), e2020GC009085.

Hsu, S. K., Wang, S. Y., Liao, Y. C., Yang, T. F., Jan, S., Lin, J. Y., & Chen, S. C. (2013). Tide-modulated gas emissions and tremors off SW Taiwan. *Earth and Planetary Science Letters*, *369*, 98–107.

Julian, B. R. (1994). Volcanic tremor: nonlinear excitation by fluid flow. *Journal of Geophysical Research: Solid Earth*, *99* (B6), 11859–11877.

Korneev, V. (2008). Slow waves in fractures filled with viscous fluid. *Geophysics*, *73*(1), N1–N7.

Korneev, V. A. (2011). Krauklis wave in a stack of alternating fluid-elastic layers. *Geophysics*, *76*(6), N47–N53.

Kovachev, S. A., Demidova, T. A., & Son'kin, A. V. (1997). Properties of noise registered by pop-up ocean-bottom seismographs. *Journal of Atmospheric and Oceanic Technology*, *14*(4), 883–888.

Kumagai, H., & Chouet, B. A. (2000). Acoustic properties of a crack containing magmatic or hydrothermal fluids. *Journal of Geophysical Research: Solid Earth*, *105*(B11), 25493–25512.

Kuşçu, İ., Okamura, M., Matsuoka, H., Gökaşan, E., Awata, Y., Tur, H., et al. (2005). Seafloor gas seeps and sediment failures triggered by the August 17, 1999 earthquake in the Eastern part of the Gulf of İzmit, Sea of Marmara, NW Turkey. *Marine Geology*, *215*(3–4), 193–214.

Kwiatek, G., & Ben-Zion, Y. (2016). Theoretical limits on detection and analysis of small earthquakes. *Journal of Geophysical Research: Solid Earth*, *121*(8), 5898–5916.

Lee, W. H., and S. W. Stewart (1981). *Principles and applications of micro-earthquake network* (Advanced Geophysics, Supplement 2). Academic Press.

Liang, C., O'Reilly, O., Dunham, E. M., & Moos, D. (2017). Hydraulic fracture diagnostics from Krauklis-wave resonance and tube-wave reflections. *Geophysics*, *82*(3), D171–D186.

Lipovsky, B. P., & Dunham, E. M. (2015). Vibrational modes of hydraulic fractures: Inference of fracture geometry from resonant frequencies and attenuation. *Journal of Geophysical Research: Solid Earth*, *120*(2), 1080–1107.

Lokmer, I., & Bean, C. J. (2010). Properties of the near-field term and its effect on polarisation analysis and source locations of long-period (LP) and very-long-period (VLP) seismic events at volcanoes. *Journal of Volcanology and Geothermal Research*, *192*(1–2), 35–47.

Maeda, Y., & Kumagai, H. (2017). A generalized equation for the resonance frequencies of a fluid-filled crack. *Geophysical Journal International*, *209*(1), 192–201.

McDonald, M. A., Hildebrand, J. A., & Wiggins, S. M. (2006). Increases in deep ocean ambient noise in the Northeast Pacific west of San Nicolas Island, California. *The Journal of the Acoustical Society of America*, *120*(2), 711–718.

Olofsson, B. (2010). Marine ambient seismic noise in the frequency range 1–10 Hz. *The Leading Edge*, *29*(4), 418–435.

Ostrovsky, A. A. (1989). On the nature of microshocks recorded by ocean bottom seismographs. *Marine geophysical researches*, *11*(2), 113–118.

Prothero, W. A., & Reid, I. D. (1982). Microearthquakes on the East Pacific Rise at 21 N and the Rivera fracture zone. *Journal of Geophysical Research: Solid Earth*, *87*(B10), 8509–8518.

Römer, M., Riedel, M., Scherwath, M., Heesemann, M., & Spence, G. D. (2016). Tidally controlled gas bubble emissions: A comprehensive study using long-term monitoring data from the NEPTUNE cabled observatory offshore Vancouver Island. *Geochemistry, Geophysics, Geosystems*, *17*(9), 3797–3814.

Rust, A. C., N. J. Balmforth, & S. Mandre (2008), The feasibility of generating low-frequency volcano seismicity by flow through a deformable channel. In S. J. Lane and J. S. Gilbert (Eds.), *Fluid motions in volcanic conduits: a source of seismic and acoustic signals* (Special Publication 307, pp. 45–56). Geological Society of London.

Scholz, C. H., Tan, Y. J., & Albino, F. (2019). The mechanism of tidal triggering of earthquakes at mid-ocean ridges. *Nature Communications*, *10*(1), 1–7.

Sgroi, T., Monna, S., Embriaco, D., Giovanetti, G., Marinaro, G., & Favali, P. (2014). Geohazards in the western Ionian Sea: Insights from non-earthquake signals recorded by the NEMO-SN1 seafloor observatory. *Oceanography*, *27*(2), 154–166.

Sohn, R. A., Hildebrand, J. A., Webb, S. C., & Fox, C. G. (1995). Hydrothermal microseismicity at the megaplume site on the southern Juan de Fuca Ridge. *Bulletin of the Seismological Society of America*, *85*(3), 775–786.

Stein, S., & Wysession, M. (2009). *An introduction to Seismology, Earthquakes, and Earth Structure*. John Wiley & Sons.

Sultan, N., Plaza-Faverola, A., Vadakkepuliyambatta, S., Buenz, S., & Knies, J. (2020). Impact of tides and sea-level on deep-sea Arctic methane emissions. *Nature Communications*, *11*(1), 1–10.

Sultan, N., Riboulot, V., Ker, S., Marsset, B., Geli, L., Tary, J. B., et al. (2011). Dynamics of fault-fluid-hydrate system around a shale-cored anticline in deepwater Nigeria. *Journal of Geophysical Research: Solid Earth*, *116*(B12).

Tary, J. B., 2011. *Case studies on fluids and seismicity in submarine environments based on ocean bottom seismometers (OBS) recordings from the Sea of Marmara and application to the Niger Delta* (PhD thesis). Université de Bretagne Occidentale.

Tary, J. B., Geli, L., Guennou, C., Henry, P., Sultan, N., Çağatay, N., & Vidal, V. (2012). Microevents produced by gas migration and expulsion at the seabed: a study based on sea bottom recordings from the Sea of Marmara. *Geophysical Journal International*, *190*(2), 993–1007.

Tary, J. B., Herrera, R. H., Han, J., & van der Baan, M. (2014a). Spectral estimation—what is new? What is next? *Reviews of Geophysics*, *52*(4), 723–749.

Tary, J. B., Van der Baan, M., Sutherland, B., & Eaton, D. W. (2014b). Characteristics of fluid-induced resonances observed during microseismic monitoring. *Journal of Geophysical Research: Solid Earth*, *119*(11), 8207–8222.

Tary, J. B., Geli, L., Lomax, A., Batsi, E., Riboulot, V., & Henry, P. (2019). Improved detection and Coulomb stress computations for gas-related, shallow seismicity, in the Western Sea of Marmara. *Earth and Planetary Science Letters*, *513*, 113–123.

Tsang-Hin-Sun, E., Batsi, E., Klingelhoefer, F., & Géli, L. (2019). Spatial and temporal dynamics of gas-related processes in the Sea of Marmara monitored with ocean bottom seismometers. *Geophysical Journal International*, *216* (3), 1989–2003.

Ugalde, A., Gaite, B., Ruiz, M., Villaseñor, A., & Ranero, C. R. (2019). Seismicity and noise recorded by passive seismic monitoring of drilling operations offshore the eastern Canary Islands. *Seismological Research Letters*, *90*(4), 1565–1576.

van der Elst, N. J., & Brodsky, E. E. (2010). Connecting near-field and far-field earthquake triggering to dynamic strain. *J. Geophys. Res.*, *115*, B07311.

Vavryčuk, V. (1992). Polarization properties of near-field waves in homogeneous isotropic and anisotropic media: numerical modelling. *Geophysical Journal International*, *110*(1), 180–190.

Vavryčuk, V. (2001). Inversion for parameters of tensile earthquakes. *Journal of Geophysical Research: Solid Earth*, *106* (B8), 16339–16355.

Vidal, V., Géminard, J. C., Divoux, T., & Melo, F. (2006). Acoustic signal associated with the bursting of a soap film which initially closes an overpressurized cavity. *The European Physical Journal B-Condensed Matter and Complex Systems*, *54*(3), 321–339.

Walter, W. R., & Brune, J. N. (1993). Spectra of seismic radiation from a tensile crack. *Journal of Geophysical Research: Solid Earth*, *98*(B3), 4449–4459.

Webb, S. C. (1998). Broadband seismology and noise under the ocean. *Reviews of Geophysics*, *36*(1), 105–142.

Zhu, L., & Rivera, L. A. (2002). A note on the dynamic and static displacements from a point source in multilayered media. *Geophysical Journal International*, *148*(3), 619–627.

Zitter, T. A. C., Henry, P., Aloisi, G., Delaygue, G., Çagatay, M. N., De Lepinay, B. M., et al. (2008). Cold seeps along the main Marmara Fault in the Sea of Marmara (Turkey). *Deep Sea Research Part I: Oceanographic Research Papers*, *55*(4), 552–570.

9

Short-Duration Events Associated With Active Seabed Methane Venting: Scanner Pockmark, North Sea

Gaye Bayrakci[1], Jonathan M. Bull[2], Timothy A. Minshull[2], Adam H. Robinson[2], Aude Lavayssiere[2,3], Finnigan Illsley-Kemp[2,4], Timothy J. Henstock[2], Calum Macdonald[5], and Mark Chapman[5]

ABSTRACT

The process by which methane propagates through near-surface sediments and is vented into the ocean is uncertain. Here, we describe seismic events (short-duration events [SDEs]) recorded on ocean-bottom seismometers (OBSs) close to an actively venting pockmark at a water depth of 150–165 m. Analysis of 120 hours of data recorded by 10 OBSs showed natural events with consistent characteristics: monochromatic, with a peak frequency of 4 Hz and relatively short duration (less than 6 seconds); an absence of clear body- or surface-wave arrivals; spectral content that decreases in frequency over the event duration; and larger amplitudes on horizontal components than on the vertical component, indicating a shallow subsurface origin. Only the SDEs with the largest geophone amplitudes are recorded on the hydrophones, suggesting that the events originate in the seabed rather than in the water column. In the pockmark, clusters of events are recorded simultaneously on two OBSs 20 m apart; but more generally, events are not recorded on multiple instruments, indicating the very local nature of the events. The rate of occurrence of the SDEs is negatively correlated with tidal height and total water depth, indicating a hydrostatic pressure control.

9.1. INTRODUCTION

The release of greenhouse gases from natural (Dean et al., 2018) and manmade (Pörtner et al., 2022) sources has been identified as a major cause of global climate change. Marine sediments host large amounts of methane gas. Methane within marine sediments can be found in free gas, hydrate, or dissolved states (Ruppel & Kessler, 2017) and may have a biogenic or thermogenic origin (Hovland et al., 1993). Subsurface gas is commonly located in high-permeability reservoirs capped by a low-permeability overburden. It can migrate from the deeper strata to the seabed using anthropogenic pathways such as abandoned wells (Ebigbo et al., 2007) or natural pathways such as up-dip migration along permeable stratigraphic horizons, and focused fluid-escape structures such as faults, chimneys, and pipes (Løseth et al., 2009).

Pockmarks are local depressions in soft sediments, assumed to be formed by gas or liquid expulsion through

[1]*National Oceanography Centre, Southampton, UK*

[2]*School of Ocean and Earth Science, University of Southampton, Southampton, UK*

[3]*Institut de Physique du Globe de Paris, Paris, France*

[4]*School of Geography, Environment and Earth Sciences, Victoria University of Wellington, Wellington, New Zealand*

[5]*School of Geosciences, University of Edinburgh Edinburgh, UK*

Noisy Oceans: Monitoring Seismic and Acoustic Signals in the Marine Environment, Geophysical Monograph 284, First Edition.
Edited by Gaye Bayrakci and Frauke Klingelhoefer.

DOI:10.1002/9781119750925.ch09

the seafloor (King & MacLean 1970; Hovland et al., 1987; Cathles et al., 2010). These circular depressions are meters to hundreds of meters in diameter and meters to tens of meters in depth, and they are common features in North Sea sediments (Hovland et al., 1984). They often form on top of focused fluid conduits that bring the deeper gas to the sea bottom (Cartwright et al., 2007; Løseth et al., 2011; Andresen, 2012; Karstens & Berndt, 2015; Callow et al., 2021). Focused fluid conduits such as chimneys (Hustoft et al., 2010) and pipes (Moss & Cartwright, 2010) are vertical to subvertical anomalies of a circular shape observed on seismic reflection images. Amplitude blanking and discontinuous and chaotic reflections are common seismic observations within these structures (e.g., Karstens & Berndt, 2015).

In near-surface marine sediments, the pore space is usually saturated with seawater. In coarse-grained sediments, gas moves in the pores by capillary intrusion, with displacement of the pore water but without grain displacement. Controlled subseafloor gas-release experiments (e.g., Cevatoglu et al., 2015; Taylor et al., 2015; Flohr et al., 2021; Roche et al., 2021) have shown that in fine-grained sediments, gas migration occurs instead by fracture generation and/or reactivation of existing fractures, forming fluid-escape structures (Cevatoglu et al., 2015; Roche et al., 2021). Gas bubbles in fine-grained sediments studied with CT scans (Boudreau et al., 2005) and within transparent gelatine (Johnson et al., 2002) that mimics fine-grained sediments have small inverse aspect ratios (b/a; a = major half-axis and b=minor half-axis). These observations also suggest that tensile fracturing is the gas-migration mechanism in fine-grained sediments (Boudreau, 2012). This second type of gas migration results in ground displacement and can therefore be recorded by seismometers and studied using seismology (e.g., Tary et al., 2012).

Seismology is widely used to study active fluid flow at mid-ocean ridges (e.g., Johnson et al., 2000) and oceanic transform faults (Géli et al., 2014), and hydraulic fracturing (e.g., Li et al., 2019). Monochromatic seismologic events with durations of only a few seconds and without clear body-wave arrivals, called short-duration events (SDEs), are observed worldwide in various geodynamic environments (e.g., Tary, this volume). There is no consensus on the definition (e.g., duration, frequency range, waveform, etc.) or origin of these events. Proposed origins for these short-duration signals include marine biologic impacts on the instruments (e.g., fish hitting the instruments; Buskirk et al., 1981), the resonance of fluid conduits (Diaz et al., 2007), and gas migration in shallow sediments and expulsion from the seabed (Tary et al., 2012).

In this chapter, we characterize SDEs recorded on a network of ocean-bottom seismometers (OBSs) deployed in the neighborhood of the Scanner pockmark complex, an active fluid flow structure in the northern North Sea (Fig. 9.1). The vigorous methane seepage from this pockmark (272 to 456 L/min derived acoustically; Li et al., 2020) and the availability of a variety of previous studies make this location an ideal location to study the relationship between active seafloor gas emission and SDE occurrence. These studies include seismic stratigraphic characterization (Böttner et al., 2019; Callow et al., 2021), sediment sampling, coring (Robinson et al., 2021), fracture characterization via S-wave splitting (Bayrakci et al., 2021; Robinson et al., 2022), P-wave velocity distribution (Bayrakci et al., 2020), porosity and free-gas estimation from controlled source electromagnetics (Gehrmann et al., 2021) and numerical simulation of gas migration trough sediments (Marín-Moreno et al., 2019).

9.2. SCANNER POCKMARK COMPLEX

Scanner pockmark complex comprises two seafloor depressions 15–20 m deep and extends over a ~900 × 450 m area of relatively flat seafloor at ~150 m water depth (Fig. 9.1a). The complex is located in a ~600 m thick Quaternary sediment succession that is subdivided into five units, S5 (shallowest) to S1 (deepest) (Fig. 9.1c,d) (Böttner et al., 2019; Callow et al., 2021). It erodes down to the base of the shallowest unit, S5, which is also called Witch Ground Formation (Fig. 9.1b). Quaternary sediments are underlain by the Nordland and Hordaland groups of Neogene and Paleogene ages, respectively (Judd et al., 1994).

The pockmark complex is on top of an elliptical and diffuse seismic reflection anomaly: a seismic chimney structure (Fig. 9.1c,d). Seismic chimney structures are interpreted as a series of near-vertical fractures in fine-grained sediments acting as fluid pathways, hydraulically connecting deeper sedimentary groups to the seafloor (Callow et al., 2021). Methane venting is observed in both depressions of the Scanner pockmark complex (Fig. 9.1a) on both multibeam and single-beam acoustic data (Li et al., 2020).

9.3. CHIMNEY SEISMIC EXPERIMENT

The CHIMNEY seismic experiment took place in September 2017 around the Scanner pockmark in the North Sea, using the *RRS James Cook* (cruise JC152) (Bull et al., 2018; Robinson et al., 2021). We acquired wide-angle and

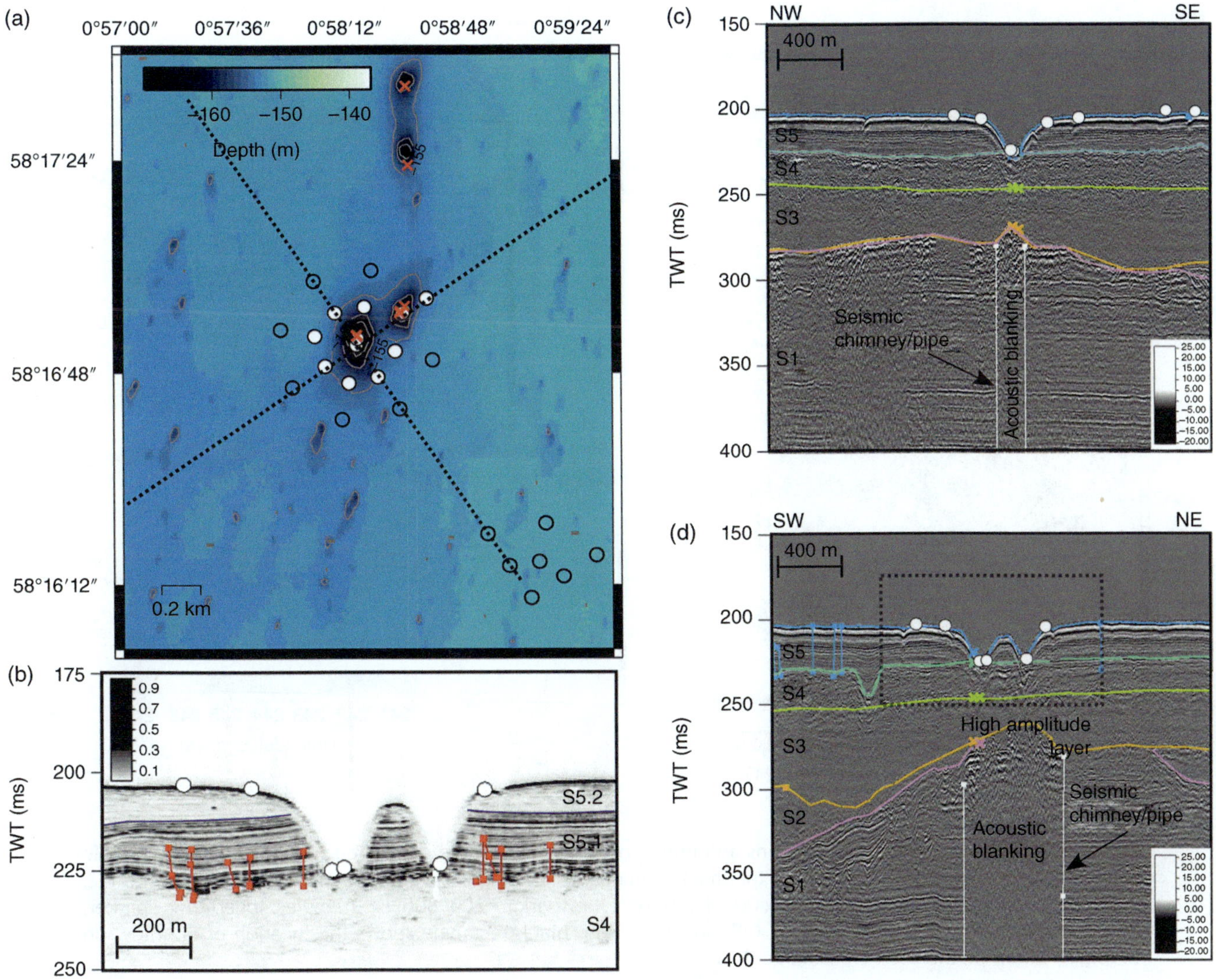

Figure 9.1 (a) Circles mark locations of the OBSs deployed around the Scanner pockmark, with white-filled circles marking those used in this study. Crosses mark locations of gas seeps observed on multibeam and multifrequency single-beam echosounders. Dashed black lines show the location of NW–SE and SW–NE seismic profiles in (c) and (d), respectively. (b) Sub-bottom profiler (SBP) seismic reflection data. Subvertical lines mark fault picks. (c) NW–SE and (d) SW–NE Squid 2000 sparker seismic reflection profiles. S5 = Witch Ground Formation (WGF). S5.2 = Witch member of WGF. S5.1 = Fladen member of WGF. S4 = Last Glacial Maximum (LGM) deposits. S3 = Coal Pit Formation (CPF). S2 = Ling Bank Formation (LBF). S1 = Aberdeen Ground Formation (AGF). Interpreted edges of the chimney are displayed with subvertical white lines. Detailed structural interpretation is adapted from Callow et al. (2021). The dashed black rectangle shows the extent of the SBP profile in (b). White circles in (b–d) mark OBS positions.

multichannel seismic and high-frequency acoustic data. A total of 25 four-component OBSs were deployed by free-fall at water depths varying between 150 and 160 m (Figs. 9.1a and 9.2). Eighteen OBSs were deployed in and around the Scanner pockmark along with seven at a reference site that showed no sign of gas presence. OBS spacing was generally ~200 m, with closer spacing (20 m) in the pockmark. Instruments were equipped with three orthogonal 4.5 Hz geophones and a hydrophone. The sampling rate was 4 kHz to record the high-frequency sources used for the active source seismic experiment.

All 25 OBSs recorded the active-source experiment, which used five different seismic sources (Bolt and GI air-guns, Squid 2000 and Duraspark surface sparkers, and a

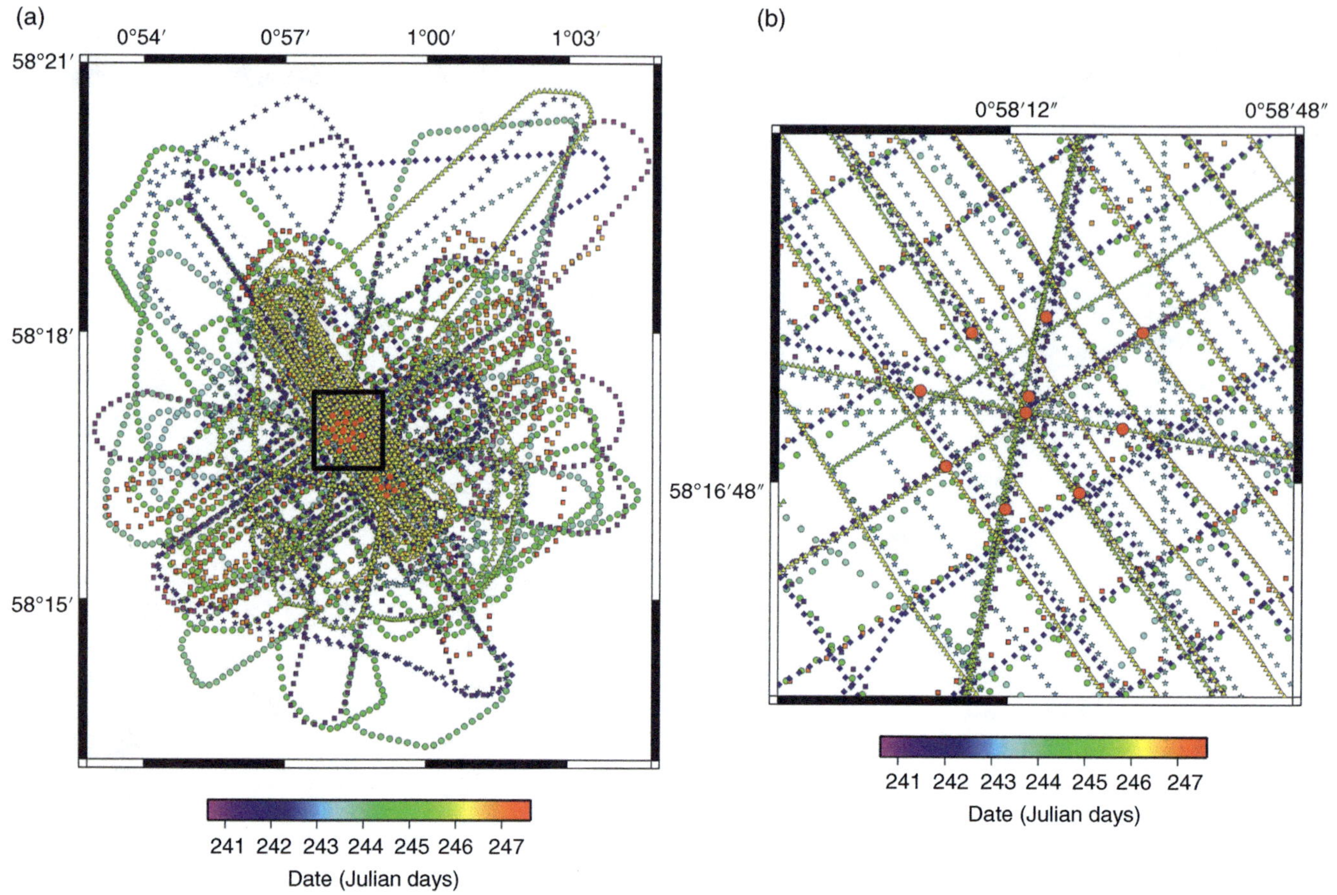

Figure 9.2 (a) Map showing the locations and timing of controlled source shots fired with five different seismic sources during the CHIMNEY seismic experiment (squares: BOLT airguns, <200 Hz; circles: GI guns, <400 Hz; diamonds: Squid 2000 sparker, ~50–800 Hz; stars: Duraspark, ~200–800Hz; triangles: deep-tow sparker, ~1,000–3,000 Hz). Larger circles show OBS locations. The black rectangle shows the location of the blow-up in (b). (b) Zoom on the used OBSs (large dots).

deep-towed sparker; Fig. 9.2) to study fluid-migration pathways within sediments (Bayrakci et al., 2021; Robinson et al., 2021; Robinson et al., 2022). In addition to seismic sources used in our experiment, analysis of the data set showed that the OBSs recorded shots from a neighboring seismic experiment (Fig. 9.3a; best visible on the hydrophone component) with unknown origin times and locations but with a constant shot interval of 8 s and a peak recorded frequency of 3 Hz. These shots covered the entire duration of our seismic experiment. The five seismic sources from our cruise and the one from the neighboring seismic cruise covered a frequency range of ~3–1,600 Hz, the maximum frequency we could record with our sampling rate of 4 kHz. OBSs also recorded ambient noise throughout their deployment period of 7 days (Julian days 241–248). During our survey, the only nonshooting periods were at the beginning, the end of the seismic survey, and when the seismic source was changed (Fig. 9.2). However, shots from the neighboring seismic experiment were present during the silent periods of our experiment, which, together with the use of five seismic sources covering a wide range of frequencies during our active-source survey, makes the study of natural seismic signals challenging with this data set.

Here, we used the continuous data recorded on 10 OBSs deployed in or close to the Scanner pockmark complex (Figs. 9.1a, 9.2b, and 9.4) where vigorous gas venting is observed. First we analyzed the entire recording period, including the active source shooting periods. Then the periods of shooting with airguns (Bolt airguns and GI guns) were removed from the data set due to their high amplitudes and lower-frequency content that is similar to that of SDEs. The shots from the neighboring seismic experiment could not be removed from the data set because the shooting period covers the entire record length of our experiment.

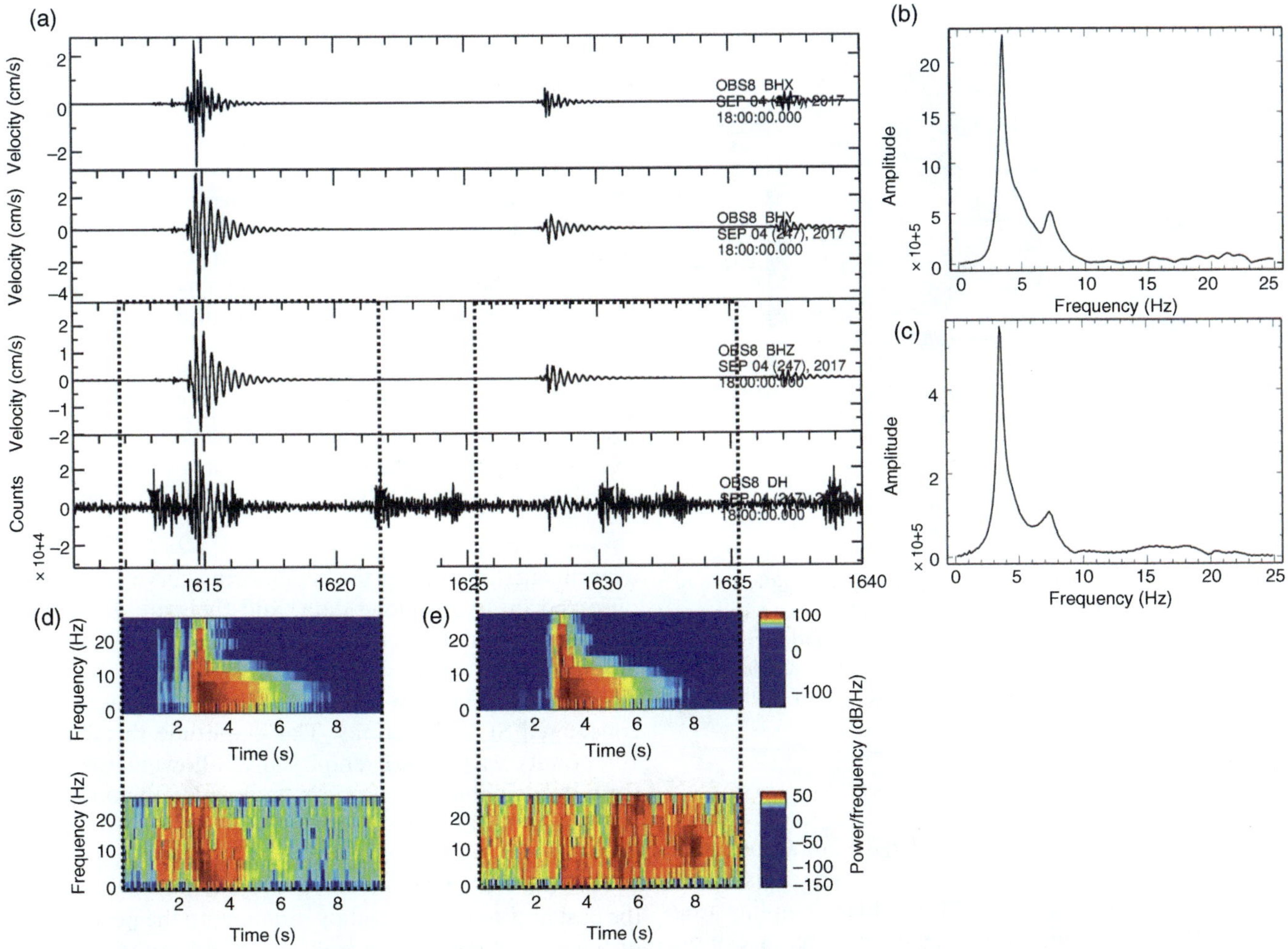

Figure 9.3 (a) Two consecutive SDEs were observed during an SDE swarm recorded by OBS-8 (see location in Fig. 9.4) on Julian day 247 at 18:26:54.75 and 18:27:07.99, respectively. Top to bottom, BHX, BHY, BHZ, and DH are the geophone *x*, *y*, and *z* components and the hydrophone, respectively. The hydrophone signal is high-pass filtered with a Butterworth filter using a corner frequency of 3 Hz. Black arrows on the hydrophone signal show the neighboring seismic shots with a constant time interval of 8 s. Note that the higher-amplitude SDE is clearly observed on the hydrophone as well as on the geophones, whereas the hydrophone signal of the lower-amplitude SDE is masked by the noise. The dashed black frames show the windows used for the fast Fourier transform (FFT). Frequency content of the (b) higher-amplitude and (c) lower-amplitude SDEs calculated on the Y component. (d, e) Spectrograms of vertical and hydrophone components of the high- and low-amplitude SDEs, respectively.

9.4. METHODS

Instruments were relocated using the controlled-source shots (Bayrakci et al., 2021). A grid search algorithm was used to find the optimal average water velocity (1,490 m/s), receiver locations, and delay times that account for the variations in the depth of the airgun and feathering of the airgun tow cable. The sum of the squared residuals (the difference between observed and predicted direct wave arrival times) was computed at every point of the grid, and the location of the minimum was chosen. We did not invert for the depths of the receivers. Instead, we extracted the seafloor depth at the relocated positions of the instruments from the multibeam bathymetry acquired during the survey. Due to the shallow seafloor depth (~150 m) and small seafloor dips, the differences between deployment and relocated depths were small (<0.5 m).

Continuous data was reduced in compact MiniSEED format (data-only version of Standard for the Exchange of Earthquake Data [SEED] of the Incorporated Research Institutions for Seismology [IRIS]). The hour-long MiniSEED files were converted to Seismic Analysis Code (SAC; Helffrich et al., 2013) format, and an automatic event-identification algorithm was applied.

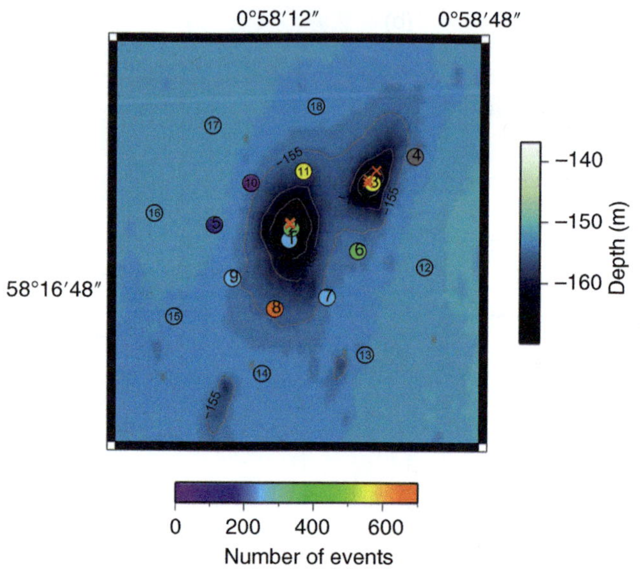

Figure 9.4 Number of events observed on studied OBSs. Empty circles represent OBSs not used in this study. OBS-4 did not have usable records. Crosses represent the seabed location of gas seeps observed on multibeam and single-beam echosounder data. Black lines show the location of the SBP profiles shown in Fig. 9.9.

The frequency range of SDEs found in the literature varies between 3 and 30 Hz (Tary, this volume). The initial analysis of the original data set with a 4 kHz sampling rate showed that the SDEs occurring in this region exhibit frequencies lower than 10 Hz (Fig. 9.3b,c). Therefore, the original 4 kHz sampling frequency was reduced to 50 Hz to speed up the automatic SDE detection. Downsampling was carried out using SAC software (Helffrich et al., 2013). The anti-alias filter used in downsampling to 50 Hz discarded the high-frequency sparker sources (frequencies >50 Hz).

The geophone counts were converted to ground velocity by multiplying the counts by the analog-to-digital conversion factor of 409 nV/count and dividing this value by the factor of the applied gain (16) and the geophone sensitivity of 31.1 V/(m/s).

Automatic event detection was carried out with a short-term average to long-term average ratio (STA/LTA) algorithm (https://github.com/csdms-contrib/slepian_oscar/blob/master/stalta.m). Initial visual analysis of the SDEs showed that the SDE amplitudes in counts were one order of magnitude larger on the horizontal components than the vertical. Therefore, the detection was run on both horizontal components (X and Y) and was repeated several times until satisfactory parameters were reached for detecting the SDEs with the highest amplitudes and discarding the shots from the neighboring seismic

Table 9.1 Parameters used for the final SDE detection.

DT	Sampling interval (s)	0.02
BE	Beginning and end time of signal (s)	0,
	3600	STA
	Short-term average window length (s)	0.3
LTA	Long-term average window length (s)	10
TR	Value of STA/LTA ratio that triggers	10
DTR	Value of STA/LTA ratio that untriggers	1
PEM	Time buffer added before triggering time (s)	2
PET	Time buffer added after triggering time (s)	2
PNL	Minimum window length of any triggered section (s)	5
ATL	Minimum time between trigger and detrigger	2
BTL	Maximum time between trigger and detrigger	6

Note: Parameters were selected to discard the neighboring seismic experiment shots. They may not be appropriate for elsewhere.

experiment. Shots from the neighboring survey were best observed on hydrophone data, and their amplitudes on the geophone data were smaller than most of the SDE amplitudes. Events with amplitudes larger than 10^5 counts (0.082 mm/s) in either the X or Y component were considered SDE detections. The amplitude threshold of 10^5 counts was chosen empirically, allowing the shots from the neighboring seismic experiment with amplitudes lower than SDEs to be discarded (see Section 9.6.4). The final detection parameters are given in Table 9.1.

The rotation angle to orient the X and Y components to the east and south is found by minimizing the power ratio of the amplitude of transverse and radial components of the direct wave arrivals of controlled-source shots (Haacke & Westbrook, 2006; Bayrakci et al., 2021). The data was rotated using the SAC software.

9.5. RESULTS

A total of 3,135 events were detected using the final detection parameters shown in Table 9.1 in our data set from 10 OBSs that covered five days of recording time outside periods of GI-gun and Bolt airgun shooting. Of these, 2,937 detected events were SDEs, and 198 were identified visually as residual active source shots from the neighboring seismic experiment.

The maximum duration of the detected SDEs was 6 s. Detected SDEs show similar frequency content (Fig. 9.3). They are monochromatic, with a peak frequency of 4 Hz and a smaller peak at 8 Hz. Body- or surface-wave arrivals cannot be distinguished clearly, but an initial higher frequency (up to >20 Hz) phase is followed by a lower-frequency phase of less than 10 Hz that decays quickly with time (Fig. 9.3d,e). SDEs show high amplitudes (between 10^5 counts or 0.082 mm/s, the cut-off amplitude,

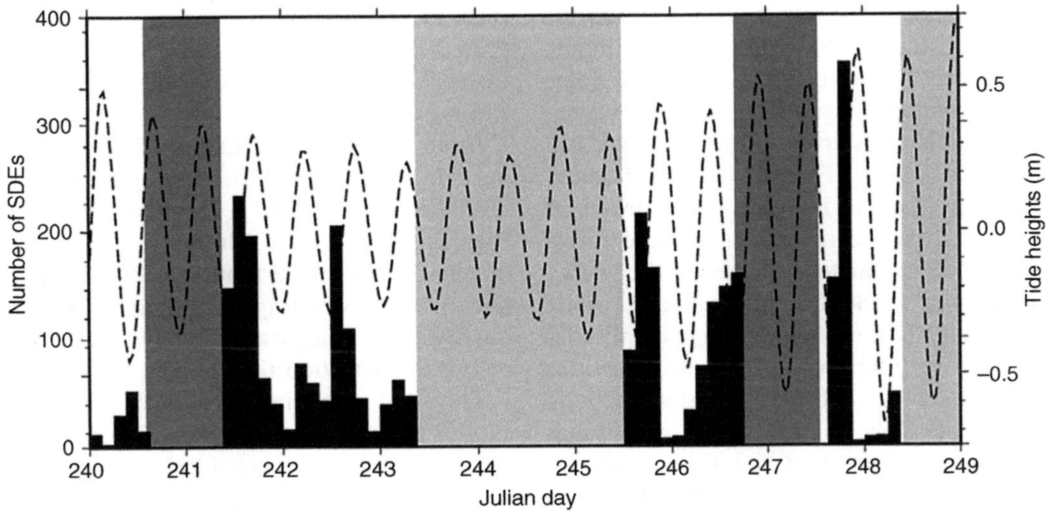

Figure 9.5 Comparison of tide heights (dashed line) and cumulative three hourly SDE occurrences recorded on all OBSs (black bars). Correlation R = −0.25; likelihood of noncorrelation (Pearson value) P = 0.03. The two dark-gray areas show the Bolt airgun shooting periods removed from the SDE-detection data set. The two pale gray areas show the GI-gun shooting period removed from the data set and the end of the survey where no data are recorded, but Julian day 248 is plotted as a full day.

and 10^7 counts or 8.22 mm/s), systematically higher than the shots from the neighboring seismic survey with frequencies similar to the SDEs. Often, their amplitudes are higher on one of the horizontal components and similar on the second horizontal and the vertical components. Systematically, SDEs with higher amplitudes on the geophones were also recorded on the hydrophones, and the hydrophone signals were absent for smaller-amplitude events. No SDE is observed simultaneously on multiple instruments, even when they are only 20 m apart, suggesting that SDEs occur in the shallow subsurface (down to a maximum 20 m depth).

Over the entire record length, the maximum number of SDEs is observed on OBS-11 (582 SDEs) located at the NNE corner of the Scanner pockmark complex, followed by OBS-8 (568 SDEs) at the SSW corner, whereas the minimum number of SDEs (79 SDEs) is observed on OBS-10 located at the NW corner (Fig. 9.4). OBS-4 did not record any signal lower than 50 Hz and could not be used for detection.

Ocean tides are the semidiurnal vertical rise and fall of ocean waters due to the gravitational attraction of the Moon and Sun on the Earth (e.g., Schwiderski, 1980). A decrease in pressure during low tide promotes gas emission from the seafloor. Preferential release of bubbles at low tide has been linked to the total load drop on the bubbles due to the hydrostatic pressure drop, which allows the total internal bubble pressure to become larger than the reduced total load plus the fracture resistance of the sediments (Boudreau, 2012; Algar et al., 2011).

The local ocean tides show a periodicity of ~12.6 hours, leading to sea-level variations of ±0.7 m. The cumulative three hourly distribution of SDEs shows a visual negative correlation with the local low-tide times (Fig. 9.5). Although the anticorrelation is clear visually, the coefficient of correlation is small (R = −0.25), potentially due to no-data periods (Bolt airgun and GI gun) that are removed from our data set. The low Pearson value (P = 0.03) indicates a low probability of no correlation.

9.6. DISCUSSION

9.6.1. Characteristics of SDEs

The peak frequency of SDEs observed in the Scanner pockmark is 4 Hz, with a smaller 8 Hz peak (Fig. 9.3b, c). Observed frequencies remain at the lower end of the frequency range suggested for SDEs in the literature (3–30 Hz; Tary et al., 2012; Tary, this volume). The spectrograms of the events recorded on different OBSs similarly indicate a singular mechanism of occurrence. SDEs exhibit a monochromatic nature. The frequency content of seismic waveforms is related to their source mechanism, the path between the source and receiver (Li et al., 2020), and the receiver coupling and transfer function. The observed monochromatic behavior may be explained by resonant emissions from fractures, such as Scholte waves propagating at the wall-fluid interface without interference between the waves propagating along two crack walls (Korneev, 2009), or Krauklis waves, where the movement of two crack walls interfere with each other. (Korneev, 2011; Tary, this volume). When Krauklis waves reach the tip of a crack, some energy is reflected and propagates as standing waves, whereas some energy is transferred to the host rock and propagates as body waves (Frehner, 2014; Korneev et al., 2012). The amplitude of Krauklis waves decreases exponentially away from the crack, so they are rarely recorded directly by seismometers. However, the energy transmitted to the host rock that propagates as body waves is recorded by the OBSs, often with high amplitudes due to the constructive interference between reflections from the tips of the cracks before conversion to body waves in the host rock. The relatively high amplitudes of SDEs (sometimes higher than the GI-gun shots), accompanied by the absence of simultaneous SDE observations on multiple instruments, is consistent with the Krauklis wave hypothesis.

We report SDEs up to 6 s in duration, which is longer than SDEs usually reported in other regions. The longer the waveform duration, the lower the frequency, and the greater the event magnitude (Lomax et al., 2007; Tsumura, 1967). Here, the amplitude of the SDEs varies broadly between 10^5 counts (minimum amplitude chosen for the detection algorithm) and 10^7, which correspond to 0.082 mm/s and 8.22 mm/s of ground velocity, respectively. We observe higher-amplitude SDEs on the hydrophone component; the hydrophone signals of lower-amplitude SDEs are absent, which is consistent with observations of shorter-duration SDEs in other regions. This observation suggests that the SDEs observed in the Scanner pockmark are of larger magnitude than elsewhere. Tsang-Hin-Sun et al. (2019) report ground velocities of ±0.02 mm/s, whereas Hsu et al. (2013) report ±0.6 mm/s for observed SDEs, which are smaller in magnitude than or similar to the range (0.082–8.22 mm/s) we observe in the Scanner pockmark.

SDEs at the Scanner pockmark are observed as single events and as swarms. The single or swarmed nature of SDEs was also reported by Batsi et al. (2019), Tsang-Hin-Sun et al. (2019), and Tary (this volume), sometimes with some periodicity. Although the amplitudes of SDEs were higher than for the sparker sources, there is no evidence of an SDE observed simultaneously on more than one instrument, suggesting that SDEs propagate as a phase that attenuates over small distances (<20 m, which is the smallest distance between instruments in our OBS network). It was not possible to pick the body- or surface-wave phases of SDEs. However, the spectrograms of the SDEs observed in the Scanner pockmark (Fig. 9.3d,e) are similar to those observed in SW Taiwan (Hsu et al., 2013) and in Chile (Batsi et al., 2019) and show a clear higher-frequency train followed by a lower-frequency one. In SW Taiwan (Hsu et al., 2013), this frequency drop has been interpreted as a body-wave train (P- and S-waves) followed by a 7 Hz damped harmonic motion following the P-wave direction.

In an attempt to distinguish different phases via their particle motion, we plotted hodograms for a series of periods (Figs. 9.6 and 9.7). The first rows of hodograms in Figs. 9.6b1 and 9.67b1 do not show any coherent (linear or elliptical) particle motion and correspond to seismic noise with a lower amplitude than the signal in the following rows. No smaller-amplitude P-wave phase is observed before the detection time of SDEs.

The onset of the signals (Figs. 9.6b2 and 9.7b2) does not exhibit linear particle motion. The particle motion of a P-wave is usually expected to be linear in all three panels (E–N, E–Z, and N–Z; e.g., Greenhalgh et al., 2018). However, in certain cases, such as with an anelastic medium, the particle motion of body waves may also not appear linear (see Tary, this issue, for further discussion). For the larger-amplitude SDE (Fig. 9.6a), the complex particle motion between 170 and 190 ms (Fig. 9.6b2) could be due to the composite movement of a potential P-wave phase affected by the shot of unknown azimuth from the neighboring experiment that is clearly observed on the hydrophone component (Fig. 9.6a, black arrow). The beginning of the second and smaller-amplitude SDE is not affected by neighboring shots (Fig. 9.7a), but the observed particle motion is still nonlinear (Fig. 9.7b2).

The observed ellipticity in the E–N panels (Figs. 9.6b2 and 9.7b2) can be due to S-wave splitting, with S1 (fast) and S2 (slow) arriving almost simultaneously at a 50 Hz sampling rate. In the Scanner pockmark, we have evidence of S-wave splitting (Bayrakci et al., 2021; Robinson et al., 2022), so we expect to observe S1 and S2 waves. When S-wave splitting is observed due to azimuthal anisotropy, the S1 and S2 waves are polarized parallel and perpendicular to the cracks. In that case, instead of the source back azimuth, the particle motions depend on the orientations of the cracks (e.g., Bale, 2005; Bayrakci et al., 2021; Robinson et al., 2022). Below OBS-8, in the shallowmost sediments (<5m depth), the anisotropic symmetry axes lie along 70/160° (Bayrakci et al., 2021; Robinson et al., 2022). Two other S-wave splitting events are observed at depths of 0.45–0.50 s and >0.82 s two-way travel time, displaying a 45/135° orientation of the symmetry plains (Robinson et al., 2022), which is coherent with the observed particle motions in Figs. 9.6b2 and 9.7b2.

For both SDEs, the later and lower-frequency parts of the waveforms (Figs. 9.6b4–6 and 9.7b4–6) show linear particle motion in all panels. The particle motions of both SDEs indicate the same direction in all panels. The observed damped harmonic motion may correspond to

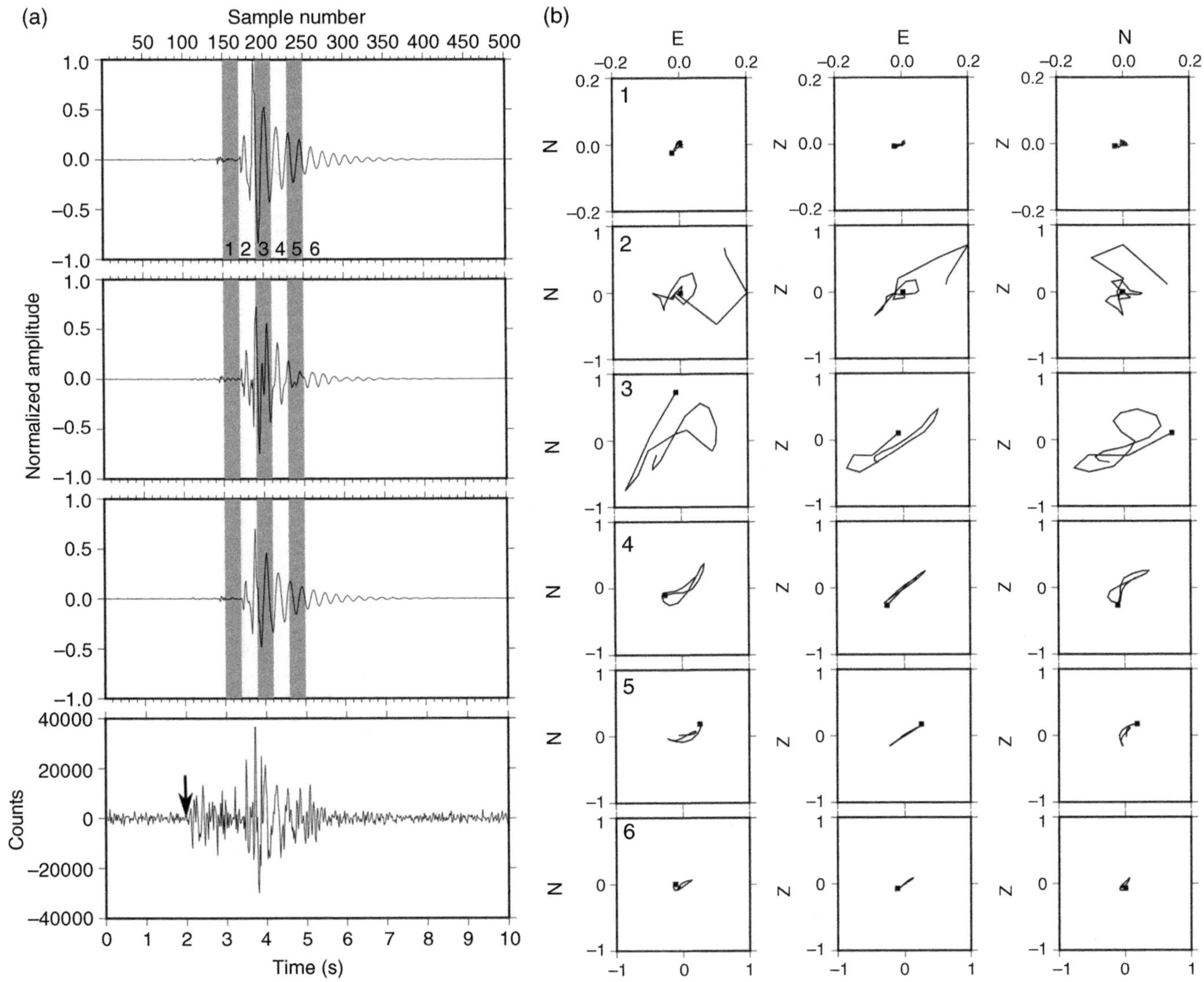

Figure 9.6 (a) High-amplitude SDE shown in Fig. 9.3 rotated (61°) to the geographic coordinate frame. Top to bottom: east–west (E), north–south (N), and vertical (Z) geophone components and the hydrophone (H). Alternating grayed areas marked by numbers (1–6) of 400 ms duration (20 samples) show the samples used in the particle motion panel in (b). The black arrow shows the beginning of a shot from the neighboring seismic experiment. (b) Corresponding particle motion panels. Left to right: E–N, E–Z, and N–Z panels. The first row starts 400 ms (20 samples) earlier than the beginning of the SDE to visualize a potential earlier phase. The second row is the beginning of the SDE, and each row represents 400 ms (20 samples) of the signal. Note that the scale of the first row differs from that of the other rows.

body waves initiated from Krauklis waves that propagate back and forth along the walls of the fluid-filled crack, which are transferred to the host rock as body waves before being recorded by OBSs (Frehner, 2014). The observed particle motion could correspond to an S-wave polarized along an ~40–45° azimuth with an inclination of 45–50° (Figs. 9.6b4–6 and 9.7b4–6). The fracture azimuth 45° matches the anisotropic symmetry axes observed via S-wave splitting within the shallow sediments (Robinson et al., 2022).

We cross-correlated hourly histograms of the number of SDEs recorded on different instruments (Fig. 9.8) to determine where similar SDE crisis patterns are observed on different OBSs (Fig. 9.9). Although simultaneous observation of a given SDE on multiple OBSs seems to be absent, simultaneous event crises are observed on OBS-1 and -2, which showed a correlation of 0.7. Apart from these two OBSs that are only 20 m apart, only OBSs 8 and 11, located at the SSW and NNE corners of the western Scanner pockmark, respectively, showed a

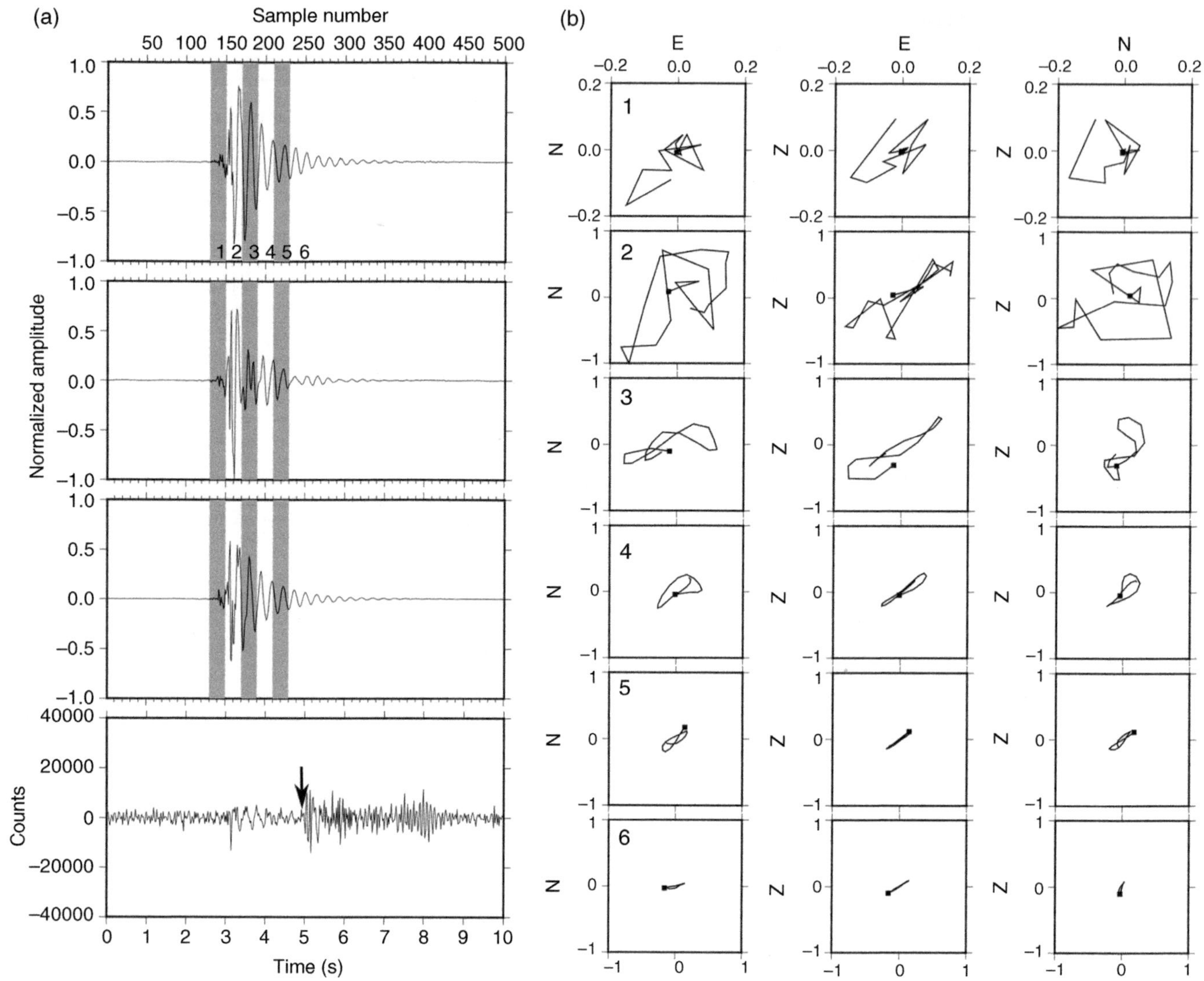

Figure 9.7 (a) Low-amplitude SDE shown in Fig. 9.3 rotated (61°) to the geographic coordinate frame. Top to bottom: east–west (E), north–south (N), and vertical (Z) geophone components and the hydrophone (H). Alternating grayed areas marked by numbers (1–6) of 400 ms duration (20 samples) show the samples used in the particle motion panel in (b). The black arrow shows a shot from the neighboring seismic experiment. (b) Corresponding particle motion panels. Left to right: E–N, E–Z, and N–Z panels. The first row starts 400 ms (20 samples) earlier than the beginning of the SDE to visualize a potential earlier phase. The second row is the beginning of SDE, and each row represents 400 ms (20 samples) of the signal. Note that the scale of the first row differs from that of the other rows.

correlation higher than 0.5 (Fig. 9.9). These two OBSs recorded the highest number of SDEs over the entire record period (Fig. 9.4). Only one large SDE peak (more than 300 SDEs in three hours, on Julian day 247 at 18:00–21:00) is observed on the cumulative distribution of three hourly occurrence of SDEs recorded by all 10 OBSs (Figs. 9.5 and 9.10). This SDE crisis is mainly recorded by OBS-8 and -11, although these instruments were 400 m apart.

9.6.2. Spatial Distribution of SDEs

The largest numbers of SDEs are observed on OBS-11 and -8 located at the NNE and SSW corners of the western Scanner pockmark, respectively (Fig. 9.4). The orientation of the long axis of elliptical western Scanner pockmark is also in the SSW–NNE direction, which suggests a relationship between the seafloor morphology and the spatial distribution of SDEs. OBS-2, which is the

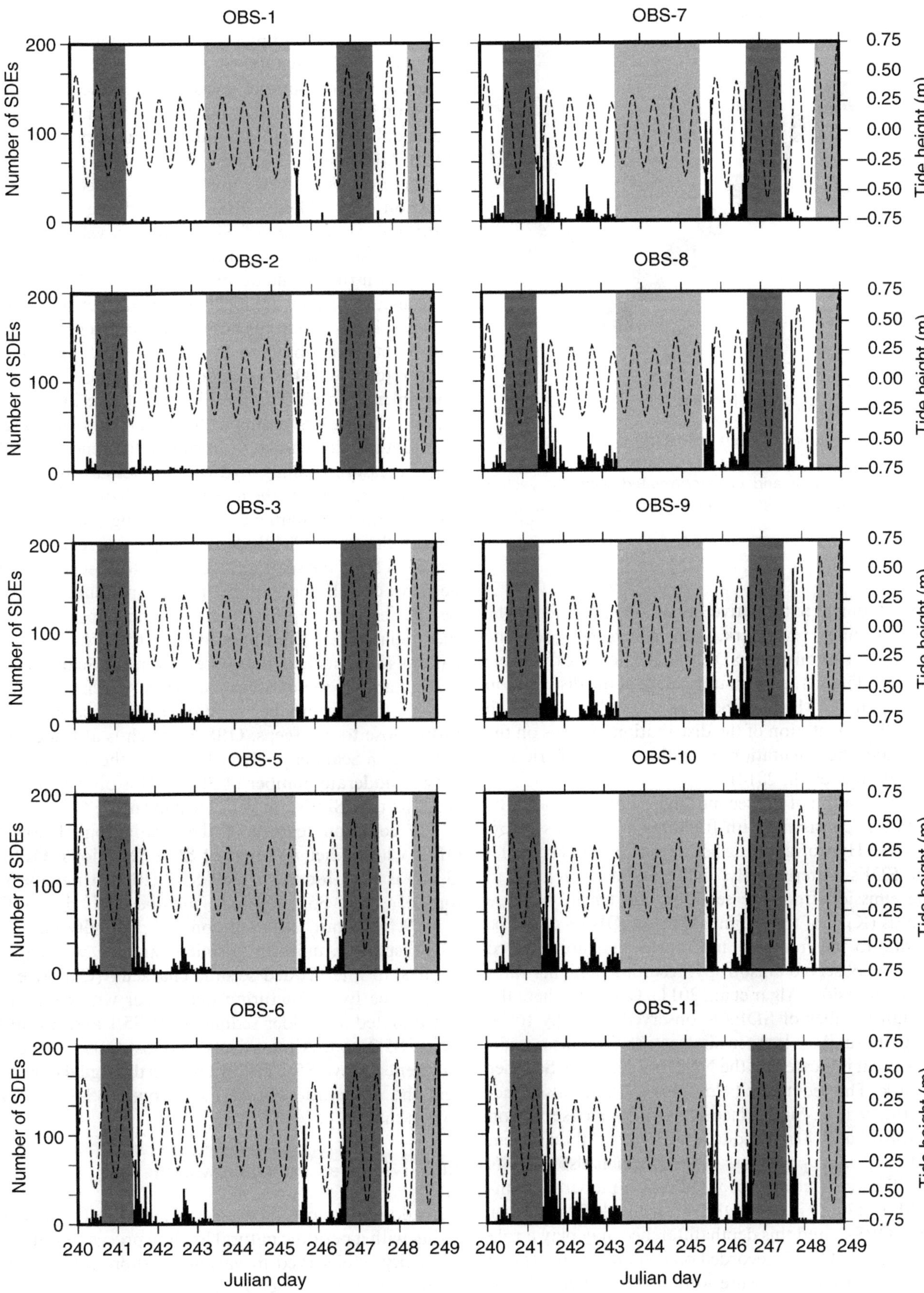

Figure 9.8 Hourly histograms of the number of SDEs recorded on different instruments.

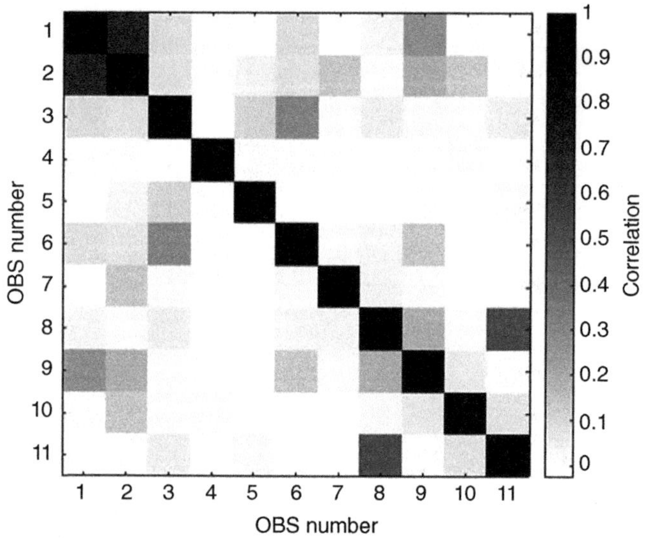

Figure 9.9 Correlation between histograms of events observed on different OBSs. The hourly histograms of each OBS are converted to a vector and cross-correlated with the vectors observed on other OBSs. Higher values (dark) show good correlation.

closest instrument to the active gas venting observed in the western Scanner pockmark, and which recorded the fourth-highest number of SDEs, is located on the same SSW–NNE line. An earlier study imaged the distribution of gas seeps in the Marmara Sea and showed a clear match between the orientation of the distribution of seeps on the seafloor and the orientation of seafloor bathymetric features (Bayrakci et al., 2014).

The observation of the second-highest number of SDEs on OBS-8 is consistent with fractures and cracks below these OBSs. Hydrofracturing below the OBS-8 may nucleate the SDEs. Alternatively, if the SDEs are related to bubble emission from the seafloor, the tensile fracture toughness (K_{IC}) of the sediments below OBS-8 could be lower due to the fractures and cracks that are inferred to cause the observed anisotropy, resulting in higher rates of bubble emission (Algar et al., 2011). OBS-11, where the maximum number of SDEs is observed, is likely to be located on similar, low-tensile-fracture-toughness sediments since it is located on the NNE border of the Scanner pockmark. The sediments below this OBS are also fractured (Fig. 9.11a). Simultaneous SDE crisis observed on both OBS-11 and -8 may indicate a connected fluid system at depth below the pockmark. OBS-1, which also exhibits shallow S-wave splitting, recorded only a moderate number of SDEs (201; Fig. 9.4). OBS-1 is located ~20 m deeper than the OBS-8, and this depth difference may play a role in the number of SDEs recorded on these instruments due to the hydrostatic pressure difference. Alternatively, aligned fractures and cracks below OBS-1 may be inactive or sealed, or the fluid pressure close to this OBS may not be as high as below OBS-8.

Active and continuous gas seeps are observed in western and eastern Scanner pockmark on the multibeam water-column data collected during the survey (Li et al., 2020). OBS-2 and -3 in the western and eastern Scanner pockmark, respectively, are close to these active gas-emission sites. These OBSs recorded the fourth- and third-highest number of SDEs, respectively (Fig. 9.4). The hydrostatic pressure at the depth of these instruments may play a role in the occurrence of SDEs close to these instruments, as they are located 20 m deeper than others. A general decrease in the number of occurrences of SDEs with depth was observed by Buskirk et al. (1981), who interpreted the SDEs as evidence for biological activity and linked this observation to the decrease in the abundance of living organisms with depth. However, here we interpret this decrease as the result of differences in hydrostatic pressure. Falling hydrostatic pressure reduces the effective stress below the tensile strength of the sediments, which leads to dynamic conduit opening and gas release (Scandella et al., 2011). If the SDEs are related to gas emission from the seafloor, we expect to observe a similar negative correlation with hydrostatic pressure controlled by the seafloor depth at the instrument location. After the initiation of fluid-migration pathways, continuous gas emission, such as that observed in the Scanner pockmark, can use the existing pathways, which can explain why the highest number of SDEs is not observed on the OBSs close to gas seeps. OBS-1, which is also located in the western Scanner pockmark, close to the seep, recorded only a moderate number of SDEs. This observation also suggests that SDEs are very local events.

The common feature of the stratigraphy below the OBSs with higher numbers of SDE detections (OBS-11, -3, -2) is the absence of unit S5.2 (Fladen member) on top of unit S5.1 (Witch member of the WGF) (Fig. 9.11). Both units are composed of silty clay: unit S5.2 displays a uniform seismic character, whereas unit 5.1 has an interbedded seismic character (Callow et al., 2021). The hydrofracturing may occur within the more consolidated and older sediments of S5.1 and not in the recent S5.2 unit. Alternatively, at most OBSs, thick S5.2 sediments above S5.1 may attenuate the signals and prevent the observation of SDEs. Below OBS-8, however, S5.2 is thin.

9.6.3. Negative Correlation with the Tide

In addition to tectonic triggering, tidal earthquake triggering with a periodic natural loading process modulating seismicity is observed in various offshore settings: mid-ocean ridges (e.g., Scholz et al., 2019; Sahoo et al.,

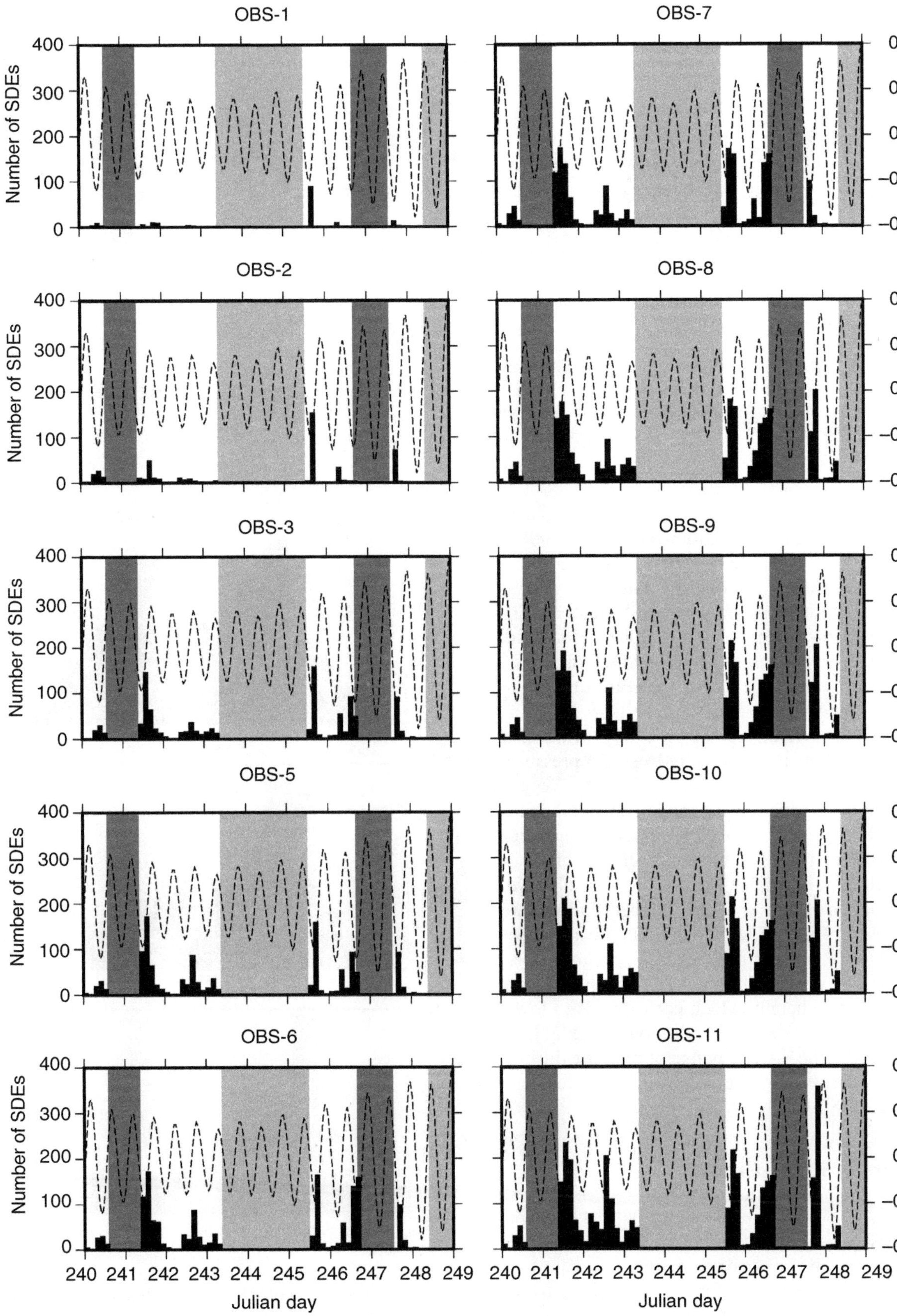

Figure 9.10 Cumulative distribution of SDEs recorded by all OBSs.

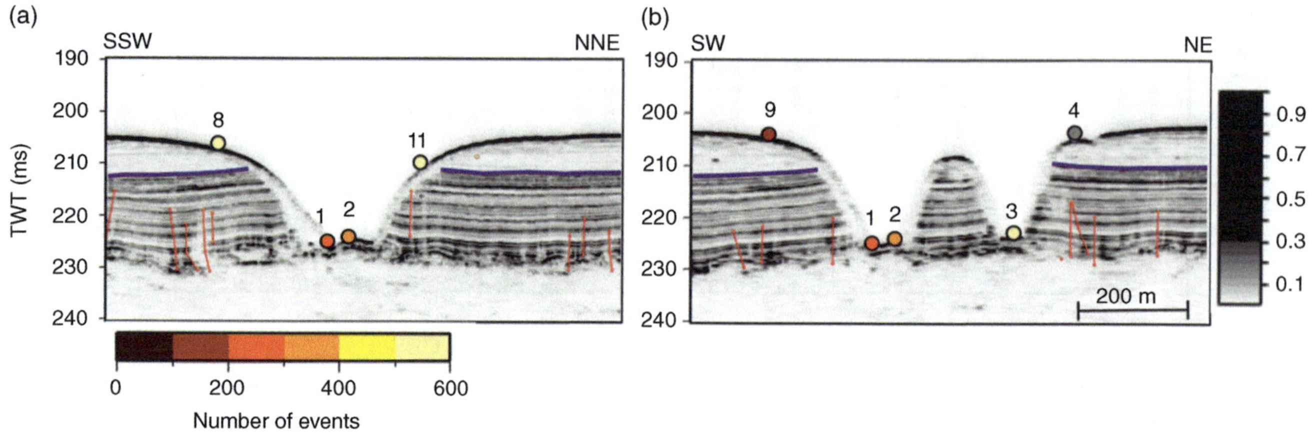

Figure 9.11 (a, b) Two sub-bottom profiler profiles showing the shallow structure beneath the OBSs with higher SDE observations. Circles show the location of OBSs along the lines, filled based on the number of observed SDEs. The numbers in (b) correspond to the sedimentary units. S5 = WGF; S5.2 = Fladen member; S5.1 = Witch member of the WGF; S4 = LGM deposits. Detailed structural interpretation Ben Callow et al. 2021 / with permission from Elsevier.

2021), hydrothermal systems (e.g., Glasby & Kasahara 2001), subduction zones (e.g., Rubinstein et al., 2008), and mud volcanoes (e.g., Hsu et al., 2013). Tidal triggering is more significant for smaller and shallower earthquakes due to less confining pressure and greater tidal stress for these events (Métivier et al., 2009). Here, we can also expect a similar observation for SDEs originating in the shallow subsurface where the confining pressure is small, especially with the shallow seafloor depth (~150 m) and relatively large tidal variations, which would translate into large hydrostatic pressure variations.

The distribution of SDEs as a function of tide height (Fig. 9.12) shows a greater number of SDEs during periods of low tide. When the distribution of SDEs is weighted by the duration of tide heights (Fig. 9.12c), we observe that the number of SDEs is clearly higher during the low-tide slack water (Fig. 9.12c, –0.8 m) than the high-tide slack water (0.6 m). A clear peak in SDE occurrence is observed at –0.4 m during the low-tide periods but not during the lowest tide height, which corresponds to low-tide slack water, suggesting that in addition to the hydrostatic pressure drop, the rate of change of pressure during the ascending or descending tide plays a role in determining the number of SDEs.

9.6.4. Efficiency of SDE Detection

In this study, we used a challenging data set where active source shooting was continuous and broadband with five different seismic sources covering a frequency range of 3–1,600 Hz (Fig. 9.2). In addition, shots from another seismic experiment with unknown origin times and location, a peak frequency of 3 Hz, and a time interval of 8 s—discovered after an initial analysis of the data set—

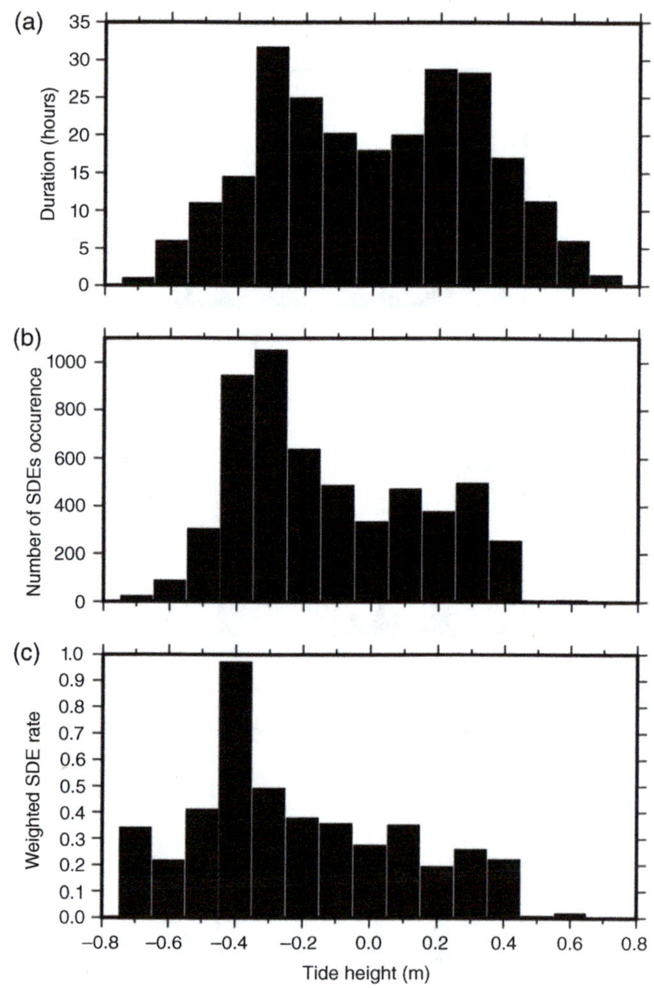

Figure 9.12 Distribution of SDEs as a function of tide height. (a) Histogram of tidal height durations. (b) Number of SDEs as a function of tidal height. (c) Normalized number of SDEs as a function of tidal height weighted by the duration of tidal height.

make the detection of SDEs even more challenging. These foreign shots are best observed on the hydrophones (Fig. 9.3a), and their geophone records are usually several orders of magnitude lower than those of SDEs. Rarely, and only on a few instruments, potentially those closest to the neighboring shooting vessel, these shots are recorded with amplitudes similar to those of SDEs. However, simultaneous data on longer offset instruments was clean enough to be used for SDE detection. The neighboring seismic shooting covers the entire record period of our experiment. To discard the neighboring shots recorded on short-offset OBSs, we applied an amplitude criterion in our detection algorithm. The amplitude of the SDEs spans several orders of magnitude. Smaller-amplitude SDEs ($<10^5$ counts) have low signal-to-noise ratios, and they are difficult to detect reliably either automatically with the STA/LTA algorithm or manually. Therefore, an amplitude criterion was also useful to obtain a homogeneous data set. With the parameters used for the detection, and the additional minimum amplitude criterion ($>10^5$ counts, 0.082 mm/s of ground velocity), the algorithm detected only 198 residual neighbor shots. These false detections were removed manually from the data set.

There are other false detections at the beginning of shot profiles, triggered due to a low long-term average and a high short-term average at the start of shooting. Similarly, the presence of shots, even if they are not detected as SDEs, may prevent the detection of SDEs by lowering the STA/LTA ratio by increasing the long-term average, especially if the shot amplitudes are high. Correlation-based matched filtering analysis (MFA) techniques with a master signal detection could improve SDE detection by allowing events to be distinguished from artificial shots. The classification of waveforms via, for example, principal component analysis would be required for MFA, which would allow analysis of the distribution of waveforms related to different mechanisms (Caffagni et al., 2016).

9.7. CONCLUSION

We observed more than 3000 SDEs on 10 OBSs over a duration of 120 h. Observed SDEs are up to 6 s long and monochromatic, with a peak frequency of 4 Hz. All observed SDEs show similar frequency content, suggesting a similar mechanism of occurrence. SDEs with higher amplitudes on geophone components show a hydrophone signal, whereas the hydrophone signal is obscured by noise for smaller-amplitude SDEs, indicating a subsurface origin. SDEs have higher horizontal geophone amplitudes than vertical geophone amplitudes, suggesting a shallow subsurface origin. SDEs do not exhibit clearly identifiable body and surface waves. However, the presence of a higher-frequency train followed by a quickly decreasing and lower-frequency train suggests a change in wave type. The hodograms of SDEs suggest that the beginning of the signal can be an S-wave phase propagating in an anisotropic medium, with slow and fast S-wave phases observed almost simultaneously. We speculate that the later monochromatic part of the signal can be related to Krauklis waves, generated by the interaction of body waves created by a hydrofracturing event and the walls of the fluid-filled cracks or fractures.

Although simultaneous SDE crises are observed on multiple instruments, no SDE is simultaneously observed on multiple instruments even when the instruments are only 20 m apart. Two OBSs located at the SSW and NNE corners of the pockmark recorded the greatest numbers of SDEs, followed by two OBSs located close to the known gas-venting sites in the eastern and western Scanner pockmarks, respectively. The western Scanner pockmark is elliptical with an SSW–NNE major axis, suggesting a link between seafloor morphology and SDE occurrence. We observed a clear negative correlation between tidal heights and the number of SDEs, with the greatest number of SDEs occurring during low tide, but not during the low-tide slack water. This observation indicates that, in addition to the hydrostatic pressure control, there is a pressure gradient control on the rate of SDE occurrence and gas flux.

ACKNOWLEDGMENTS

Data acquisition and analysis was funded by the UK Natural Environment Research Council (grants NE/N016130/1 and NE/N015762/1). We thank the officers, engineers, crew, and scientific party of research cruise JC152 for their support at sea. OBSs and associated technical support were provided by the NERC Ocean-Bottom Instrumentation Facility (Minshull et al., 2005). We thank Sarah Bazin, Jean-Baptiste Tary, and the anonymous reviewer for their constructive comments.

REFERENCES

Algar, C. K., Boudreau, B. P., & Barry, M. A. (2011). Release of multiple bubbles from cohesive sediments. *Geophysical Research Letters*, *38*(8).

Andresen, K. J. (2012). Fluid flow features in hydrocarbon plumbing systems: what do they tell us about the basin evolution? *Marine Geology*, *332*, 89–108.

Bale, R., Gratacos, B., Mattocks, B., Roche, S., Poplavskii, K., & Li, X. (2009). Shear wave splitting applications for fracture analysis and improved imaging: some onshore examples. *First Break*, *27*(9). https://doi.org/10.3997/1365-2397.27.1304.32448

Batsi, E., Tsang-Hin-Sun, E., Klingelhoefer, F., Bayrakci, G., Chang, E. T., Lin, J. Y., et al. (2019). Nonseismic signals in the ocean: indicators of deep sea and seafloor processes on ocean-bottom seismometer data. *Geochemistry, Geophysics, Geosystems*, *20*(8), 3882–3900.

Bayrakci, G., Callow, B., Bull, J., Minshull, T. A., Provenzano, G., North, L., et al. (2021). Seismic anisotropy within an active fluid flow structure: Scanner pockmark, *North Sea. Frontiers in Earth Science*, *9*, 626416.

Bayrakci, G., Minshull, T. A., Bull, J. M., Henstock, T. J., Provenzano, G., Birinci, H., et al. (2020). P-wave velocity anisotropy in an active methane venting pockmark: the Scanner pockmark, northern North Sea. In *EGU General Assembly Conference Abstracts*.

Bayrakci, G., Scalabrin, C., Dupré, S., Leblond, I., Tary, J.-B., Lanteri, N., et al. (2014). Acoustic monitoring of gas emissions from the seafloor. Part II: a case study from the Sea of Marmara. *Marine Geophysical Research*, *35*(3), 211–229.

Boudreau, B. P., Algar, C. K., Johnson, B. D., Croudace, I., Reed, A., Furukawa, Y., et al. (2005). Bubble growth and rise in soft sediments. *Geology*, *33*(2005), 517–520.

Boudreau, B. P. (2012). The physics of bubbles in surficial, soft, cohesive sediments. *Marine and Petroleum Geology*, *38*(1), 1–18.

Böttner, C., Berndt, C., Reinardy, B. T., Geersen, J., Karstens, J., Bull, J. M., et al. (2019). Pockmarks in the Witch Ground Basin, central North Sea. *Geochemistry, Geophysics, Geosystems*, *20*(4), 1698–1719.

Bull, J., Berndt, C., Minshull, T., Henstock, T., Bayrakci, G., Gehrmann, R., et al. (2018). *Constraining leakage pathways through the overburden above sub-seafloor CO2 storage reservoirs*. Paper presented at American Geophysical Union, Fall Meeting 2018.

Burrell, A., Evans, J., & De Kauwe, M. (2020). Anthropogenic climate change has driven over 5 million km^2 of drylands towards desertification. *Nature Communications*, *11*(1), 1–11.

Buskirk, R. E., Frohlich, C., Latham, G. V., Chen, A. T., & Lawton, J. (1981). Evidence that biological activity affects ocean bottom seismograph recordings. *Marine Geophysical Researches*, *5*(2), 189–205.

Caffagni, E., Eaton, D. W., Jones, J. P., & van der Baan, M. (2016). Detection and analysis of microseismic events using a matched filtering algorithm (MFA). *Geophysical Journal International*, *206*(1), 644–658.

Callow, B., Bull, J. M., Provenzano, G., Böttner, C., Birinci, H., Robinson, A. H., et al. (2021). Seismic chimney characterisation in the North Sea—implications for pockmark formation and shallow gas migration. *Marine and Petroleum Geology*, *105301*.

Cartwright, J., Huuse, M., & Aplin, A. (2007). Seal bypass systems. *AAPG Bulletin*, *91*(8), 1141–1166.

Cathles, L., Su, Z., & Chen, D. (2010). The physics of gas chimney and pockmark formation, with implications for assessment of seafloor hazards and gas sequestration. *Marine and Petroleum Geology*, *27*(1), 82–91.

Cevatoglu, M., Bull, J. M., Vardy, M. E., Gernon, T. M., Wright, I. C., & Long, D. (2015). Gas migration pathways, controlling mechanisms and changes in sediment acoustic properties observed in a controlled sub-seabed CO2 release experiment. *International Journal of Greenhouse Gas Control*, *38*, 26–43.

Chand, S., & Minshull, T. (2003). Seismic constraints on the effects of gas hydrate on sediment physical properties and fluid flow: a review. *Geofluids*, *3*(4), 275–289.

Dean, J. F., Middelburg, J. J., Röckmann, T., Aerts, R., Blauw, L. G., Egger, M., et al. (2018). Methane feedbacks to the global climate system in a warmer world. *Reviews of Geophysics*, *56*(1), 207–250.

Diaz, J., Gallart, J., & Gaspà, O. (2007). Atypical seismic signals at the Galicia Margin, North Atlantic Ocean, related to the resonance of subsurface fluid-filled cracks. *Tectonophysics*, *433*(1–4), 1–13.

Ebigbo, A., Class, H., & Helmig, R. (2007). CO2 leakage through an abandoned well: problem-oriented benchmarks. *Computational Geosciences*, *11*(2), 103–115.

Flohr, A., Schaap, A., Achterberg, E. P., Alendal, G., Arundell, M., Berndt, C., et al. (2021). Towards improved monitoring of offshore carbon storage: a real-world field experiment detecting a controlled sub-seafloor CO2 release. *International Journal of Greenhouse Gas Control*, *106*, 103237.

Frehner, M. (2014). Krauklis wave initiation in fluid-filled fractures by seismic body waves. *Geophysics*, *79*(1), T27–T35.

Gehrmann, R. A., Provenzano, G., Böttner, C., Marín-Moreno, H., Bayrakci, G., Tan, Y. Y., et al. (2021). Porosity and free gas estimates from controlled source electromagnetic data at the Scanner pockmark in the North Sea. *International Journal of Greenhouse Gas Control*, *109*, 103343.

Géli, L., Piau, J.-M., Dziak, R., Maury, V., Fitzenz, D., Coutellier, Q., et al. (2014). Seismic precursors linked to highly compressible fluids at oceanic transform faults. *Nature Geoscience*, *7*(10), 757–761.

Glasby, G., & Kasahara, J. (2001). Influence of tidal effects on the periodicity of earthquake activity in diverse geological settings with particular emphasis on submarine hydrothermal systems. *Earth-Science Reviews*, *52*(4), 261–297.

Greenhalgh, S., Sollberger, D., Schmelzbach, C., & Rutty, M. (2018). Single-station polarization analysis applied to seismic wavefields: A tutorial. *Advances in Geophysics*, *59*, 123–170.

Helffrich, G., Wookey, J., & Bastow, I. (2013). *The Seismic Analysis Code: A Primer and User's Guide*. Cambridge University Press.

Hovland, M., Judd, A. G., & Burke Jr, R. (1993). The global flux of methane from shallow submarine sediments. *Chemosphere*, *26*(1–4), 559–578.

Hovland, M., Judd, A. G., & King, L. H. (1984). Characteristic features of pockmarks on the North Sea Floor and Scotian Shelf. *Sedimentology*, *31*(4), 471–480.

Hovland, M., Talbot, M. R., Qvale, H., Olaussen, S., & Aasberg, L. (1987). Methane-related carbonate cements in pockmarks of the North Sea. *Journal of Sedimentary Research*, *57* (5), 881–892.

Hsu, S.-K., Wang, S.-Y., Liao, Y.-C., Yang, T. F., Jan, S., Lin, J.-Y., & Chen, S.-C. (2013). Tide-modulated gas emissions and tremors off SW Taiwan. *Earth and Planetary Science Letters*, *369*, 98–107.

Hustoft, S., Bünz, S., & Mienert, J. (2010). Three-dimensional seismic analysis of the morphology and spatial distribution of chimneys beneath the Nyegga pockmark field, offshore mid-Norway. *Basin Research*, *22*(4), 465–480.

Johnson, B. D., Boudreau, B. P., Gardiner, B. S., & Maass, R. (2002). Mechanical response of sediments to bubble growth. *Marine Geology*, *187*(2002), 347–363.

Johnson, H. P., Hutnak, M., Dziak, R. P., Fox, C. G., Urcuyo, I., Cowen, J. P., et al. (2000). Earthquake-induced changes in a hydrothermal system on the Juan de Fuca mid-ocean ridge. *Nature*, *407*(6801), 174–177.

Judd, A., Long, D., & Sankey, M. (1994). Pockmark formation and activity, UK block 15/25, North Sea. *Bulletin of the Geological Society of Denmark*, *41*(1), 34–49.

Karstens, J., & Berndt, C. (2015). Seismic chimneys in the Southern Viking Graben—Implications for paleo fluid migration and overpressure evolution. *Earth and Planetary Science Letters*, *412*, 88–100.

Kettlety, T., & Verdon, J. P. (2021). Fault triggering mechanisms for hydraulic fracturing-induced seismicity from the Preston New Road, UK case study. *Frontiers in Earth Science*, *9*, 382.

King, L. H., & MacLean, B. (1970). Pockmarks on the Scotian shelf. *Geological Society of America Bulletin*, *81*(10), 3141–3148.

Kopf, A. J. (2003). Global methane emission through mud volcanoes and its past and present impact on the Earth's climate. *International Journal of Earth Sciences*, *92*(5), 806–816.

Korneev, V. (2009). Resonant seismic emission of subsurface objects. *Geophysics*, *74*(2), T47–T53.

Korneev, V. A. (2011). Krauklis wave in a stack of alternating fluid-elastic layers. *Geophysics*, *76*(6), N47–N53.

Korneev, V., Goloshubin, G., Kashtan, B., Troyan, V., Shigapov, R., Bakulin, A., et al. (2012). Krauklis wave—half a century after. Paper presented at Saint Petersburg 2012–5th Saint Petersburg International Conference & Exhibition 2012: Making the Most of the Earth's Resources.

Lee, M. W. (2004). Elastic velocities of partially gas-saturated unconsolidated sediments. *Marine and Petroleum Geology*, *21*(6), 641–650.

Li, J., Roche, B., Bull, J. M., White, P. R., Leighton, T. G., Provenzano, G., et al. (2020). Broadband acoustic inversion for gas flux quantification—application to a methane plume at Scanner pockmark, central North Sea. *Journal of Geophysical Research: Oceans*, *125*(9), e2020JC016360.

Li, L., Tan, J., Wood, D. A., Zhao, Z., Becker, D., Lyu, Q., et al. (2019). A review of the current status of induced seismicity monitoring for hydraulic fracturing in unconventional tight oil and gas reservoirs. *Fuel*, *242*, 195–210.

Lomax, A., Michelini, A., & Piatanesi, A. (2007). An energy-duration procedure for rapid determination of earthquake magnitude and tsunamigenic potential. *Geophysical Journal International*, *170*(3), 1195–1209.

Løseth, H., Wensaas, L., Arntsen, B., Hanken, N.-M., Basire, C., & Graue, K. (2011). 1000 m long gas blow-out pipes. *Marine and Petroleum Geology*, *28*(5), 1047–1060.

Løseth, H., Gading, M., & Wensaas, L. (2009). Hydrocarbon leakage interpreted on seismic data. *Marine and Petroleum Geology*, *26*(7), 1304–1319.

Marín-Moreno, H., Bull, J. M., Matter, J. M., Sanderson, D. J., & Roche, B. J. (2019). Reactive transport modeling insights into CO2 migration through sub-vertical fluid flow structures. *International Journal of Greenhouse Gas Control*, *86*, 82–92.

Métivier, L., de Viron, O., Conrad, C. P., Renault, S., Diament, M., & Patau, G. (2009). Evidence of earthquake triggering by the solid earth tides. *Earth and Planetary Science Letters*, *278* (3–4), 370–375.

Minshull, T.A., Sinha, M.C. & Peirce, C. (2005). Multi-disciplinary, sub-seabed geophysical imaging—a new pool of 28 seafloor instruments in use by the United Kingdom Ocean Bottom Instrument Consortium. *Sea Technology*, *46*(10), 27–31.

Moss, J. and J. Cartwright (2010). 3D seismic expression of km-scale fluid escape pipes from offshore Namibia. *Basin Research*, *22*(4), 481–501.

Pörtner, H. O., Roberts, D. C., Adams, H., Adler, C., Aldunce, P., Ali, E., et al. (2022). *Climate change 2022: impacts, adaptation and vulnerability* (IPCC Sixth Assessment Report).

Robinson, A. H., Callow, B., Böttner, C., Yilo, N., Provenzano, G., Falcon-Suarez, I. H., et al. (2021). Multiscale characterization of chimneys/pipes: fluid escape structures within sedimentary basins. *International Journal of Greenhouse Gas Control*, *106*, 103245.

Roche, B., Bull, J. M., Marin-Moreno, H., Leighton, T. G., Falcon-Suarez, I. H., Tholen, M., et al. (2021). Time-lapse imaging of CO2 migration within near-surface sediments during a controlled sub-seabed release experiment. *International Journal of Greenhouse Gas Control*, *109*, 103363.

Rubinstein, J. L., La Rocca, M., Vidale, J. E., Creager, K. C., & Wech, A. G. (2008). Tidal modulation of nonvolcanic tremor. *Science*, *319*(5860), 186–189.

Ruppel, C. D., & Kessler, J. D. (2017). The interaction of climate change and methane hydrates. *Reviews of Geophysics*, *55*(1), 126–168.

Sahoo, S., Senapati, B., Panda, D., Tiwari, D. K., Santosh, M., & Kundu, B. (2021). Tidal triggering of micro-seismicity associated with caldera dynamics in the Juan de Fuca ridge. *Journal of Volcanology and Geothermal Research, 107319.*

Sawyer, J. S. (1972). Man-made carbon dioxide and the "greenhouse" effect. *Nature*, *239*(5366), 23–26.

Scandella, B. P., Varadharajan, C., Hemond, H. F., Ruppel, C., & Juanes, R. (2011). A conduit dilation model of methane venting from lake sediments. *Geophysical Research Letters*, *38*(6).

Scholz, C. H., Tan, Y. J., & Albino, F. (2019). The mechanism of tidal triggering of earthquakes at mid-ocean ridges. *Nature communications*, *10*(1), 1–7.

Schwiderski, E. W. (1980). On charting global ocean tides. *Reviews of Geophysics*, *18*(1), 243–268.

Tary, J.-B., Geli, L., Guennou, C., Henry, P., Sultan, N., Çağatay, N., & Vidal, V. (2012). Microevents produced by gas migration and expulsion at the seabed: a study based on sea bottom recordings from the Sea of Marmara. *Geophysical Journal International*, *190*(2), 993–1007.

Taylor, P., Stahl, H., Vardy, M. E., Bull, J. M., Akhurst, M., Hauton, C., et al. (2015). A novel sub-seabed CO2 release experiment informing monitoring and impact assessment for geological carbon storage. *International Journal of Greenhouse Gas Control*, *38*, 3–17.

Tsumura, K. (1967). Determination of earthquake magnitude from total duration of oscillation. *Bulletin of the Earthquake Research Institute*, *45*, 7–18.

Tsang-Hin-Sun, E., Batsi, E., Klingelhoefer, F., & Géli, L. (2019). Spatial and temporal dynamics of gas-related processes in the Sea of Marmara monitored with ocean bottom seismometers. *Geophysical Journal International*, *216*(3), 1989–2003.

10

Ambient Bubble Acoustics: Seep, Rain, and Wave Noise

Ben Roche[1], Timothy G. Leighton[2], Paul R. White[2], and Jonathan M. Bull[1]

ABSTRACT

This chapter discusses the sounds emitted by gas bubbles when they are generated underwater. Here, we define bubbles to be volumes of gas surrounded by liquid (in this case, taken to be water), having surface-tension forces (the so-called *Laplace pressure*) generated by a single wall; they are distinguished from the soap bubbles familiar in children's games, where the volume of gas is surrounded by two gas/liquid boundaries. Compared to other acoustic sources, such as marine mammals, ships, and tectonic events, a single bubble may seem insignificant. Indeed, without ideal conditions, it can be difficult to observe the sound of a single bubble from a distance of more than a few tens of centimeters. However, natural processes rarely produce single bubbles and can generate them by the million, at which point the sound generated is significant. The formation of bubbles due to gas seeps, rainfall, and breaking waves is a major component of ambient noise in the marine environment and can even alter the propagation of sound waves from other sources. This chapter focuses on the passive emissions of bubbles as they are formed, released, or injected into water, with linear volume pulsations.

10.1. INTRODUCTION

In this chapter, we will discuss the mechanics behind an individual bubble's acoustic signature, particularly the Minnaert equation and other relevant properties, before discussing the formation of bubbles from subsurface gas migration, rainfall, and wave action, characterizing the acoustic nature of each process. The primary focus will be on the sound resulting from bubble generation from each source. Several different units are used to define each acoustic source; although this may appear confusing and make direct comparison difficult, we do this to be consistent with the literature. The topics covered here are broad, so the approach taken is to summarize the key principles and state of the field while providing substantial linkage to the literature.

[1]*School of Ocean and Earth Science, University of Southampton, Southampton, UK*

[2]*Institute of Sound and Vibration Research, University of Southampton, Southampton, UK*

10.2. BUBBLES AS ACOUSTIC SOURCES

Although bubbles may be found throughout the water column and produced in all manner of ways, from fish flatulence to volcanic emissions, it is only the initial formation of the bubble near the source that is of interest in passive acoustics. Additionally, only a few sources of bubble production are common and large enough to warrant a full discussion: bubbles released from gas seeps on the seabed and those produced by either rainfall or breaking waves at the surface. The following section will discuss the initial release of gas bubbles into a body of water and the resulting acoustic signal. These are the fundamental principles behind bubble acoustics and are directly applicable to all initial sources of bubble production.

Noisy Oceans: Monitoring Seismic and Acoustic Signals in the Marine Environment, Geophysical Monograph 284, First Edition.
Edited by Gaye Bayrakci and Frauke Klingelhoefer.

DOI:10.1002/9781119750925.ch10

10.2.1. The Injection of a Gas Bubble

A bubble does not instantly appear fully formed in the water column. The gas is injected into the body of water over a very short time. Although there are several processes by which this injection can happen, the core principles remain the same: a small volume of gas from a larger reservoir encroaches into a body of water, with the two volumes of gas connected via a thin neck. As the small volume of gas extends further and further into the body of water, the neck is stretched thinner and thinner, eventually snapping and releasing the gas into the water as a distinct bubble. The snapping of the bubble neck is of most interest to us as it results in a jet of water being momentarily propelled into the bubble, triggering an initial volume oscillation. This volume oscillation ultimately results in the acoustic signal of a bubble release (Leighton, 1994; Longuet-Higgins et al., 1991; Czerski & Deane, 2010).

The easiest way to understand the process of bubble release is to study gas being injected into a body of water with a needle, as shown in Fig. 10.1. Theoretical calculations have been used to deduce the stages of bubble injection via a needle, reinforced by lab observations (Longuet-Higgins et al., 1991). These stages are best described in relation to the radius of curvature (the radius of a circular arc that best approximates the curve) of the meniscus at the top of the bubble; the scales are dimensionless. The bubble initially grows from the surface of the nozzle as gas flows through it, the radius of curvature decreasing from 1 to <0.5 with volume increasing steadily (t = –830 ms in Fig. 10.1). The bubble profile changes from nearly horizontal to semicircular in shape. Near the moment *the tangent to the meniscus at the point of attachment to the nozzle* becomes vertical (t = −730 ms in Fig. 10.1), the volume increases rapidly while the radius of curvature remains roughly constant (t = –480 ms in Fig. 10.1). Subsequently, the volume and radius of curvature increase steadily. Here, a neck begins to form; this is the narrowest part of the bubble profile, located between the nozzle and the main body of the bubble (t = –80 ms in Fig. 10.1). Once the radius of curvature equals ~0.655, the *tangent to the meniscus at the point of attachment* (now the neck of the bubble) becomes near vertical again, and there is a second sharp increase in volume. The bubble now has a distinct diapir-like shape (t = –2 ms in Fig. 10.1). The volume of gas in the bubble reaches a maximum; beyond this point, the bubble is considered unstable. Further air forced into the bubble causes it to detach, and the snapping of the neck releases it, allowing it to rise upward (t = 0 ms in Fig. 10.1). (Longuet-Higgins et al., 1991) The upper half of the neck recedes back into the bubble as a jet of water propels itself inward (t = 2 ms cross-section in Fig. 10.1); this decreases the volume of the bubble, resulting in a volume oscillation (Leighton, 1994; Longuet-Higgins et al., 1991; Czerski & Deane, 2010).

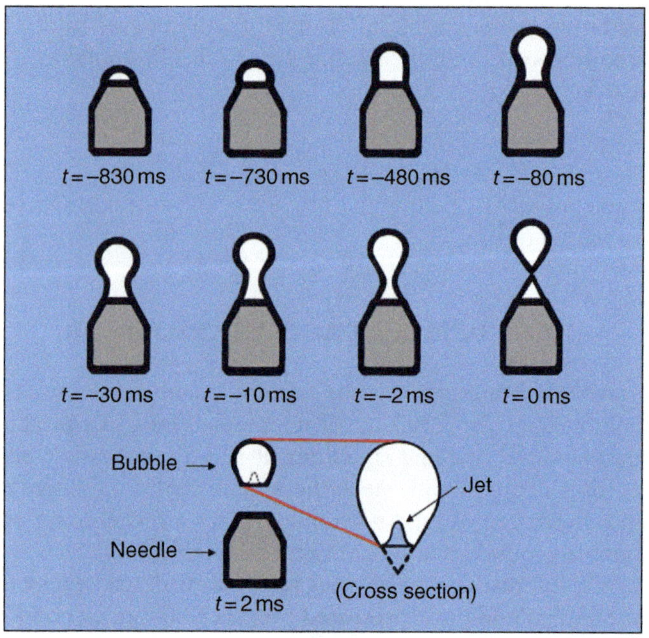

Figure 10.1 A bubble emerging from an underwater nozzle with internal diameter 4.00 mm. As the bubble grows, a neck forms between it and the injection nozzle; the neck eventually snaps, releasing the bubble and propelling a jet of water into it. This jet decreases the volume of the bubble and causes it to undergo simple harmonic motion. Times are given in milliseconds relative to the moment the bubble is detached (i.e., the neck snaps); note that these timings change with nozzle size and gas flow rate. Adapted from Longuet-Higgins et al. (1991) and Czerski and Deane (2010).

10.2.2. Bubbles as Simple Harmonic Oscillators

Immediately after its release into the water column, regardless of the means of production, a bubble begins to pulsate. The bubble itself may undergo a wide range of oscillatory changes in shape, but these can be decomposed into a summation of spherical harmonic pulsations; only one of these (the zeroth order) changes the bubble volume to first order and hence changes the gas pressure to first order (at low Mach numbers), and so couples to acoustic fields (Leighton, 1994). Therefore, even though the bubble often departs from sphericity, with a few notable exceptions that will not be discussed further in this chapter, it is appropriate when discussing the interaction with sound fields at low Mach numbers to refer to the pulsations of a spherical bubble. This oscillation approximates a simple harmonic oscillator at low amplitudes, occurring at the natural frequency of the bubble (Leighton, 1994). It is possible to derive the relationship between the radius of a bubble and the frequency of its initial free oscillations by assuming there are no dissipative losses—for example, through viscosity or thermal conduction—via consideration of the flow between potential and internal energy. The natural frequency of bubble oscillation is known as the Minnaert frequency (Minnaert, 1933).

As a simple harmonic oscillator, the pulsation of a bubble is analogous to the classic bob on a spring system of unloaded length l_0 and loaded length l. The water around the bubble is the bob weight, and the gas within the bubble is the spring, as shown in Fig. 10.2a. Note that the water's contribution to the system's effective mass declines with distance from the bubble wall, so the mass is in effect finite. The displacement ε from the equilibrium position corresponds to displacement of the bubble wall R_ε between its equilibrium radius R_0 (the bob at $l - l_0$) and its present radius R at any given moment (the bob at $l + \varepsilon - l_0$), see Fig. 10.2. The gas pressure following compression or expansion restores the bubble to its equilibrium position, which is analogous to the spring stiffness in the spring–bob example. However, it is important to note that the gas in the bubble is less dense than the surrounding medium (unlike the bob in the air). So, whereas in the spring–bob system, inertia and momentum are dominated by the bob, the inertia of the water surrounding a bubble dominates in the bubble system, the mass of the gas being negligible.

This final stage of the bubble formation triggers the simple harmonic motion of a bubble (Leighton, 1994; Czerski & Deane, 2010). The neck snapping triggers an initial volume oscillation that acts as an exciting force, causing the bubble to emit sound at its natural frequency. We assume this initial driving impulse is of infinitesimally small duration, meaning that while the bubble undergoes subsequent oscillation, it effectively experiences no driving force.

With this idea of a bubble as a simple harmonic oscillator, we can describe the shape of the bubble over time. Imagine a bubble (Fig. 10.2b) with a mean radius R_0 that remains spherical at all times while undergoing a volume oscillation at a frequency ω_0. The expressed maximum displacement of the bubble wall is $R_{\varepsilon 0}$ so that $R_{max} = R_0 + R_{\varepsilon 0}$ and $R_{min} = R_0 - R_{\varepsilon 0}$. Thus the bubble radius R at any time t can be expressed as the real part of

$$R = R_0 + R_\varepsilon(t) = R_0 + R_{\varepsilon 0}e^{i\omega_0 t}. \tag{10.1}$$

The displacement of the bubble wall R_ε from equilibrium over time describes a motion

$$R_\varepsilon = -R_{\varepsilon 0}e^{i\omega_0 t}. \tag{10.2}$$

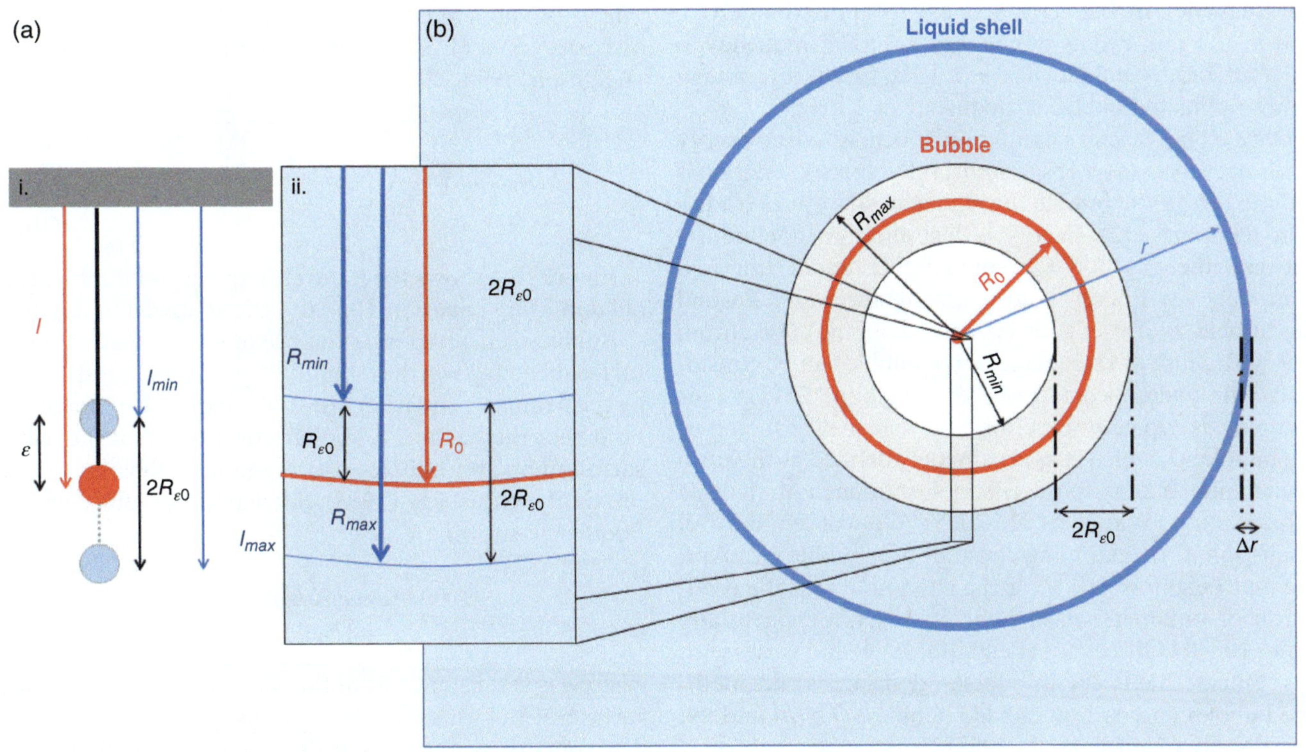

Figure 10.2 (a) Diagram comparing simple harmonic oscillators i. A spring bob system. ii. A bubble wall moving. (b) Diagram of a bubble of radius R_0, the wall of which is undergoing small-amplitude oscillations of amplitude $R_{\varepsilon 0}$. It is surrounded by spherical shells of liquid, one of which has a radius of r and a thickness of Δr.

10.2.3. Minnaert Frequency

With a description of the motion of the bubble wall over time, we can describe the flow between kinetic and potential energy. From here, we can apply the concept to the conservation of energy to derive the Minnaert (or natural) frequency of a bubble (Leighton, 1994; Minnaert, 1933):

$$f_M = \frac{1}{2\pi R_0}\sqrt{\frac{3\kappa p_0}{\rho}}, \tag{10.3}$$

where ρ is the density of water, p_0 is the hydrostatic liquid pressure outside the bubble, and κ is the polytropic index (which takes a value equal to unity when the gas behaves isothermally and equals the ratio of the specific heats of the gas at constant pressure to that at constant volume when the gas behaves adiabatically). A full derivation of the Minnaert frequency can be found in the appendix of this chapter.

The Minnaert equation demonstrates that the frequency of a bubble's oscillation is inversely proportional to its equilibrium radius R_0. As the other factors are fairly consistent or easily predictable (polytropic constant, density of water, water pressure outside the bubble), it is relatively easy to measure the size of a bubble based on its acoustic signal. As a general rule of thumb for bubbles near the surface, the radius in millimeters multiplied by the frequency in kilohertz is equal to approximately 3: that is, a 1 mm radius bubble has a 3 kHz frequency, a 1.5 mm radius bubble has a 2 kHz frequency, and a 3 mm radius bubble has a frequency of 1 kHz.

Once a bubble starts oscillating, it begins to lose energy in three ways. (1) Most importantly for us, energy is radiated from the bubble through acoustic waves (radiation damping). (2) Energy is lost through conduction between the gas and the surrounding liquid (thermal damping). (3) Energy is lost moving the water around the bubble as it oscillates (viscous damping) (Leighton, 1994). Because of these factors, the bubble can be considered lightly damped (Ainslie & Leighton, 2011). This damping is typically described by the quality factor of the bubble, Q, which is approximately defined as the ratio of the initial energy to the energy lost in one radian cycle of oscillation (Walton et al., 2005). We will avoid a full discussion of the damping constant of a bubble (see Ainslie and Leighton [2011] for this) and note that the oscillation of millimeter-sized bubbles decays exponentially over ~10–30 milliseconds (Leighton, 1994) and varies with gas content: that is, for air bubbles, $Q = 34$; for pure methane bubbles and carbon dioxide bubbles, $Q = 24$ and 29, respectively (Walton et al., 2005).

The polytropic adaptation of the Minnaert equation was first used in the 1980s to infer the size distribution and number of bubbles formed in the natural world in waterfalls and streams (Leighton & Walton, 1987). In subsequent years, this method was extended to do the same for bubbles entrained by breaking waves (Medwin & Beaky, 1989) and rainfall (Pumphrey & Crum, 1990). It works well when the signature passive emission from each bubble is clearly separated in time from others; however, this method of counting and sizing bubbles becomes more difficult as the signatures from each bubble get closer in time and overlap. Although signal-processing techniques can alleviate the problem (Leighton et al., 1998), eventually the degree of overlap becomes so great that this technique must be replaced by a spectral approach (discussed later) (Leighton & White, 2012).

A recording of a bubble signature (Fig. 10.3) shows a sinusoidal wave that decays exponentially, indicative of a lightly damped oscillator with a frequency consistent with that predicted by the Minnaert equation (equation 10.3). Note that because the sound generated by a bubble is an exponentially decaying sinusoid, the sound contains a range of frequencies, and the spectral profile of each bubble is Lorentzian (Leighton, 1988), centered around the natural frequency (Leighton, 1994). The bubble in Fig. 10.3 was released at a water depth of 2.5 m and has a frequency of 0.38 kHz. Using equation 10.3 (or the rule of thumb $R_{0(\text{mm})} = 3/f_{M(\text{kHz})}$), this corresponds to a radius of 7.9 mm.

The Minnaert equation was later adapted to include the effects of vapor pressure p_v, surface tension σ, and shear viscosity η and so is more correctly presented as (Leighton, 1994, 2004)

$$f_M = \frac{1}{2\pi R_0\sqrt{\rho_0}}\sqrt{3\kappa\left(p_0 - p_v + \frac{2\sigma}{R_0}\right) - \frac{2\sigma}{R_0} + p_v - \frac{4\eta^2}{\rho_0 R_0^2}}. \tag{10.4}$$

Figure 10.4 shows the natural frequency of bubbles calculated using equation 10.4 at various sizes and depths.

Bubbles generated near the ocean surface have a natural frequency f_0, which is slightly higher than that specified by the Minnaert equation due to the reduced inertia of the fluid near the surface. A similar effect occurs if the water surrounding the bubble also contains bubbles in close proximity. Strasberg (1953) calculated the effect on the frequency, showing that

$$f_0 = \frac{f_M}{\sqrt{1 - (R_0/2h) - (R_0/2h)^4}}, \tag{10.5}$$

where h is the distance from the *center* of the bubble to the surface of the water. As h is always greater than R_0, the denominator is always less than 1, so the oscillation of a bubble in the near surface is always slightly greater than that predicted by the Minnaert equation. For example, a bubble at depth $h = 4R_0$ has a frequency ~7% higher. Alternatively, one should also be able to see that when

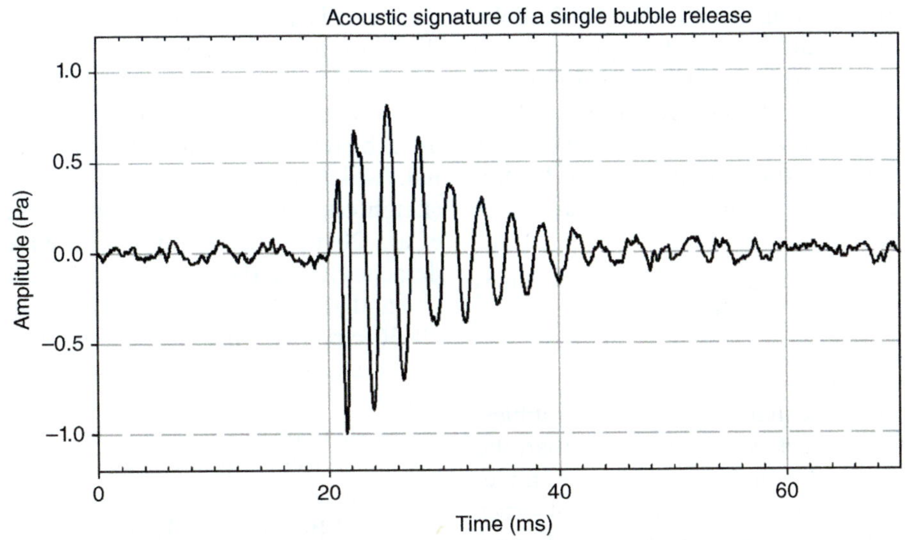

Figure 10.3 Sonogram displaying the typical acoustic emission of a bubble as recorded by a hydrophone. The bubble was released from sediment at a water depth of 2.5 m, 25 cm from the hydrophone. It became detached at around 20 ms, triggering simple harmonic oscillation resulting in an exponentially decaying sinusoidal wave. The bubble oscillates with a frequency of 0.38 kHz (t = 2.6 ms), which we can invert via the Minnaert equation (equation 10.3) to indicate a radius of 7.9 mm.

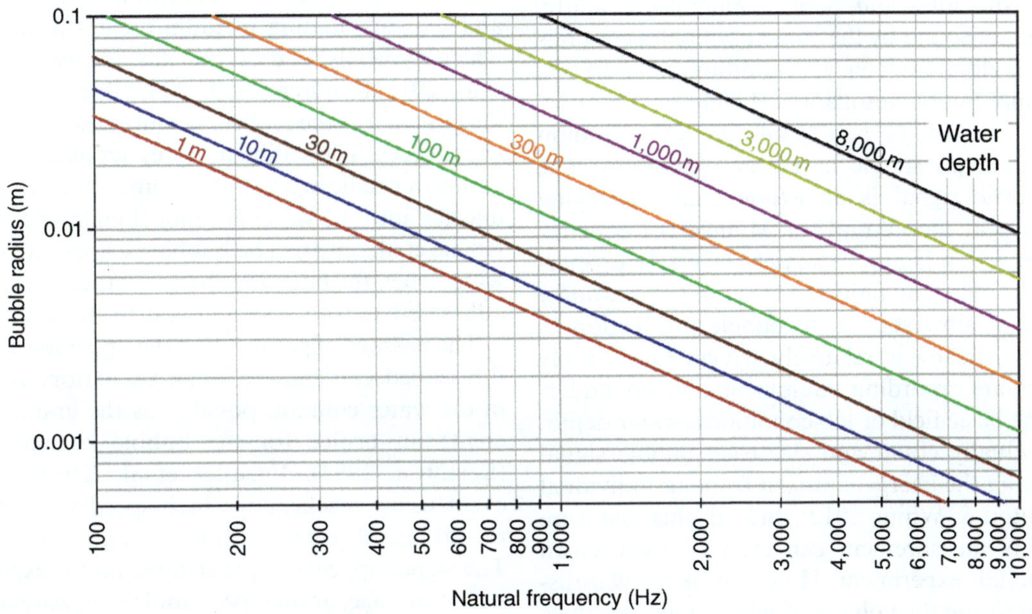

Figure 10.4 Graph displaying the natural frequency of bubbles of various sizes according to the refined Minnaert in equation 10.4 at a range of water depths (1, 10, 30, 100, 300, 1,000, 3,000, and 8,000 m). Calculated assuming ρ = 1,000 kg/m^3 and κ = 1.4 at a temperature of 10 °C and a salinity of 0%.

$h \gg R_0$—that is, the bubble is a few tens of centimeters or more beneath the water's surface—the frequency of the bubble is equal to the Minnaert frequency:

$$f_0 = f_M \text{ for } h \gg R_0. \quad (10.6)$$

Another notable deviation from the Minnaert equation occurs when bubbles are generated (nearly) simultaneously in close proximity, as is the case with wave-generated bubbles. The bubbles are linked by acoustic and hydrodynamic interactions, resulting in coupled oscillator systems that tend to oscillate at much lower frequencies than the natural frequency of any individual bubble within the system. In effect, a cloud of small bubbles can emit an acoustic signature similar to that of a much larger bubble. A region of bubbly water containing a total of N_b identical bubbles (each having a radius R_0 and a natural frequency of f_0) forms an air–water mixture with a void fraction VF. If this bubbly water is submerged in bubble-free water and the boundary between the two is assumed to be rigid (a poor assumption but a useful starting point), the modal frequency f_n of a bubble cloud can be given by (Lu et al., 1990)

$$f_n = f_0 \frac{n}{N_b^{1/3} \cdot \{VF\}^{1/6}}, \tag{10.7}$$

where n is the mode of oscillation. It should be apparent that for any cloud with more than a few hundred bubbles, the lower-order modal frequencies will be lower than the natural frequency of the individual bubbles. For example, a 10 cm cloud of 1,000 bubbles will have a first-order modal frequency one-third that of the bubble oscillations (Leighton, 1994). The greater the number of bubbles in the same space, the lower the modal frequency. Obviously, bubble clouds do not have rigid walls, but the general trend holds true, with complexities in cloud geometry and bubble-size distribution being a greater source of error. In practice, this means if bubbles exist in clouds, the emission—and, perhaps more prominently, the scattering—of sound by the cloud of bubbles contains elements at this cloud frequency, in addition to the signals of the individual bubble resonances (Leighton, 1994).

In summary, releasing a bubble into the water column causes it to undergo simple harmonic oscillation. The resulting acoustic signal is an exponentially decaying sinusoidal wave at the natural (or Minnaert) frequency of the bubble, which is approximately inversely proportional to its equilibrium radius. Measuring the volume of gas release at slow sources of bubble production (a few hertz) acoustically is relatively trivial. One simply needs appropriate recording equipment and an understanding of the basic field or lab conditions (water depth, etc.) to individually count and size each bubble signal without knowing the energy emission from an individual bubble (Leighton & White, 2012). Indeed, this signature method of flux measurement can even be used as an undergraduate lab experiment. However, as we will discuss later, measuring the volume of released gas becomes increasingly difficult as the rate of bubble production increases.

10.3. SUBSURFACE GAS RELEASE

Gas can be generated below the seabed from several sources: thermogenic, biogenic, and anthropogenic. When this gas reaches the seabed, it escapes upward into the water column by the formation of bubbles (Judd, 2003; Knittel & Boetius, 2009; McGinnis et al., 2006). This can have a major impact on ocean chemistry via dissolution and is a poorly understood part of the global carbon cycle (Roche et al., 2020; Johnson et al., 2012; Johnson et al., 2002; Jain & Juanes, 2009). Additionally, the release of each bubble produces an acoustic signal, which can affect the soundscape in local areas (Leighton, 1994; Leighton & White, 2012; Maksimov et al., 2016). The sound is emitted as soon the bubble detaches from the seafloor and lasts ~20 ms, which, given that bubbles tend to rise upward at a speed of 20–30 cm/s, confines the production to within ~5 mm of the seafloor. We will first describe the passage and release of a single bubble before discussing localized seeps, their resultant signal, and flux inversion techniques. The following applies to bubbles of any gas type, with the Minnaert frequency varying only slightly as described by equation 10.4.

In a typical near-surface marine sediment, the pores between grains are saturated with water. If gas is introduced, such as from an underlying fault, it slowly invades the surrounding pores, displacing the water. This intrusion can occur either by capillary invasion or fracture opening (Johnson et al., 2012; Johnson et al., 2002; Jain & Juanes, 2009). The difference in density between the gas and the surrounding medium creates a buoyancy force, which typically causes the gas to rise upward (Algar et al., 2011; Boudreau et al., 2005; Boudreau, 2012). When the gas reaches the seabed, it continues to rise due to buoyancy forces. The sediment pores act like a kind of nozzle, akin to a needle in a test tank, through which the bubble is injected into the water column (Leighton & White, 2012; Roche et al., 2020). The bubble escapes into the overlying water when the buoyancy forces acting on it overcome the adhesive-like forces attaching it to the sediment.

The passage of gas through the upper few centimeters of the seabed can cause a weak oscillatory signal, audible in the water column, possibly as the grains rearrange to create an orifice for the bubble beneath the surface (Leighton, 1994). Vazquez et al. (2015) observed this event using synchronous high-speed video and acoustic recordings of gas migrating through granular sediment. The signal appears unpredictable and is expected to vary with grain size, grain type, bubble size, water pressure, etc. Indeed, some experimental evidence indicates that the sound is absent for fine silts and coarse pebbles. As the magnitude of this precursor signal is smaller than that of subsequent bubble oscillation, the phenomenon remains largely unexplored. Thus, the acoustic signature of a single bubble released from sediment can be defined as an exponentially decaying sinusoidal wave resulting from bubble oscillation (Leighton & Walton, 1987), potentially preceded by a weak, unpredictable oscillatory signal in certain sediment types (Vazquez et al., 2015), as shown in Fig. 10.5.

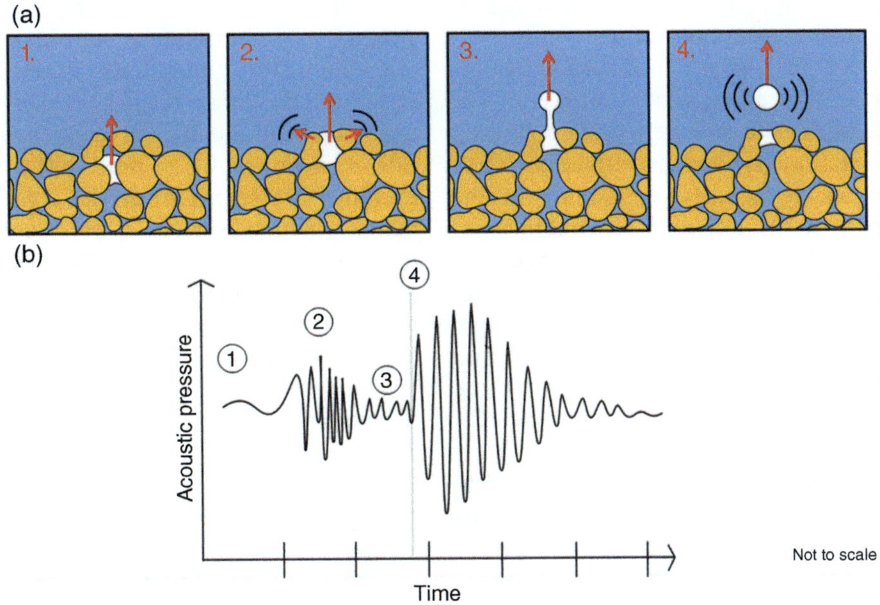

Figure 10.5 (a) The release of a bubble from granular sediment. (b) The corresponding acoustic signal. Note (2) the chaotic weak signal resulting from the rearrangement of grains as the gas reaches the seabed and (4) the stronger distinct acoustic signature of the bubble being released into the water column. Adapted from Vazquez et al. (2015).

10.3.1. Gas-Seep Acoustics

The continuous passage of gas through the same area may cause the development of open channels (or chimneys) in the sediment, which direct the flow of gas to a single localized point on the seabed, forming a seep (Suess, 2014; Coughlan, et al., 2021; Hovland, 2002). A *subsea gas seep* is broadly defined as the *continuous* release of gas from the seabed into the water column. There is no set magnitude for the flux of gas from a seep (Suess, 2014; Coughlan, et al., 2021; Hovland, 2002), meaning the term encompasses seeps that release tens to millions of bubbles per minute. Similarly, there are no strictly defined time scales for being continuous; some seeps are born and die within a few hours, some are only active for certain times of the day or year, and others have been active for centuries (Coughlan et al., 2021; Li et al., 2020b; Böttner et al., 2019). It is also worth noting that a seep does not have to be in sediment; it may also come from exposed bedrock or even manmade features such as leaking pipes. The term *pockmark* is often synonymous with gas seeping from sediment, although strictly speaking, the term only defines the depressions created in the seabed by the gas release (Coughlan et al., 2021; Böttner et al., 2019).

The size of a bubble released from the seabed is difficult to predict. In a lab, the size of a bubble released from a needle is generally considered a factor of the size of the nozzle, the gas injection pressure, and the overlying water pressure, although even these have limited control of bubble size (Longuet-Higgins et al., 1991; Leighton et al., 1991). Even in a controlled setting, it is difficult to regularly produce identically sized bubbles, making bubble size highly variable in the field. Assuming the pores between grains act as nozzles, one might anticipate that larger pore spacings, which are generally associated with larger and more rounded grains, would generate larger bubbles. However, when open conduits form in sediment, pore size becomes less important than conduit size, which can vary significantly based on numerous factors, including the chimney's age. Consequently, bubble-size distributions are unique for each seep. Indeed, the exact bubble-size distribution (and thus gas flux) is likely to change over time as the underlying conduits evolve and the overlying water pressure fluctuates with tidal and seasonal variations (Römer et al., 2016; Bergès et al., 2015; Sultan et al., 2020; Boles et al., 2001; Scandella et al., 2016; Klaucke et al., 2010; Riedel et al., 2018; Leifer, 2015; Wiggins et cl., 2015). Gas flux from underwater seeps can also vary due to underlying causes such as seabed temperature and seismic or volcanic activity (Leifer, 2015; Leifer & Patro, 2002; Muyakshin & Sauter, 2010; Ostrovsky et al., 2008; Li et al., 2020a; Esposito et al., 2006). In deep marine settings, bubble sizes have commonly been observed between 1 and 6 mm in radius (Bergès et al., 2015; Li et al., 2020a); shallower waters (<10 m) have been observed to contain larger bubbles greater than 10 mm in radius (Leighton, 1994; Leighton & White,

2012; Li et al., 2020b; Li et al., 2021), although this trend is far from a rule and exceptions are plentiful.

The acoustic signature of a seep is thus defined by its bubble-generation rate: the rate at which bubbles of different sizes are released. Unfortunately, as every seep has a unique bubble-generation rate, it is difficult to define a general rule for passive acoustic emissions. In the simplest case of a slow seep releasing a few bubbles per second, its acoustic emission can be defined by the continuous release of bubbles: a continuous repetition of the signal seen in Fig. 10.3 (Leighton & White, 2012; Greene & Wilson, 2012). Ultimately, these signals are weak and have little impact on the marine soundscape due to the low flux rates, meaning they are of little interest to many researchers.

Larger seeps with higher gas-flux rates generate stronger signals, which can be observed at greater distances and may have a noticeable impact on ocean chemistry. However, as the frequency of bubble release increases with flux rate, eventually the acoustic signals of each release begin to overlap, making it impossible to distinguish individual bubble oscillations (Bergès et al., 2015; Roche et al., 2019; Chen et al., 2016).

By considering the combined signal of multiple bubble releases, Leighton and White (2012) derived the power spectral density $S(\omega)$ of the far-field acoustic signature of a gas seep at some distance r:

$$S(\omega) = \int_{0}^{\infty} B(R_0)|X_b(\omega, R_0)|^2 dR_0, \tag{10.8}$$

where ω is the angular frequency and $B(R_0)$ is the bubble-size distribution as a function of R_0 defined such that $\Psi(n) = \int_{R_1}^{R_2} B(R_0)dR_0$ represents the number of bubbles generated per second with a radius between R_1 and R_2: that is, the bubble generation rate, δ_{tot}, is the total damping constant for pulsation at resonance, and

$$|X_b(\omega, R_0)|^2 = R_{\epsilon 0i}^2 \frac{\omega_0^4 R_0^4 \rho^2}{r^2} \left(\frac{4\left[(\omega_0\delta_{tot})^2 + 4\omega^2\right]}{\left[(\omega_0\delta_{tot})^2 + 4(\omega_0 - \omega)^2\right]\left[(\omega_0\delta_{tot})^2 + 4(\omega_0 + \omega)^2\right]} \right). \tag{10.9}$$

The important unknown in this equation is the initial amplitude of displacement of the bubble wall at the start of the emission ($R_{\epsilon 0i}$) (not to be confused with $R_{\epsilon 0}$ the maximum displacement of bubble wall from the equilibrium radius). There is strong evidence to suggest this is a function of the equilibrium bubble radius (R_0). However, this exact relationship is yet to be defined. Consequently, many studies have elected to treat $R_{\epsilon 0i}/R_0$ as a constant somewhere between 1.4×10^{-4} and 5.6×10^{-4}, based on experimental observations (Leighton & White, 2012. However, it should be stressed that this is a pragmatic choice with no theoretical foundation. It is important to note that the above formulation excludes the signal from the rearrangement of grains before a bubbles release. However, this is reasonable because (1) the signal is very weak (Vazquez et al., 2015), and (2) seeps with a higher flux contain open conduits that do not require grains to be rearranged to facilitate the migration of gas (Roche et al., 2020).

This spectral approach allowed Leighton and White to invert the signal from a given gas seep (at a known distance from a hydrophone) to determine the number of bubbles of various sizes released within a given period, providing them with an estimate of gas flux (Leighton & White, 2012). In replacing the signature method for counting and sizing bubbles in circumstances where their passive acoustic emissions overlapped, Leighton and White drew particular attention to the lack of knowledge about the energy of an individual bubble's emission. Although the signature method had managed to bypass this unknown, their spectral method could not. Although the spectral method had the power to count and size bubbles when the signatures overlapped, Leighton and White noted that the reliance on literature values for the energy released by a bubble was the greatest source of uncertainty, particularly as the energy associated with the release of a given bubble is likely to vary with the mode by which it is entrained (injected by a needle, through sediment, via a gas pipe leak, or entrained in the upper ocean by rainfall or breaking waves) and the depth at which it is entrained. In simple terms, if the count of bubbles of a certain size is based on the energy detected in a given frequency band, then if the acoustic energy in that band is divided between the bubbles contributing to it, the estimation calculates fewer bubbles entrained, the more acoustic energy is contained in each bubble signature (Leighton & White, 2012). Further complications were identified in that a given injection process can cause a bubble to fragment after release or merge with other bubbles, which can lead to the injection of a single bubble generating multiple signatures (Leighton et al., 1991). Despite all this, the use of the spectral method in the field has proven effective to date, providing continuous estimates of gas flux over extended periods as validated by intermittent physical measurements. However, a need to reduce the uncertainty in these measurements is continually noted (Bergès et al., 2015; Li et al., 2021; Roche et al., 2019).

Given the highly variable nature of seafloor gas seeps, particularly regarding the size of the bubbles and their rate of release, it is difficult to give a general impression of

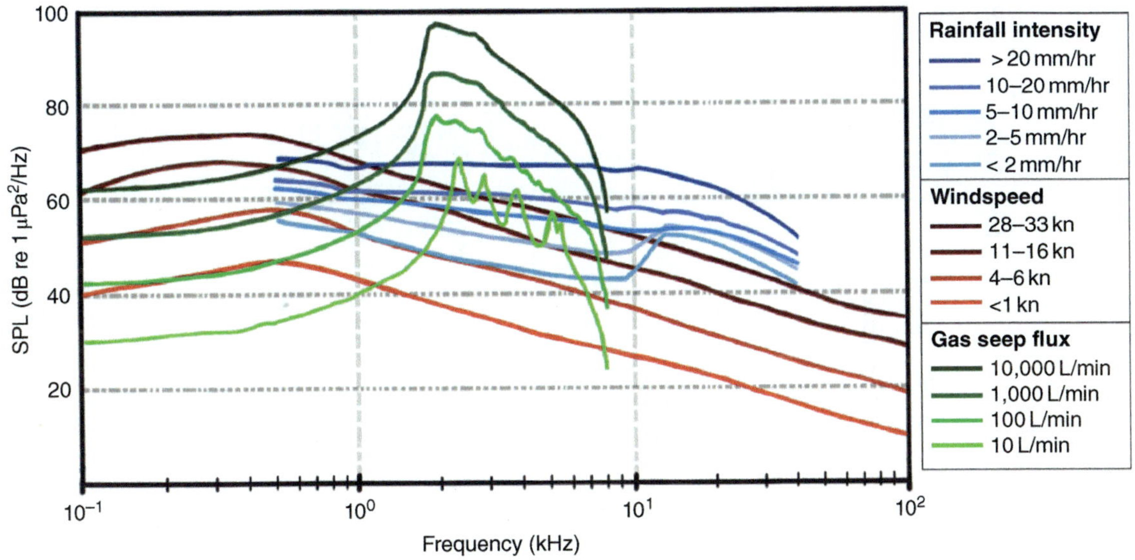

Figure 10.6 Ambient noise spectral density from 0.1 to 100 kHz for common bubble-production sources in the marine environment, including gas seeps, rainfall, and breaking waves. Gas seepage is simulated at different flux rates in L/min at 100 m water depth, assuming a Gaussian bubble size distribution with radius between 0.5 and 10 mm. Rainfall data at different intensity levels in mm/hr is from Ma and Nystuen (2004), and breaking wave data observations at different wind speeds in knots is from Wenz (1962).

their contribution to the marine soundscape. To do so here, we use the above equations to simulate the sound pressure level (SPL) of a single point-focused seep venting gas at the rate of 10, 100, 1,000, and 10,000 L/min, assuming a log-normal bubble-size distribution with radius between 0.5 and 10 mm. The results are displayed in Fig. 10.6 alongside the SPL of various intensities of wind- and rain-generated bubble noise. We emphasize that these graphs are meant to serve only as an approximate guide to the potential effect natural seeps can have on the marine soundscape. Here, we see that the signal is confined mainly between 1 to 10 kHz (a result of the selected bubble-size distribution), with the magnitude of the signal increasing in line with the rate of gas flux. A maximum amplitude of 97 dB re 1 $\mu Pa^2/Hz$ is seen at 10,000 L/min, well in excess of wind- and rain-generated bubble noise.

In summary, the release of gas from the seabed generates an acoustic signal at the natural frequency of the resulting bubble. As the gas flow out of a seep increases, the acoustic signals of each bubble released begin to overlap, making the resultant signal a summation of each individual bubble's natural frequency. Consequently, it is currently impossible to predict the sound resulting from a gas seep (or a field of seeps) without understanding the bubble-size distribution. However, by observing the acoustic signature of a known seep, it is possible to quantify the size of bubbles being released and thus estimate the flux, observing tidal and seasonal variations.

10.4. RAINFALL ACOUSTICS

When a rain droplet impacts a body of water, it forms an impact crater and may entrain a bubble (Fig. 10.7). Consequently, falling rain produces two sounds in the marine environment: (1) the initial impact of the droplet on the body of water, which generates a compressional wave; and (2) the simple harmonic motion of a bubble following its release into the water, once again at the Minnaert frequency (Leighton, 1994). This scenario is directly comparable to a leaky tap dripping water into a sink, producing a distinct "plinking" noise (Leighton, 1994; Pumphrey & Elmore, 1990; Pumphrey & Walton, 1988). Although some incorrectly assume that this sound is a consequence of the initial collision between the drop and the water surface, the entrainment of bubbles produces the majority of the acoustic signature. We will first discuss the sound of the initial droplet impact before discussing the processes of entrainment, the resulting acoustic signature of rainfall, and methods of rainfall quantification.

The impact of the rain droplet on the water's surface initially produces a sharp acoustic pulse with a duration of 10–40 μs, as a result of the water-hammer effect (a pressure surge caused when the motion of a fluid is stopped). The pressure radiated by the impact is given by Pumphrey & Crym, 1990)

$$p_{\text{impact}} = \frac{\rho u_d^3 L_d}{2c} \frac{\cos\theta}{r} u, \tag{10.10}$$

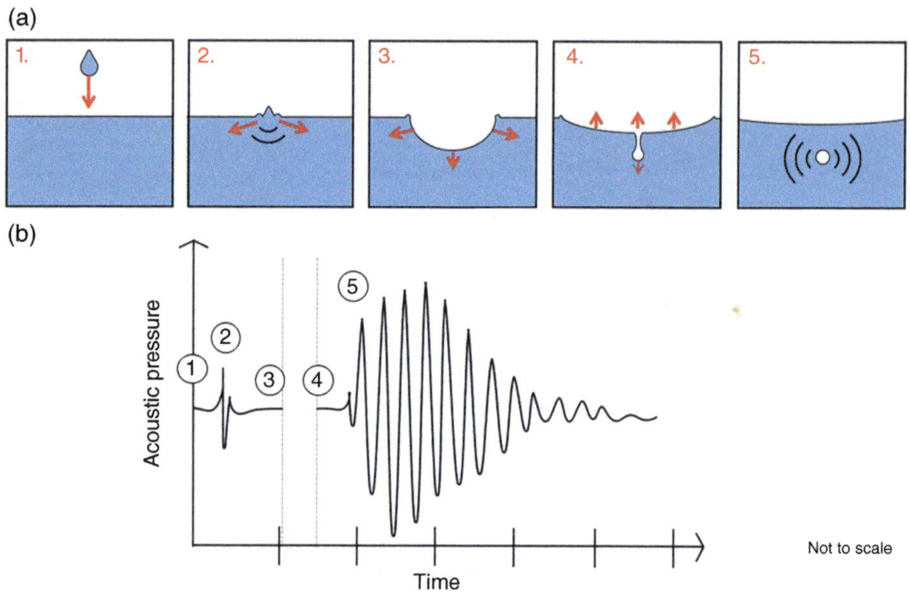

Figure 10.7 (a) A diagram of regular entrainment of a bubble following the impact of a water droplet. (b) The acoustic signature of a raindrop. Note the weak, sharp signal from the impact (2) followed by the larger oscillation of the bubble (5) once it detaches from the crater. Adapted from Medwin and Nystuen (1990).

where ρ is the water density, u_d is the impact speed of the droplet, L_d is the diameter of the droplet, c is the speed of sound in water, θ is the angle between the observer and sound source relative to the z axis, and u is the impact Mach number. Raindrops typically have a diameter between 0.5 and 5.0 mm (larger droplets tending to break up), resulting in an impact velocity between ~2.0 and 9.0 m/s (Nystuen, 2001). This means that although for individual droplets, it is easy to identify the impact signature, this sound is dwarfed by the later oscillation of a bubble by a factor as large as 200:1. During rainfall (where bubbles are continuously oscillating), the sound of impact has very little effect on the overall acoustic signature and is responsible only for a weak broadband signal (Pumphrey & Elmore, 1990).

The entrainment of a bubble by a droplet of water is a dynamic process, much more complex than the injection of gas through sediment pores. The exact mechanism by which this occurs varies based on several factors, mainly impact velocity and droplet diameter. These processes are (Pumphrey & Elmore, 1990) as follows:

1. *Irregular entrainment:* The complex and unpredictable details of a splash somehow entrain a bubble(s).
2. *Regular entrainment:* A retreating impact crater leaves behind a small volume of gas connected via a narrow neck that is eventually pinched off, leaving behind a single bubble (see Fig. 10.7a).
3. *Entrainment of large bubbles:* Most of the volume of the crater is trapped as a bubble.
4. *Mesler entrainment:* Many tiny bubbles are trapped in the early stages of the impact process, possibly between the crest of capillary waves on the droplet and body of water.

Bubbles produced by entrainment act identically to examples discussed previously, oscillating to produce an exponentially decaying sinusoidal wave in the near field (Pumphrey & Elmore, 1990). The only notable difference from seabed gas release is that the bubbles are much closer to a free surface (the water surface), meaning the mass of water regulating the oscillations is lower, and thus the natural frequency of the bubbles is slightly higher than the Minnaert frequency (see equation 10.5).

Given a large enough area and a sufficient number of raindrops (of a consistent size distribution), one can assume that a constant number of raindrops are impacting the water per second, and therefore a constant number of bubbles are being entrained. Consequently, rainfall results in a constant ambient noise. Using this principle, and quantifying the number of bubbles entrained per second $n(f)$ in a 1 Hz frequency band over a 1 m^2 area of water, Pumphrey and Elmore (1990) were able to show that the intensity below the surface of the oscillating bubbles at any given frequency f is

$$I_{\text{Rain}} = \frac{n(f)D^2Q}{4f\rho c}, \quad (10.11)$$

where Q is the quality factor and D is the initial dipole strength of the bubble. From this, the intensity spectrum level is given by

$$ISL_{\text{Rain}} = 10\log\frac{I_{\text{Rain}}\cdot\rho c}{1\ \mu\text{Pa}^2/\text{Hz}} = 10\log\frac{n(f)D^2Q/4f}{1\ \mu\text{Pa}^2/\text{Hz}}. \quad (10.12)$$

It is important to note that whereas D increases with increasing f, Q decreases. these two effects cancel each other out, so the spectrum is dominated by the number of bubbles entrained per second (Pumphrey & Elmore, 1990). Additionally, it has been observed that, neglecting refraction and absorption, 90% of the rain signal arrives from a sample area with a radius equal to three times the observer's depth (Pumphrey & Elmore, 1990). Thus, using equation 10.12 and the size and number of bubbles produced per second by entrainment, one can calculate the acoustic spectrum produced by rainfall or vice versa.

It is difficult to precisely predict the relative occurrence of each entrainment process during a rainstorm. Although regular entrainment is by far the most well-understood process, its name is primarily a consequence of being the easiest to comprehend and predict. Indeed, when Pumphrey and Elmore (1990) mapped which process occurs at which ratio of impact velocity to drop diameter, the plot is dominated by Mesler entrainment. Additionally, if one were to assume all impacts occurred at terminal velocity, the entrainment of large bubbles would never occur, and irregular entrainment would occur only during storms with particularly large droplets (Pumphrey & Elmore, 1990; Nystuen, 2001; Ma & Nystuen, 2005; Black et al., 1997; Serra, 2018; Yang et al., 2015; Pensieri et al., 2015; Taylor et al., 2020; Nystuen et al., 2015; Ashokan et al., 2015; Anagnostou et al., 2008; Nystuen et al., 2008). It is logical to assume that the splashing of water will produce some droplets that impact below terminal velocity, meaning all entrainment processes are likely to occur at some point during a rainstorm. However, it is reasonable for now to assume only regular and Mesler entrainment dominate and justify further consideration.

Mesler entrainment produces multiple very small bubbles ~25 μm in radius regardless of the size and velocity of the droplet. This results in a natural frequency of approximately 1.3 MHz. The high frequency and small size of Mesler bubbles ultimately means they produce very little noise, with high levels of attenuation. Consequently, they have little to no impact on the acoustic signature of rainfall, especially in the far field, and thus regular entrainment is responsible for the majority of bubble-oscillation sound during rainfall (Pumphrey & Elmore, 1990; Nystuen, 1986, 2001; Ma & Nystuen, 2005; Black et al., 1997; Serra, 2018; Yang et al., 2015; Pensieri et al., 2015; Taylor et al., 2020; Nystuen et al., 2000, 2015; Ashokan et al., 2015; Scrimger et al., 1987; Pumphrey, 1989; Nystuen & Selsor, 1997).

Regular entrainment produces different bubble sizes for different droplet sizes and impact velocities. If we consider only the bubbles produced by raindrops traveling at terminal velocity, regular entrainment is the result of droplets 0.40 to 0.55 mm in radius (with larger and smaller droplets resulting in Mesler entrainment) (Ma & Nystuen, 2005; Ashokan et al., 2015; Nystuen et al., 2000). Bubbles produced by regular entrainment of droplets of this size are predicted to be in the range of 0.16 to 0.33 mm in radius, resulting in frequencies between 10 and 20 kHz. Laboratory and field data has consistently shown a general increase in the number of bubbles entrained with bubble radius, peaking at ~0.23 mm in radius and dropping off rapidly above ~0.27 mm (Pumphrey & Elmore, 1990; Ma & Nystuen, 2005; Pensieri et al., 2015; Ashokan et al., 2015; Nystuen et al., 2008).

Consequently, the spectral content of rainfall on a body of water is expected to gradually increase in intensity with decreasing frequency, leading to a large peak at around 14–15 kHz followed by a sharp decline below 10–12 kHz. The exact intensity of the signal depends on the number of bubbles entrained per second (a consequence of the number of raindrops impacting per second) (Nystuen, 2001; Ma & Nystuen, 2005; Ashokan et al., 2015). This prediction fits well with field observations (see Fig. 10.8). Data collected from lakes, land-based water tanks, brackish ponds, and deep marine environments all show a distinctive peak at 14–15 kHz with a sudden drop-off below 10–12 kHz (Nystuen, 2001).

One consistent observation in repeat studies is a decrease in the prominence of the 14–15 kHz peak with increased rates of rainfall, with the peak being almost indistinguishable above 30 mm/hr (see Figs. 10.6 and 10.8) (Medwin et al., 1990; Ma & Nystuen, 2007). This

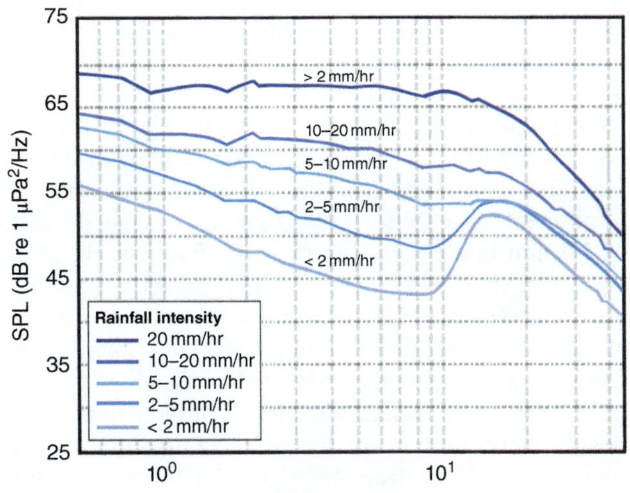

Figure 10.8 The average SPL spectra of rainfall acoustic signals at various rates as recorded by buoys in lakes and seas worldwide over a collective total of 30 months. Note the distinct peak at 14 kHz caused by the regular entertainment of bubbles from rainfall that becomes less prominent as rainfall becomes more intense. Adapted from Ma and Nystuen (2004).

is because at higher rainfall rates, more bubbles oscillating between 10 and 20 kHz are generated per second, resulting in increased signal intensity. However, as can be seen in equation 10.12, this is a logarithmic increase with diminishing returns: although the surrounding frequencies increase in intensity, the 14–15 kHz peak is relatively unmoved, flattening the spectrum. Additionally, in the field, increased rainfall tends to be accompanied by increased windspeed, which, as discussed next, also affects the rain spectrum. For this reason, the rain spectrum is best observed during a drizzle or light rain (Cavaleri & Bertotti, 2018).

The rain signature above 10 kHz is known to be affected by wind speed (Pumphrey & Elmore, 1990; Ma & Nystuen, 2005; Black et al., 1997; Anagnostou et al., 2008). Ma and Nystuen (2004) noted that as wind speed increased from 0.6 to 3.3 m/s, the 15 kHz peak became less prominent and broader, shifting up by a few kilohertz. The increased wind speed drives waves on the surface of the water, which has two effects. First, it alters the angle of incidence of raindrops on the water; this reduces the probability that an individual droplet will produce a bubble from 100% at normal incidence to 10% at a deviation of 20° (Ma & Nystuen, 2005). Additionally, a deviation of 20° causes a 30% decrease in the energy emitted by the initial impact (Ma & Nystuen, 2005). This means the dominance of the bubble noise over the impact noise reduces by a factor of 10, making the peak less prominent. Second, as we will discuss later, at high wind speeds, breaking waves (white caps) can also produce bubbles of similar magnitude, which interfere with the sound of rainfall. However, it has been observed that under certain conditions, rainfall can prevent the formation of breaking waves (Leighton, 1994; Wu, 1979; Cavaleri, 2020; Holthuijsen, 2007).

Some studies have noted a secondary rise in the rain spectrum starting at 2–3 kHz and peaking around 5 kHz. This is believed to be a consequence of irregular entrainment of bubbles caused by very large droplets 2.0–3.5 mm in diameter (Serra, 2018; Yang et al., 2015; Pensieri et al., 2015; Taylor et al., 2020; Anagnostou et al., 2008). We had previously dismissed irregular entrainment and the entrainment of large bubbles by assuming all the droplets impacted at terminal velocity and that larger droplets were less common than small ones. However, it appears that when a rainstorm consists of particularly large droplets (>2.0 mm), the frequency of irregular and large entrainment events is significant enough to cause a recognizable spike in the spectrum, possibly as a result of accompanying wave action lowering the impact velocity. This secondary 5 kHz peak, although less conspicuous than the 14–15 kHz peak, may be more useful, as it exists in the part of the spectrum less affected by wind and wave noise (2–10 kHz) (Ma & Nystuen, 2005). Thus, observations of the intensity of the 5 kHz peak can be used for rainfall quantification regardless of wind speeds. Using comparative rain gauge data, Ma and Nystuen (2001) proposed the following equation for calculating the rainfall rate S_{rain} in mm/hr based on the SPL at 5 kHz ($SPL_{5\text{kHZ}}$):

$$10 \log_{10} S_{\text{rain}}/10 = (SPL_{5\text{kHZ}} - 42.4)/15.4. \qquad (10.13)$$

Although the exact relationship varies from location to location based on local conditions and ambient noise levels, acoustic inversion of rainfall (with sufficient calibration) is a highly promising technique for use in meteorological and oceanographic research and is becoming increasingly common (Ma & Nystuen, 2005, 2007; Yang et al., 2015; Pensieri et al., 2015; Taylor et al., 2020; Anagnostou et al., 2008; Nystuen et al., 2000; Deane & Stokes, 2002).

In summary, the acoustic signature of rainfall in the marine environment is caused by the entrainment of bubbles, not the impact of the droplets themselves. The rain spectrum is a distinctive peak at 14–15 kHz with a sudden drop-off below 10–12 kHz, caused by regular entrainment, and occasionally a secondary smaller peak at 5 kHz, caused by irregular entrainment when droplets are particularly large. The intensity of these peaks is dictated by the number of raindrops impacting the water per second. As the intensity of the rainfall increases, the peaks become broader and less well-defined. Increasing wind speeds also mute the 14–15 kHz peak by altering the impact angle of droplets and interference from wave noise; however, the 5 kHz peak is less affected by wind and can be used for rainfall quantification.

10.5. ACOUSTICS OF BREAKING WAVES

In the natural marine environment, sufficiently high wind speeds can cause surface gravity waves to break as whitecaps (or white horses) (Leighton, 1994). Unsurprisingly, this process entrains a large number of bubbles that oscillate near the surface, as described by equation 10.5 (Deane & Stokes, 2002; Bass & Hay, 1997; Manasseh et al, 2006; Farmer & Vagle, 1988). Not only do these bubbles have a noticeable effect on the ambient noise of the ocean (via oscillation) (Deane & Stokes, 2002), but they also may affect the passage of other acoustic signals by altering the propagation of sound waves near the sea surface (Deane & Stokes, 2002). First we will discuss waves as acoustic sources before discussing the effect wave-generated bubbles have on the speed of sound. Then we will discuss ambient noise generated by wave action and how this can be related to wind speed. We will not discuss in detail the hydrodynamic controls behind breaking waves, other than to note that in general, strong winds result in larger breaking waves (which would demand a full chapter) (Leighton, 1994; Deane & Stokes, 2002).

The entrainment of bubbles from wave action is a highly dynamic process (even more so than rainfall), with the exact minutia of bubble generation being poorly understood (Deane & Stokes, 2002). We know, however, from laboratory and field data that distinct bubbles are initially generated during one of two phases: jet entrainment and cavity collapse (Deane & Stokes, 2002).

Jet entrainment begins as soon as the wave starts to break. The crest of the wave overturns and plunges into the wave face, forming a plunging jet with a cavity of air trapped between the two bodies of water. This chaotic collision of the jet generates bubbles, generally between 0.1 and 2.0 mm in radius (2 to 30 kHz) (Leighton, 1994). Additionally, the impact of the jet causes the water to splash and a number of droplets to also entrain bubbles. Toward the end of a breaking wave's life cycle, the air cavity trapped between the wave face and the plunging jet collapses. This forms many bubbles, most of which are between 2.0 and 10.0 mm in radius (0.4 to 2 kHz) (Deane & Stokes, 2002). Given the high density of bubbles, the remnants of the cavity act as a bubble cloud. As discussed earlier, bubbles in a cloud tend to act as coupled oscillators with normal modes of oscillation much lower than that of individual bubbles (Deane & Stokes, 2002; Carey & Bradley, 1985). The cavity-collapse phase is thus responsible for frequencies between 0.1 and 0.5 kHz due to the bubble cloud and higher frequencies up to ~2 kHz from individual oscillations (Deane & Stokes, 2002; Carey & Bradley, 1985). At this time, the plunging jet also forms a shear zone along the wave surface, which encircles the cavity remnants. Some bubbles are pulled through this shear zone, which can cause a bubble to fragment into two or more smaller fragments that once again oscillate although at a higher frequency than their parent bubble (Leighton, 1994; Deane & Stokes, 2002, 2010). For example, a large bubble oscillating at 3.1 kHz may produce two daughters, one at 50 kHz and the other at 32.3 kHz (Deane & Stokes, 2002; Thorpe, 1982). The intensity of the cavity-collapse signature is far greater than that of the jet period (or later shearing); thus, when waves continuously break in the marine environment, the sound of these bubble clouds dominates. Therefore, the acoustic signature of a breaking wave near the surface can most easily be recognized by a low-frequency signal between approximately 0.2 and 2 kHz (Deane & Stokes, 2002; Carey & Bradley, 1985), distinct from that of rain and gas seeps (Bergès et al., 2015; Ma & Nystuen, 2005).

Deane and Stokes (2002) presented the average acoustic signature of 10 cm plunging breakers (Fig. 10.9). Here, one can clearly see the jet period—continuous throughout the breaking of the wave—responsible for the signal above 2 kHz, with the majority of the sound generated below 10 kHz. The cavity-collapse period can also be clearly identified as a quick (~0.3 s) low-frequency burst centered around 0.3 kHz (Deane & Stokes, 2002; Thorpe, 1982). It should be intuitive that the acoustic signature (or rather, the resulting bubble cloud) of a wave is a consequence of its size (and the style of breaking), which is typically a function of wind speed. Thus, by observing the breaking of a wave, one can infer the acoustic signal or vice versa.

Given the energetic and variable nature of breaking waves, it is difficult to predict exactly what happens to the resulting bubbles postproduction. The exception is during Langmuir circulation—the slow, shallow, counter-rotational vortices aligned with wind direction that develop when the wind blows steadily over a body of water—which has been extensively analyzed. After formation, Langmuir circulation can carry wave-generated bubbles up to 10 m below the surface (Monahan & Lu, 1990). In wind speeds >7 m/s, this has been known to result in linear bubble clouds orientated parallel to the wind direction (Leighton, 1994). These *Langmuir bubble clouds* consist of bubbles produced throughout the lifecycle of a wave (jet and cavity-collapse phase) and potentially those entrained by rainfall. Individual bubbles in the cloud naturally shrink due to dissolution and eventually disappear (Liang et al., 2011); however, a high concentration of bubbles may delay this process. A continued supply of freshly generated bubbles can allow the cloud to persist as long as circulation is active. (Leighton, 1994; Liang et al., 2011; Clay & Medwin, 1977) The clouds generally have void fractions between 10^{-4} and 10^{-5}% that are assumed to be uniform in the horizontal plane but fall off exponentially with depth (Leighton, 1994; Farmer & Vagle, 1988; Vagle et al., 1990; Farmer et al., 2001).

As the speed of sound depends on the inertia and stiffness of the material it is passing through, the speed of sound in a bubble cloud (or bubbly liquid) differs from that of pure water. Gas is less dense than water, so sound waves travel more slowly through bubble clouds, becoming slower the larger and more numerous the bubbles are—that is, the larger the void fraction (Garrett et al., 2000; Wang et al., 2016). If there is a distribution of bubble sizes within a cloud, such that $n_n^{gr}(z, R)dR_0$ is the number of bubbles per unit volume at depth z having radii between R_0 and $R_0 + dR_0$, the speed of sound in a cloud c_c is given by (Clay & Medwin, 1977)

$$c_c(z,\omega)= c\left\{1-(2\pi c^2)\int_{R_0=0}^{\infty}\frac{R_0}{\omega^2}\left(\frac{(\omega_0/\omega)^2-1}{\{(\omega_0/\omega)^2-1\}^2+d^2}\right)n_n^{gr}(z,R)\,dR_0\right\}, \tag{10.14}$$

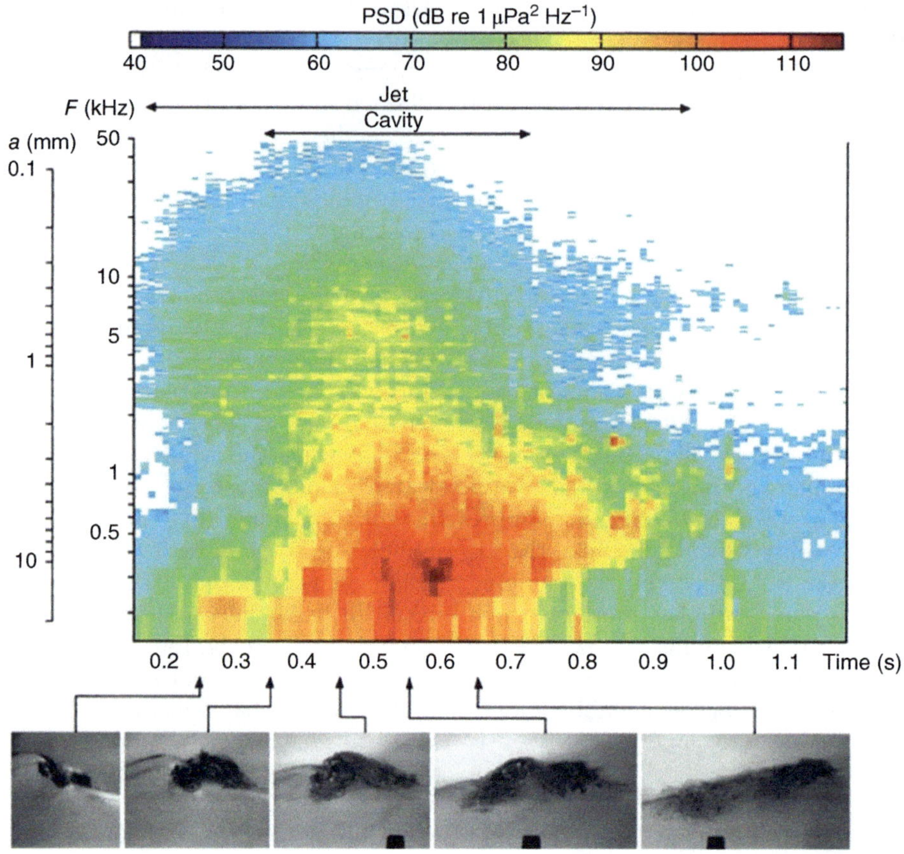

Figure 10.9 Spectrogram of wave noise calculated from an average of 17 breaking events. Note the jet and cavity phases. The color contours represent sound intensity plotted on a decibel scale (the intensity if referenced to 1 μPa^2/Hz) versus frequency and time. The log scale labeled *a* on the left-hand side indicates the radius of a bubble resonant at the corresponding frequency of the frequency scale (*F*). The wave noise was measured by a hydrophone mounted in the wave flume beneath the bubble plume. The images plotted below the spectrogram show the sequence of flow features at different times during the acoustic emissions. Deane et al. (2002) / Reproduced from Springer *Nature*.

where *d* is the dimensionless damping constant for a single bubble, and ω_0 is the resonant circular frequency of the bubble given by $\omega_0 = 2\pi f_0$.

Given the above and the fact that surface-generated bubble clouds tend to decrease in concentration with depth, one can see how the presence of breaking waves can result in an ocean model where sound speed increases noticeably with depth in the upper ~10 m (Leighton, 1994; Knudsen et al., 1948; Wenz, 1962; Farmer & Vagle, 1989; Buckingham, 1991; Zhao et al., 2014). In such a scenario, downward-propagating sound waves tend to turn, due to refraction, bending upward back toward the surface. Similarly, upward-propagating waves also turn, refracting downward. Repeating this cycle can result in the horizontal propagation of sound waves, trapping acoustic energy in the near surface (Deane & Stokes, 2010; Farmer et al., 2001). In terms of wave acoustics, Farmer and Vagle (1989) and Buckingham (1991) both suggested that for a given mode, the signal becomes evanescent (unable to propagate further) below a certain extinction depth. They suggest that the trapping of sound in such a waveguide may influence the ambient acoustic spectra of wave noise and that by observing certain drop-out frequencies, one could infer the bubble-size population generated by breaking waves, although Buckingham argues the loss of signal alone is not sufficient for a full analysis (Farmer & Vagle, 1989; Buckingham, 1991). Unfortunately, the latter appears correct, as despite numerous attempts in the following years, little progress has been made inverting bubble populations from wave acoustics (Garrett et al., 2000; Wang et al., 2016).

Accounting for bubble-cloud effects, Deane and Stokes (2010) presented a model for calculating the underwater noise of a single breaking wave at a distance r with good agreement with experimental observations. Here, assuming wave noise is a superposition of oscillations from generated bubbles, with the creation times of bubbles being uniform and randomly distributed throughout the breaking period, the Power spectrum is given by

$$P(\omega, r) = \int_V \int_{a_{min}}^{a_{max}} \lambda(a, r)|\gamma(\omega, a)\alpha(\omega, a)|^2 da dV, \quad (10.15)$$

where a_{min} and a_{max} are the minimum and maximum bubble sizes generated, V is the plume volume, $\lambda(a, r)$ is the rate at which bubbles are generated, and $\gamma(\omega, a)$ and $\alpha(\omega, a)$ are the Fourier transforms of the convolution of free-space bubble pulses and Green's function for the medium of propagation, respectively (Deane & Stokes, 2010).

With an understanding of the individual acoustic signal of a breaking wave and how bubble clouds affect the near surface, one might assume that calculating the resulting signal of multiple breaking waves would be straightforward. After all, at a given depth, an observer will record a signal that is the superposition of all the waves breaking above it at any given moment. Given a large enough area and a sufficient number of waves—that is, an ocean—one can assume a constant number of waves are breaking, resulting in a constant ambient noise, as is the case with rainfall. Indeed, if all the waves were identical and occurring in some symmetrical pattern around the recorder, we could attempt to estimate the signal via theoretical calculations. However, this is unrealistic and would be of little practical use; a range of breaking wave sizes and styles will always exist, distributed erratically along the sea surface (Bass & Hay, 1997; Manasseh et al., 2006; Knudsen et al., 1948; Wenz, 1962). Additionally, a detailed understanding of the size distribution of bubbles generated in a breaking wave $\lambda(a, r)$ in equation 10.15 is required, something lacking outside easily replicable waves (Deane & Stokes, 2010).

For simplicity's sake, the seminal work of Knudsen et al. (1948) and Wenz (1962) describing the ambient SPL in the ocean at different wind speeds using field observations, seen in Fig. 10.6, is still relevant (Medwin & Beaky, 1989; Vagle et al., 1990; Zhao et al., 2014). It starts at around 0.20 kHz and rises 3–5 dB re 1 μPa^2/Hz to a peak at approximately 0.5 kHz (consistent with the above discussion) before dropping off slowly, ~25 dB re 1 μPa^2/Hz by 10 kHz, with peak SPLs of 60 and 73 dB re 1 μPa^2/Hz, respectively, for wind speeds of 3.4–5.5 m/s and 17.2–20.7 m/s (Farmer & Vagle, 1988; Vagle et al., 1990; Zhao et al., 2014; Felizardo & Melville., 1995). This does not generally cover strong gales (wind speed >20.8 m/s), as during higher wind speeds, it becomes difficult to identify periods of pure wind noise (i.e., non-rain-contaminated).

Despite the complexity of the task, however, many still wish to be able to calculate the ambient noise of breaking waves: for example, for storm monitoring (Yang et al., 2015; Pensieri et al., 2015; Anagnostou et al., 2008; Zhao et al., 2014; Zhao et al., 2017) or studying ocean atmospheric mixing (Deane & Stokes, 2002; Vagle et al., 2010). The most widespread approach is via WOTAN (Wind Observations Through Ambient Noise) calculations. Here, observations of the ambient noise from breaking waves have been correlated with wind speed through numerous studies to empirically map their relationship (Vagle et al., 1990; Zhao et al., 2014). Originally this work was done to estimate wind speed based on ambient wave noise, but the reverse should also be possible (calculating ambient wave noise based on wind speed).

Using past studies and their data, Vagle et al. (1990) determined that the source sound level at a depth of 1 m from breaking waves was given by

$$SSL_0 = q \log f + G, \quad (10.16)$$

where q is the slope of the logarithmic spectrum of the wind-generated sound, which they find to be equal to –19.0 dB/decade (in good agreement with past estimates (Farmer & Vagle, 1988)). G is a variable function of wind speed. Vagle et al. (1990) determined values for G between set wind speeds, which we note approximately follow $G = 1.3U_{10}+56$ (U_{10} being the wind speed 10 m above water level). Unfortunately, this sound-level equation only holds true for low wind speeds ($U_{10} \leq 15$ m/s) and below a certain critical frequency, $\log f_c = 1.9 - 0.07U_{10}$ (Vagle et al., 2010).

Zhao et al. (2014) expanded on this work by studying typhoons' underwater acoustics using Lagrangian floats. Figure 10.10 displays the spectral content they observed at a range of high wind speeds from a number of floats. They noted that low-frequency sound (<1 kHz) monotonically increased with wind speed, whereas while intermediate and higher frequencies initially increased and then decreased with wind speed. They presented the following empirical equation to calculate the sound pressure level of a given frequency in wind speeds up to 50 m/s (Zhao et al., 2014):

$$SPL = S_{noise} + S_{10} \frac{\left(U_{10}/10\right)^{n_{lf}}}{1 + \left(U_{10}/f_{peak}\right)^{n_{hf}}}, \quad (10.17)$$

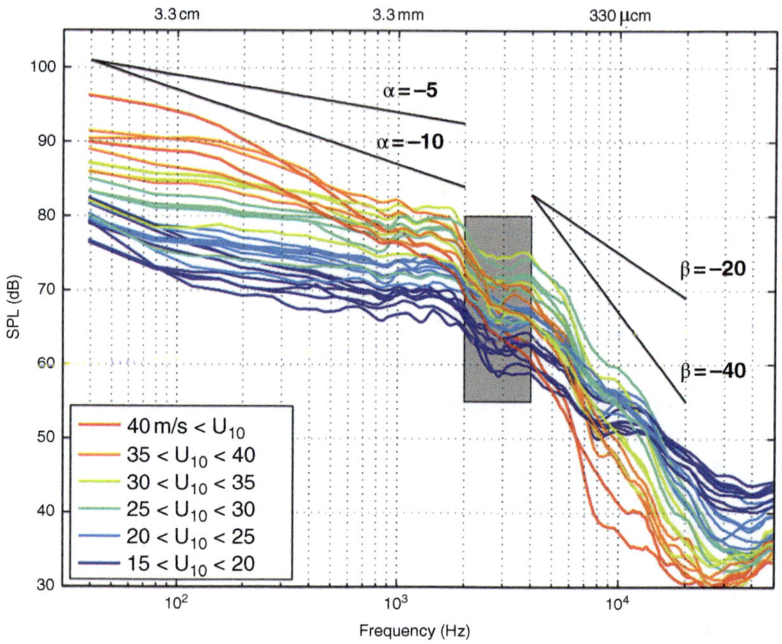

Figure 10.10 Spectrogram of breaking-wave noise at various wind speeds recorded at sea during tropical cyclones. Each curve averages all spectra in 5 m/s wind speed bins from measurements at depth >2 m of a single float; multiple curves of the same color denote observations from multiple floats. The four black lines show representative spectral slopes at low (α) and high (β) frequencies. The gray box shows the transition frequency (2–4 kHz) between the two slope regions. The dip in sound level near 3 kHz may be an instrumental effect. Zhao et al. (2014) / American Meteorological Society.

where S_{noise} is the noise floor, S_{10} is the sound level at 10 m/s wind, f_{peak} controls the wind speed with the sound level maximum, and n_{lf} and n_{hf} are values that control the increasing/decreasing in lower/higher wind conditions. All of these values are a function of the target frequency, although the exact relationship is not fully understood or linear and is determined instead by least-square fitting of observations (Zhao et al., 2014). Consequently, real-time analysis of wind speed via acoustics is still an emerging field. Although highly promising, this approach needs to be tested in multiple environments many more times, especially if the underlying variables are to be better understood (Pensieri et al., 2015; Vagle et al., 1990; Zhao et al., 2014; Cazau et al., 2019; Cazau et al., 2017; Farrell et al., 2016; Cauchy et al., 2018).

It should be apparent from the above that accurately predicting the acoustic signal recorded at a hydrophone as a result of breaking waves at any given wind speed is an exceedingly difficult task, especially in gale-force winds (>17.6 m/s) (Vagle et al., 1990; Zhao et al., 2014), and particularly without some prior observations for calibration. Furthermore, many empirical WOTAN studies are intrinsically flawed for the purpose of calculating noise levels at depth as they only study the effects of wind on specific frequencies. Additionally, without a better understanding of the effect bubble clouds have on the downward propagating of the signal, estimations at depth (>100 m) are highly speculative.

In summary, the acoustic signature of a breaking wave is primarily the result of bubble-cloud generation during the final cavity collapse phase of a wave's life cycle. Although individual bubble frequencies range from 0.4 to 2.0 kHz, the frequency of the bubble cloud itself is typically lower, at 0.1 to 0.5 kHz. The exact frequency spectrum depends on the properties of the wave, which are usually a consequence of wind speed and can be highly variable even under laboratory conditions. Bubbles generated by breaking waves can be pulled down as much as 10 m by Langmuir currents, where they can create steep sound-speed profiles with depth, possibly trapping select acoustic signals. The highly dynamic and unpredictable nature of breaking waves makes predicting ambient noise from multiple breaking waves difficult, especially in gale-force winds. WOTAN allow for measurements of wind speed via the ambient noise of wave action based on empirical observations but is insufficient for calculating ambient noise at depth. Knudsen curves are still the most commonly used prediction of ambient noise from wave action, with a positively skewed peak at around 0.5 kHz increasing in intensity with wind speed.

10.6. CONCLUSION

Bubbles have subtle yet far-reaching effects on marine acoustics. The initial formation of a bubble triggers simple harmonic motion at a natural frequency known as the Minnaert frequency, which is approximately inversely proportional to the bubble's radius. Thus, by observing the acoustic signature of a bubble, one can determine its size. Although the sound of a single bubble is low energy, the continuous release of multiple bubbles can significantly impact the ambient marine soundscape. To accurately predict the ambient noise produced by a gas seep, rainfall, or breaking waves, one must have a detailed understanding of the size distribution of bubbles being generated. Unfortunately, it is all but impossible to predict the size of bubbles released. However, it is possible to use observations of ambient noise to infer the characteristics of these sources: the flux from a gas seep, the intensity of rainfall, and the wind speed resulting in breaking waves.

FURTHER READING

The following sources are recommended for continuing research in each area introduced in this chapter:

- Bubble physics:
  - Leighton, T. G. (1994). The freely-oscillating bubble. In *The Acoustic Bubble* (pp. 129–286). Elsevier. https://doi.org/10.1016/B978-0-12-441920-9.50008-0
- Subsea gas-seep acoustics:
  - Leighton, T. G., & White, P. R. (2012). Quantification of undersea gas leaks from carbon capture and storage facilities, from pipelines and from methane seeps, by their acoustic emissions. *Proceedings of the Royal Society A: Mathematical, Physical and Engineering Sciences 468*, 485–510.
  - Bergès, B. J. P., Leighton, T. G., & White, P. R. (2015). Passive acoustic quantification of gas fluxes during controlled gas release experiments. *International Journal of Greenhouse Gas Control, 38*, 64–79.
- Rainfall acoustics:
  - Serra, Y. L. (2018). Precipitation measurements from the Tropical Moored Array: a review and look ahead. *Quarterly Journal of the Royal Meteorological Society, 144*, 221–234.
  - Ma, B. B., & Nystuen, J. A. (2005). Passive acoustic detection and measurement of rainfall at sea. *Journal of Atmospheric and Oceanic Technology, 22*, 1225–1248.
  - Pumphrey, H. C., & Crum, L. A. (1990). Free oscillations of near-surface bubbles as a source of the underwater noise of rain. *The Journal of the Acoustical Society of America, 87*, 142–148.
  - Pumphrey, H. C., & Elmore, P. A. The entrainment of bubbles by drop impacts. *Journal of Fluid Mechanics, 220*, 539–567 (1990).
  - Pumphrey, H.C., & Walton, A.J., 1988. Experimental study of the sound emitted by water drops impacting on a water surface. *European Journal of Physics, 9*, 225–231. https://doi.org/10.1088/0143-0807/9/3/011
- Breaking wave acoustics:
  - Deane, G. B., & Stokes, M. D. (2002). Scale dependence of bubble creation mechanisms in breaking waves. *Nature, 418*, 839–844.
  - Deane, G. B., & Stokes, M. D. (2010). Model calculations of the underwater noise of breaking waves and comparison with experiment. *The Journal of the Acoustical Society of America, 127*, 3394–3410.
  - Zhao, Z., D'Asaro, E. A., & Nystuen, J. A. (2014). The sound of tropical cyclones. *Journal of Physical Oceanography, 44*, 2763–2778.
  - Leighton, T. G. (1994). The freely-oscillating bubble. In *The Acoustic Bubble* (pp. 129–286). Elsevier. https://doi.org/10.1016/B978-0-12-441920-9.50008-0
  - Vagle, S., Large, W. G., & Farmer, D. M. An evaluation of the WOTAN technique of inferring oceanic winds from underwater ambient sound. *Journal of Atmospheric and Oceanic Technology, 7*, 576–595 (1990).

APPENDIX

Minnaert Frequency Derivation

The following is a derivation of the Minnaert equation following Leighton (1994).

We can find the kinetic energy, φ_K, of the water surrounding a bubble by integrating over shells of liquid from the bubble wall to infinity. A shell of radius r and a thickness dr has a mass of $4\pi r^2 \rho dr$ (where ρ is the density of water), and thus the kinetic energy of the surrounding water is

$$\varphi_K = \int_R^\infty \left(4\pi r^2 \rho \, dr\right) \dot{r}^2. \tag{A1}$$

The mass of liquid flowing in time dt through any spherical surface around the bubble is $4\pi r^2 \dot{r} \rho dt$. Assuming the liquid is incompressible, then by conservation of mass, this general flow can be equated to the flow at the bubble surface, which can be shown to be $\dot{r}/\dot{R} = R^2/r^2$. Substituting this into the above gives

$$\varphi_K = 2\pi R^3 \rho \dot{R}^2 \tag{A2}$$

Kinetic energy is a maximum at the equilibrium position (as with any harmonic oscillator) when $R = R_0$ and $\dot{R} = i\omega_0 R_{\varepsilon 0} e^{i\omega_0 t}$, implying that $|\dot{R}|^2 = (\omega_0 R_{\varepsilon 0})^2$. Thus, the maximum value of the kinetic energy is

$$\varphi_{\mathrm{KMax}} = \frac{1}{2} m_{\mathrm{RF}}^{\mathrm{rad}} \left(\omega_0 R_{\varepsilon 0}\right)^2 = 2\pi R_0{}^3 \rho (\omega_0 R_{\varepsilon 0})^2, \tag{A3}$$

where $m_{\mathrm{RF}}^{\mathrm{rad}}$ is the radiation mass of the bubble in the radius-force frame. This mass is the effective inertia of the liquid component of the oscillating system the pulsating bubble represents: that is, $m_{\mathrm{RF}}^{\mathrm{rad}} = 4\pi R_0{}^3 \rho$. It arises from the liquid that is transmitting acoustic waves and is the only inertia considered by the Minnaert derivation.

Through conservation of energy, the maximum kinetic energy φ_{KMax} must equal the maximum internal energy φ_{PMax}, which occurs when $R = R_0 \pm R_{\varepsilon 0}$ and $\dot{R} = 0$. The work done compressing the bubble from equilibrium volume V_0 (at radius R_0) to minimum volume $V_{\min}$ (at radius $R_0 - R_{\varepsilon 0}$) is the integral of $-(p_g - p_0)dV$, where p_g is the gas pressure and p_0 is the hydrostatic liquid pressure outside the bubble:

$$\varphi_{\mathrm{PMax}} = -\int_{V_{\max}}^{V_{\min}} \left(p_g - p_0\right) dV = -\int_{R_0}^{R_0 - R_{\varepsilon 0}} \left(p_g - p_0\right) 4\pi r^2 \, dr \tag{A4}$$

Minnaert derived his equation assuming that the gas behaved adiabatically: that is, that there was no heat flow across the bubble wall. This was adapted by the introduction of the polytropic index κ (which takes a value equal to unity when the gas behaves isothermally and equals the ratio of the specific heat of the gas at constant pressure to that at constant volume when the gas behaves adiabatically) (Leighton & Walton, 1987). Assuming the gas behaves polytropically so that $p_g V^\kappa = constant$, then since $R_\varepsilon = R - R_0$, equating the pressure and volume condition at equilibrium to those when the bubble attains minimum volume gives

$$p_g (R_0 + R_\varepsilon)^{3k} = p_0 R_0{}^{3k}. \tag{A5}$$

For small displacements, the binomial expansion of this is

$$p_0 - p_g \approx \frac{3\kappa R_\varepsilon p_0}{R_0}. \tag{A6}$$

Substituting this into the maximum internal energy φ_{PMax} with the use of first-order $R_\varepsilon = R - R_0$ coordinates gives

$$\varphi_{PMax} = \int_0^{R_{\varepsilon 0}} \frac{3\kappa R_\varepsilon p_0}{R_0} 4\pi R_0{}^2 \, dR_\varepsilon = 6\pi\kappa p_0 R_0 R_{\varepsilon 0}{}^2. \tag{A7}$$

This allows us to equate the maximum kinetic energy and maximum potential energy:

$$\varphi_{KMax} = 2\pi R_0{}^3 \rho (\omega_0 R_{\varepsilon 0})^2 = 6\pi\kappa p_0 R_0 R_{\varepsilon 0}{}^2 = \varphi_{PMax}, \tag{A8}$$

which can be solved for the resonance circular frequency ω_0:

$$\omega_0 = \frac{1}{R_0}\sqrt{\frac{3\kappa p_0}{\rho}} \tag{A9}$$

Finally, using $\omega_0 = 2\pi f_M$ gives us the Minnaert frequency equation:

$$f_M = \frac{1}{2\pi R_0}\sqrt{\frac{3\kappa p_0}{\rho}}. \tag{A10}$$

SYMBOLOGY

Symbol	Definition
$a_{\min}$	Minimum bubble size generated
$a_{\max}$	Maximum bubble size generated
$B(R_0)$	Bubble-size distribution
c	Speed of sound in water
c_c	Speed of sound through a bubble cloud
d	Dimensionless damping constant for a single bubble $=2\beta/\omega$
D	Initial dipole strength of the bubble
f	Frequency
f_{peak}	Peak frequency
f_M	Minnaert frequency: oscillation frequency of a bubble as predicted by the Minnaert equation
f_0	Natural frequency of a bubble oscillation
G	A variable function of wind speed

Symbol	Definition
h	Distance from the center of the bubble to the surface of the water
I_{Rain}	Intensity of rainfall beneath the surface at a given frequency
ISL_{Rain}	Intensity spectrum level of rainfall beneath the surface at a given frequency
l	Loaded length of spring
l_0	Unloaded length of spring
L_d	Diameter of water droplet
m_{RF}^{rad}	Radiation mass of bubble in radius force frame
n	Mode number
$n(f)$	Number of bubbles entrained per second by rainfall
n_{lf}	A quantity controlling the increasing slope of wave noise in lower wind conditions
n_{hf}	A quantity controlling the decreasing slope of wave noise in higher wind conditions
$n_n^{gr}(z, R)dR_0$	Number of bubbles per unit volume at depth z having a radii between R_0 and $R_0 + dR_0$
N_b	Number of identical bubbles in a bubble cloud
p_g	Gas pressure inside the bubble
p_0	Hydrostatic liquid pressure outside the bubble
p_v	Vapor pressure
$P(\omega, r)$	Power spectrum of a breaking wave
q	Quality factor
Q_w	Slope of the logarithmic spectrum of the wind-generated sound
r	Radial distance
R	Radius of Bubble wall
R_0	Equilibrium radius of bubble
R_ϵ	Displacement of the bubble wall from equilibrium radius
$R_{\epsilon 0}$	Maximum displacement of bubble wall from equilibrium radius
$R_{\epsilon 0i}$	Initial amplitude of displacement of the bubble wall
R_{max}	Maximum radius of bubble
R_{min}	Minimum radius of bubble
S	Column vector containing the measured spectrum $S(\omega_k)$
$S(\omega)$	Power spectral density of a marine gas seep
S_{10}	Sound level at 10 ms^{-1} wind
S_{noise}	Noise floor
S_{rain}	Rainfall rate
SPL	Sound pressure level
SPL_{5kHZ}	Sound pressure level at 5 kHz
SSL_0	Source sound level of breaking waves at a depth of 1m
t	Time
u	Impact Mach number
u_d	Impact speed of water droplet
U_{10}	Wind speed 10 m above water level
V	Volume of bubble plume
V_0	Equilibrium bubble volume
V_{min}	Minimum bubble volume
VF	Void Fraction

Symbol	Definition
z	Depth below sea surface
$\alpha(\omega, a)$	Fourier transform of the Greens function for the medium of propagation, respectively
$\gamma(\omega, a)$	Fourier transform of the convolution of free-space bubble pulses
δ_{tot}	Total damping constant for bubble pulsation at resonance
ε	Displacement from equilibrium
θ	Polar angle, angle between observer and source relative to the z axis
φ_K	Kinetic energy
φ_{KMax}	Maximum kinetic energy
φ_{PMax}	Maximum internal energy
κ	Polytropic index
σ	Surface tension
η	Shear viscosity
ρ	Density of water
ω	Angular frequency $=2\pi f$
ω_0	Angular resonate frequency
$\Psi(n)$	Bubble generation rate (for marine gas seep)

REFERENCES

Ainslie, M. A., & Leighton, T. G. (2011). Review of scattering and extinction cross-sections, damping factors, and resonance frequencies of a spherical gas bubble. *The Journal of the Acoustical Society of America*, *130*(5), 3184–3208.

Algar, C. K., Boudreau, B. P., & Barry, M. A. (2011). Initial rise of bubbles in cohesive sediments by a process of viscoelastic fracture. *Journal of Geophysical Research*, *116*, B04207.

Anagnostou, M. N., Nystuen, J. A., Anagnostou, E. N., Nikolopoulos, E. I., & Amitai, E. (2008). Evaluation of underwater rainfall measurements during the ionian sea rainfall experiment. *IEEE Transactions on Geoscience and Remote Sensing*, *46*, 2936–2946.

Ashokan, M., Latha, G., & Ramesh, R. (2015). Analysis of shallow water ambient noise due to rain and derivation of rain parameters. *Applied Acoustics*, *88*, 114–122.

Bass, S. J., & Hay, A. E. (1997). Ambient noise in the natural surf zone: wave-breaking frequencies. *IEEE Journal of Oceanic Engineering*, *22*, 411–424.

Bergès, B. J. P., Leighton, T. G., & White, P. R. (2015). Passive acoustic quantification of gas fluxes during controlled gas release experiments. *International Journal of Greenhouse Gas Control*, *38*, 64–79.

Black, P. G., Proni, J. R., Wilkerson, J. C., & Samsury, C. E. (1997). Oceanic rainfall detection and classification in tropical and subtropical mesoscale convective systems using underwater acoustic methods. *Monthly Weather Review*, *125*, 2014–2042.

Boles, J. R., Clark, J. F., Leifer, I., & Washburn, L. (2001). Temporal variation in natural methane seep rate due to tides, Coal Oil Point area, California. *Journal of Geophysical Research: Oceans*, *106*, 27077–27086.

Böttner, C., Berndt, C., Reinardy, B. T. I., Geersen, J., Karstens, J., Bull, J. M., et al. (2019). Pockmarks in the Witch Ground Basin, Central North Sea. *Geochemistry, Geophysics, Geosystems, 20*, 1698–1719.

Boudreau, B. P. (2012). The physics of bubbles in surficial, soft, cohesive sediments. *Marine and Petroleum Geology, 38*(1), 1–18.

Boudreau, B. P., Algar, C. K., Johnson, B. D., Croudace, I., Reed, A., Furukawa, Y., et al. (2005). Bubble growth and rise in soft sediments. *Geology, 33*(6), 517.

Buckingham, M. J. (1991). On acoustic transmission in ocean-surface waveguides. *Philosophical Transactions: Physical Sciences and Engineering, 335*, 513–555.

Carey, W. M., & Bradley, M. P. (1985). Low-frequency ocean surface noise sources. *The Journal of the Acoustical Society of America, 78*, S1–S2.

Cauchy, P., Heywood, K. J., Merchant, N. D., Queste, B. Y., & Testor, P. (2018). Wind speed measured from underwater gliders using passive acoustics. *Journal of Atmospheric and Oceanic Technology, 35*, 2305–2321.

Cavaleri, L. (2020). Rain, wave breaking and spray. In *Recent advances in the study of oceanic whitecaps* (pp. 65–75). Springer International Publishing. https://doi.org/10.1007/978-3-030-36371-0_5

Cavaleri, L., & Bertotti, L. (2018). Rain on generative seas. *Geophysical Research Letters, 45*, 7049–7056.

Cazau, D., Bonnel, J., & Baumgartner, M. (2019). Wind speed estimation using acoustic underwater glider in a near-shore marine environment. *IEEE Transactions on Geoscience and Remote Sensing, 57*, 2097–2106.

Cazau, D., Bonnel, J., Jouma'a, J., le Bras, Y., & Guinet, C. (2017). Measuring the marine soundscape of the Indian Ocean with southern elephant seals used as acoustic gliders of opportunity. *Journal of Atmospheric and Oceanic Technology, 34*, 207–223.

Chen, L., Trinh, V., Yang, W., & Mohanangam, K. (2016). Prediction of bubble generation based on acoustic emission. *Acoustics Australia, 44*, 325–331.

Clay, C. S., & Medwin, H. (1977). Acoustical oceanography: principles and applications. *Journal of the Marine Biological Association of the United Kingdom, 58*, 543–543.

Coughlan, M., Roy, S., O'Sullivan, C., Clements, A., O'Toole, R., & Plets, R. (2021). Geological settings and controls of fluid migration and associated seafloor seepage features in the north Irish Sea. *Marine and Petroleum Geology, 123*, 104762.

Czerski, H., & Deane, G. B. (2010). Contributions to the acoustic excitation of bubbles released from a nozzle. *The Journal of the Acoustical Society of America, 128*(5), 2625–2634.

Deane, G. B., & Stokes, M. D. (2002). Scale dependence of bubble creation mechanisms in breaking waves. *Nature, 418*, 839–844.

Deane, G. B., & Stokes, M. D. (2010). Model calculations of the underwater noise of breaking waves and comparison with experiment. *The Journal of the Acoustical Society of America, 127*, 3394–3410.

Esposito, A., Giordano, G., & Anzidei, M. (2006). The 2002–2003 submarine gas eruption at Panarea volcano (Aeolian Islands, Italy): volcanology of the seafloor and implications for the hazard scenario. *Marine Geology, 227*, 119–134.

Farmer, D. M., & Vagle, S. (1988). On the determination of breaking surface wave distributions using ambient sound. *Journal of Geophysical Research, 93*, 3591.

Farmer, D. M., & Vagle, S. (1989). Waveguide propagation of ambient sound in the ocean-surface bubble layer. *The Journal of the Acoustical Society of America, 86*, 1897–1908.

Farmer, D. M., Deane, G. B., & Vagle, S. (2001). The influence of bubble clouds on acoustic propagation in the surf zone. *IEEE Journal of Oceanic Engineering, 26*, 113–124.

Farrell, W. E., Berger, J., Bidlot, J.-R., Dzieciuch, M., Munk, W., Stephen, R. A., & Worcester, P. F. (2016). Wind sea behind a cold front and deep ocean acoustics. *Journal of Physical Oceanography, 46*, 1705–1716.

Felizardo, F. C., & Melville, W. K. (1995). Correlations between ambient noise and the ocean surface wave field. *Journal of Physical Oceanography, 25*, 513–532.

Garrett, C., Li, M., & Farmer, D. (2000). The connection between bubble size spectra and energy dissipation rates in the upper ocean. *Journal of Physical Oceanography, 30*, 2163–2171.

Greene, C. A., & Wilson, P. S. (2012). Laboratory investigation of a passive acoustic method for measurement of underwater gas seep ebullition. *The Journal of the Acoustical Society of America, 131*, EL61–EL66.

Holthuijsen, L. H. (2007). *Waves in oceanic and coastal waters*. Cambridge University Press. https://doi.org/10.1017/CBO9780511618536

Hovland, M. (2002). On the self-sealing nature of marine seeps. *Continental Shelf Research, 22*, 2387–2394.

Jain, A. K., & Juanes, R. (2009). Preferential mode of gas invasion in sediments: grain-scale mechanistic model of coupled multiphase fluid flow and sediment mechanics. *Journal of Geophysical Research, 114*, B08101.

Johnson, B. D., Barry, M. A., Boudreau, B. P., Jumars, P. A., & Dorgan, K. M. (2012). In situ tensile fracture toughness of surficial cohesive marine sediments. *Geo-Marine Letters, 32*(1), 39–48.

Johnson, B. D., Boudreau, B. P., Gardiner, B. S., & Maass, R. (2002). Mechanical response of sediments to bubble growth. *Marine Geology, 187*(3-4), 247–363.

Judd, A. G. (2003). The global importance and context of methane escape from the seabed. *Geo-Marine Letters, 23*(2), 147–154.

Klaucke, I., Weinrebe, W., Petersen, C. J., & Bowden, D. (2010). Temporal variability of gas seeps offshore New Zealand: multi-frequency geoacoustic imaging of the Wairarapa area, Hikurangi margin. *Marine Geology, 272*, 49–58.

Knittel, K., & Boetius, A. (2009). Anaerobic oxidation of methane: progress with an unknown process. *Annual Review of Microbiology, 63*, 311–334.

Knudsen, V. O., Alford, R. S., & Emling, L. W. (1948). Underwater ambient noise. *Marine Research, 7*.

Leifer, I. (2015). Seabed bubble flux estimation by calibrated video survey for a large blowout seep in the North Sea. *Marine and Petroleum Geology, 68*, 743–752.

Leifer, I., & Patro, R. K. (2002). The bubble mechanism for methane transport from the shallow sea bed to the surface:

a review and sensitivity study. *Continental Shelf Research, 22*, 2409–2428.

Leighton, T. G. (1988). The frequency analysis of transients. *European Journal of Physics, 9*, 69–70.

Leighton, T. G. (1994). The freely-oscillating bubble. In *The acoustic bubble* (pp. 129–286). Elsevier. https://doi.org/10.1016/B978-0-12-441920-9.50008-0

Leighton, T. G. (2004). From seas to surgeries, from babbling brooks to baby scans: the acoustics of gas bubbles in liquids. *International Journal of Modern Physics B, 18*(18), 3267–3314.

Leighton, T. G., & Walton, A. J. (1987). An experimental study of the sound emitted from gas bubbles in a liquid. *European Journal of Physics, 8*, 98–104.

Leighton, T. G., & White, P. R. (2012). Quantification of undersea gas leaks from carbon capture and storage facilities, from pipelines and from methane seeps, by their acoustic emissions. *Proceedings of the Royal Society A: Mathematical, Physical and Engineering Sciences, 468*(2140), 485–510.

Leighton, T. G., K.J., F., & Field, J.E. (1991). Acoustic and photographic studies of injected bubbles. *European Journal of Physics, 12*, 77–85.

Leighton, T. G., White, P. R., & Schneider, M. F. (1998). The detection and dimension of bubble entrainment and comminution. *The Journal of the Acoustical Society of America, 103*(4), 1825–1835.

Li, J., Roche, B., Bull, J. M., White, P. R., Davis, J. W., Deponte, M., Gordini, E., & Cotterle, D. (2020a). Passive acoustic monitoring of a natural CO2 seep site – implications for carbon capture and storage. *International Journal of Greenhouse Gas Control, 93*, 102899.

Li, J., Roche, B., Bull, J. M., White, P. R., Leighton, T. G., Provenzano, G., et al. (2020b). Broadband acoustic inversion for gas flux quantification—application to a methane plume at Scanner pockmark, Central North Sea. *Journal of Geophysical Research: Oceans, 125*.

Li, J., White, P. R., Roche, B., Bull, J. M., Leighton, T. G., Davis, J. W., & Fone, J. W. (2021). Acoustic and optical determination of bubble size disruptions - quantification of undersea gas emissions. *International Journal of Greenhouse Gas Control.*

Liang, J.-H., McWilliams, J. C., Sullivan, P. P., & Baschek, B. (2011). Modeling bubbles and dissolved gases in the ocean. *Journal of Geophysical Research, 116*, C03015.

Longuet-Higgins, M. S., Kerman, B. R., & Lunde, K. (1991). The release of air bubbles from an underwater nozzle. *Journal of Fluid Mechanics, 230*, 365–390.

Lu, N. Q., Prosperetti, A., & Yoon, S. W. (1990). Underwater noise emissions from bubble clouds. *IEEE Journal of Oceanic Engineering, 15*(3), 275–281.

Ma, B. B., & Nystuen, J. A. (2005). Passive acoustic detection and measurement of rainfall at sea. *Journal of Atmospheric and Oceanic Technology, 22*, 1225–1248.

Ma, B. B., & Nystuen, J. A. (2007). Detection of rainfall events using underwater passive aquatic sensors and air–sea temperature changes in the tropical Pacific Ocean. *Monthly Weather Review, 135*, 3599–3612.

Maksimov, A. O., Burov, B. A., Salomatin, A. S., & Chernykh, D. V. (2016). Sounds of undersea gas leaks. In *Underwater acoustics and ocean dynamics* (pp. 107–116). Springer Singapore. https://doi.org/10.1007/978-981-10-2422-1_15

Manasseh, R., Babanin, A. V., Forbes, C., Rickards, K., Bobevski, I., & Ooi, A. (2006). Passive acoustic determination of wave-breaking events and their severity across the spectrum. *Journal of Atmospheric and Oceanic Technology, 23* (4), 599–618.

McGinnis, D. F., Greinert, J., Artemov, Y., Beaubien, S. E., & Wüest, A. (2006). Fate of rising methane bubbles in stratified waters: how much methane reaches the atmosphere? *Journal of Geophysical Research, 111*, C09007.

Medwin, H., & Beaky, M. M. (1989). Bubble sources of the Knudsen sea noise spectra. *The Journal of the Acoustical Society of America, 86*(3), 1124–1130.

Medwin, H., Kurgan, A., & Nystuen, J. A. (1990). Impact and bubble sound from raindrops at normal and oblique incidence. *The Journal of the Acoustical Society of America, 88*, 413–418.

Minnaert, M. (1933). On musical air-bubbles and the sounds of running water. *The London, Edinburgh, and Dublin Philosophical Magazine and Journal of Science, 16*, 235–248.

Monahan, E. C., & Lu, M. (1990). Acoustically relevant bubble assemblages and their dependence on meteorological parameters. *IEEE Journal of Oceanic Engineering, 15*, 340–349.

Muyakshin, S. I., & Sauter, E. (2010). The hydroacoustic method for the quantification of the gas flux from a submersed bubble plume. *Oceanology, 50*, 995–1001.

Nystuen, J. A. (1986). Rainfall measurements using underwater ambient noise. *The Journal of the Acoustical Society of America, 79*, 972–982.

Nystuen, J. A. (2001). Listening to raindrops from underwater: an acoustic disdrometer. *Journal of Atmospheric and Oceanic Technology, 18*, 1640–1657.

Nystuen, J. A., & Selsor, H. D. (1997). Weather classification using passive acoustic drifters. *Journal of Atmospheric and Oceanic Technology, 14*, 656–666.

Nystuen, J. A., Amitai, E., Anagnostou, E. N., & Anagnostou, M. N. (2008). Spatial averaging of oceanic rainfall variability using underwater sound: Ionian sea rainfall experiment 2004. *The Journal of the Acoustical Society of America, 123*, 1952–1962.

Nystuen, J. A., Anagnostou, M. N., Anagnostou, E. N., & Papadopoulos, A. (2015). Monitoring Greek seas using passive underwater acoustics. *Journal of Atmospheric and Oceanic Technology, 32*, 334–349.

Nystuen, J. A., McPhaden, M. J., & Freitag, H. P. (2000). Surface measurements of precipitation from an ocean mooring: the underwater acoustic log from the South China Sea. *Journal of Applied Meteorology, 39*, 2182–2197.

Ostrovsky, I., McGinnis, D. F., Lapidus, L., & Eckert, W. (2008). Quantifying gas ebullition with echosounder: The role of methane transport by bubbles in a medium-sized lake. *Limnology and Oceanography: Methods, 6*, 105–118.

Pensieri, S., Bozzano, R., Nystuen, J., Anagnostou, E., Anagnostou, M., & Bechini, R. (2015). Underwater acoustic measurements to estimate wind and rainfall in the Mediterranean Sea. *Advances in Meteorology, 2015*, 1–18.

Phelps, A. D., Ramble, D. G., & Leighton, T. G. (1997). The use of a combination frequency technique to measure the surf zone bubble population. *The Journal of the Acoustical Society of America, 101*(4), 1981–1989.

Pumphrey, H. C. (1989). *Sources of ambient noise in the ocean: an experimental investigation* (PhD thesis). Mississippi University.

Pumphrey, H. C., & Crum, L. A. (1990). Free oscillations of near-surface bubbles as a source of the underwater noise of rain. *The Journal of the Acoustical Society of America, 87* (1), 142–148.

Pumphrey, H. C., & Elmore, P. A. (1990). The entrainment of bubbles by drop impacts. *Journal of Fluid Mechanics, 220*, 539–567.

Pumphrey, H. C., & Walton, A. J. (1988). Experimental study of the sound emitted by water drops impacting on a water surface. *European Journal of Physics, 9*, 225–231.

Riedel, M., Scherwath, M., Römer, M., Veloso, M., Heesemann, M., & Spence, G. D. (2018). Distributed natural gas venting offshore along the Cascadia margin. *Nature Communications, 9*, 3264.

Roche, B., Bull, J. M., Marin-Moreno, H., Leighton, T. G., Suarez, I. F., Tholen, M., et al. (2020). Time-lapse imaging of CO2 migration within near-surface sediment during a controlled sub-seabed release experiment. *International Journal of Greenhouse Gas Control, 109*, 103363.

Roche, B., Li, J., White, P., Bull, J., Davis, J., Deponte, M., et al. (2019). Validating passive acoustic methods for gas flux quantification, offshore Panarea, Mediterranean Sea. *The Journal of the Acoustical Society of America, 146*, 2965–2965.

Römer, M., Riedel, M., Scherwath, M., Heesemann, M., & Spence, G. D. (2016). Tidally controlled gas bubble emissions: A comprehensive study using long-term monitoring data from the NEPTUNE cabled observatory offshore Vancouver Island. *Geochemistry, Geophysics, Geosystems, 17*, 3797–3814.

Scandella, B. P., Pillsbury, L., Weber, T., Ruppel, C., Hemond, H. F., & Juanes, R. (2016). Ephemerality of discrete methane vents in lake sediments. *Geophysical Research Letters, 43*, 4374–4381.

Scrimger, J. A., Evans, D. J., McBean, G. A., Farmer, D. M., & Kerman, B. R. (1987). Underwater noise due to rain, hail, and snow. *The Journal of the Acoustical Society of America, 81*, 79–86.

Serra, Y. L. (2018). Precipitation measurements from the Tropical Moored Array: a review and look ahead. *Quarterly Journal of the Royal Meteorological Society, 144*, 221–234.

Strasberg, M. (1953). The pulsation frequency of nonspherical gas bubbles in liquids. *The Journal of the Acoustical Society of America, 25*(3), 536–537.

Suess, E. (2014). Marine cold seeps and their manifestations: Geological control, biogeochemical criteria and environmental conditions. *International Journal of Earth Sciences, 103*, 1889–1916.

Sultan, N., Plaza-Faverola, A., Vadakkepuliyambatta, S., Buenz, S., & Knies, J. (2020). Impact of tides and sea-level on deep-sea Arctic methane emissions. *Nature Communications, 11*, 5087.

Taylor, W. O., Anagnostou, M. N., Cerrai, D., & Anagnostou, E. N. (2020). Machine learning methods to approximate rainfall and wind from acoustic underwater measurements. *IEEE Transactions on Geoscience and Remote Sensing, 1–12*. https://doi.org/10.1109/TGRS.2020.3007557

Thorpe, S. A. (1982). On the clouds of bubbles formed by breaking wind-waves in deep water, and their role in air--sea gas transfer. *Philosophical Transactions of the Royal Society of London. Series A, Mathematical and Physical Sciences, 304*, 155–210.

Vagle, S., Large, W. G., & Farmer, D. M. (1990). An evaluation of the WOTAN technique of inferring oceanic winds from underwater ambient sound. *Journal of Atmospheric and Oceanic Technology, 7*, 576–595.

Vagle, S., McNeil, C., & Steiner, N. (2010). Upper ocean bubble measurements from the NE Pacific and estimates of their role in air-sea gas transfer of the weakly soluble gases nitrogen and oxygen. *Journal of Geophysical Research, 115*, C12054.

Vazquez, A., Manasseh, R., & Chicharro, R. (2015). Can acoustic emissions be used to size bubbles seeping from a sediment bed? *Chemical Engineering Science, 131*, 187–196.

Walton, A. J., Gunn, M. G., & Reynolds, G. T. (2005). The quality factor of oscillating bubbles as an indication of gas content with particular reference to methane. *IEEE Journal of Oceanic Engineering, 30*(4), 924–926.

Wang, D. W., Wijesekera, H. W., Jarosz, E., Teague, W. J., & Pegau, W. S. (2016). Turbulent diffusivity under high winds from acoustic measurements of bubbles. *Journal of Physical Oceanography, 46*, 1593–1613.

Wenz, G. M. (1962). Acoustic ambient noise in the ocean: Spectra and sources. *The Journal of the Acoustical Society of Tommy America, 34*, 1936–1956.

Wiggins, S. M., Leifer, I., Linke, P., & Hildebrand, J. A. (2015). Long-term acoustic monitoring at North Sea well site 22/4b. *Marine and Petroleum Geology, 68*, 776–788.

Wu, J. (1979). Oceanic whitecaps and sea state. *Journal of Physical Oceanography, 9*, 1064–1068.

Yang, J., Riser, S., Nystuen, J., Asher, W., & Jessup, A. (2015). Regional rainfall measurements using the passive aquatic listener during the SPURS field campaign. *Oceanography, 28*, 124–133.

Zhao, B., Qiao, F., Cavaleri, L., Wang, G., Bertotti, L., and Liu, L. (2017). Sensitivity of typhoon modeling to surface waves and rainfall. *Journal of Geophysical Research: Oceans, 122*, 1702–1723.

Zhao, Z., D'Asaro, E. A., & Nystuen, J. A. (2014). The sound of tropical cyclones. *Journal of Physical Oceanography, 44*, 2763–2778.

11

Baleen Whale Vocalizations

Paul R. White[1] and Victoria L.G. Todd[2]

ABSTRACT

Many baleen (mysticete) whales produce high-intensity calls, the frequency content of which resides primarily in a band below 1 kHz. Consequently, there is a reasonable probability of vocalizing animals being detected on hydrophones designed to collect marine-acoustic-geophysical data, such as towed and bottom-mounted seismic streamers. This represents an opportunity to collect potentially untapped data on whale distribution and behavior over broad spatial scales. The objectives of this chapter are to present those with geophysical-research backgrounds with an opportunity to recognize and classify frequently encountered whale sounds in the field. The chapter introduces the general classification and description of marine mammals, a brief physical description of sound, and the range of vocalizations produced by baleen whales on a species-by-species basis.

11.1. INTRODUCTION

The collection of acoustic data is integral to many marine geophysical surveys. The incidental collection of recordings of biogenic sounds from the marine environment is almost inevitable. This chapter aims to describe the sounds of baleen whales, the species group most likely to be encountered in such surveys. The goal is not to provide a complete compendium of all the known vocalizations of these animals but rather to provide an overview, presenting some of the most common and characteristic calls that might be recorded, providing the interested analyst the opportunity to appropriately attribute the sounds.

The chapter begins with a summary of the taxonomy of marine mammals, focusing on the baleen whales. This is followed by a review of some basic principles of acoustics, as they pertain to bioacoustics (the study of animal sounds). The core of the discussion then follows and consists of a species-by-species description of the most relevant sounds produced by baleen whales.

11.1.1. Marine Mammal Classification

The classification of all organisms, both living and extinct, is organized using seven distinct levels, referred to commonly as taxonomic hierarchical categories: kingdom, phylum, class, order, family, genus, and species. Scientific names and groups that marine mammals belong to change as new information comes to light (especially from molecular-genetic studies), but it is important to have some understanding of the basics.

All large multicellular, mobile organisms belong to the kingdom Animalia (animals), and marine mammals are classified as chordates (any organism possessing a backbone)—phylum Chordata, class Mammalia. These include any group of vertebrates that, inter alia, give birth to live young nourished with milk secreted by mammary glands. At the order level, marine mammals can be

[1]*Institute of Sound and Vibration Research, University of Southampton, Southampton, UK*
[2]*Ocean Science Consulting Ltd., Dunbar, UK*

Noisy Oceans: Monitoring Seismic and Acoustic Signals in the Marine Environment, Geophysical Monograph 284, First Edition.
Edited by Gaye Bayrakci and Frauke Klingelhoefer.

DOI:10.1002/9781119750925.ch11

Table 11.1 Families within the suborder mysticete.

Family	Visual features	Member species
Balaenopteridae (rorquals)	Sleek, streamlined body; throat grooves extending from chin to navel (except for sei whale); a substantial increase in mouth volume during feeding because of throat grooves	Sei (*Balaenoptera borealis*), Bryde's (*B. brydei*), fin (*B. physalus*), blue (*B. musculus*), Omura's (*B. omurai*), common minke (*B. acutorostrata*), Antarctic minke (*B. bonaerensis*), humpback (*Megaptera novaeangliae*)
Eschrichtiidae (gray whale)	Mottled dark skin, heavily parasitized (barnacles, whale lice); barnacles especially on top of the head, around blowhole, and on the anterior part of the back; lack of dorsal fin; dorsal ridge (8–9 humps) present in last third of back; short, heavy baleen; two longitudinal throat grooves	Gray whale (*Eschrichtius robustus*)
Balaenidae (right whales)	Arched rostrum (upper jaw); long, slender baleen plates (up to 4 m); absence of dorsal fin and throat grooves; southern right has barnacles in callosities	North Atlantic right whale (*Eubalaena glacialis*), north Pacific right whale (*E. japonica*), southern right (*E. australis*), bowhead (*Balaena mysticetus*)
Neobalaenidae (pigmy right whale)	Arched rostrum; long, slender baleen plates (up to 4 m); small, triangular dorsal fin; two throat grooves	Pygmy right whale (*Caperea marginata*)

Taxonomy is complicated, and subspecies are not included here for simplicity. Adapted from Macdonald (2006).

subdivided into three groups: Cetacea (whales, dolphins, and porpoises), Sirenia (dugongs and manatees), and Carnivora (seals, sea lions, the walrus, sea otters, and polar bears). Unlike Sirenia and Carnivora, the order Cetacea is further divided into two distinct suborders: Mysticeti (mysticetes, or baleen whales) and Odontoceti (odontocetes, or toothed whales) (Folkens et al., 2011).

Although this chapter focuses primarily on the baleen whales, as the species most likely to be detected on geophysical-survey hydrophones, some discussion of other marine mammal species is included.

Mysticeti (Baleen Whales)

The suborder Mysticeti encompasses relatively few living species (Table 11.1), which range in size from a minimum of 6.4 m (pigmy right whale) to nearly 33 m (blue whale). Baleen whales are characterized by the presence of *baleen plates*, which are bristle-like keratin structures (analogous to human fingernails) that have evolved as a substitute for teeth. The plates occur in two racks situated along either side of the upper jaw and serve the function of straining small fish and crustaceans. A further diagnostic feature is that they possess paired blow holes, in contrast to the toothed whales, which have a single blow hole.

The fast-swimming, sleeker rorquals, such as blue whales (Sears & Perrin, 2009), are the largest group of baleen whales, characterized by, among other things, pleated throats and (with the exception of humpback) a general absence of encrusting barnacles. The pleats allow their throats to expand massively to gulp large amounts of water, which they sieve through the plates as they expel the contents of their mouths to extract small food items (Lambertsen, 1983; Pyenson et al., 2012). Rorquals comprise two genera: *Balaenoptera* and *Megaptera* (Würsig et al., 2017), the latter consisting of a single species, the humpback. Updated taxonomy is published regularly by the Committee on Taxonomy (https://marinemammalscience.org/science-and-publications/list-marine-mammal-species-subspecies/#list).

Odontoceti (Toothed Whales)

The suborder Odontoceti encompasses about 90% of living cetacean species, some of the major families being the dolphins (*Delphinidae*), porpoises (*Phocoenidae*), beaked whales (*Ziphiidae*), and the sperm whale (*Physeteridae*). Most toothed whales tend not to exceed 4.5 m in length, although they can reach a maximum length of 18 m (sperm whales). Unlike mysticetes, odontocetes have a set of identical teeth in the upper and/or lower jaw and feed on squids, fish, and other marine mammals. Most vocalizations from these species are higher in frequency than are typically captured using equipment designed for geophysical purposes, so they are not described here.

11.2. PHYSICAL DESCRIPTION OF SOUND AND ITS CONVENTIONS

Before expanding on vocalization types in marine mammals, it is important to note that the measurement of sound underwater can be the source of considerable confusion; therefore, this section aims to explain conventional acoustic terminologies. A sound wave is characterized by a small number of basic measures. The two most familiar properties are the amplitude of the pressure wave (which relates to human perception of loudness) and frequency of the signal (which relates to human perception of pitch).

11.2.1. Sound Pressure Level (SPL)

An underwater sound wave generates local pressure fluctuations: that is, compression and rarefaction of the water. Along with these pressure fluctuations, there is cyclical motion of water particles.

The amplitude of an acoustic wave is usually expressed in logarithmic units called decibels (dB) and is referred to as the sound pressure level (SPL). The dB is a relative quantity, and in underwater acoustics, the modern convention is to base the dB on a reference pressure of 1 μPa, in contrast with 20 μPa, which is used in air. It is good practice to cite SPLs using the full (but rather long-winded) unit notation dB re 1 μPa. The use of different reference units is not the sole distinction between measurements made in water and air, so making comparisons between SPLs in the two media can lead to misleading conclusions and should generally be avoided (Finfer et al., 2008).

An additional source of ambiguity when considering SPLs can arise depending on whether the sound is short-duration, transient, or continuous. For short-duration and transient sounds, it is normal to measure SPL based on peak pressure levels, most commonly the peak-to-peak level (p-p). For continuous sounds, however, it is conventional to use the root mean square (rms) value. In the context of cetacean vocalizations, short-duration sounds, like echolocation clicks, are typically reported using SPLs based on peak measurements. For longer vocalizations, such as most communication signals (e.g., whistles and pulsed calls), it is more appropriate to base SPLs on rms levels. Throughout the chapter, the subscript "rms" is used to emphasize when a metric is based on root mean squared values, and the subscript "p-p" is used when the metric represents a peak-to-peak measurement.

Note that in general, it is not possible to convert SPLs based on measurements of peak values to those based on rms. However, one can say that peak-based SPLs yield higher values than those based on rms. Most baleen whale vocalizations are of longer duration, so they are suited to measurements based on rms metrics; however, there are some instances in the literature where only peak SPLs are reported and occasions where the basis of SPL calculations is not stated.

If $p(t)$ is defined as the acoustic pressure at time t, then the SPL based on the rms level is defined as

$$SPL_{rms} = 20\log_{10}\left(\frac{p_{rms}}{p_{ref}}\right) \qquad p_{rms} = \sqrt{\frac{1}{T}\int_0^T p(t)^2 dt}$$

where T is the chosen integration time, and p_{ref} is the reference pressure (i.e., 1 μPa for underwater applications). The integration time can be a source of variation: it needs to be chosen to encompass at least several cycles of the waveform, but the amplitude of observed calls typically varies throughout their duration, so the choice of integration period affects the reported value of SPL_{rms}.

11.2.2. Source Level (SL)

One important characteristic of a sound source is the acoustic power it radiates, which, loosely speaking, relates to its absolute loudness. In underwater acoustics, this is usually represented using the source level (SL). One practical use of a SL is to predict the distance at which a sound source could be detected in specific environmental conditions. The SL can be defined by considering a hypothetical scenario in which the source is replaced by an idealized, equivalent monopole. This monopole is located at the acoustic center of the physical source, and its acoustic power is selected to account for the measured SPL at the observer's location. The SL is then defined as the SPL at a reference distance, R_0, from this hypothetical monopole source, with the reference distance being (almost) universally selected as 1 m. The reader is cautioned not to mistake this definition of SL as being equivalent to the SPL measured 1 m from the actual source. Close to a source—that is, in the acoustic *near field*—the properties of the sound field exhibit complex behaviors. Consequently, measuring at a single point in the near field of the source is not generally representative of the source characteristics observed at distance (i.e., in the *far field*) (Kinsler et al., 1999).

SL is commonly expressed in dB, but based on a different reference unit, and two equivalent notations are in common usage. One notation employs "dB re 1 μPa @ 1 m," with "@ 1 m" being shorthand for the hypothetical-monopole-process described in the preceding paragraph; however, we believe this notation invites misunderstanding, implicitly implying that the SL measurement is made 1 m from the source. Consequently, to reduce the risk of misunderstanding, the alternative notation "dB re 1 μPa.m" is preferred herein.

SL estimation of a whale vocalization is performed based on a measuring received level (RL): that is, the SPL of the source at a known distance (range) from the animal. To compute the SL from the RL, a correction is applied to account for the effect of the sound propagating from the reference point, R_0, to the measurement location. This reduction in level as sound travels through the water from reference point to receiver is called the transmission loss (TL), expressed in dB:

$$RL = SL - TL$$

Consequently, the SL is estimated by adding an estimate of the TL to the RL.

The TL comprises two elements: absorption and spreading (geometric) losses. Absorption is the irreversible loss of acoustic energy, ultimately through heating, as the sound wave propagates through the water: for example, as a consequence of the viscosity of the water. Losses due to absorption increase with increasing frequency. At frequencies considered in this chapter (<1 kHz), the effect of absorption is usually very small and can be neglected: for example, at 1 kHz, sound must travel over 16 km for absorption to account for a 1 dB loss (Urick, 1975).

Spreading losses account for geometric effects; specifically, as the sound wave moves away from the source, the energy it carries is distributed over a greater surface area; so, on average, the energy density reduces, decreasing the RL. A simple heuristic model used widely to model for spreading losses has the form

$$TL = K \log_{10}(R/R_0)$$

where R represents the range between source and receiver, and the factor K is a constant that characterizes the environment. Two extreme idealized scenarios are commonly cited. (1) In the case where there are no reflecting boundaries and the speed of sound is constant throughout the medium, the energy leaving the source spreads out on the surface of a sphere, a process called *spherical spreading* and corresponding to the choice $K = 20$. (2) In a medium bounded by upper and lower interfaces that reflect or refract sound in a lossless manner, a so-called *cylindrical spreading* model, $K = 10$, may be considered. Using a spherical spreading model may be considered at short ranges in deep water, but the straightforward application of a cylindrical spreading is inappropriate in most practical (field) applications. A suitable intermediate value of K is sometimes determined by employing numerical-propagation models run over various scenarios, which account for uncertainties in environmental characterization. The resulting values of K typically lie between 15 and 20. An example review of computer modeling of underwater acoustics can be found in Jensen et al. (2011).

This chapter chooses to report rms SLs for vocalizations of different baleen whale species, as this is a key factor in determining the detectability of a call. The SLs recorded for the same species differ by very significant margins. This is a consequence of the innate variability in call level, allied to methodological differences in the approaches to the calculation of the SL. For instance, some species change the level of their vocalizations in response to an external anthropogenic noise stimulus or elevated natural ambient noise, known as the *Lombard effect* (Dunlop et al., 2010; Helble et al., 2020; Thode et al., 2020). Some investigations into the Lombard effect have reported a nearly 1 dB increase in SL for every dB increase in the background noise level (e.g., Dunlop, 2016). There is also evidence of interference between signals propagating along direct and surface-reflected paths (the *Lloyd's mirror effect*), which should be accounted for (Charif et al., 2002). The degree to which the two factors (biological and methodological) contribute to the observed variation in SL is unclear.

11.2.3. Whale-Sound Analysis

The analysis of cetacean vocalizations is usually conducted by visually inspecting calls in the form of a spectrogram augmented by listening to recordings. Quantitative descriptions of the frequency content of sound are usually constructed by inspection of the spectrogram. The spectrogram is an image-based representation of a sound created using a short-time Fourier transform. Over many years, across various applications, the spectrogram has proven to be an effective tool for revealing the character of signals. To compute a spectrogram, the signal is divided into overlapping segments of a fixed duration. Each segment is then windowed, and the squared magnitude of the Fourier transform of the windowed segment is computed. This provides an estimate of the spectrum of the signal for different points in time, yielding a representation of signal energy as a function of two variables: time and frequency. The spectrogram is plotted herein with time on the horizontal (x axis) and frequency on the vertical (y axis), with energy on a dB scale conveyed through the use of a color scale. The scale selected here is ordered such that darker hues correspond to lower energies, and lighter hues represent higher energies.

Despite its great utility, using a spectrogram is not without problems. Display settings, such as the dynamic range of the color scale or choice of color map, can render signal features harder or easier to discern. Normalization of the spectrogram can also be undertaken, with the goal of emphasizing some signal features at the expense of others. More fundamentally, the choice of segment duration—that is, window length (used to compute the spectrogram)—affects the nature of the visual result. Specifically, the ability of a spectrogram to separate two components within a signal—that is, its resolution—can be considered with respect to both time and frequency, with each of these resolutions being defined by the duration of the window. Use of a long-duration window yields a spectrogram with good frequency resolution but poor time resolution, whereas a short-duration window offers the converse. This phenomenon is encountered in speech processing, where one approach is to consider spectrograms with different window durations to observe the differing properties of human speech (Rabiner, 1978). An illustration of

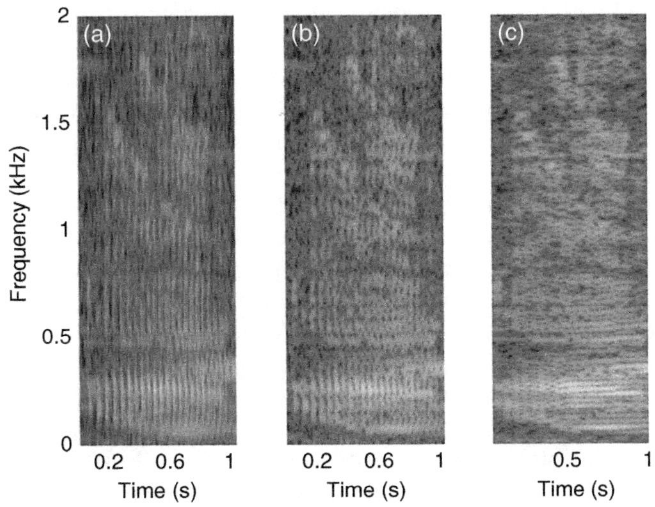

Figure 11.1 Spectrograms of an amplitude-modulated call, analyzed using different window durations: (a) 23 ms, (b) 46 ms, (c) 93 ms.

how window duration can change the nature of the spectrogram is provided in Fig. 11.1. This is an example of an amplitude-modulated call, which consists of a series of regular pulses spaced approximately 40 ms apart (i.e., repeating roughly 25 times per second, corresponding to a pitch frequency of approximately 25 Hz). When the short window duration (23 ms) is used, individual pulses appear as vertical lines, representing a burst of energy at the epoch of each pulse. Conversely, when the longer window is used (93 ms), horizontal lines are visible, which constitute the harmonics of the pitch frequency. In general, the choice of optimal spectrogram-window duration depends on the signal type (e.g., a pulsed call or a high-frequency-continuous whistle) and the specific features of the signal that are of interest: there is no universally optimal choice of window duration.

11.3. MARINE MAMMAL VOCALIZATIONS

A range of soniferous (sound-producing) aquatic animals contributes to the biological soundscapes of the ocean. These include over 700 species of fish (Luczkovich et al., 2008) and invertebrates (Au & Banks, 1998); however, marine mammals, particularly cetaceans, are acoustically reliant animals that utilize sound for prey detection, navigation, and communication. In fact, sounds from fin whales provided the stimulus for much of the early progress in the design of equipment and techniques for acoustic observations at sea (Watkins, 1981).

All species of baleen whales produce low-frequency sounds (e.g., moans, thumps, and knocks) in the 10–200 Hz range (Au, 2000; Thompson et al., 1979). Some mysticete species also produce pulses, chirps, and "songs" at frequencies ≤10 kHz (Thompson et al., 1979). Toothed whales (i.e., dolphins and porpoises), on the other hand, produce communication and echolocation sounds at higher frequencies, including (1) narrow-band, frequency-modulated (FM) continuous tonal sounds referred to as *whistles* (0.5–80 kHz), and (2) broadband sonar clicks (0.25–220 kHz) including burst-pulse sounds (Au, 2000; Gordon & Tyack, 2002).

Cetaceans use different sound frequency bands for several activities, including communication, navigation, foraging, and a range of activities within the wider social group, such as group bonding, warnings, and maternal relationships (André et al., 2010; Erbe et al., 2018; NMFS, 2018; Southall et al., 2007; Southall et al., 2019). Sound perception and production are also important for seals and sea lions (Mikkelsen et al., 2019; Reichmuth et al., 2013; Schusterman & van Parijs, 2003; Vincenzi et al., 2019). In most cases, the hearing range of cetaceans is poorly understood, but it is assumed that generally they hear over frequency ranges similar to the sounds they produce. Toothed whales are considered to be more sensitive to underwater sound, producing sound at higher frequencies than all other marine mammals, especially large whales (NMFS, 2018; Southall et al., 2019), which are the focus of this chapter.

In recent years, recognizing the importance of sound for aquatic fauna has led to the development of methods for characterizing marine soundscapes (Miksis-Olds et al., 2018). Such studies use long-term recordings of underwater sound to provide information on the behavioral patterns of marine life over temporo-spatial scales. An appreciation of the role of sound within aquatic ecosystems has raised concerns about the impacts of anthropogenic noise on marine life. This has led to greater regulation of many anthropogenic activities in the marine environment. In the context of geophysical applications, primary concern has focused on the impact of seismic surveys, one consequence of which is the increasing levels of marine-mammal monitoring required as part of the licensing processes (Todd et al., 2015). Nonetheless, there are synergies between geophysical work and the study of marine life, which can and should be exploited usefully. For example, ocean-bottom seismometers (OBSs) collect data that can inadvertently (or possibly intentionally) contain data from marine mammals, as discussed in Fisher et al. in this volume, permitting monitoring of animals over large temporo-spatial scales. Conversely, whale calls can also be used as a source of opportunity to extract geophysical measurements of the subsurface structure in a noninvasive manner (Kuna & Nábělek, 2021).

The purpose of this chapter is to describe vocalizations of species that may be encountered during geophysical survey data collection, and hence our focus on frequencies below 1 kHz (Ampilov et al., 2019). Soniferous fish species produce sounds in this band; individual fish calls tend to be much quieter than those of baleen whales. Aggregations of fish occur at certain locations: for example, at coral reefs, where choruses from many fish can be a dominant feature in the soundscape (Parsons et al., 2016). For the purposes of this chapter, sounds of baleen whales (mysticetes) will be discussed in detail, as these are the animals most readily detected and identified, together with information about the whale's movements and behaviors having the potential to be of scientific utility.

11.3.1. Sirenia and Carnivora

Sirenians (manatees and dugongs) produce low-frequency underwater vocalizations (Sousa-Lima et al., 2002), but these are relatively quiet and are produced mainly in shallow water. Thus, they are unlikely to be recorded during the vast majority of geophysical deployments. Polar bears and sea otters appear to produce sounds in the air only (Reidenberg & Laitman, 2010), and consequently, most studies have focused on the hearing range in the air (e.g., Reichmuth, 2008). Some pinniped species are vocal underwater and can be heard widely, especially in polar regions (see the review by Miksis-Olds et al., 2016, for examples). These calls commonly contain energy in the frequency band up to 1 kHz.

11.3.2. Toothed Whales

Unlike baleen whales, toothed whales produce powerful, high-frequency, highly directional echolocation clicks (sonar) to navigate and locate prey by personifying their surroundings (Thomas et al., 2004). Echolocation clicks have some distinctive features that coincide with the characteristics of manmade sonar systems designed to form high-resolution images. Echolocation pulses tend to be short-duration, narrowband, highly directional, and high frequency (in most cases, center frequencies >10 kHz). As a result of their high-frequency content, such sounds are unlikely to be recorded on conventional geophysical recording systems.

Although most toothed whale vocalizations are high frequency in nature, a few exceptions contain some energy below 1 kHz. Examples include sperm whales (*Physeter macrocephalus*) (Madsen et al., 2002), killer whales (*Orcinus orca*) (Ford, 2011), and long-finned pilot whales (*Globicephala melas*) (Courts et al., 2020); however, most low-frequency-cetacean (<1 kHz) calls arise from baleen whales, and these form the focus for the remainder of this chapter.

11.3.3. Baleen Whales

Baleen communication and vocalizations differ from those used for echolocation by toothed whales. Calls tend to be lower in frequency, longer in duration, and much less directional. These properties are consistent with use for long-distance communication since they tend to be detectable at longer ranges. There is, however, some evidence that humpback whales use low-frequency acoustic trains of pulses and buzzes called *megapclicks* produced during night-time foraging (Stimpert et al., 2007).

The link between vocal repertoire and taxonomy is inconsistent across baleen whale species (Reeves et al., 2004). For example, biogeographical characterization of blue-whale (*Balaena musculus*) songs worldwide has been achieved to identify populations (McDonald et al., 2006). There are also clear and apparent differences in sounds emitted by humpback whale populations in different hemispheres/ocean basins. However, animals are not sufficiently genetically distinct to be currently regarded as subspecies (Winn et al., 1981). Conversely, right whales comprise three different species, but their vocalizations overlap to the extent that no clear differentiation is available (Clark, 1982; McDonald & Moore, 2002; Parks & Tyack, 2005). Knowledge gaps in genetics and acoustic behavior are likely responsible for some of this inconsistency.

General Behavior

Many baleen whales undertake significant migrations, best documented in species such as humpback and gray (*Eschrichtius robustus*) whales, which favor coastal waters where seasonal presence has long been recognized and is most easily studied (Brown et al., 1995; Chittleborough,

1953; Modest et al., 2021; Rasmussen et al., 2007; Sommerville, 1994). Movements of more arcane, deep-water species, such as sei whales (*Balaenoptera borealis*), are less well understood (Olsen et al., 2009; Prieto et al., 2014). Typically, animals spend summer months feeding in the highly productive waters of higher latitudes, overwintering in the more benign conditions of the lower latitudes, where they calve and mate (Ramp et al., 2015). For species that have stocks in both hemispheres, such as humpback and fin (*Balaenoptera physalus*) whales, this pattern of movement means there may be limited opportunity for northern and southern populations to mix (Aguilar & García-Vernet, 2018; Reeves et al., 2004). In many species, such as bowhead whales (*Balaena mysticetus*), movement tends to be constrained within a single ocean basin so that populations in different basins in the same hemisphere may also remain largely isolated (Clapham et al., 1999). These barriers to intermixing can create genetic and/or vocal divergence in populations (McDonald et al., 2006; Rosel et al., 2021).

Before embarking on a species-level description of baleen-whale vocal characteristics, the following two subsections provide an overview of some common behaviors and acoustic features shared by many of the species.

General Sound Types

Whales produce both (intentional) vocal and nonvocal sounds and unintentional (nonvocal) noise. For example, surface activities such as breaching tail/fin slapping in humpback whales produces loud percussive sound (Dunlop et al., 2008). Other noises include expelling air at the surface (blowing), such as by North Atlantic right whales (Parks & Tyack, 2005); and bubble production, or expelling air while underwater, such as by humpback whales (Reidenberg & Laitman, 2007).

The term *call* is used generally to describe vocalizations that do not form part of a song (see the following subsection); however, not all calls and songs are distinct in that some vocalizations may be used both as components in a song or as calls. The *vocal repertoire* of a species constitutes all intentionally emitted sounds. The observed repertoire may be context-dependent; for example, blue whales typically exhibit different vocal behaviors on feeding grounds, during migration, and on breeding grounds (Leroy et al., 2016; Oleson et al., 2007). Consequently, temporospatial recordings yield differences in call types, meaning that to capture the full repertoire of a species, recordings need to be made at multiple locations, during various seasons, and over long periods.

Vocalizations of cetaceans are frequently described using subjective or onomatopoeic terminology, employing descriptions such as "moans," "groans," "thumps," "clanks," "grunts," "creaks," etc. (Edds-Walton, 1997; Erbe et al., 2017; Todd et al., 2015). Descriptions that try to represent a sound's timbre include terms such as "metallic," as in sei whales (Nieukirk et al., 2016; Thompson et al., 1979). Although these descriptions can be effective, they are also open to misinterpretation since it is not clear that one researcher's "clang" differs from another researcher's "bang." Quantitative descriptions of calls are favored since they can facilitate comparison between different studies. Nonetheless, limited use of subjective terms can help to convey a sense of call quality and are especially useful when describing sounds that are more challenging to develop quantitative metrics for.

Stereotyped calls are vocalizations that are highly repeatable; their structure remains constant over long time scales and is consistent between animals within a population. Such calls are frequently species-specific, such as in North Pacific right (Mellinger et al., 2004) and minke (Mellinger et al., 2000; Moore et al., 2006) whales, and so can be used reliably for identification (Halkias et al., 2013). Conversely, a wide range of calls from several species defy classification, lying on a continuum (Payne & McVay, 1971). In these instances, to facilitate analysis, it is common practice to group such calls into broad classes, which tend to follow similar structures between species. Nonetheless, details vary, and there is commonly a lack of universal agreement between researchers as to how many—and which classes—are appropriate for a given species.

Calls can be analyzed by capturing a few salient summary statistics, usually extracted from a spectrogram or time-series data (Matthews et al., 1999). In the time domain, commonly considered statistics include call duration and interval between calls. These are usually supplemented with some frequency-domain metrics, which tend to be call-specific but try to capture the band of frequencies present and clarify at what frequency most of the energy is focused.

One distinction often drawn between vocalizations is whether the call is *narrowband* or *broadband*, although even these terms can present challenges. Narrowband vocalizations are considered (at any point in time) to comprise a small number of discrete-frequency components, such as harmonically related components, with the lowest-frequency element termed the *fundamental frequency*. In FM narrowband calls, the temporal evolution of the fundamental frequency is commonly measured. The nature of this frequency modulation is an additional basis on which to classify sounds, specifically considering whether the frequency increases with time (up-sweeping), reduces with time (down-sweeping), or stays approximately constant (constant frequency [CF]). Somewhat perversely, narrowband signals can contain energy across a wide band of frequencies because of their frequency modulations and/or harmonics.

The term *broadband call* is reserved for those that at any point in time distribute energy across a continuum of frequencies instead of a small number of discrete frequencies (harmonics). Within this class, it is possible to recognize specific subtypes, including transients: short-duration pulses, moan/groan type calls, and pulsed (or amplitude-modulated) sounds that are sequences of closely spaced pulses that are very close together, such as those presented in Fig. 11.1. Broadband sounds tend to be more difficult to characterize using a small number of quantitative metrics, making them harder for researchers to study. It is when studying these sound types that onomatopoeic terminology has the most benefit. These calls also tend to be harder to detect than equivalent narrowband calls because the energy emitted is distributed across a wider range of frequencies and is more readily masked by noise. Presently, there is little information about the species-specific characteristics of most broadband sounds produced by baleen whales.

Songs

In bioacoustics, the term *song* is considered widely in the context of birds, where the distinction is drawn between songs and calls. For this discussion, this categorization has been adopted for cetacean vocalizations, and songs are defined as "long, complicated, vocalizations produced by males in the breeding season" (Catchpole & Slater, 2003).

As in the avian world, the complexity of baleen whale song varies greatly between species. Singing occurs seasonally, primarily on breeding grounds and migration routes (Homfeldt et al., 2022). Any seasonally produced, repeated-patterned signals are often termed *song* (Stafford et al., 2008). Blue, fin, and minke whales (the three rorquals that sing) all share aspects of their singing behavior, including conservation of most song structures over time (Stafford et al., 2008). Of all baleen whale species, only humpbacks and bowheads sing varying and complex songs that change rapidly over time (Stafford et al., 2008).

This section describes the vocal behavior of baleen whale species by species. In several instances, distinct vocal characteristics of subspecies are also discussed, but the aim is not to provide a complete catalog of baleen whale vocalizations.

Humpback (*Megaptera novaeangliae*)

Humpback song has been detected successfully during geophysical surveys through seismic streamers in the Barreirinhas basin of Brazil (L'Her et al., 2017). Humpbacks exhibit one of the most interesting vocal behaviors of any animal species, and their song has infused modern culture to a point where it has become synonymous with the song of all whales. However, humpback song is much more of an exception than a rule when it comes to baleen whale vocalizations. The species also make a wide range of calls (Dunlop et al., 2007), which are not restricted to a single sex or age, seemingly produced throughout the year. In contrast, many other rorqual whale species (like fin and blue whales) have a limited vocal repertoire, characterized by a small number of stereotyped, simple-repetitive-call units (Tyack & Clarke, 2000).

The humpback whale is highly migratory, with significant populations in both the Northern and Southern hemispheres (Fleming & Jackson, 2011). Song emanates predominantly from males in the breeding season (Winn & Winn, 1978), on breeding grounds, and during migration (Norris et al., 1999), with some evidence of singing on feeding grounds (Tyarks et al., 2021; Vu et al., 2012).

Humpback song is formed around a complex hierarchical structure (Cholewiak et al., 2013; Payne & McVay, 1971), and it is extraordinary that all singing males in a local population sing the same song. Each "ballad" develops with time, and song elements are transmitted between different populations (Garland et al., 2011; Noad et al., 2000; Payne, 2000; Payne & Payne, 1985); although sharing of calls between populations primarily happens within ocean basins (Garland et al., 2011; Winn et al., 1981), there is also evidence of transmission between basins (Rekdahl et al., 2018).

The structure of humpback song was first described in the seminal work of Payne and McVay (1971). Songs comprise *units*, defined as "vocalizations with a period of silence between them," and these units form the building blocks of humpback song. The concept of a unit is common throughout baleen-whale vocalizations, in that individuals typically produce one sound and then pause before producing the next. This contrasts, for instance, with some avian species, whose vocal apparatus allows them to produce song continuously (Greenewalt, 1969). Humpback-song units are usually organized into *phrases* comprising a small number of units (typically two or three) repeated together. Several phrases are recited to form a *theme*, and a *song* consists of a set of themes that are repeated cyclically. The duration of an individual unit is variable but usually lasts 0.1–4 s. Song-cycle duration varies considerably, usually between 8 and 16 min (Payne, 2000). The song structure evolves over multiyear time scales, with a general trend for year-on-year increases in complexity, followed by rapid simplification (Allen et al., 2018; Garland et al., 2011; Payne, 1983).

Humpback song units comprise a few basic forms that have been described in several broad categories. For instance, one can use the following categories (originally proposed for social calls): low-frequency-narrowband sounds, mid-frequency-narrowband sounds, high-frequency narrowband sounds, amplitude-modulated sounds, broadband sounds, and repetitive sounds (Dunlop et al., 2007). Spectrograms of example song units from each of these groups are shown in Fig. 11.2. The first three call types shown in Fig. 11.2 comprise narrowband units. In

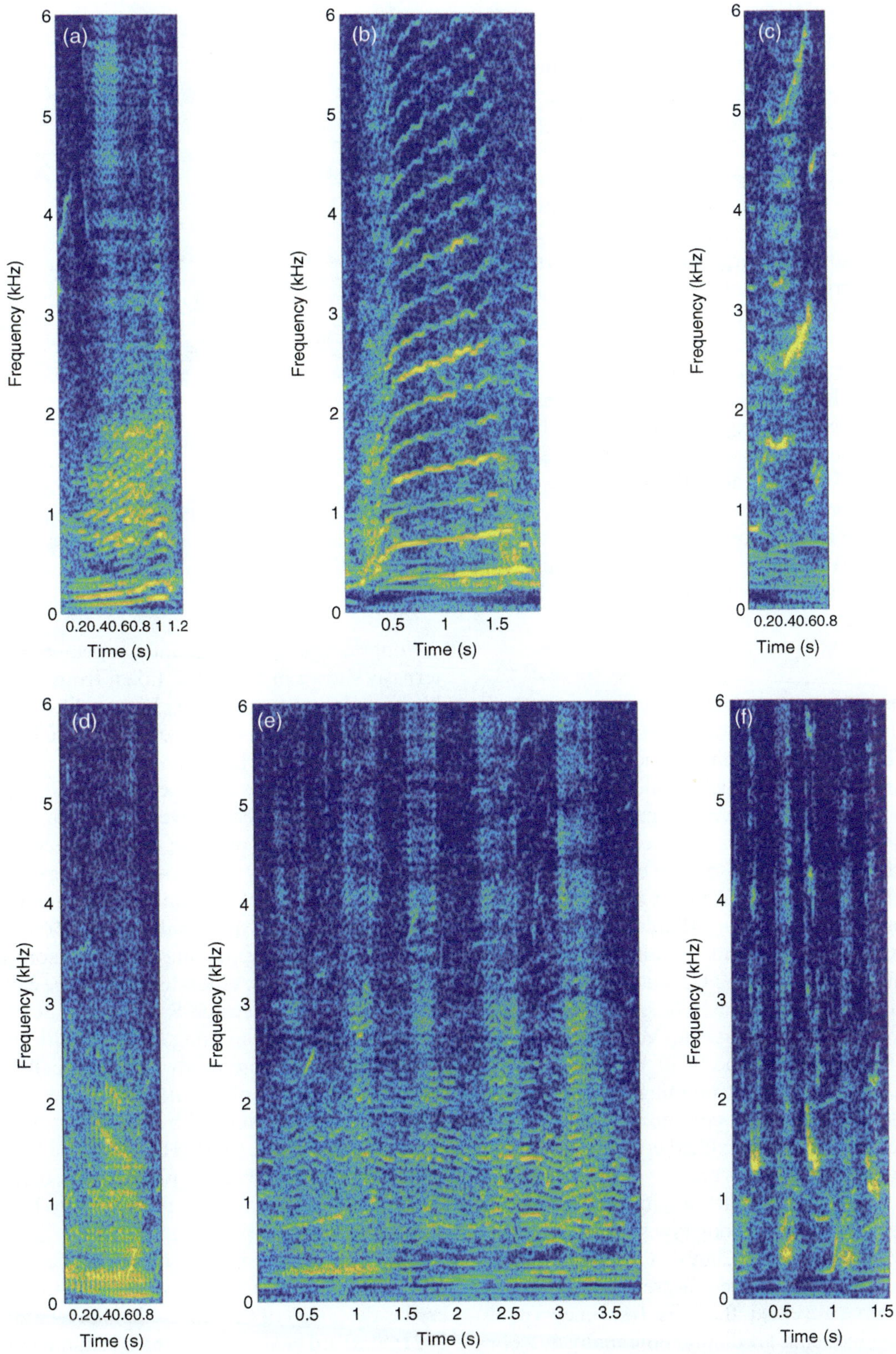

Figure 11.2 Spectrograms of humpback whale vocalizations (46 ms Hanning window, 90% overlap, 4096-point FFT, 22.05 kHz sample rate). Time, frequency, and amplitude (color) scales are common to all plots. Data includes units taken from a single song recorded in the St. Marie Channel, Madagascar, and is typical of the broad classes of humpback whale call types: (a) low-frequency call, (b) mid-frequency harmonic call, (c) high-frequency tonal call, (d) amplitude-modulated call, (e) complex-broadband call, and (f) repetitive call.

humpbacks, this type of call typically has several harmonics, as seen in Fig. 11.2a–c. Figure 11.2e presents an example of a broadband unit: in this case, an amplitude-modulated, or pulsed, call, which sounds like a modulated buzzing sound, becoming softer and louder throughout its duration. The final call, Fig. 11.2f, is an example of a repetitive unit—specifically, a sequence of short pulses that alternate in pitch—in this case, three high-pitched pulses interleaved with two of lower pitch.

Vocalizations contain energy between 100 Hz and 4 kHz (Tyack & Clarke, 2000). However, the precise range is ill-defined, and it is common to observe frequency components outside of this range in data recorded at good signal-to-noise ratios (SNRs) (Au et al., 2006).

Source levels within a humpback whale song vary considerably between units (e.g., 25 dB difference between the loudest and quietest units in a song) and between individuals (e.g., up to 10 dB spread in the mean SL_{rms} for one unit measured from different individuals) (Au et al., 2006). Overall, the rms source levels for song units range from 145 to 173 dB re 1 μPa.m. The SL_{rms} of social calls show a similarly wide spread of values (total range 124–184 dB re 1 μPa.m) (Fournet et al., 2018).

Blue (Balaenoptera musculus)

The blue whale is famed for being the largest animal to have existed on the earth, with individuals reported at lengths of 31.7–32.6 m (Mizroch et al., 1984a; Sears & Perrin, 2009). At least five accepted geographically distinct subspecies of blue whales are recognized (NOAA, 2020). Blue whale populations exist in both hemispheres, being found in the Atlantic (Davis et al., 2020), Pacific (Reilly & Thayer, 1990), Indian (Samaran et al., 2013; Torterotot et al., 2020), and Southern (McDonald et al., 2006; Širović et al., 2009) Oceans.

As befits its size, the blue whale produces loud, extremely low-frequency (infrasonic) vocalizations, with fundamental frequencies in the 5–30 Hz range; energy over 100 Hz is infrequent (Cummings & Thompson, 1971; Edds, 1982). The most commonly reported blue whale calls are narrowband (Stafford et al., 1999). Individuals sing, however, their songs are very different to those of humpback whales, with males singing all year round. There are at least 10 different song types, which vary geographically (McDonald et al., 2009) and appear to maintain a stable structure over time; however, observations over decades have revealed that the frequency (pitch) of calls has been reducing gradually, potentially as a consequence of stocks recovering from the effects of whaling and requisite increase in acoustic competition (McDonald et al., 2009). Alternatively, it has been posited that calls have changed in frequency because of elevated levels of anthropogenic noise (Leroy et al., 2018).

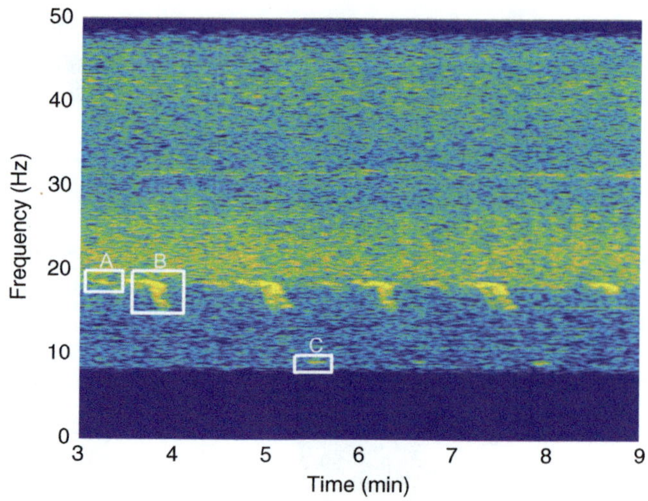

Figure 11.3 Spectrogram of a blue whale song recorded in the North Atlantic (10 s Hanning window, 90% overlap, 4096-point FFT, 100 Hz sample rate). Examples of the three units making up the song are labeled and shown in white boxes. Source: www.mobysound.org.

Songs consist of several units, repeated in a regular pattern, as shown in Fig. 11.3, taken from a North Atlantic blue-whale song sequence. Figure 11.3 shows a three-component structure, as described in Mellinger and Clark (2003), with units labeled A, B, and C. Units A and B typically last up to 20 s (Lewis et al., 2018). The pattern AB can be seen repeated five times, with a C call at 9 Hz, evident after three of these. Such AB combinations are observed frequently in blue-whale songs. Similar repetitions of a small number of units constitute the song of blue whales worldwide (McDonald et al., 2009), but in the various regions, units appearing within the song can differ. In addition to the song, blue whales produce calls (E. M. Oleson et al., 2007; Recalde-Salas et al., 2014). These can include units used in songs with additional call types. One of the most commonly observed calls is referred to as a *D call* (Berchok et al., 2006; Mellinger & Clark, 2003) and is characterized by a rapid (< 5 s) down-sweep from about 80 Hz to 30 Hz (McDonald et al., 2001), although the observed frequencies for these calls seems quite variable. D-calls are produced by both males and females in many populations and are believed to be associated with social interactions and feeding behaviors (Oleson et al., 2007). A further call type frequently considered is the *Z call* of the Antarctic blue whale (Leroy et al., 2021; Stafford et al., 2004), which is named after the shape of the call observed in a spectrogram.

There are a variety of estimates for mean blue-whale call SL, with SL_{rms} estimates ranging from 180 to 189 dB re 1 μPa.m (Cummings & Thompson, 1971; Davis et al., 2020; Samaran et al., 2010; Thode et al., 2020), suggesting that it is, at least on average, the loudest of all baleen whales.

Fin (Balaenoptera physalus)

The fin whale, or *finback,* as it is sometimes referred to (especially in North America), is second in size to the blue whale (Mizroch et al., 1984b). The fin whale is found in a large proportion of the world's oceans, favoring temperate and cooler waters, and is more scarce in equatorial regions than in extreme northerly or southerly latitudes (Wursig et al., 2008). The number of subspecies recognized is a matter of contention, with populations in the Northern and Southern hemispheres accepted widely as separate subspecies, and with the Pacific population constituting a third (Archer et al., 2019; Edwards et al., 2015). The fin whale is a cosmopolitan species, traveling great distances and swimming at high speed (Aguilar & García-Vernet, 2018).

The most frequently encountered fin whale vocalizations is the so-called *20 Hz pulsed call* (P. O. Thompson et al., 1992; Watkins, 1981a; Watkins et al., 1987). This sound was familiar to military observers for many years, although its source remained a mystery until the early 1960s when it was identified as originating from a fin whale (Watkins et al., 1987). In deep-water environments, these calls are one of the most frequently encountered (and readily identifiable) of the baleen whales; consequently, it is a fin that has been associated with most work in the context of OBS measurements and other seismic or geophysical-related studies (Iwase, 2015; Matias & Harris, 2015; McDonald et al., 1995; Wilcock, 2012).

The 20 Hz fin whale call is a narrowband, frequency modulated, down-swept pulse, lasting approximately 1 s, with a center frequency in the region of 20 Hz (Watkins et al., 1987); see Fig. 11.4. They are typically emitted in long pulse trains. Songs comprise regular sequences of pulses with occasional lower-frequency (and lower-amplitude) pulses called *back-beats* (Clark et al., 2002), emitted only by males (Croll et al., 2002). Interpulse intervals range between 10 and 20 sec and appear stable over time (Hatch & Clark, 2004), varying geographically, a feature which may assist in discriminating between stocks (Delarue et al., 2009b).

In addition to the 20 Hz and associated back-beat call, fin produces other vocalizations, including down-swept and tonal calls in the band between 30 and 100 Hz and a selection of other amplitude-modulated calls (P. M. Thompson et al., 1992; Watkins, 1981b).

Several measurements of SL_{rms} for the 20 Hz pulsed calls are available (Charif et al., 2002; Cholewiak et al., 2018; Širović et al., 2007; Wang et al., 2016), with estimated values covering a large (40 dB) range from 159 to 189 dB re 1 μPa.m.

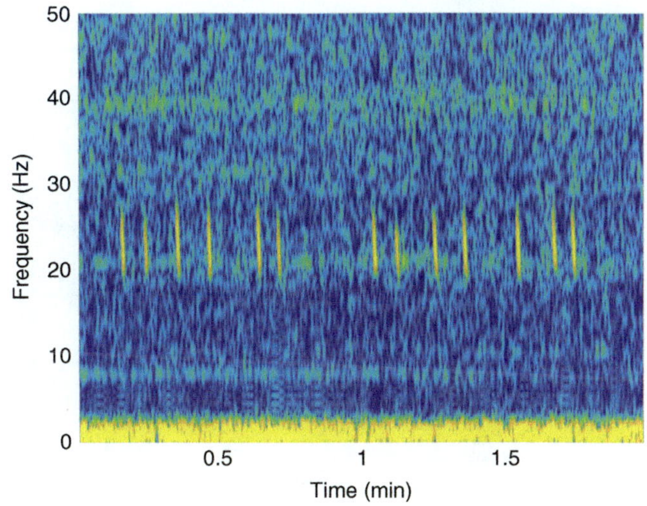

Figure 11.4 Spectrogram of sequence of fin whale 20 Hz calls (1 s Hanning window, 90% overlap, 1024-point FFT, 256 Hz sample rate) recorded from an OBS off the Galician coast in the Northeastern Atlantic. See Fisher et al. in this volume.

Bryde's (Balaenoptera edeni)

Bryde's whale is one of the smallest rorquals. Taxonomy remains a matter of contention (Rosel et al., 2021), with suggestions of two subspecies, Eden's whale (*B. edeni edeni*) and Bryde's whale (*B. edeni brydei*), or possibly separate species (Constantine et al., 2018; Sasaki et al., 2006; Wada et al., 2003). These species/subspecies are thought to represent a smaller in-shore and a larger off-shore form. Moreover, recent discoveries suggest that the population resident in the northern Gulf of Mexico should also be regarded as a separate species (Rosel et al., 2021). To complicate matters, there is insufficient acoustic data to draw distinctions between vocal repertoires of the different species/subspecies; therefore, for the purposes of this chapter, Bryde's are regarded as a *species complex* as a whole, accepting that this may be revised in the future.

Bryde's whale is found primarily between latitudes of 40° N and 40° S and was thought not to undertake large-scale movements (E. M. Oleson et al., 2003); however, more recent findings suggest that there is a seasonal poleward migration associated with ocean warming and climate change (Kerosky et al., 2012). This does not appear to correlate directly between whale presence and sea surface temperature but indirectly due to increased prey availability (Kerosky et al., 2012).

Bryde's whale known vocal repertoire consists of around 12 call types (Cummings et al., 1986; Edds et al., 1993; Heimlich et al., 2005; Kerosky et al., 2012; E. M. Oleson et al., 2003; Širović et al., 2014; Viloria-Gomora et al., 2015) consisting of tonal- and frequency-modulated (swept) calls, as well as broadband vocalizations (Širović et al., 2014). Tonal calls are

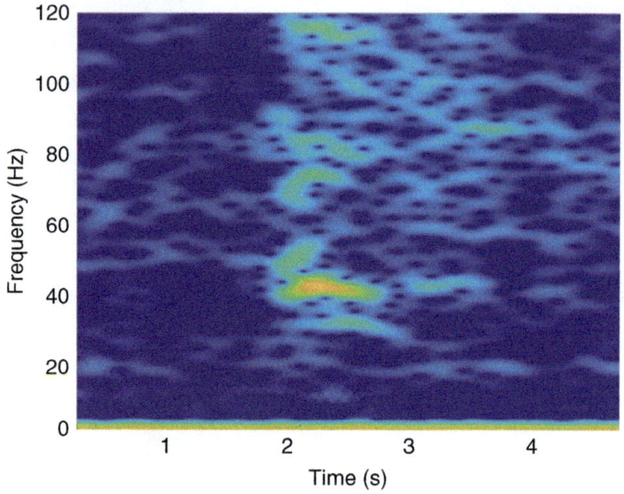

Figure 11.5 Spectrogram of a Bryde's whale vocalization (0.5 s Hanning window, 90% overlap, 128-point FFT, 249 Hz sample rate) similar to the Be2 call (Oleson et al., 2003). Source: www.mobysound.org.

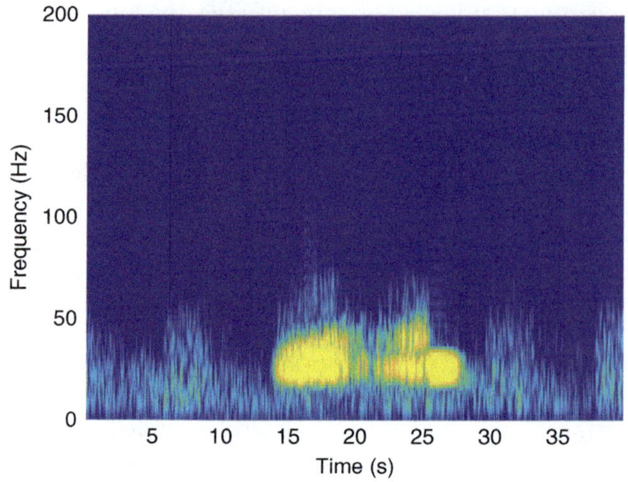

Figure 11.6 Putative spectrogram of an Omura's-whale call not verified by visual confirmation (0.2 s Hanning window, 90% overlap, 4096-point FFT, 6 kHz sample rate). Taken from a sequence of calls recorded off the coast of Western Australia in November 2012. Source: https://acoustic.aodn.org.au/acoustic/.

generally higher frequency than the 20 Hz fin whale call (Watkins et al., 1987) or the low-frequency (12.5 to 200 Hz) blue-whale calls (Cummings & Thompson, 1971), with some call types having fundamental frequencies above 100 Hz. Calls are lower frequency than most humpback vocalizations, typically comprising most energy between 100 Hz and 4 kHz (Tyack & Clarke, 2000). Figure 11.5 shows one example of a Bryde's tonal call, which forms part of a sequence recorded in the Eastern Tropical Pacific. Like the fin, some vocalizations are emitted in "pulsed pairs" as described in Širović et al. (2014). The vocal repertoire of Bryde's whale exhibits notable geographical variations (Heimlich et al., 2005; Viloria-Gomora et al., 2015).

Calls of Bryde's whale are quieter than other baleen-whale species, with p-p source levels ($SL_{p\text{-}p}$) estimated to be 155 ± 14 dB re 1 μPa.m for one particular down-swept (Be9) call (Širović et al., 2014) with the equivalent SL_{rms} lower than this. This data is broadly consistent with earlier observations for moans, which have SLs of 152–174 dB re 1 μPa.m, although it is unclear if these measurements are based on rms or p-p levels (Cummings et al., 1986).

Omura's Whale (Balaenoptera omurai)

Omura's whale was only discovered in 2003 (Wada et al., 2003) and is similar in appearance to Bryde's and sei; this species was first regarded as a dwarf subspecies of Bryde's whale. Consequently, there remains a large knowledge gap surrounding this species' taxonomy and ecology (Cerchio & Yamada, 2018). Omura's is believed to occur primarily in the Indo-Pacific region, derived from a clear physical description of live specimens (Cerchio & Yamada, 2018). Sighting rates are increasing, leading to more information about distribution and behavior (Cerchio et al., 2019). Much data is derived from strandings of animals in the Atlantic, from equatorial (Moreira et al., 2020), South (Cypriano-Souza et al., 2017), and West Africa areas. (Jung et al., 2015).

There is limited data on the vocal behavior of Omura's whale, and the fact that the animal is difficult to identify visually casts some doubt on the veracity of this data. It is assumed that Omura's produce low-frequency (15–50 Hz), amplitude-modulated calls composed of two distinct components (Madhusudhana et al., 2020), emitted in sequences with intervals of between two and three minutes (Erbe et al., 2017). There is no source-level information available at present. Figure 11.6 shows a spectrogram of what is tentatively identified as an Omura's call, recorded on an archival system without an associated visual confirmation.

Sei Whale (Balaenoptera borealis)

The sei whale closely resembles Bryde's; consequently, it was not until the early twentieth century that the two species were recognized as separate (Gambell, 1985). The sei is found in all ocean basins, with northern and southern populations sometimes regarded as subspecies (Erbe et al., 2017). Migrations are less well understood than those of, for example, humpback (Stevick et al., 2011) or blue (Abrahms et al., 2019) whales, but movements probably follow similar patterns for most other large baleen whale species. There is a possible migratory corridor between wintering grounds in the Azores and feeding grounds in the Labrador Sea (Prieto et al., 2014).

As with most whale species, the full sei vocal repertoire is unknown. Animals emit various broadband sounds, such as moans and grunts (McDonald et al., 2005; Rankin & Barlow, 2007), with higher-frequency tonal calls centered above 400 Hz (McDonald et al., 2005). The most frequently reported calls comprise downward-sweeping tonal signals, with a start and end frequencies of about 100 Hz and 40 Hz, respectively, lasting roughly 1 s (Baumgartner et al., 2008; Calderan et al., 2014; Johnson et al., 2014; Newhall et al., 2012; Rankin & Barlow, 2007; Romagosa et al., 2015). Calls are heard commonly in short sequences of 2–4 (Davis et al., 2020; Newhall et al., 2012), with an average-call interval of 3.5 s (Baumgartner et al., 2008). Up-sweeping vocalizations in the same frequency band have also been recorded and associated with sei whales (Calderan et al., 2014); however, a spectrogram of a sei call was not available for illustration here because of their rarity.

Efforts to compute sei whale SLs have focused on down-swept calls; most studies suggest SL_{rms} in the range of 170–183 dB re 1 μPa.m (Newhall et al., 2012; Romagosa et al., 2015; Wang et al., 2016). One study conducted in Antarctic feeding grounds estimated an anomalous SL_{rms} of 147–156 dB re 1 μPa.m (McDonald et al., 2005).

Common Minke (Balaenoptera acutorostrata)

The common minke is the smallest rorqual species (Perrin et al., 2018). Taxonomy is complex and ambiguous. Until the 1990s, common and Antarctic minke (see section 3.3) were considered the same species, but they are now regarded as separate (Rice, 1998); however, to obfuscate this neat dichotomy, both species can hybridize to produce fertile offspring (Glover et al., 2013). To add further uncertainty, according to the National Oceanic and Atmospheric Administration (NOAA), there are two other subspecies of common minke in the Northern Hemisphere: the North Pacific (*Balaenoptera acutorostrata scammoni*) and the North Atlantic (*B. a. acutorostrata*) minke. To complicate taxonomy further, there are supposed to be two forms of minke whales in East Australian waters. Finally, in tropical, temperate, and polar waters of the South Atlantic, South Pacific, and Indian Oceans, there is also a subspecies of the common minke, referred to as the dwarf minke whale (Arnold et al., 1987), which is currently considered an unnamed subspecies of the common minke whale.

All minke whales likely migrate, but since taxonomy is, putting it mildly, complex, verified movements are restricted to a handful of studies, which are mostly inferred (e.g., Pastene et al., 2010). What is regarded as a common minke is found throughout the oceans of the Northern Hemisphere in both shallower shelf seas and deeper waters, and again, probably follows the typical pattern of whale migrations.

It follows that full confidence in the species' acoustic repertoire is similarly problematical. The common minke is the source of a "boing" sound, detected first in the northern Pacific Ocean, reported by Wenz (1964); after mystifying scientists for 40 years, it was only identified recently as minke (Rankin & Barlow, 2005). Boings generally contain most energy in the 1–4 kHz range and "consist of a brief pulse followed by a long call that is both frequency and amplitude modulated" (Rankin & Barlow, 2005). Common minke whales in the North Atlantic produce sequences of pulses in a band up to about 500 Hz (Mellinger et al., 2000; Risch et al., 2013; Risch et al., 2014). Interpulse intervals (IPIs) are variable, with durations of tens of seconds on average, and intervals between pulses of around 0.5 s (Mellinger et al., 2000); IPIs can be constant, accelerating, or decelerating (Mellinger et al., 2000; Winn & Perkins, 1976). This long sequence of rapid pulses differs in character from pulse sequences observed from other rorquals. A spectrogram of one example of such a pulse train is shown in Fig. 11.7. The acoustic repertoire of the Southern Hemisphere subspecies, the dwarf minke whale, includes down-sweeps and a complex call, sometimes referred to as the *star-wars* (SW) call (Gedamke et al., 2001). More information on minke vocal behavior can be found in Martin et al. (2013), Oswald et al. (2011), and Risch et al. (2013, 2014, 2019).

The SL for some other sounds from common minke has been reported for high-frequency-isolated pulses (Beamish & Mitchell, 1973) and the North Pacific boing

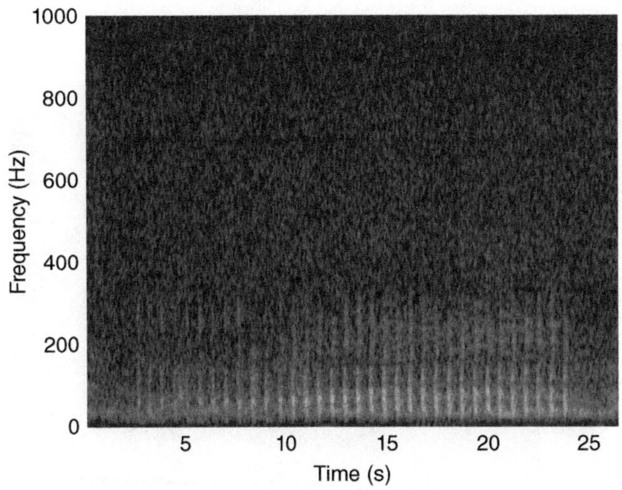

Figure 11.7 Spectrogram of a pulse train from a common minke whale (0.13 s Hanning window, 90% overlap, 1024-point FFT, 8 kHz sample rate). Recorded in the North Atlantic off the western coast of Scotland, collected as part of the COMPASS Interreg Project.

call (Helble et al., 2020); however, applicable here for low-frequency pulsed calls, SL_{rms} has been estimated to range between 159 and 178 dB re 1 μPa.m when computed across the whole pulse sequence, individual pulses being 2–3 dB greater than this (Cholewiak et al., 2018; Risch et al., 2014; Wang et al., 2016). The SL_{rms} of dwarf minke complex and stereotyped sound sequence star-wars vocalization are estimated to be somewhat lower than this, specifically 148–165 dB re 1 μPa.m (Gedamke et al., 2001).

Antarctic Minke Whale (Balaenoptera bonaerensis)

Consistent with its name, the Antarctic minke whale spends much of the summer in the Southern Ocean (Ainley et al., 2012). In the winter, it migrates to breeding grounds in warmer northerly waters (Kasamatsu et al., 1995), and it has also been reported to migrate from pole to pole (Glover et al., 2010). The Antarctic minke is distinguished from the dwarf minke (the Southern Hemisphere subspecies of the common minke) by its greater size, being roughly 2 m longer (Arnold et al., 1987); nonetheless, like so many of its ilk, its physical similarity means that it can be challenging to discriminate between these two species in field conditions.

The lower polar accessibility of waters occupied by the Antarctic minke whale means its vocal behavior is more poorly described than that of many other baleen whales. Fifty years ago, the pioneering whale acoustician duo Bill Schevill and Bill Watkins from the Woods Hole Oceanographic Institution (WHOI) recorded Antarctic minke whales making down-swept calls (Schevill & Watkins, 1972). More recently, Risch et al. (2014) reported that the Antarctic minke was the source of the so-called *bio-duck* vocalization. Both calls have been observed to have significant diel (24-hour) and seasonal variability (Dominello & Sirovic, 2016; Shabangu et al., 2019).

Figure 11.8 shows a spectrogram of a typical bio-duck call from an Antarctic minke whale. In Fig. 11.8a, the structure of the call can be seen across timescales of tens of seconds, with bursts of sounds lasting approximately 2 s and silences of about 1 s, with energy between 100–200 Hz. Note the unrelated fin-whale 20 Hz calls marked in the white square of this plot. Fig. 11.8b expands the dotted area in frame Fig. 11.8a, revealing that each burst of sound consists of a sequence of pulses, with IPIs of 100–200 ms.

SL estimates for the Antarctic minke are scarce but appear to overlap with the SL_{rms} estimated for common minke; specifically, the down-sweeps are reported in the range 160–165 dB re 1 μPa.m (Schevill & Watkins, 1972).

Right Whales (Eubalaenidae)

There are three species of right whale (Gaines et al., 2005; Rosenbaum et al., 2000), which are all highly endangered (Meyer-Gutbrod & Greene, 2017; Pennisi, 2017) and on the International Union of the Conservation of Nature (IUCN) red list. These are the North Atlantic right whale (*Eubalaena glacialis*), the North Pacific right whale, *E. japonica* (Marques et al., 2011), and the Southern right whale, *E. australis* (Crespo et al., 2018). All species are restricted (mostly) to specific geographical regions and rarely intermix (Gaines et al., 2005; Rosenbaum et al., 2000).

All right whale populations were devastated by whaling activities of the last century (Carroll et al., 2014). This is

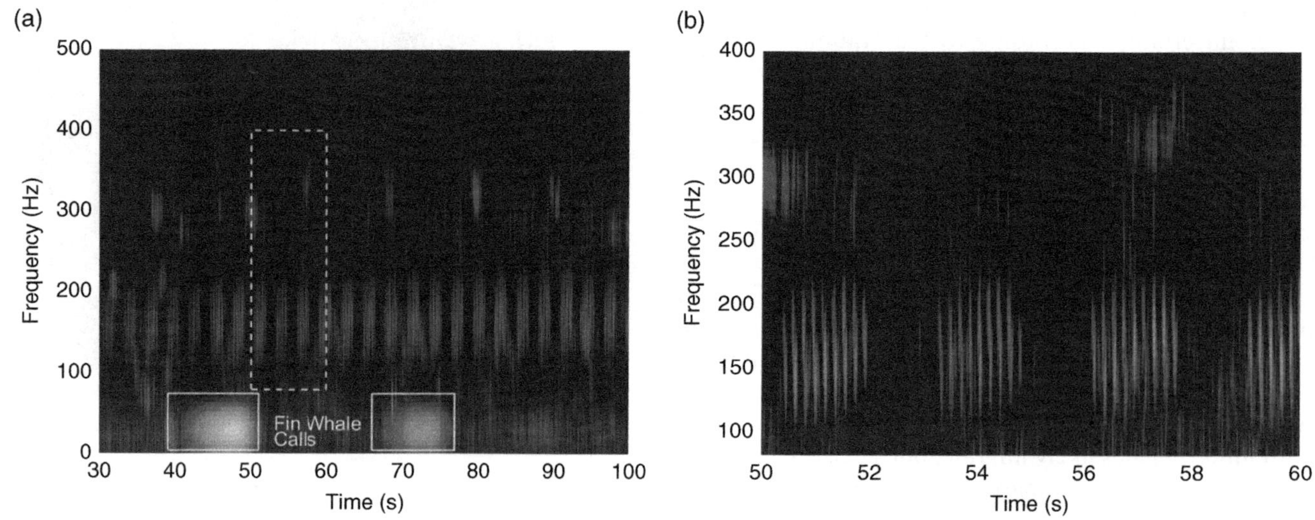

Figure 11.8 Antarctic minke whale bio-duck call (40 ms Hanning window, 90% overlap, 1024-point FFT, 6 kHz sample rate). (a) A section of 70 s of recorded data, with incidental fin-whale calls indicated in white boxes close to 20 Hz. (b) Expanded view of the area in a dotted box in (a). Source: https://acoustic.aodn.org.au/acoustic/.

because they are slow swimming and favor accessible coastal waters, and their bodies are positively buoyant post mortem as a consequence of being rich in oil (Cummings, 1985)—factors that contributed to them being the "right" whale to hunt. Numbers have been slow to recover (Crespo et al., 2018), at least in part because of low reproduction rates (Kraus et al., 2001) and continued threats from entanglement and ship strikes (Knowlton & Kraus, 2001). Consequently, these animals' survival remains precarious.

The three right whale species are closely related (Gaines et al., 2005), and their vocal repertoires are thought to be qualitatively similar; consequently, sounds from the three species have not been differentiated for the purpose of this chapter. Moreover, since there is little geographical overlap between the three species, detection on a hydrophone provides sufficient information to identify the species with high confidence.

Right-whale calls are usually divided into two broad categories: *stereotyped* calls—in this context, discrete calls—and *variable* calls, lying on a continuum (Clark, 1982; Payne & McVay, 1971). The right whales are also known to sing, insomuch they can produce regular patterned sequences of calls (Crance et al., 2019).

The most useful calls by which to identify the right whale are stereotyped calls (Matthews & Parks, 2021), the two most commonly researched of which are the *up* (Calderan et al., 2021) call (see Fig. 11.9) and the *gunshot* call (Parks et al., 2012).

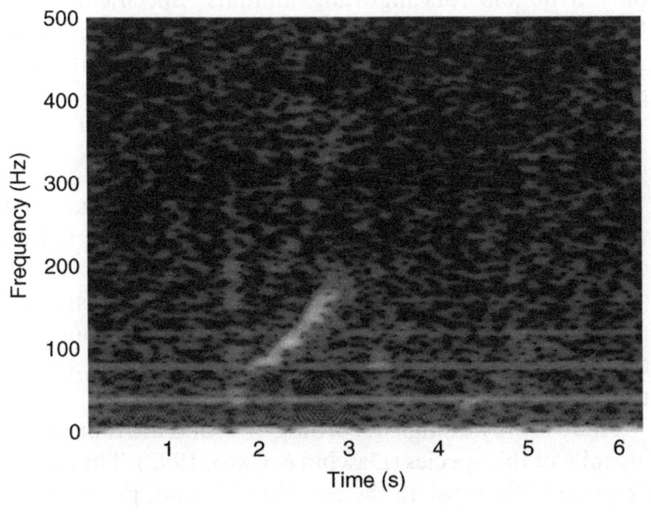

Figure 11.9 Spectrogram of an up call from a North Atlantic right whale recorded off the coast of Massachusetts (256 ms Hanning window, 90% overlap, 512-point FFT, 2 kHz sample rate). Note the tonal self-noise horizontal lines below 150 Hz. Source: Detection, Classification, Localization and Density Estimation of Marine Mammals using Passive Acoustics (DCLDE) Workshop database 2013 (https://soi.st-andrews.ac.uk/dclde2013/).

The up call is a narrow-band vocalization, believed to be a contact call, demonstrated through up-call playback experiments when animals were attracted to an underwater speaker (Clark & Clark, 1980). Up calls are emitted frequently by animals of either sex or age. Frequency sweep starts at approximately 70 Hz and finishes at roughly 170 Hz, although this range appears to be evolving with time, likely in response to changes in anthropogenic noise (Parks et al., 2007). Call duration is 2–3 s. The characteristic nature of the up call means it is well suited for monitoring animal presence: for instance, to reduce risks associated with ship strikes (Soldevilla et al., 2014). The broadband, short-duration, transient gunshot call appears to be related to breeding behavior (Parks et al., 2005). These calls span a frequency range from 20 Hz up to about 20 kHz and last approximately 200 ms.

The SL_{rms} of the up call has been estimated to be between 150 and 178 dB re 1 μPa.m, whereas gunshot calls have been estimated to have $SL_{p\text{-}p}$ of 174–192 dB re 1 μPa.m, with SL_{rms} estimates for the other call types in the range 137 to 162 dB re 1 μPa.m (Munger et al., 2011; Parks & Tyack, 2005). Note that the gunshot is reported in terms of a p-p SL, a choice consistent with its transient nature, but this means it should not be directly compared to other SLs given here.

Bowhead Whale (Balaena mysticetus)

The bowhead whale is an arctic/subarctic species, with populations found throughout the northern polar waters, following the pattern of the advancing and retreating ice shelf. This species is similar physically to the right whale and, as such, was originally called the Greenland right whale; the two were not separated from each other as independent species until the mid-nineteenth century (Reeves & Leatherwood, 1985).

The bowhead whale is vocally distinct from right whales. Animals produce songs in the breeding season, with a structure not dissimilar to that of the humpback whale, albeit much simpler in form. Songs are shorter in duration (roughly 1 minute), and each cycle contains fewer (up to 10) units (Cummings & Holliday, 1987), with bouts of singing lasting hours (Stafford et al., 2008). Like that of humpbacks, bowhead song is repeated cyclically, involves patterns of calls (Julien Delarue et al., 2009), and song structure evolves over time, with individuals in one group singing the same song.

Bowhead vocalizations cover a wide frequency range from 25 Hz to 4 kHz. Call form is graded—that is, lies on a continuum—and so is often placed into a small number of broad categories that are not classified further (Ljungblad et al., 1982; Stafford et al., 2008). Classes resemble those defined for other whale species and are classed typically as simple moans (near-tonal FM calls), complex moans (broadband calls, including amplitude-modulated

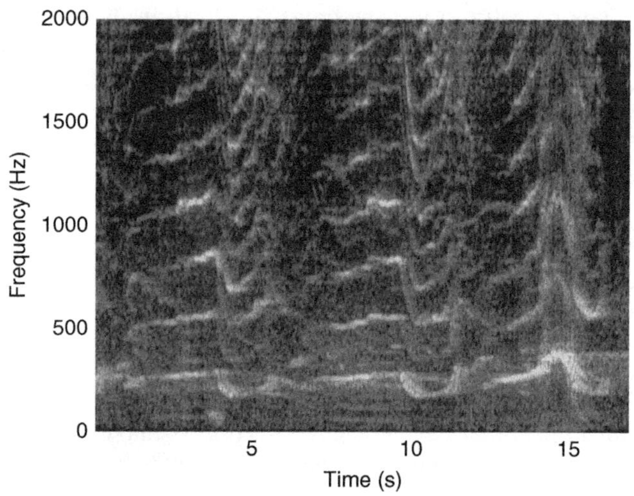

Figure 11.10 Spectrogram of bowhead song clip containing three frequency-modulated units (100 ms Hanning window, 90% overlap, 1024-point FFT, 10.24 kHz sample rate). Source: WHOI Watkins Library (https://cis.whoi.edu/science/B/whalesounds/index.cfm).

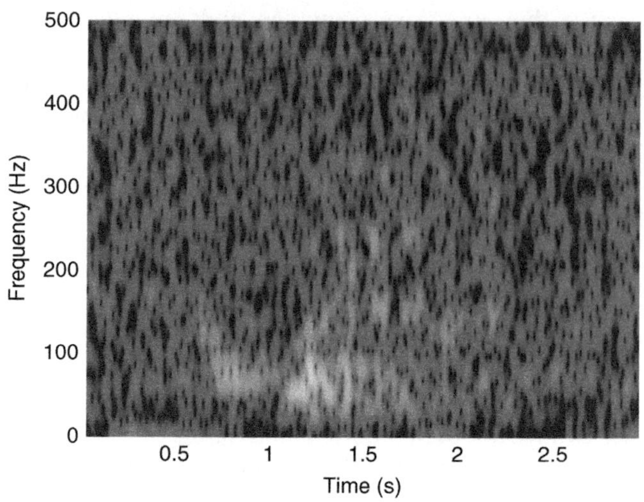

Figure 11.11 Spectrogram of M3 call for a gray whale (64 ms Hanning window, 90% overlap, 4096 point FFT, 64 kHz sample rate), data supplied as supplementary material for (Burnham et al., 2018).

sounds), gunshots (Delarue et al., 2009), and miscellaneous calls, including a *warbling call* (Fig. 11.10). Alternatively, some authors classify calls into a larger number of smaller categories (Cummings & Holliday, 1987).

The SL_{rms} for different bowhead whale call types vary between 130 and 165 dB re 1 μPa.m and have been shown to be affected by the ambient noise environment: that is, bowheads exhibit the Lombard effect (Thode et al., 2020).

Gray Whale (Eschrichtius robustus)

The gray whale is found primarily in the coastal waters of both sides of the northern Pacific Ocean; it became extinct in the Atlantic in the eighteenth century (Bryant, 1995). The western Pacific population is classed as endangered, while the eastern population is close to full carrying capacity (Moore et al., 2001). These animals undertake extremely long annual migrations, the eastern population feeding in the Bering and Chukchi seas in summer and breeding in Baja, California in winter (Pike, 1962; Rugh et al., 2001).

Gray whale vocal behavior changes considerably depending on whether animals are feeding, migrating, or on breeding grounds (Guazzo et al., 2017). Fewer vocalizations are produced on northern feeding grounds than at other times (Dahlheim, 1987). Whales produce sounds in the 20 Hz to 2 kHz range, and call nomenclature stems from the work of Dahlheim (1987), wherein an alphanumeric notation is used. Letters (N, S, or M) signify whether the call is heard in the North, in the South, or on Migration, and the number represents the call type. Five vocalizations are recognized (and two sounds are associated with surface activities). These sounds are stereotyped except for the S1/M1/N1 call, which consists of "knocks" and pulses, usually observed in sequences, with an average duration of about 2 s and a mean of 9.4 pulses per sequence. Other sounds produced are rapid-frequency sweeps, amplitude-modulated calls, and sequences of clicks (Fish et al., 1974). The spectrogram of an example M3 call is shown in Fig. 11.11.

Reliable SL information for gray whales is currently only available for migrating animals. Specifically, the M3 call, one of the amplitude-modulated calls, has been studied, and SL_{rms} are estimated as 150–165 dB re 1 μPa.m (Guazzo et al., 2017).

Pygmy Right Whale (Caperea marginata)

The pygmy right whale is not related closely to right whales, despite their name. This is the smallest baleen whale species, measuring up to 6.5 m in length (Fordyce & Marx, 2012; Kemper, 2018). A cryptic and understudied animal, it is believed to reside throughout the Southern Ocean (Kemper, 2002) and does not appear to migrate (Kemper, 2009).

There is only a single recording of sound attributed to a juvenile of this species (Dawbin & Cato, 1992). That study described "thumps" in the 50–200 Hz band, produced in short sequences of 2 or 3 calls. SLs are reported as p-p values ($SL_{p\text{-}p}$) and are given as 165–179 dB re 1 μPa.m.

11.4. CONCLUSIONS

Whale calls most likely to be recorded on instruments designed for geophysical surveys are those of fin and blue whales since these are the loudest and lowest-frequency of

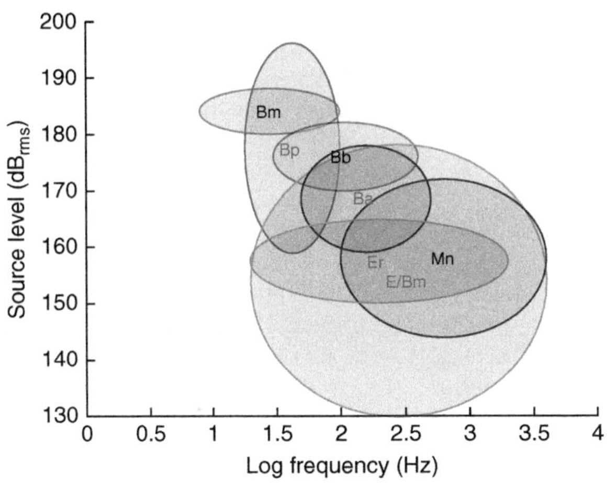

Figure 11.12 Plot of baleen whale range of rms source levels and frequencies: Mn = humpback whale, Bm = blue whale, Bp = fin whale, Bb = sei whale, Ba = combined common and Antarctic minke whale, Er = gray whale, and, E/Bm = combined right and bowhead whale.

the baleen whale species. The stereotyped nature of most calls from these two species makes them readily identifiable. More generally, a variety of other sounds may be observed from baleen whales <1 kHz.

Figure 11.12 illustrates the range of SLs and approximate bandwidths for baleen whales. Ellipses show limits in terms of (logarithmic) frequency and rms source level reported for each species or species group.

To maintain clarity in the construction of Fig. 11.12, some compromises were made, such as the grouping of species: minke whales and bowheads with rights. The spread of SLs is only slightly exacerbated by the combining of species. Labels for each group at the center of the ellipses provide a simple measure of typical SL and representative frequencies for these species' groups. Figure 11.12 should not be mistaken for a full meta-analysis, from which a true mean SL could be estimated; rather, it is intended as a visualization of some available data. Generally, balaenopterids emit sounds with greater SL_{rms} than other baleen whales, with SLs for these species ordered according to physical size. Data on typical values reported in the remaining baleen-whale species are clustered in a similar area of the plot, albeit with considerable variation in reported metrics.

In general, baleen-whale vocalizations are distinctive when recorded at good SNRs but can be somewhat more challenging when animals are detected at greater ranges. In most circumstances, if information on location and time of year is associated with a recording, species identification can be assigned to a few candidate species. This is because repertoires of many baleen-whale species include stereotyped vocalizations, which are diagnostic at the species level. Although recent advances in research and software have furthered automated detection and classification methods—particularly approaches that build on accelerated development of machine learning and artificial intelligence—classification typically still requires considerable manual intervention, and processing long time series is currently unrealistic. Nonetheless, there are clear opportunities for geophysical surveys to supplement marine-mammal-acoustic data collection, adding to knowledge of the behavior and distribution of these magnificent creatures.

ACKNOWLEDGMENTS

PRW would like to acknowledge the tireless efforts of the people who generously collected, curated, and provided access to the recordings used as examples in this work. Data were sourced from Australia's Integrated Marine Observing System (IMOS), enabled by the National Collaborative Research Infrastructure Strategy (NCRIS), operated by a consortium of institutions as an unincorporated joint venture, with the University of Tasmania as lead agent. Data was also collected as part of the Collaborative Oceanography and Monitoring for Protected Areas and Species (COMPASS) project, supported by the European Union's (EU) Interreg VA Programme, managed by the Special EU Programmes Body. The charity Cetamada provided support for further data collection in Madagascar. Individuals who have been of direct help in these endeavors include Tim Minshull, Gaye Bayrakci, Ellen White, Federica Pace, Kate Reid, Olivier Adam, and Denise Risch. VLGT would like to thank Ocean Science Consulting's Research Assistants, Laura Park, Zoe Hodgson, Tierney Carter, and Katie Rapson for sourcing additional citations and Laura Williamson for photo contributions.

REFERENCES

Abrahms, B., Hazen, E. L., Aikens, E. O., Savoca, M. S., Goldbogen, J. A., Bograd, S. J., et al. (2019). Memory and resource tracking drive blue whale migrations. *Proceedings of the National Academy of Sciences*, *116*(12), 5582. https://doi.org/10.1073/pnas.1819031116

Aguilar, A., & García-Vernet, R. (2018). Fin whale: *Balaenoptera physalus*. In B. Würsig, J. G. M. Thewissen, & K. M. Kovacs (Eds.), *Encyclopedia of marine mammals* (pp. 368–371). Elsevier. https://doi.org/10.1016/B978-0-12-804327-1.00128-X

Ainley, D. G., Jongsomjit, D., Ballard, G., Thiele, D., Fraser, W. R., & Tynan, C. T. (2012). Modeling the relationship of Antarctic minke whales to major ocean boundaries. *Polar Biology*, *35*(2), 281–290. https://doi.org/10.1007/s00300-011-1075-1

Allen, J. A., Garland, E. C., Dunlop, R. A., & Noad, M. J. (2018). Cultural revolutions reduce complexity in the songs of humpback whales. *Proceedings of the Royal Society B: Biological Sciences, 285*(1891). https://doi.org/10.1098/rspb.2018.2088

Ampilov, Y. P., Vladov, M., & Tokarev, M. Y. (2019). Broadband marine seismic acquisition technologies: challenges and opportunities. *Seismic Instruments, 55*(4), 388–403.

André, M., van der Schaar, M., Zaugg, S., Houégnigan, L., Sánchez, A., Mas, A., & Castell, J. V. (2010). Real-time monitoring of noise in cetacean acoustic niches. Paper presented at the 2nd International Conference on the Effects of Noise on Aquatic Marine Life, 15–20 August, Cork.

Archer, F. I., Brownell, R. L., Hancock-Hanser, B. L., Morin, P. A., Robertson, K. M., Sherman, K. K., et al. (2019). Revision of fin whale *Balaenoptera physalus* (Linnaeus, 1758) subspecies using genetics. *Journal of Mammalogy, 100*(5), 1653–1670. https://doi.org/10.1093/jmammal/gyz121

Arnold, P., Marsh, H., & Heinsohn, G. (1987). The occurrence of two forms of minke whales in East Australian waters with a description of external characters and skeleton of the diminutive or dwarf form. *The Scientific Reports of the Whales Research Institute, 38*, 1–46.

Au, W. W. (2000). Hearing in whales and dolphins: an overview. In R. R. Fay, A. N. Popper, & R. R. Fay (Eds.), *Hearing by whales and dolphins: Springer handbook of auditory research* (pp. 1–42). Springer.

Au, W. W., & Banks, K. (1998). The acoustics of the snapping shrimp *Synalpheus parneomeris* in Kaneohe Bay. *Journal of the Acoustical Society of America, 103*(1), 41–47. http://dx.doi.org/10.1121/1.423234

Au, W. W., Pack, A. A., Lammers, M. O., Herman, L. M., Deakos, M. H., & Andrews, K. (2006). Acoustic properties of humpback whale songs. *Journal of the Acoustical Society of America, 120*(2), 1103–1110. https://doi.org/DOI: 10.1121/1.2211547

Baumgartner, M. F., Van Parijs, S. M., Wenzel, F. W., Tremblay, C. J., Carter Esch, H., & Warde, A. M. (2008). Low frequency vocalizations attributed to sei whales (Balaenoptera borealis). *Journal of the Acoustical Society of America, 124* (2), 1339–1349.

Beamish, P., & Mitchell, E. (1973). Short pulse length audio frequency sounds recorded in the presence of a minke whale (*Balaenoptera acutorostrata*). *Deep Sea Research, 20*(4), 375–386.

Berchok, C. L., Bradley, D. L., & Gabrielson, T. B. (2006). St. Lawrence blue whale vocalizations revisited: characterization of calls detected from 1998 to 2001. *Journal of the Acoustical Society of America, 120*(4), 2340–2354. http://dx.doi.org/10.1121/1.2335676

Brown, M. R., Corkeron, P. J., Hale, P. T., Schultz, K. W., & Bryden, M. M. (1995). Evidence for a sex-segregated migration in the humpback whale (*Megaptera novaeangliae*). *Proceeding of the Royal Society B Biological Sciences, 259* (1355), 229–234.

Bryant, P. J. (1995). Dating remains of gray whales from the Eastern North Atlantic. *Journal of Mammalogy, 76*(3), 857–861.

Burnham, R., Duffus, D., & Mouy, X. (2018). Gray whale (Eschrictius robustus) call types recorded during migration off the west coast of Vancouver Island [original research]. *Frontiers in Marine Science, 5*. https://doi.org/10.3389/fmars.2018.00329

Calderan, S., Miller, B., Collins, K., Ensor, P., Double, M., Leaper, R., & Barlow, J. (2014). Low-frequency vocalizations of sei whales (*Balaenoptera borealis*) in the Southern Ocean. *Journal of the Acoustical Society of America, 136*(6), EL418–EL423. https://doi.org/doi:http://dx.doi.org/10.1121/1.4902422

Calderan, S. V., Leaper, R. C., Miller, B. S., Andriolo, A., Buss, D. L., Carroll, E. L., et al. (2021). Southern right whale vocalizations on foraging grounds in South Georgia. *JASA Express Letters, 1*(6), 061202. https://doi.org/10.1121/10.0005433

Carroll, E. L., Jackson, J. A., Paton, D., & Smith, T. D. (2014). Two intense decades of 19th century whaling precipitated rapid decline of right whales around New Zealand and East Australia. *PLoS ONE, 9*(4), e93789. https://doi.org/10.1371/journal.pone.0093789

Catchpole, C. K., & Slater, P. J. B. (2003). *Bird song: biological themes and variations* (2nd ed.). Cambridge University Press.

Cerchio, S., & Yamada, T. K. (2018). Omura's whale: *Balaenoptera omurai*. In B. Würsig, J. G. M. Thewissen, & K. M. Kovacs (Eds.), *Encyclopedia of marine mammals* (3rd ed., pp. 656–659). Academic Press. https://doi.org/10.1016/B978-0-12-804327-1.00186-2

Cerchio, S., Yamada, T. K., & Brownell Jr, R. L. (2019). Global distribution of Omura's whales *(Balaenoptera omurai)* and assessment of range-wide threats. *Frontiers in Marine Science, 6*(67), 19. https://doi.org/10.3389/fmars.2019.00067

Charif, R. A., Mellinger, D. K., Dunsmore, K. J., Fristrup, K. M., & Clark, C. W. (2002). Estimated source levels of fin whale (Balaenoptera physalus) vocalizations: Adjustments for surface interference. *Marine Mammal Science, 18*(1), 81–98. https://doi.org/10.1111/j.1748-7692.2002.tb01020.x

Chittleborough, R. G. (1953). Aerial observations on the humpback whale *Megaptera nodosa* (Bonnaterre), with notes on other species. *Australian Journal of Marine and Freshwater Research, 4*(2), 219–226.

Cholewiak, D., Clark, C. W., Ponirakis, D., Frankel, A., Hatch, L. T., Risch, D., et al. (2018). Communicating amidst the noise: modeling the aggregate influence of ambient and vessel noise on baleen whale communication space in a national marine sanctuary *Endangered Species Research, 36*, 18. https://doi.org/10.3354/esr00875

Cholewiak, D. M., Sousa-Lima, R. S., & Cerchio, S. (2013). Humpback whale song hierarchical structure: historical context and discussion of current classification issues. *Marine Mammal Science, 29*(3), E312–E332. https://doi.org/10.1111/mms.12005

Clapham, P. J., Young, S. B., & Brownell, R. B. J. (1999). Baleen whales: conservation issues and the status of the most endangered populations. *Mammal Review, 29*(1), 35–60.

Clark, C. W. (1982). The acoustic repertoire of the southern right whale, a quantitative analysis. *Animal Behaviour, 30*, 508–511.

Clark, C. W., Borsani, J. F., & Notarbartolo-di-Sciara, G. (2002). Vocal activity of fin whales *(Balaenoptera physalus)*, in the Liguarian sea. *Marine Mammal Science 18*(1), 10. https://doi.org/10.1111/J.1748-7692.2002.TB01035.X

Clark, C. W., & Clark, J. M. (1980). Sound playback experiments with southern right whales (*Eubalaena australis*). *Science, 207*(4431), 663–665.

Constantine, R., Iwata, T., Nieukirk, S. L., & Penry, G. S. (2018). Future directions in research on Bryde's whales. *Frontiers in Marine Science, 5*(333). https://doi.org/10.3389/fmars.2018.00333

Courts, R., Erbe, C., Wellard, R., Boisseau, O., Curt Jenner, K., & Jenner, M.-N. (2020). Australian long-finned pilot whales *(Globicephala melas)* emit stereotypical, variable, biphonic, multi-component, and sequenced vocalisations, similar to those recorded in the northern hemisphere. *Scientific Reports 10*, 14. https://doi.org/10.1038/s41598-020-74111-y

Crance, J. L., Berchok, C. L., Wright, D. L., Brewer, A. M., & Woodrich, D. F. (2019). Song production by the North Pacific right whale, *Eubalaena japonica. The Journal of the Acoustical Society of America, 145*(6), 3467. https://doi.org/10.1121/1.5111338

Crespo, E. A., Pedraza, S. N., Dans, S. L., Svendsen, G. M., Degrati, M., & Coscarella, M. A. (2018). The southwestern Atlantic southern right whale, *Eubalaena australis*, population is growing but at a decelerated rate. *Marine Mammal Science, 35*(1), 93–107. https://doi.org/10.1111/mms.12526

Croll, D. A., Clark, C. W., Acevedo, A., Tershy, B., Flores, S., Gedamke, J., & Urban, J. (2002). Only male fin whales sing loud songs. *Nature, 417*, 3. https://www.nature.com/articles/417809a.pdf

Cummings, W. C. (1985). Right whales - *Eubalaena glacialis* and *Eubalaena australis*. In S. H. Ridgway & S. R. Harrison (Eds.), *Handbook of marine mammals* (Vol. *3*, pp. 30). Academic Press.

Cummings, W. C., & Holliday, D. V. (1987). Sounds and source levels from bowhead whales of Pt. Barrow, Alaska. *Journal of the Acoustical Society of America, 82*, 814–821.

Cummings, W. C., & Thompson, P. O. (1971). Underwater sounds from the blue whale, *Balaenoptera musculus. Journal of the Acoustical Society of America, 50*(1193–1198).

Cummings, W. C., Thompson, P. O., & Ha, S. J. (1986). Sounds from Bryde, *Balaenoptera edeni,* and finback, *B. physalus,* whales in the Gulf of California. *Fishery Bulletin, 84*(2), 359–370.

Cypriano-Souza, A. L., de Meirelles, A. C. O., Carvalho, V. L., & Bonatto, S. L. (2017). Rare or cryptic? The first report of an Omura's whale (Balaenoptera omurai) in the South Atlantic Ocean. *Marine Mammal Science, 33*(1), 80–95. https://doi.org/10.1111/mms.12348

Dahlheim, M. E. (1987). *Bio-acoustics of the gray whale (Eschrichtius robustus)*. University of British Columbia.

Davis, G. E., Baumgartner, M. F., Corkeron, P. J., Bell, J., Berchok, C., Bonnell, J. M., et al. (2020). Exploring movement patterns and changing distributions of baleen whales in the western North Atlantic using a decade of passive acoustic data. *Global Change Biology*. https://doi.org/10.1111/gcb.15191

Dawbin, W. H., & Cato, D. H. (1992). Sounds of a pygmy right whale (*Caperea marginata*). *Marine Mammal Science, 8*(3), 213–219. https://doi.org/10.1111/j.1748-7692.1992.tb00405.x

Delarue, J., Laurinolli, M., & Martin, B. (2009a). Bowhead whale (*Balaena mysticetus*) songs in the Chukchi Sea between October 2007 and May 2008. *Journal of the Acoustical Society of America, 126*(6), 3319–3328. http://link.aip.org/link/?JAS/126/3319/1

Delarue, J., Todd, S. K., Van Parijs, S. M., & Iorio, L. (2009b). Geographic variation in Northwest Atlantic fin whale (*Balaenoptera physalus*) song: Implications for stock structure assessment. *Journal of the Acoustical Society of America, 125*(3), 1774–1782. https://doi.org/10.1121/1.3068454

Dominello, T., & Sirovic, A. (2016). Seasonality of Antarctic minke whale *(Balaenoptera bonaerensis)* calls off the western Antarctic Peninsula *Marine Mammal Science 32*(3), 13. https://doi.org/10.1111/mms.12302

Dunlop, R. A. (2016). The effect of vessel noise on humpback whale, *Megaptera novaeangliae*, communication behaviour. *Animal Behaviour, 111*, 13–21. https://doi.org/10.1016/j.anbehav.2015.10.002

Dunlop, R. A., Cato, D. H., & Noad, M. J. (2008). Non-song acoustic communication in migrating humpback whales (*Megaptera novaeangliae*). *Marine Mammal Science, 24*(3), 613–629. https://doi.org/10.1111/j.1748-7692.2008.00208.x

Dunlop, R. A., Cato, D. H., & Noad, M. J. (2010). Your attention please: increasing ambient noise levels elicits a change in communication behaviour in humpback whales (*Megaptera novaeangliae*). *Proceedings of the Royal Society B: Biological Sciences, 277*(1693), 2521–2529.

Dunlop, R. A., Noad, M. J., Cato, D. H., & Stokes, D. (2007). The social vocalization repertoire of east Australian migrating humpback whales (*Megaptera novaeangliae*). *Journal of the Acoustical Society of America, 122*(5), 2893–2905. http://dx.doi.org/10.1121/1.2783115

Edds-Walton, P. L. (1997). Acoustic communication signals of mysticete whales. *Bioacoustics, 8*(1&2), 47–60.

Edds, P. L. (1982). Vocalizations of the blue whale, *Balaenoptera musculus*, in the St. Lawrence River. *Journal of Mammalogy, 63*(2), 345–347.

Edds, P. L., Odell, D. K., & Tershy, B. R. (1993). Vocalizations of a captive juvenile and free-ranging adult-calf pairs of Bryde's whales, *Balaenoptera edeni. Marine Mammal Science, 9*(3), 269–284.

Edwards, E. F., Hall, C., Moore, T. J., Sheredy, C., & Redfern, J. V. (2015). Global distribution of fin whales *Balaenoptera physalus* in the post-whaling era (1980–2012). *Mammal Review, 45*(4), 197–214. https://doi.org/10.1111/mam.12048

Erbe, C., Dunlop, R., & Dolman, S. (2018). Effects of noise on marine mammals. In H. Slabbekoorn et al. (Eds.), *Effects of anthropogenic noise on animals* (pp. 277–309). Springer New York. https://doi.org/10.1007/978-1-4939-8574-6_10

Erbe, C., Dunlop, R., Jenner, K. C., S., Jenner, M.-N. M., McCauley, R. D., Parnum, I., et al. (2017). Review of underwater and in-air sounds emitted by Australian and Antarctic marine mammals. *Acoustics Australia, 45*(2), 179–241. https://doi.org/10.1007/s40857-017-0101-z

Finfer, D. C., Leighton, T. G., & White, P. R. (2008). Issues relating to the use of a 61.5 dB conversion factor when comparing airborne and underwater anthropogenic noise levels *Applied Acoustics, 69*, 8. https://doi.org/10.1016/j.apacoust.2007.05.008

Fish, J. F., Sumich, J. L., & Lingle, G. L. (1974). Sounds produced by the gray whale, *Eschrichtius robustus. Marine Fisheries Review, 36*(4), 38–45.

Fleming, A., & Jackson, J. (2011). Global review of humpback whales (Megaptera novaeangliae) (NOAA-TM-NMFS-SWFSC-474).

Folkens, P. A., Reeves, R. R., Stewart, B. S., Clapham, P. J., & Powell, J. A. (2011). *Guide to marine mammals of the world* (1st ed.). Alfred A. Knopf.

Ford, J. K. B. (2011). Acoustic behaviour of resident killer whales *(Orcinus orca)* off Vancouver Island, British Columbia. *Canadian Journal of Zoology 67*, 20.

Fordyce, R. E., & Marx, F. G. (2012). The pygmy right whale *Caperea marginata*: the last of the cetotheres. *Proceedings Of The Royal Society B*, 280.

Fournet, M. E. H., Matthews, L. P., Gabriele, C. M., Mellinger, D. K., & Klinck, H. (2018). Source levels of foraging humpback whale calls. *The Journal of the Acoustical Society of America 143*(2), 7. https://doi.org/10.1121/1.5023599

Gaines, C. A., Hare, M. P., Beck, S. E., & Rosenbaum, H. C. (2005). Nuclear markers confirm taxonomic status and relationships among highly endangered and closely related right whale species. *Proceedings of the Royal Society B: Biological Sciences, 272*(1562), 533–542. https://doi.org/10.1098/rspb.2004.2895

Gambell, R. (1985). Sei whale—*Balaenoptera borealis*. In S. H. Ridgway & S. R. Harrison (Eds.), *Handbook of marine mammals* (Vol. *3*, pp. 16). Academic Press.

Garland, E. C., Goldizen, A. W., Rekdahl, M. L., Constantine, R., Garrigue, C., Hauser, N. D., et al. (2011). Dynamic horizontal cultural transmission of humpback whale song at the ocean basin scale. *Current Biology, 21*(8), 687–691. https://doi.org/10.1016/j.cub.2011.03.019

Gedamke, J., Costa, D. P., & Dunstan, A. (2001). Localization and visual verification of a complex minke whale vocalization. *The Journal of the Acoustical Society of America, 109*(6), 3038–3047.

Glover, K. A., Kanda, N., Haug, T., Pastene, L. A., Øien, N., Goto, M., et al. (2010). Migration of Antarctic minke whales to the Arctic. *PLoS ONE, 5*(12), e15197. https://doi.org/10.1371/journal.pone.0015197

Glover, K. A., Kanda, N., Haug, T., Pastene, L. A., Øien, N., Seliussen, B. B., et al. (2013). Hybrids between common and Antarctic minke whales are fertile and can back-cross. *BMC Genetics, 14*(1), 25. https://doi.org/10.1186/1471-2156-14-25

Gordon, J., & Tyack, P. L. (2002). Sounds and cetaceans. In P. G. H. Evans & J. A. Raga (Eds.), *Marine mammals: biology and conservation* (pp. 139–196). Kluwer Academic/Plenum Press.

Greenewalt, C. H. (1969). How birds sing. *Scientific American, 221*(5), 126–139.

Guazzo, R. A., Helble, T. A., D'Spain, G. L., Weller, D. W., Wiggins, S. M., & Hildebrand, J. A. (2017). Migratory behavior of eastern North Pacific gray whales tracked using a hydrophone array. *PLoS ONE, 12*(10), e0185585. https://doi.org/10.1371/journal.pone.0185585

Halkias, X. C., Paris, S., & Glotin, H. (2013). Classification of mysticete sounds using machine learning techniques. *Journal of the Acoustical Society of America, 134*(5), 3496–3505. https://doi.org/doi:http://dx.doi.org/10.1121/1.4821203

Hatch, L., & Clark, C. (2004). Acoustic differentiation between fin whales in both the North Atlantic and North Pacific Oceans, and integration with genetic estimates of divergence. Unpublished paper to the IWC Scientific Committee.

Heimlich, S. L., Mellinger, D. K., Nieukirk, S. L., & Fox, C. G. (2005). Types, distribution, and seasonal occurrence of sounds attributed to Bryde's whales (*Balaenoptera edeni*) recorded in the eastern tropical Pacific, 1999–2001. *Journal of the Acoustical Society of America, 118*(3), 1830–1837. http://dx.doi.org/10.1121/1.1992674

Helble, T. A., Guazzo, R. A., Alongi, G. C., Martin, C. R., Martin, S. W., & Henderson, E. E. (2020). Fin whale song patterns shift over time in the central north Pacific. *Frontiers in Marine Science, 7*(907). https://doi.org/10.3389/fmars.2020.587110

Homfeldt, T. N., Risch, D., Stevenson, A., & Henry, L.-A. (2022). Seasonal and diel patterns in singing activity of humpback whales migrating through Bermuda. *Frontiers in Marine Science, 2*.

Iwase, R. (2015). Fin whale vocalizations observed with ocean bottom seismometers of cabled observatories off east Japan Pacific Ocean. *Japanese Journal of Applied Physics, 54*(7S1), 07HG03. https://doi.org/10.7567/jjap.54.07hg03

Jensen, F., Kuperman, W., Porter, M., & Schmidt, H. (2011). *Computational ocean acoustics* (2nd ed.). Springer. https://doi.org/10.1007/978-1-4419-8678-8

Johnson, S. C., Sirovic, A., Buccowich, J. S., Debich, A. J., Roche, L. K., Thayre, B., et al. (2014). *Passive acoustic monitoring for marine mammals in the Jacksonville Range Complex 2010*. M. P. Laboratory. https://citeseerx.ist.psu.edu/viewdoc/download?doi=10.1.1.721.1572&rep=rep1&type=pdf

Jung, J.-L., Mullié, W. C., Van Waerebeek, K., Wagne, M. M., Ould Bilal, A. S., Ould Sidaty, Z. E. A., et al. (2015). Omura's whale off West Africa: autochthonous population or inter-oceanic vagrant in the Atlantic Ocean? *Marine Biology Research, 12*(1), 66–75. https://doi.org/10.1080/17451000.2015.1084424

Kasamatsu, F., Nishiwaki, S., & Ishikawa, H. (1995). Breeding areas and southbound migrations of southern minke whales *Balaenoptera acutorostrata*. *Marine Ecology Progress Series, 119*, 1–10. https://www.int-res.com/abstracts/meps/v119/p1-10/

Kemper, C. M. (2002). Distribution of the pygmy right whale, *Caperea marginata*, in the Australasian region. *Marine Mammal Science, 18*(1), 99–111. https://doi.org/10.1111/j.1748-7692.2002.tb01021.x

Kemper, C. M. (2009). Pygmy right whale: *Caperea marginata*. In *Encyclopedia of marine mammals* (pp. 939–941). Elsevier.

Kemper, C. M. (2018). Pygmy right whale: *Caperea marginata*. In *Encyclopedia of marine mammals* (3rd ed., pp. 790–792). Academic Press. https://doi.org/10.1016/B978-0-12-804327-1.00211-9

Kerosky, S. M., Širović, A., Roche, L. K., Baumann-Pickering, S., Wiggins, S. M., & Hildebrand, J. A. (2012). Bryde's whale seasonal range expansion and increasing presence in the Southern California Bight from 2000 to 2010. *Deep Sea Research Part I: Oceanographic Research Papers, 65*, 125–132. https://doi.org/10.1016/j.dsr.2012.03.013

Kinsler, L. E., Frey, A. R., Coppens, A. B., & Sanders, J. V. (1999). *Fundamentals of acoustics* (4th ed.). Wiley.

Knowlton, A. R., & Kraus, S. D. (2001). Mortality and serious injury of northern right whales (*Eubalaena glacialis*) in the western North Atlantic Ocean. *Journal of Cetacean Research and Management*, *2*(193–208).

Kraus, S. D., Hamilton, P. K., Kenney, R. D., Knowlton, A. R., & Slay, C. K. (2001). Reproductive parameters of the North Atlantic right whale. *Journal of Cetacean Research and Management*, *2*, 231–236.

Kuna, V. M., & Nábělek, J. L. (2021). Seismic crustal imaging using fin whale songs. *Science*, *371*, 4. https://www.gwern.net/docs/technology/2021-kuna.pdf

L'Her, C., Vallez, S., & Gouriou, M. (2017). Automatic detection of humpback whales and localization of sperm whales during marine seismic survey. Paper presented at the 15th International Congress of the Brazilian Geophysical Society & EXPOGEF, Rio de Janeiro, Brazil, 31 July–3 August.

Lambertsen, R. H. (1983). Internal mechanism of rorqual feeding. *Journal of Mammology*, *64*(1), 76–88.

Leroy, E. C., Royer, J. Y., Alling, A., Maslen, B., & Rogers, T. L. (2021). Multiple pygmy blue whale acoustic populations in the Indian Ocean: whale song identifies a possible new population. *Scientific Reports*, *11*(1), 8762. https://doi.org/10.1038/s41598-021-88062-5

Leroy, E. C., Royer, J. Y., Bonnel, J., & Samaran, F. (2018). Long-term and seasonal changes of large whale call frequency in the Southern Indian Ocean. *Journal of Geophysical Research: Oceans*, *123*(11), 8568–8580.

Leroy, E. C., Samaran, F., Bonnel, J., & Royer, J. Y. (2016). Seasonal and diel vocalization patterns of Antarctic blue whale (*Balaenoptera musculus intermedia*) in the Southern Indian Ocean: a multi-year and multi-site study. *PLoS ONE*, *11*(11), e0163587. https://doi.org/10.1371/journal.pone.0163587

Lewis, L. A., Calambokidis, J., Stimpert, A. K., Fahlbusch, J., Friedlaender, A. S., McKenna, M. F., et al. (2018). Context-dependent variability in blue whale acoustic behaviour. *Royal Society Open Science*, *5*(8), 180241. https://doi.org/10.1098/rsos.180241

Ljungblad, D. K., Thompson, P. O., & Moore, S. E. (1982). Underwater sounds recorded from migrating bowhead whales, *Balaena mysticetus*, in 1979. *Journal of the Acoustical Society of America*, *71*, 477–482.

Luczkovich, J. J., Mann, D. A., & Rountree, R. A. (2008). Passive acoustics as a tool in fisheries science. *Transactions of the American Fisheries Society* *137*(1), 9. https://doi.org/10.1577/T06-258.1

Macdonald, D. W. (Ed.). (2006). *The encyclopedia of mammals*. Oxford University Press.

Madhusudhana, S., Murray, A., & Erbe, C. (2020). Automatic detectors for low-frequency vocalizations of Omura's whales, *Balaenoptera omurai:* a performance comparison. *The Journal of the Acoustical Society of America*, *147*(5), 3078–3090. https://doi.org/10.1121/10.0001108

Madsen, P. T., Payne, R., Kristiansen, N. U., Wahlberg, M., Kerr, I., & Møhl, B. (2002). Sperm whale sound production studied with ultrasound time/depth-recording tags. *Journal of Experimental Biology*, *205*(13), 1899–1906.

Marques, T. A., Munger, L., Thomas, L., Wiggins, S., & Hilderbrand, J. A. (2011). Estimating North Pacific right whale *Eubalaena japonica* density using passive acoustic cue counting. *Endangered Species Research*, *13*, 163–172.

Martin, S. W., Marques, T. A., Thomas, L., Morrissey, R. P., Jarvis, S., DiMarzio, N., et al. (2013). Estimating minke whale (*Balaenoptera acutorostrata*) boing sound density using passive acoustic sensors. *Marine Mammal Science*, *29*(1), 142–158. https://doi.org/10.1111/j.1748-7692.2011.00561.x

Matias, L., & Harris, D. (2015). A single-station method for the detection, classification and location of fin whale calls using ocean-bottom seismic stations. *The Journal of the Acoustical Society of America*, *138*(1), 504–520. https://doi.org/10.1121/1.4922706

Matthews, J. N., Rendell, L. E., Gordon, J. C. D., & Macdonald, D. W. (1999). A review of frequency and time parameters of cetacean tonal calls. *Bioacoustics*, *10*(1), 47–71. https://doi.org/10.1080/09524622.1999.9753418

Matthews, L. P., & Parks, S. E. (2021). An overview of North Atlantic right whale acoustic behavior, hearing capabilities, and responses to sound. *Marine Pollution Bulletin*, *173*, 113043. https://doi.org/10.1016/j.marpolbul.2021.113043

McDonald, M. A., Calambokidis, J., Teranishi, A., & Hildebrand, J. A. (2001). The acoustic calls of blue whales off California with gender data. *Journal of the Acoustical Society of America*, *109*(4), 1728–1735.

McDonald, M. A., Hildebrand, J. A., & Mesnick, S. (2009). Worldwide decline in tonal frequencies of blue whale songs. *Endangered Species Research*, *9*(1), 9. https://doi.org/10.3354/esr00217

McDonald, M. A., Hildebrand, J. A., Wiggins, S. M., Thiele, D., Glasgow, D., & Moore, S. E. (2005). Sei whale sounds recorded in the Antarctic. *Journal of the Acoustical Society of America*, *118*(6), 3941–3945. http://dx.doi.org/10.1121/1.2130944

McDonald, M. A., Hilderbrand, J. A., & Webb, S. C. (1995). Blue and fin whales observed on a seafloor array in the northeast Pacific. *Journal of the Acoustical Society of America*, *98* (2), 712–721.

McDonald, M. A., Mesnick, S. L., & Hildebrand, J. A. (2006). Biogeographic characterization of blue whale song worldwide: using song to identify populations. *Journal of Cetacean Research and Management*, *8*(1), 55–65. http://www.escholarship.org/uc/item/5r16c2mz

McDonald, M. A., & Moore, S. E. (2002). Calls recorded from North Pacific right whales (*Eubalaena japonica*) in the eastern Bering Sea. *Journal of Cetacean Research and Management*, *4* (3), 261–266.

Mellinger, D. K., Carson, C., & Clark, C. W. (2000). Characteristics of minke whale (*Balaenoptera acutorostrata*) pulse trains recorded near Puerto Rico. *Marine Mammal Science*, *16*, 739–756.

Mellinger, D. K., & Clark, C. W. (2003). Blue whale (*Balaenoptera musculus*) sounds from the North Atlantic. *Journal of the Acoustical Society of America*, *114*(2), 1108–1119. http://dx.doi.org/10.1121/1.1593066

Mellinger, D. K., Stafford, K. M., Moore, S. E., Munger, L., & Fox, C. G. (2004). Detection of North Pacific right whale (*Eubalaena japonica*) calls in the Gulf of Alaska. *Marine Mammal Science*, *20*(4), 872–879. https://doi.org/10.1111/j.1748-7692.2004.tb01198.x

Meyer-Gutbrod, E. L., & Greene, C. H. (2017). Uncertain recovery of the North Atlantic right whale in a changing ocean. *Global Change Biology*, *24*(1), 455–464. https://doi.org/10.1111/gcb.13929

Mikkelsen, L., Johnson, M., Wisniewska, D. M., van Neer, A., Siebert, U., Madsen, P. T., & Teilmann, J. (2019). Long-term sound and movement recording tags to study natural behavior and reaction to ship noise of seals. *Ecology and Evolution*, *9* (5), 2588–2601. https://doi.org/10.1002/ece3.4923

Miksis-Olds, J. L., Martin, B., & Tyack, P. L. (2018). Exploring the ocean through soundscapes. *Acoustics Today*, *14*(1), 9. https://acousticstoday.org/wp-content/uploads/2018/03/Exploring-the-Ocean-Through-Soundscapes.pdf

Miksis-Olds, J. L., Van Opzeeland, I. C., Van Parijs, S. M., & Jones, J. (2016). Pinniped sounds in the polar oceans. In W. Au & M. Lammers (Eds.), *Listening in the ocean* (pp. 257–308). Springer. https://doi.org/10.1007/978-1-4939-3176-7_11

Mizroch, S., Rice, D., & Breiwick, J. (1984a). The blue whale, *Balaenoptera musculus*. *Marine Fisheries Reviews*, *46*.

Mizroch, S., Rice, D. W., & Breiwick, J. M. (1984b). The fin whale, *Balaenoptera physalus*. *Marine Fisheries Review*, *46*, 20–24.

Modest, M., Irvine, L., Andrews-Goff, V., Gough, W., Johnston, D., Nowacek, D., et al. (2021). First description of migratory behavior of humpback whales from an Antarctic feeding ground to tropical calving ground. *Animal Biotelemetry*, *9*(42).

Moore, C. J., Moore, S. L., Leecaster, M. K., & Weisberg, S. B. (2001). A comparison of plastic and plankton in the North Pacific central gyre. *Marine Pollution Bulletin*, *42*(12), 1297–1300. http://www.sciencedirect.com/science/article/pii/S0025326X0100114X

Moore, S. E., Stafford, K. M., Mellinger, D. K., & Hildebrand, J. A. (2006). Listening for large whales in the offshore waters of Alaska. *BioScience*, *56*(1), 49–55.

Moreira, S. C., Weksler, M., Sousa-Lima, R. S., Maia, M., Sukhovich, A., Royer, J.-Y., et al. (2020). Occurrence of Omura's whale, *Balaenoptera omurai* (Cetacea: Balaenopteridae), in the Equatorial Atlantic Ocean based on passive acoustic monitoring. *Journal of Mammalogy*, *101*(6), 1727–1735. https://doi.org/10.1093/jmammal/gyaa130

Munger, L. M., Wiggins, S. M., & Hildebrand, J. A. (2011). North Pacific right whale up-call source levels and propagation distance on the southeastern Bering Sea shelf. *Journal of the Acoustical Society of America*, *129*(6), 4047–4054. http://dx.doi.org/10.1121/1.3557060

Newhall, A. E., Lin, Y.-T., Lynch, J. F., Baumgartner, M. F., & Gawarkiewicz, G. G. (2012). Long distance passive localization of vocalizing sei whales using an acoustic normal mode approach. *Journal of the Acoustical Society of America*, *131* (2), 1814–1825. http://link.aip.org/link/?JAS/131/1814/1

Nieukirk, S. L., Fregosi, S., Mellinger, D. K., & Klinck, H. (2016). A complex baleen whale call recorded in the Mariana Trench Marine National Monument. *The Journal of the Acoustical Society of America*, *140*(3), EL274. https://doi.org/10.1121/1.4962377

NMFS. (2018). *2018 Revisions to: Technical guidance for assessing the effects of anthropogenic sound on marine mammal hearing (Version 2.0): underwater thresholds for onset of permanent and temporary threshold shifts.*

NOAA. (2020). *Recovery plan for the blue whale (Balaenoptera musculus)* O. *o. P. R. a. W. C. Region.*

Noad, M. J., Cato, D. H., Bryden, M. M., Jenner, M. N., & Jenner, K. C. S. (2000). Cultural revolution in whale songs. *Nature*, *408*(6812), 537–537. https://doi.org/10.1038/35046199

Norris, T. F., McDonald, M., & Barlow, J. (1999). Acoustic detections of singing humpback whales (*Megaptera novaeangliae*) in the eastern North Pacific during their northbound migration. *Journal of the Acoustical Society of America*, *106* (1), 506–514.

Oleson, E. M., Barlow, J., Gordon, J., Rankin, S., & Hildebrand, J. A. (2003). Low frequency calls of Bryde's whales. *Marine Mammal Science*, *19*(2), 407–419.

Oleson, E. M., Calambokidis, J., Burgess, W. C., McDonald, M. A., LeDuc, C. A., & Hildbrand, J. A. (2007). Behaviourial context of call productions by eastern North Pacific blue whales. *Marine Ecology Progress Series*, *330*(1), 16. https://www.int-res.com/articles/meps_oa/m330p269.pdf

Oleson, E. M., Wiggins, S. M., & Hildebrand, J. A. (2007). Temporal separation of blue whale call types on a southern California feeding ground. *Animal Behaviour*, *74*(4), 881–894. https://doi.org/10.1016/j.anbehav.2007.01.022

Olsen, E., Budgell, P., Head, E., Kleivane, L., Nøttestad, L., Prieto, R., et al. (2009). First satellite-tracked long-distance movement of a sei whale (*Balaenoptera borealis*) in the North Atlantic. *Aquatic Mammals*, *35*(3), 313–318.

Oswald, J. N., Au, W. W. L., & Duennebier, F. (2011). Minke whale (*Balaenoptera acutorostrata*) boings detected at the Station ALOHA Cabled Observatory. *Journal of the Acoustical Society of America*, *129*(5), 3353–3360. http://link.aip.org/link/?JAS/129/3353/1

Parks, S. E., Clark, C. W., & Tyack, P. L. (2007). Short-and long-term changes in right whale calling behavior: The potential effects of noise on acoustic communication. *The Journal of the Acoustical Society of America*, *122*(6), 3725–3731.

Parks, S. E., Hamilton, P. K., Kraus, S. D., & Tyack, P. L. (2005). The gunshot sound produced by male North Atlantic right whales (*Eubalaena glacialis*) and its potential function in reproductive advertisement. *Marine Mammal Science*, *21*(3), 458–475. https://doi.org/10.1111/j.1748-7692.2005.tb01244.x

Parks, S. E., Hotchkin, C. F., Cortopassi, K. A., & Clark, C. W. (2012). Characteristics of gunshot sound displays by North Atlantic right whales in the Bay of Fundy. *Journal of the Acoustical Society of America*, *131*(4), 3173–3179.

Parks, S. E., & Tyack, P. L. (2005). Sound production by North Atlantic right whales (*Eubalaena glacialis*) in surface active groups. *Journal of the Acoustical Society of America*, *117* (5), 3297–3306.

Parsons, M., Erbe, C., McCauley, R., McWilliam, J., Marley, S., Gavrilov, A., & Parnum, I. (2016). Long-term monitoring of soundscapes and deciphering a usable index: examples of fish choruses from Australia. *Proceedings of Meetings on Acoustics*, *27*, 010023.

Pastene, L. A., Acevedo, J., Goto, M., Zerbini, A. N., Acuna, P., & Aguayo-Lobo, A. (2010). Population structure and possible migratory links of common minke whales, *Balaenoptera acutorostrata*, in the Southern Hemisphere. *Conservation Genetics*, *11*(4), 1553–1558.

Payne, K. (2000). The progressively changing songs of humpback whales: A window on the creative process in a wild animal In N. L. Wallin, B. Merker, & S. Brown (Eds.), *The origins of music* (pp. 15). The MIT Press.

Payne, R. S. (1983). *Communication and behavior of whales* (Vol. *76*). Westview Press.

Payne, R. S., & McVay, S. (1971). Songs of humpback whales. *Science*, *173*(3997), 585–597.

Payne, R. S., & Payne, K. (1985). Large scale changes over 19 years in songs of humpback whales in Bermuda. *Zeitschrift für Tierpsychologie*, *68*(2), 89–114.

Pennisi, E. (2017). North Atlantic right whale faces extinction. *Science*, *358*(6364), 703–704.

Perrin, W. F., Mallette, S. D., & Brownell Jr, R. L. (2018). Minke whales: *Balaenoptera acutorostrata* and *B. bonaerensis*. In B. Würsig, J. G. M. Thewissen, & K. M. Kovacs (Eds.), *Encyclopedia of marine mammals* (pp. 608–613). Elsevier.

Pike, G. C. (1962). Migration and feeding of the gray whale (*Eschrichtius gibbosus*). *Journal of the Fisheries Research Board of Canada*, *19*(5), 815–838.

Prieto, R., Silva, M. A., Waring, G. T., & Gonçalves, J. M. A. (2014). Sei whale movements and behaviour in the North Atlantic inferred from satellite telemetry. *Endangered Species Research*, *26*(2), 103–113. https://www.int-res.com/abstracts/esr/v26/n2/p103-113/

Pyenson, N. D., Goldbogen, J. A., Vogl, A. W., Szathmary, G., Drake, R. L., & Shadwick, R. E. (2012). Discovery of a sensory organ that coordinates lunge feeding in rorqual whales. *Nature*, *485*(7399), 498–501. https://doi.org/10.1038/nature11135

Rabiner, L. R. (1978). *Digital processing of speech signals*. Pearson Education India.

Rabiner, L. R., & Schafer, R. W. (Eds.). (1978). *Digital processing of speech signals*. Prentice Hall.

Ramp, C., Delarue, J., Palsboll, P. J., Sears, R., & Hammond, P. S. (2015). Adapting to a warmer ocean–seasonal shift of baleen whale movements over three decades. *PLoS ONE*, *10*(3), e0121374. https://doi.org/10.1371/journal.pone.0121374

Rankin, S., & Barlow, J. (2005). Source of the North Pacific "boing" sound attributed to minke whales. *Journal of the Acoustical Society of America*, *118*(5), 3346–3351. http://dx.doi.org/10.1121/1.2046747

Rankin, S., & Barlow, J. (2007). Vocalizations of the sei whale *Balaenoptera borealis* off the Hawaiian Islands. *Bioacoustics*, *16*, 137–145.

Rasmussen, K., Palacios, D. M., Calambokidis, J., Saborío, M. T., Dalla Rosa, L., Secchi, E. R., et al. (2007). Southern hemisphere humpback whales wintering off central America: insights from high water temperature into the longest mammalian migration. *Biology Letters*, *3*, 302–305.

Recalde-Salas, A., Salgado Kent, C. P., Parsons, M. J. G., Marley, S. A., & McCauley, R. D. (2014). Non-song vocalizations of pygmy blue whales in Geographe Bay, Western Australia. *Journal of the Acoustical Society of America*, *135*(5), EL213-EL218. https://doi.org/doi:http://dx.doi.org/10.1121/1.4871581

Reeves, R. R., & Leatherwood, S. (1985). Bowhead whale—*Balaena mysticetus*. In S. H. Ridgway & S. R. Harrison (Eds.), *Handbook of marine mammals* (Vol. *3*, pp. 40). Academic Press.

Reeves, R. R., Perrin, W. F., Taylor, B. L., Baker, C. S., & Mesnick, S. L. (2004). *Report of the workshop on shortcomings of cetacean taxonomy in relation to needs of conservation and management. April 30–May 2, 2004* (Technical Memorandum NMFS).

Reichmuth, C. (2008). Hearing in marine carnivores. *Bioacoustics*, *17*(1–3), 89–92.

Reichmuth, C., Holt, M. M., Mulsow, J., Sills, J. M., & Southall, B. L. (2013). Comparative assessment of amphibious hearing in pinnipeds. *Journal of Comparative Physiology A*, *1999*(6), 491–507.

Reidenberg, J. S., & Laitman, J. T. (2007). Blowing bubbles: an aquatic adaptation that risks protection of the respiratory tract in humpback whales (*Megaptera novaeangliae*). *The Anatomical Record*, *290*(6), 569–580. https://doi.org/10.1002/ar.20537

Reidenberg, J. S., & Laitman, J. T. (2010). Generation of sound in marine mammals. In *Handbook of behavioral neuroscience* (Vol. *19*, pp. 451–465). Elsevier.

Reilly, S. B., & Thayer, V. G. (1990). Blue Whale (*Balaenoptera musculus*) distribution in the eastern tropical Pacific. *Marine Mammal Science*, *6*(4), 265–277.

Rekdahl, M. L., Garland, E. C., Carvajal, G. A., King, C. D., Collins, T., Razafindrakoto, Y., & Rosenbaum, H. (2018). Culturally transmitted song exchange between humpback whales (*Megaptera novaeangliae*) in the southeast Atlantic and southwest Indian Ocean basins. *Royal Society Open Science*, *5*(11), 172305.

Rice, D. W. (1998). *Marine mammals of the world: systematics and distribution*. Society for Marine Mammalogy.

Risch, D., Clark, C. W., Dugan, P. J., Popescu, M., Siebert, U., & Van Parijs, S. M. (2013). Minke whale acoustic behavior and multi-year seasonal and diel vocalisation patterns in Massachusetts Bay, USA. *Marine Ecology Progress Series*, *489*, 279–295. https://doi.org/10.3354/meps10426

Risch, D., Siebert, U., & Van Parijs, S. M. (2014). Individual calling behaviour and movements of North Atlantic minke whales (*Balaenoptera acutorostrata*). *Behaviour*, *151*(9), 1335–1360. https://doi.org/doi:10.1163/1568539X-00003187

Risch, D., Wilson, S. C., Hoogerwerf, M., van Geel, N. C. F., Edwards, E. W. J., & Brookes, K. L. (2019). Seasonal and diel acoustic presence of North Atlantic minke whales in the North Sea. *Scientific Reports*, *9*(1), 3571. https://doi.org/10.1038/s41598-019-39752-8

Romagosa, M., Boisseau, O., Cucknell, A. C., Moscrop, A., & McLanaghan, R. (2015). Source level estimates for sei whale (*Balaenoptera borealis*) vocalizations off the Azores. *The Journal of the Acoustical Society of America*, *138*(4), 2367–2372. https://doi.org/10.1121/1.4930900

Rosel, P. E., Wilcox, L. A., Yamada, T. K., & Mullin, K. D. (2021). A new species of baleen whale (*Balaenoptera*) from the Gulf of Mexico, with a review of its geographic distribution. *Marine Mammal Science*, *37*(2), 577–610. https://doi.org/10.1111/mms.12776

Rosenbaum, H. C., Brownell Jr, R. L., Brown, M. W., Schaeff, C., Portway, V., White, B. N., et al. (2000). World-wide genetic differentiation of Eubalaena: questioning the number of right whale species. *Molecular Ecology*, *9*(11), 1793–1802. https://doi.org/10.1046/j.1365-294x.2000.01066.x

Rugh, D. J., Shelden, K. E. W., & Schulan-Janiger, A. S. (2001). Timing of the gray whale southbound migration. *Journal of Cetacean Research and Management*, *3*(1), 31–40.

Samaran, F., Guinet, C., Adam, O., Motsch, J.-F., & Cansi, Y. (2010). Source level estimation of two blue whale subspecies in southwestern Indian Ocean. *Journal of the Acoustical Society of America*, *127*(6), 3800–3808. http://dx.doi.org/10.1121/1.3409479

Samaran, F., Stafford, K. M., Branch, T. A., Gedamke, J., Royer, J. Y., Dziak, R. P., & Guinet, C. (2013). Seasonal and geographic variation of southern blue whale subspecies in the Indian Ocean. *PLoS ONE*, *8*(8), e71561.

Sasaki, T., Nikaido, M., Wada, S., Yamada, T. K., Cao, Y., Hasegawa, M., & Okada, N. (2006). *Balaenoptera omurai* is a newly discovered baleen whale that represents an ancient evolutionary lineage. *Molecular Phylogenetics and Evolution*, *41*(1), 40–52. https://doi.org/10.1016/j.ympev.2006.03.032

Scheinin, A. P., Kerem, D., MacLeod, C. D., Gazo, M., Chicote, C. A., & Castellote, M. (2011). Gray whale (*Eschrichtius robustus*) in the Mediterranean Sea: anomalous event or early sign of climate-driven distribution change? *Marine Biodiversity Records*, *4*(1).

Schevill, W. E., & Watkins, W. A. (1972). Intense low-frequency sounds from an Antarctic minke whale, *Balaenoptera acutorostrata. Breviora: Museum of Comparative Zoology*, *388*, 1–8.

Schusterman, R. J., & van Parijs, S. M. (2003). Pinniped vocal communication: an introduction. *Aquatic Mammals*, *29*(2), 177–180.

Sears, R., & Perrin, W. F. (2009). Blue whale: *Balaenoptera musculus*. In W. F. Perrin, B. Würsig, & J. G. M. Thewissen (Eds.), *Encyclopedia of marine mammals* (2nd Ed.) (pp. 120–124). Academic Press. https://doi.org/10.1016/B978-0-12-373553-9.00033-X

Shabangu, F. W., Findlay, K., & Stafford, K. M. (2019). Seasonal acoustic occurrence, diel-vocalizing patterns and bioduck call-type composition of Antarctic minke whales off the west coast of South Africa and the Maud Rise, *Antarctica. Marine Mammal Science*, *36*(1), 18. https://doi.org/10.1111/mms.12669

Širović, A., Bassett, H. R., Johnson, S. C., Wiggins, S. M., & Hildebrand, J. A. (2014). Bryde's whale calls recorded in the Gulf of Mexico. *Marine Mammal Science*, *30*(1), 399–409. https://doi.org/10.1111/mms.12036

Širović, A., Hildebrand, J. A., & Wiggins, S. (2007). Blue and fin whale call source levels and propagation range in the Southern Ocean. *Journal of the Acoustical Society of America*, *122*(2), 1208–1215.

Širović, A., Hildebrand, J. A., Wiggins, S. M., & Thiele, D. (2009). Blue and fin whale acoustic presence around Antarctica during 2003 and 2004. *Marine Mammal Science*, *25*(1), 125–136.

Soldevilla, M. S., Rice, A. N., Clark, C. W., & Garrison, L. P. (2014). Passive acoustic monitoring on the North Atlantic right whale calving grounds. *Endangered Species Research*, *25*(2), 115–140.

Sommerville, D. C. (1994). Observations of the gray whale migration in the vicinity of Diablo Canyon, California: 1981–1994 migration (Report).

Sousa-Lima, R. S., Paglia, A. P., & Da Fonseca, G. A. (2002). Signature information and individual recognition in the isolation calls of Amazonian manatees, *Trichechus inunguis* (Mammalia: Sirenia). *Animal Behaviour*, *63*(2), 301–310.

Southall, B. L., Bowles, A. E., Ellison, W. T., Finneran, J. J., Gentry, R. L., Greene, C. R., et al. (2007). Marine mammal noise exposure criteria, initial scientific recommendations. *Aquatic Mammals*, *33*(4), 411–414. http://www.ingentaconnect.com/content/eaam/am/2007/00000033/00000004/art00002

Southall, B. L., Finneran, J. J., Reichmuth, C., Nachtigall, P. E., Ketten, D. R., Bowles, A. E., et al. (2019). Marine mammal noise exposure criteria: Updated scientific recommendations for residual hearing effects. *Aquatic Mammals*, *45*(2), 125–232.

Stafford, K. M., Bohnenstiehl, D. R., Tolstoy, M., Chapp, E., Mellinger, D. K., & Moore, S. E. (2004). Antarctic-type blue whale calls recorded at low latitudes in the Indian and eastern Pacific Oceans. *Deep Sea Research Part I: Oceanographic Research Papers*, *51*(10), 1337–1346. https://doi.org/10.1016/j.dsr.2004.05.007

Stafford, K. M., Moore, S. E., Laidre, K. L., & Heide-Jorgensen, M. P. (2008). Bowhead whale springtime song off West Greenland. *Journal of the Acoustical Society of America*, *124*(5), 3315–3323. http://dx.doi.org/10.1121/1.2980443

Stafford, K. M., Nieukirk, S. L., & Fox, C. G. (1999). Low-frequency whale sounds recorded on hydrophones moored in the eastern tropical Pacific. *Journal of the Acoustical Society of America*, *106*(6), 3687–3698. http://dx.doi.org/10.1121/1.428220

Stevick, P. T., Neves, M. C., Johansen, F., Engel, M. H., Allen, J., Marcondes, M. C. C., & Carlson, C. (2011). A quarter of a world away: female humpback whale moves 10 000 km between breeding areas. *Biology Letters*, *7*(2), 299–302. https://doi.org/10.1098/rsbl.2010.0717

Stimpert, A. K., Wiley, D. N., Au, W. W. L., Johnson, M. P., & Arsenault, R. (2007). "Megapclicks": acoustic click trains and buzzes produced during night-time foraging of humpback whales (*Megaptera novaeangliae*). *Biology Letters*, *3*(5), 467–470. https://doi.org/DOI: 10.1098/rsbl.2007.0281

Thode, A. M., Blackwell, S. B., Conrad, A. S., Kim, K. H., Marques, T., Thomas, L., et al. (2020). Roaring and repetition: how bowhead whales adjust their call density and source level (Lombard effect) in the presence of natural and seismic airgun survey noise. *The Journal of the Acoustical Society of America*, *147*(3), 2061–2080. https://doi.org/10.1121/10.0000935

Thomas, J., Moss, C., & Vater, M. (Eds.). (2004). *Echolocation in bats and dolphins*. The University of Chicago Press.

Thompson, P. M., Cornwell, H. J. C., Ross, H. M., & Miller, D. (1992). Serologic study of phocine distemper in a population of harbor seals in Scotland. *Journal of Wildlife Diseases*, *28*(1), 21–27. https://doi.org/10.7589/0090-3558-28.1.21

Thompson, P. O., Findley, L. T., & Vidal, O. (1992). 20-Hz pulses and other vocalizations of fin whales, Balaenoptera physalus, in the Gulf of California, Mexico. *Journal of the*

Acoustical Society of America, *92*(6), 3051–3057. http://dx.doi.org/10.1121/1.404201

Thompson, T. J., Winn, H. E., & Perkins, P. J. (Eds.). (1979). *Mysticete sounds* (Vol. 3). Springer.

Todd, V. L. G., Todd, I. B., Gardiner, J. C., & Morrin, E. C. N. (2015). *Marine mammal observer and passive acoustic monitoring handbook*. Pelagic Publishing Ltd.

Torterotot, M., Samaran, F., Stafford, K. M., & Royer, J.-Y. (2020). Distribution of blue whale populations in the Southern Indian Ocean based on a decade of acoustic monitoring. *Deep Sea Research Part II: Topical Studies in Oceanography*, *179*, 104874. https://doi.org/10.1016/j.dsr2.2020.104874

Tyack, P. L., & Clarke, C. W. (2000). Communication and acoustic behaviour of dolphins and whales. In R. R. Fay, A. N. Popper, & R. R. Fay (Eds.), *Hearing by Whales and Dolphins: Springer Handbook of Auditory Research* (pp. 156–224). Springer.

Tyarks, S. C., Aniceto, A. S., Ahonen, H., Pedersen, G., & Lindstrøm, U. (2021). Humpback whale (Megaptera novaeangliae) song on a subarctic feeding ground. *Frontiers in Marine Science*, 594.

Urick, R. J. (1975). *Principles of underwater sound*. McGraw Hill.

Viloria-Gomora, L., Romero-Vivas, E., & Urban, J. R. (2015). Calls of Bryde's whale (*Balaenoptera edeni*) recorded in the Gulf of California (L). *The Journal of the Acoustical Society of America*, *138*(5), 2722–2725. https://doi.org/10.1121/1.4932032

Vincenzi, G. d., Parisi, I., Torri, M., Papale, E., Mazzola, S., Nuth, C., & Buscaino, G. (2019). Influence of environmental parameters on the use and spatiotemporal distribution of the vocalizations of bearded seals (*Erignathus barbatus*) in Kongsfjorden, *Spitsbergen. Polar Biology*, *42*(7), 1241–1254. https://doi.org/10.1007/s00300-019-02514-3

Vu, E. T., Risch, D., Clark, C. W., Gaylord, S., Hatch, L. T., Thompson, M. A., et al. (2012). Humpback whale song occurs extensively on feeding grounds in the western North Atlantic Ocean. *Aquatic Biology*, *14*(2), 175–183. https://www.int-res.com/abstracts/ab/v14/n2/p175-183/

Wada, S., Oishi, M., & Yamada, T. K. (2003). A newly discovered species of living baleen whale. *Nature*, *426*, 278–281.

Wang, D., Huang, W., Garcia, H., & Ratilal, P. (2016). Vocalization source level distributions and pulse compression gains of diverse baleen whale species in the Gulf of Maine. *Remote sensing*, *8*(1), 20. https://doi.org/10.3390/rs8110881

Watkins, W. A. (1981a). Activities and underwater sounds of fin whales. *The Scientific Reports of the Whales Research Institute*, *33*, 83–117.

Watkins, W. A. (1981b). Reaction of 3 species of whales *Balaenoptera physalus*, *Megaptera novaeangliae*, and *Balaenoptera edeni* to implanted radio tags. *Deep Sea Research Part I: Oceanographic Research Papers*, *28*(6), 589.

Watkins, W. A., Tyack, P., Moore, K. E., & Bird, J. E. (1987). The 20 Hz signals of finback whales (*Balaenoptera physalus*). *Journal of the Acoustical Society of America*, *82*(6), 1901–1912.

Wenz, G. M. (1964). Curious noises and the sonic environment in the ocean. *Marine Bio-Acoustics*, *1*(1), 18.

Wilcock, W. S. D. (2012). Tracking fin whales in the northeast Pacific Ocean with a seafloor seismic network. *Journal of the Acoustical Society of America*, *132*(4), 2408–2419. http://link.aip.org/link/?JAS/132/2408/1

Winn, H. E., & Perkins, P. J. (1976). Distribution and sounds of the minke whale, with a review of mysticete sounds. *Cetology*, *19*, 1–12.

Winn, H. E., Thompson, T. J., Cummings, W. C., Hain, J., Hudnall, J., Hays, H., & Steiner, W. W. (1981). Song of the humpback whale: population comparisons. *Behavioral Ecology and Sociobiology*, *8*(1), 41–46. https://doi.org/10.2307/4599355

Winn, H. E., & Winn, L. K. (1978). The song of the humpback whale, *Megaptera novaeangliae,* in the West Indies. *Marine Biology*, *47*, 97–114.

Würsig, B., Perrin, W. F., & Thewissen, J. G. M. (2008). *Encyclopedia of marine mammals* (2nd ed.). Academic Press.

Würsig, B., Thewissen, J., & Kovacs, K. M. (2017). *Encyclopedia of marine mammals* (3rd ed.). Academic Press.

12

Tracking and Monitoring Fin Whales Offshore Northwest Spain Using Passive Acoustic Methods

Jessica Fisher[1], Timothy A. Minshull[1], Paul R. White[2], Bai Tian[2], and Gaye Bayrakci[3]

ABSTRACT

Fin whales produce regular vocalizations with a dominant frequency of ~20 Hz, similar to that of large airgun sources used in marine seismic surveys. Thus these vocalizations may be recorded on ocean-bottom seismometers (OBSs) or hydrophones (OBHs) deployed during such surveys. We used data recorded by an array of 72 OBSs/Hs with 4–6.5 km spacing, deployed offshore northwest Spain during June–August 2013, to study fin whale movements in this area. Whale vocalizations were identified automatically using signal-processing techniques and localized using time-delay estimates between several instruments. Airgun shooting took place during the deployment period, but we found no evidence for a correlation between vocalization detection rate and the presence or absence of shooting. Our analysis focused on six fin whale tracks identified during this period. Uncertainties in depth led to poor confidence intervals, preventing effective analysis of diving behavior for the identified tracks. In the horizontal plane, the localizations had a higher degree of confidence. Use of a Kalman filter resulted in smoother tracks. Mean swim velocities range from 2 to 15 km/hr, and the animals traveled distances of 1.5–15 km in the periods analyzed.

12.1. INTRODUCTION

12.1.1. Passive Acoustic Monitoring

Passive acoustic monitoring (PAM) has been used increasingly to study cetaceans because studying marine life from visual observations can be challenging (Moore et al., 2006). Visual monitoring is limited to daylight hours and periods of good weather, and it is difficult to infer animal behavior from brief and rare sightings. Cetaceans use sound for communication, navigation, and locating prey (Nowacek et al., 2016), and it has become clear that detecting their vocalizations is an effective monitoring method that can complement visual observations (Mellinger et al., 2007). PAM can be conducted without interfering with the cetaceans in their natural environment over extended periods of time (Zimmer, 2011). Recent technological advances mean acoustic surveys can be conducted at lower expense and over greater periods of time than ever before. Here we use fixed-hydrophone PAM to study fin whales (*Balaenoptera physalus*) from their vocalizations (calls). This is a highly efficient technique because at certain times of the year fin whales vocalize frequently, and the sound can be detected and recorded at distances of tens of kilometers (McDonald et al., 1995; Stafforda et al., 2007).

[1]*School of Ocean and Earth Science, University of Southampton, Southampton, UK*
[2]*Institute of Sound and Vibration Research, University of Southampton, Southampton, UK*
[3]*National Oceanography Centre, Southampton, UK*

Noisy Oceans: Monitoring Seismic and Acoustic Signals in the Marine Environment, Geophysical Monograph 284, First Edition.
Edited by Gaye Bayrakci and Frauke Klingelhoefer.

DOI:10.1002/9781119750925.ch12

Typically, hydrophones are used for passive acoustic surveys and are either towed behind a vessel or fixed on the sea floor (Mellinger et al., 2007). The use of fixed hydrophones can be subdivided into cabled long-term deployments (Au & Lammers, 2016; Clark et al., 2019; McCarthy et al., 2011), where data is returned to shore in real time (or near real time), and shorter-term deployments of autonomous recorders that need to be retrieved for data collection (Nguyen Hong Duc et al., 2021; Wiggins & Hildebrand, 2007). In some instances, as is the case in the work detailed here, the instruments are deployed for an objective unrelated to PAM, and the analysis to obtain information regarding marine mammals is secondary (Clark, 1995). The use of multiple instruments with close spacing allows vocalizations to be detected on several hydrophones and thus be used to track the source of the sound (Zimmer, 2011). Ocean-bottom seismometers and hydrophones (OBSs and OBHs) provide one type of instrument from which to make these opportunistic observations of marine mammals (Harris et al., 2013; Harris et al., 2018; Iwase, 2015; Matias & Harris, 2015; Rebull et al., 2006). Such instruments provide a well-suited platform to observe deep-water marine mammals that produce low-frequency sounds. In particular, the distinctive, high-intensity calls of fin whales provide an obvious candidate for study.

Using techniques developed for earthquake seismology, even a single OBS can be used to estimate the range to a vocalizing animal (Harris et al., 2013). A network of devices can be used to improve localization accuracy: for example, by using the time difference of arrivals (Dunn & Hernandez, 2009). In addition, it has been demonstrated that surface reflections can be used to estimate the depth of a calling whale (Pereira et al., 2020b). This data can be used for marine management planning and to obtain ecologically significant parameters, such as animal density (Harris et al., 2018).

12.1.2. Fin Whale Vocalizations

Fin whales vocalize in the frequency range 15–142 Hz (Edds, 1988; Širović et al., 2007; Thompson et al., 1992; Watkins et al., 1987). The low frequency means their calls can propagate large distances and are likely to be used for communication or navigation (Croll et al., 2002; Edds-Walton, 1997; Sirovic et al., 2013). The most commonly studied fin whale call is the *20 Hz call* (Hatch and Clark, 2004), which here will be considered the "classic" call and has maximum energy at 21 Hz (Clark et al., 2002). These calls have a distinctive downsweep in frequency (Locke & White, 2011), which can be obscured at longer ranges where reverberation and attenuation can make this structure difficult to observe. This classic call typically has a signal energy of 40 J, a signal length of approximately 1 s, and a source level measured using OBSs of ~189 dB re 1 μPa.m (Weirathmueller et al., 2013; Zimmer, 2011).

The calls are typically emitted in trains in which there are interspersed *backbeat* calls, which are upsweeps that are shorter in duration than the classic calls, have a narrower bandwidth, and are approximately 1 Hz lower in frequency (Fig. 12.1). These sequences of calls may vary in duration and call combinations depending on the activity or social behavior of the fin whale (Edds-Walton, 1997; McDonald et al., 1995). Variations in call trains also have been observed between different fin whale populations and may develop over time. Intervals between consecutive classic calls, the so-called internote interval (INI), typically range between 7 and 26 seconds (Watkins et al., 1987). For example, mean intervals are ~13 s in the eastern North Atlantic but ~15 s in the western Mediterranean, indicating the presence of two distinct populations (Castellote et al., 2012). Intervals longer than the normal INIs are called *rests* and typically last for several minutes (Watkins et al., 1987). These rests correspond to periods when the whale surfaces

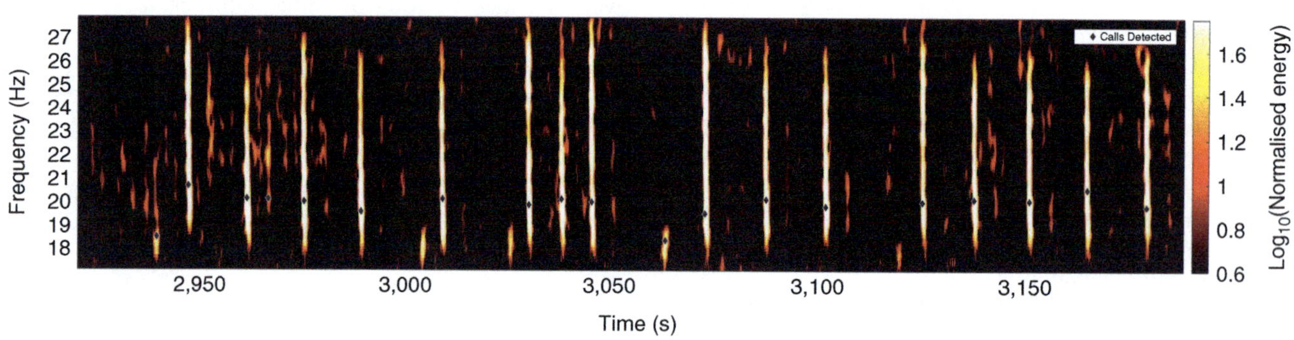

Figure 12.1 Normalized spectrogram illustrating a typical fin whale call pattern observed in our dataset. The normalization is described in the text. Broader-band signals are classic calls, and narrower-band signals are backbeats. Grey diamond markers indicate call detections by our automated algorithm.

to breathe or dives deep, and the typical rest periods for these activities are 150 s and 600 s, respectively (McDonald et al., 1995). In cases where calling ceases for more than 20 minutes, the silent period is called a *gap* (Watkins et al., 1987).

Baleen whales commonly migrate between summer feeding grounds at high latitudes and winter breeding and calving grounds at lower latitudes (Kellogg, 1929). However, a wide range of migratory behaviors has been observed (Geijer et al., 2016). Detection of vocalizations in temperate latitudes is strongly seasonal, with peak activity in the winter months and limited activity in the summer (Pereira et al., 2020a; Watkins et al., 1987). This variation may be linked both to migration patterns and to seasonal variations in behavior associated with the mating season (Oleson et al., 2014). However, a tagging study near the Azores that focused on the months of March to May in 2010–2012 suggested that tagged animals were foraging while on a northward migration toward Greenland (Silva et al., 2013).

12.1.3. Data Available for This Study

Our dataset comes from 50 four-component OBSs (Minshull et al., 2005) and 28 OBHs deployed on the seabed in ~5 km water depth during May–August 2013 as part of the Galicia 3D study (Fig. 12.2), which was focused on the structure of the crust and uppermost mantle in a region where the crust is highly stretched (Bayrakci et al., 2016; Schuba et al., 2018). The deployment covered the summer period when fewer fin whale vocalizations were expected, and indeed, fewer were observed during an OBS deployment offshore Iberia to the south of our study area (Pereira et al., 2020a). However, the deployment was within an area where fin whales have been previously detected acoustically in the summer months (Rebull et al., 2006). The instruments were distributed in an 18 × 4 grid with an east–west spacing of ~6.5 km and a north–south spacing of ~4 km to cover an area of 80 km × 25 km, with an additional six instruments on a line extending further west that provided limited useful

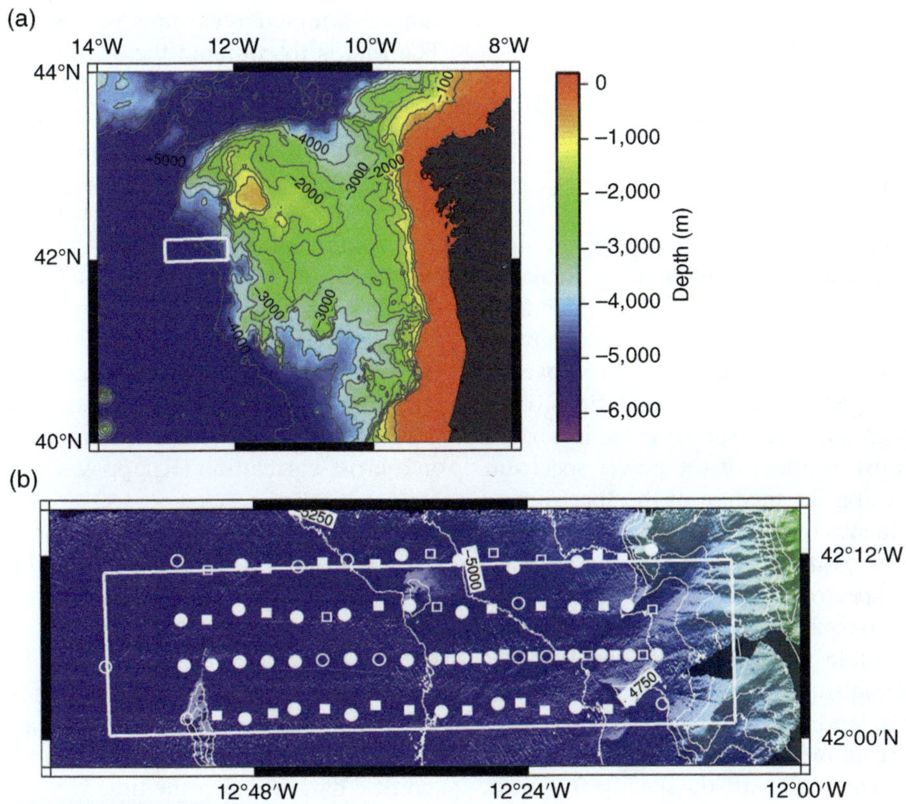

Figure 12.2 (a) White box marks the survey area on the Deep Galicia Margin. Bathymetry is from Smith and Sandwell (1997). (b) Swath bathymetric grid of survey area. Circles mark OBS locations, squares mark OBH locations, and filled symbols mark instruments that provided usable data for this study. White box marks the area of airgun shooting.

data for our study. Several OBSs and OBHs were not retrieved or returned with no usable data, leaving 44 OBSs and 26 OBHs that could be used for our study. OBS data was recorded at a sample rate of 250 Hz and OBH data at 200 Hz, with the OBS data being downsampled to 200 Hz so that a consistent sample rate was used for all data. Differences in hardware meant the OBHs and some OBSs recorded for the entire deployment period, while other OBSs lost battery power before recovery. The survey involved active airgun shooting during the periods 5–23 June (Julian days 156–174) and 16–31 July (Julian days 197–212), with the survey vessel returning to port between these periods because of an engine failure. For this study, we used hydrophone data recorded from 5 June to 12 August (Julian days 156–224). Instrument clocks were synchronized with Universal Time on deployment and recovery, and a linear clock drift was assumed between these times. Instruments used for detection and location had a mean drift rate of 4 ms/day, and the maximum rate was 25 ms/day. If such synchronization was not possible or the clock drift rate was larger, the data was not used. Instrument locations on the seabed were determined using the travel times of airgun shots (Bayrakci et al., 2016).

12.2. METHODS

12.2.1. Call Detection

The recorded data was first converted to day-long WAV files to allow for audio playback at an increased speed that renders the vocalizations audible to the human ear. Our analysis is based on the spectrogram, computed using a Hanning window of 512 samples, with an overlap of 87.5% (yielding frequency and time resolutions of 0.4 Hz and 0.32 s, respectively). The spectrogram is normalized based on a robust estimate of the power spectrum, which is computed using the median of the spectrogram values in each frequency bin (Leung & White, 1998). The normalized spectrogram is obtained by dividing each value in the original spectrogram by the power spectrum at the corresponding frequency.

In addition to fin whale calls, the instruments recorded the shots of the seismic survey, some earthquakes, noise from passing vessels, and ambient noise due to ocean waves. To reduce errors in detecting fin whale vocalizations, filtering can be used to isolate the specific frequency band to promote successful detections. In-band and out-band frequency ranges were determined to optimize the filtering of the data. To detect the whale vocalizations, the mean energies in three bands are computed based on the normalized spectrogram. The three bands are the *in-band,* which is the range of frequencies where a fin whale 20 Hz call is expected to appear—in this case, 15–30 Hz is used—and two *out-bands,* one covering the frequency range below the in-band (i.e., <15 Hz) and one above the in-band (i.e., >30 Hz). For each one-hour block, the energies in each band are calculated as the mean of the normalized spectrogram in the two bands. A detection is made only if the energy is above the threshold of 5 in the in-band but below the threshold of 5 for both out-bands. This threshold value of 5 was determined by testing various values on a small subset of data that had been manually annotated. This dual criterion is used to eliminate false alarms from sounds that occur across a range of frequencies and are not restricted to the 15–30 Hz range.

12.2.2. Delay Estimation

The localization process for a source is based on the estimation of the time differences of arrivals (TDoAs) measured between different instruments. The TDoAs represent the delays of the signals observed on the different sensors. For an array of N sensors, two broad strategies can be adopted regarding which TDoAs to compute. The first is to compute the TDoAs between all possible pairs of sensors (White et al., 2006), which results in $N(N-1)/2$ TDoAs. The second is to select a reference sensor and compute the TDoAs only between the reference sensor and each of the other sensors, which requires only $N-1$ values to be computed. The second approach has the advantage of reducing the computational load because of the smaller number of TDoAs computed. Further, in principle, one can compute all the possible TDoAs from the subset computed relative to a reference sensor. However, the first approach is more robust to estimation errors because there is some redundancy.

The calculation of the TDoAs is most commonly based on a cross-correlation (Knapp & Carter, 1976):

$$\widehat{R}_{f_1 f_2}(\tau) = \frac{1}{T-|\tau|}\int_0^T f_1(t-\tau)f_2(t)dt \qquad (12.1)$$

where f_1 and f_2 are the time series and T is their duration, and the value of the lag, τ, at which the cross-correlation is maximum, was determined and used as the estimate of the TDoA. The correlation can be computed based on the raw acoustic data or using the time series of a metric derived from the acoustic data. If the acoustic data is to be used in the correlation function, the underlying assumption is that the difference between the acoustic signals on two instruments is a simple delay and an amplitude scaling. This assumption may not be true, for various reasons, including complexities in the propagation conditions (e.g., reflections), source directivity, and noise. These approximations are more justifiable for sensors that are close to each other.

For pulsed signals, such as fin whale calls, we can regard the signal as consisting of two components. Borrowing terminology from the fields of communications, radar, and sonar, we can refer to the carrier signal as being the frequency of each pulse (~20 Hz) and the modulation signal defining the shape of the pulses at frequencies of the order of 0.1 Hz. This decomposition provides two routes for computing the delays. The first is based on correlating the carrier waveforms, which potentially provides greater levels of precision; the second is based on correlating the modulation waveform, which results in lower precision but greater robustness.

Here, we describe methods based on the modulation function, which use the envelope of the waveform as an estimate of this modulation signal. To reduce the effect of noise, two processes are adopted before computing the modulation function. First, out-of-band noise is removed by bandpass filtering the signals between 15 and 30 Hz. Then noise in the band of fin whale calls is reduced using a technique pioneered for speech enhancement: power subtraction (McAulay & Malpass, 1980). Following filtering and enhancement, the cross-correlation is calculated using the envelope of the signals. The envelope is computed using the analytic signal based on a Hilbert transform. The analytic signal, $f_a(t)$, is a complex representation of a signal $f(t)$ and is given by

$$f_a(t) = f(t) + i\tilde{f}(t) \tag{12.2}$$

where $\tilde{f}(t)$ is the Hilbert transform of $f(t)$ (Marple, 1999). The envelope is then the magnitude of the analytic signal.

A potential problem with the cross-correlation approach is that the call trains have a nearly constant INI, so the computed cross-correlation functions have multiple peaks separated by the INI. Thus, the wrong peak may be selected to compute the TDoAs. This possibility is mitigated in two ways. First, the natural variation in the INIs means averaging across longer time intervals reduces the height of the secondary (false) peaks in the cross-correlation function, favoring longer time windows, which capture more of this natural variation. Second, the ambiguities generally correspond to unphysical time delays because the interval between calls is around 14 s, corresponding to path-length differences of roughly 20 km. This observation also provides a second motivation for choosing to correlate the modulation function rather than the carrier signal since the ambiguous peaks observed for the carrier signals occur at much shorter path-length distances (~0.05 s or 75 m) and are typically physically realistic and thus much harder to eliminate.

The window length selected to compute the cross-correlation functions is a compromise. A long window reduces the potential problems of secondary peaks in the cross-correlation function and increases the signal-to-noise ratio (SNR) of the peaks. A short window allows more independent localizations to be computed, but the reduced information produces low-correlation peaks, which are more readily masked by noise. A short window also decreases the influence of the motion of the animal during the cross-correlation window, which will blur the main peak, introducing greater uncertainty in the TDoA estimates. This effect suggests a criterion for the maximum window length: the animal's motion within a window should not result in a TDoA change greater than the width of the main correlation peak. The greatest possible change in the TDoA is twice the distance traveled by the whale divided by the speed of sound. This would correspond to the whale swimming along the line connecting the two hydrophones: that is, swimming and vocalizing at the bottom of the ocean (something fin whales do not do). For that case, assuming a whale swimming at 10 km/hr and a width of the correlation peak of 1 s (the typical duration of the call), the corresponding window duration is 280 s. We used a value of 300 s, which is a realistic choice based on the extremely conservative nature of the preceding calculation. These windows were overlapped by 50%. The TDoA estimate was assumed to correspond to a position in the middle of the time window. The effects of absolute propagation time were neglected, leading to a systematic position error of up to a few tens of meters (depending on swim speed) that does not affect swim speed estimates.

The TDoA estimates were post-processed to remove outliers by examining the errors relative to a linear trend. Time delays were also displayed graphically and used to determine tracks only if they showed a smooth progression in time. We used the closest instrument to the whale as a reference and only computed the TDoAs between this instrument and all other instruments. An alternative approach would be to compute TDoAs across all pairs of instruments that detect the whale.

12.2.3. Localization and Tracking

Various methods have been used to localize the source of whale vocalizations from time-delay estimates (Baggenstoss, 2011; Rebull et al., 2006; Wilcock, 2012). We used the method of White et al. (2006) to localize and track the whales. An inversion model from the time delays needs to be applied, and this is achieved with a maximum likelihood estimator by minimizing the weighted least-squares cost function:

$$\Psi(s(n)) = \sum_{k=2}^{N} \frac{(\Delta_{1k}(n) - \mathrm{M}_{1k}(s(n)))^2}{\sigma^2} \tag{12.3}$$

where s is the estimated source location of a discrete time n, Δ_{1k} is the estimated time delay between sensor 1 and sensor k, M_{1k} is the modeled delay between sensor 1 and sensor k, and σ^2 is the weighted variance of the time delays (White et al., 2006). For efficiency, a linear sound-speed model was adopted, where the speed of sound at the ocean surface is 1505 m/s, and the vertical gradient is 0.017 /s. The gradient represents the change in hydrostatic pressure with increased depth for a constant temperature and salinity in the water column, corresponding to the deep isothermal layer (Etter, 1995). This simplification allows rapid localization and correctly reproduces the flattening of propagation paths in the deep ocean but overestimates the mean sound speed of ~1,520 m/s (e.g., Davy et al., 2018) by ~2%. When attempting the localization, it is assumed that only one whale is present, that the delay times are valid with no false detections, and that there are no surface reflections or echoes in the data. If a direct path could not be computed for a given OBS and detection, the delay time for that OBS and detection was not used. Manual checks were made on the INI to check that the tracks used involved only one animal. The extension of this method to tracking multiple animals is a significant step, leading to the field of multitarget tracking (Mahler, 2004).

A minimum of four instruments are needed to successfully locate the source. For computational convenience, the hydrophone locations were projected into Cartesian coordinates; the areas of interest were small enough to neglect the curvature of the Earth. The localization was achieved by minimizing the cost function (3) using a Nelder–Meade simplex method, implemented with random multiple initializations. A Monte Carlo method was employed to quantify the errors associated with noise processes. The depth at which fin whales vocalize is poorly characterized, with acoustic tag data suggesting depths of up to 30 m (Stimpert et al., 2015), so the algorithm was run both constrained such that the whale was assumed to be at the sea surface, and unconstrained such that its depth could correspond to anywhere in the water column. Prior knowledge of the depth distribution of vocalizing animals could usefully be incorporated into the method via a Bayesian framework.

12.2.4. Kalman Filter

By constructing a sequence of localizations from a whale, one can form tracks of the whale's motion while it is calling. Once localizations had been conducted for each sequential vocalization, the next procedure was to apply a Kalman filter to create a final animal track. Using a Kalman filter estimates the movement parameter of the whales and presents the "most probable" whale track from the data. If the stochastic processes are Gaussian and the system is linear, then the Kalman filter represents an optimal estimator, both in the minimum mean-squared sense and, more generally, as the Bayesian maximum a posteriori estimate of the state vector (Bibby & Toutenburg, 1977). The Kalman filter predicts the motion of the fin whale while reducing noise with the associated measured points. It relies on a state-space model of the whale's motion and measurement processes. A state-space model builds a representation of a system by defining a state vector; in this case, the state vector is taken as containing the whale's position and velocity. The model here describes the whale's motion in either two or three dimensions. In the case of two dimensions, the model is simplified by assuming that the animal only moves in a horizontal plane: that is, the vertical motions can be neglected, which is valid if the diving depth of the whale is small compared to the water depth. In the following, we provide details of the two-dimensional model; the extension to three dimensions follows straightforwardly. The state vector, $x_s(n)$, describing the whale's motion in two dimensions is given by

$$x_s(n) = [x(n), \dot{x}(n), y(n), \dot{y}(n)]^t \tag{12.4}$$

where $(x(n),y(n))$ is the whale's position in the plane at time n, and dot notation is used to denote differentiation with respect to time.

This state-space model has two update equations, one describing the motion of the animal (i.e., how $\mathbf{x}_s(n)$ evolves with time) and the second the measurement process. The motion model is

$$x_s(n+1) = \mathbf{A}x_s(n) + w(n+1) \tag{12.5}$$

Here, the linear transition matrix $\mathbf{A}$ describes the movement from time t_n to t_{n+1}, and (n) represents stochastic processes driving the animal's motion. To define $\mathbf{A}$, we used a "near-constant-velocity" model (Li & Jilkov, 2003) that is widely used in tracking problems: for example, to track aircraft in radar applications. We used a Gaussian white-noise process for the vector $w(n)$. According to this model, the matrix $\mathbf{A}$ has the form

$$\mathbf{A} = \begin{bmatrix} 1 & \Delta & 0 & 0 \\ 0 & 1 & 0 & 0 \\ 0 & 0 & 1 & \Delta \\ 0 & 0 & 0 & 1 \end{bmatrix} \tag{12.6}$$

where Δ is the interval between two samples. The model for measurements in this application is straightforward: specifically, the measurements are assumed to be observations of the locations corrupted by additive noise

$$y_m(n+1) = \mathbf{C}x_s(n+1) + v(n+1) \tag{12.7}$$

where $\mathbf{C}$ is the measurement matrix and $v(n)$ is the noise associated with the measurement process (Zimmer, 2011). The measurement matrix for the two-dimensional case has the form

$$\mathbf{C} = \begin{bmatrix} 1 & 0 & 0 & 0 \\ 0 & 0 & 1 & 0 \end{bmatrix} \quad (12.8)$$

The Kalman filter uses the state space model defined by equations 12.4–12.7 to produce recursive estimates of the true positions contained in $\boldsymbol{x}_s(n)$ based on the observed measurements, $\boldsymbol{y}_m(n)$. In this case, the measurement vector contains the estimated locations of the whale. In situations where the state evolution and measurement are linear and the stochastic processes are Gaussian, the Kalman filter yields the maximum a posteriori estimate of the states based on all of the measurements up to time n (Arulampalam et al., 2002). This is achieved by a two-step process: (1) a prediction step, wherein the estimates are updated based on the state model in equation 12.4; and (2) an update step in which the estimate is refined based on the new measurement. The updates for the Kalman filter can be written as follows (Arulampalam et al., 2002):

Prediction step:

$$\widehat{\boldsymbol{x}}_s(n+1|n) = \mathbf{A}\widehat{\boldsymbol{x}}_s(n)$$

$$\mathbf{P}(n+1|n) = \mathbf{A}\mathbf{P}(n|n)\mathbf{A}^t + \mathbf{Q}$$

Update step:

$$\boldsymbol{\varepsilon}(n+1) = \boldsymbol{y}_m(n+1) - \mathbf{C}\widehat{\boldsymbol{x}}(n+1|n)$$

$$\boldsymbol{K}(n) = \mathbf{P}(n+1|n)\mathbf{C}^t(\mathbf{C}\mathbf{P}(n+1|n)\mathbf{C}^t + \mathbf{R})^{-1}$$

$$\widehat{\boldsymbol{x}}_s(n+1) = \widehat{\boldsymbol{x}}_s(n+1|n) + \boldsymbol{K}(n)\boldsymbol{\varepsilon}(n+1)$$

$$\mathbf{P}(n+1|n+1) = (\mathbf{I} - \boldsymbol{K}(n)\mathbf{C})\mathbf{P}(n+1|n) \quad (12.9)$$

where, in two dimensions $\mathbf{Q}$ (4 × 4) and $\mathbf{R}$ (2 × 2) are the correlation matrices of the Gaussian noise processes $\boldsymbol{w}(n)$ and $\boldsymbol{v}(n)$, respectively, which are provided to the algorithm. The quantities of the form $\widehat{\boldsymbol{x}}_s$ are estimates of the state vector $\boldsymbol{x}_s(n)$. The notation $(n+1|n)$ is used to represent estimates of a quantity at time $n+1$ based only on measurements up to time n, sometimes referred to as a priori estimates; accordingly, notation of the form $(n|n)$ denotes estimates of a quantity at time n based on all of the available measurements up to that time. The vector $\boldsymbol{K}(n)$ is called the *Kalman gain*, and $\mathbf{P}$ (4 × 4) is an estimate of the correlation matrix for the errors in the state estimates, providing information regarding the method's accuracy. The state estimate is initialized using the first localizations and assuming a velocity of zero, whereas $\mathbf{P}(0)$ is initialized as the identity matrix $\mathbf{I}$.

12.3. RESULTS

12.3.1. Detections

The dataset was very large (~300 GB), so it was not feasible to run the location and tracking algorithms on the entire dataset. The detection algorithm was applied for periods of 1 hour each day for four representative instruments (OBSs 13, 41, 45, and 51) in different parts of the study area (Fig. 12.3) to determine peak hours for whale vocalizations. There was a large variation in the hourly detection rate (Fig. 12.3). There were periods with low call rates on all four instruments, such as Julian days 156–170 and 213–215. Our analysis of the full dataset also showed that there were hours on particular days with very high call rates on multiple instruments, including days 175, 182, and 222. In addition, there was a high density of calls during the 16th hour of day 187 on OBS 41 and the 18th hour of day 182 on OBS 51. In summary, high call rates were found on multiple OBSs in the network over 1–2 hour periods on days 175, 179, 182, 187, 197, and 222. These peak hours are indications of the presence of fin whales in the area of the OBS array and therefore likely to be identified on surrounding OBSs. These periods, when large numbers of calls are detected, presented the best opportunity of finding a sequence of vocalizations long enough to define a track. During days 156–174 and 197–212, when there was regular airgun shooting in the survey area, call-detection rates were moderate, whereas outside these periods, rates were lower for some periods and higher for others (Fig. 12.3).

12.3.2. Localization

Five tracks were constructed from stable sequences of 81–240 localizations during these periods (Fig. 12.4). On day 175, during the 80-minute period analyzed, the whale initially traveled at an azimuth of ~280° for ~1 km and then turned approximately 100° clockwise to continue traveling at an azimuth of ~020° for 1.7 km past OBS17. The mean velocity of the animal was around 2 km/hr. A track determined on day 179 covers a two-hour period, during which the whale travels backward and forward near OBS16 at an azimuth of ~50° over a length of 4.5 km (Fig. 12.4). Many successive localizations are separated by distances that would represent impossible swim velocities and/or accelerations, but a linear fit of velocity against time results in velocities between 22 and 27 km/hr. A track on day 182 also covers a two-hour period, during which the whale travels 5 km at an azimuth of ~20° before turning to an azimuth of ~80° for a further 3 km close to OBS51. The mean swim speed is 4 km/hr, but successive localizations indicate swim speeds of up to 24 km/hr for short periods. A track on day 187, close to OBS21, continues for a little less than one hour, during which the whale traveled ~15 km. This track is more discontinuous than the others, suggesting that there were some rest periods. Swim speeds were generally between 10 and 20 km/hr. Localizations on day 197 were close to OBS29 and were too scattered to construct a coherent track. These calls occurred when airgun shooting was nearby, and we infer that the shots led to either false

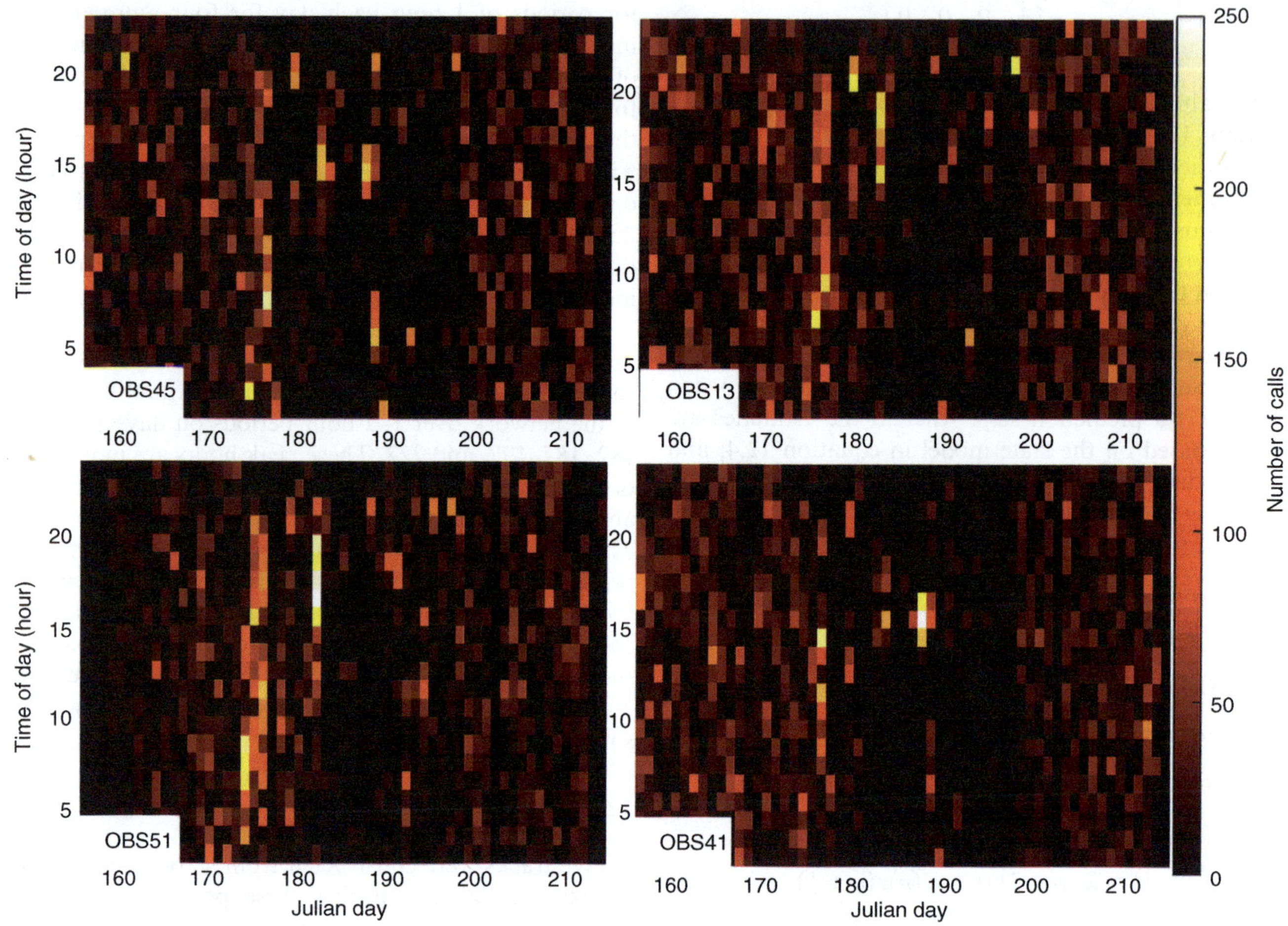

Figure 12.3 Hourly rates of call detection for four representative instruments during the period covered by all four (see Fig. 12.4 for locations). The period of recording available varies between instruments.

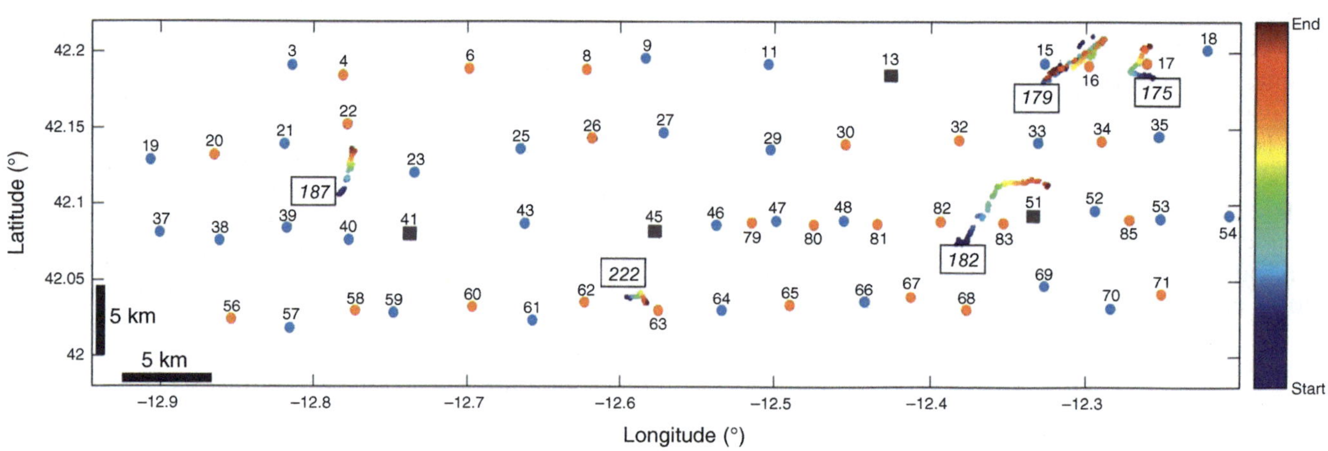

Figure 12.4 Fin whale tracks successfully followed. Symbols mark OBS/OBH positions. Instrument locations marked with grey squares are those used for detection; circles are OBSs and OBHs. Numbers in boxes mark the Julian day of each track.

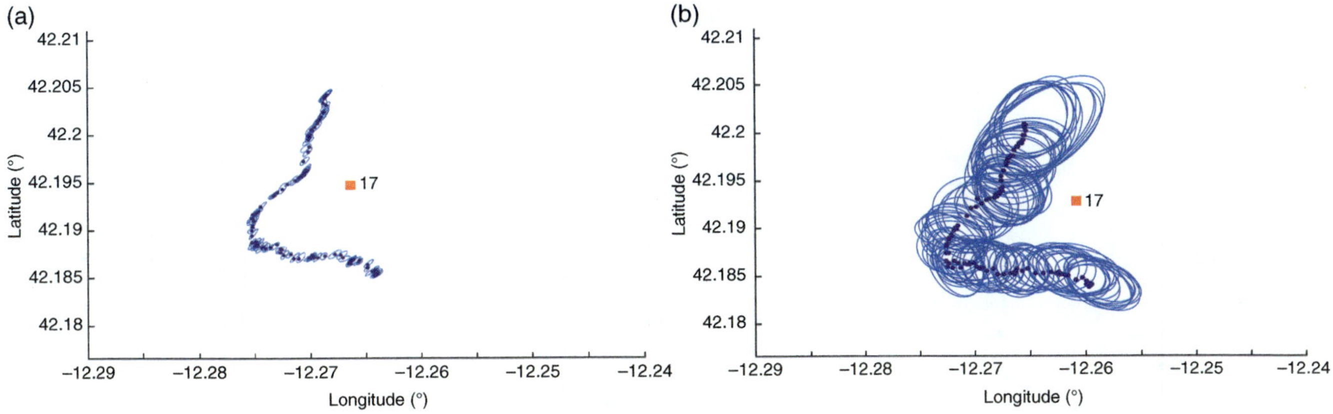

Figure 12.5 (a) Monte Carlo 95% confidence ellipses for localizations on day 175 when OBS13 is used as the reference instrument. (b) Corresponding confidence ellipses when localizations using six different OBSs/OBHs as reference instruments are combined.

detections or spurious cross-correlation maxima. Finally, a one-hour track was detected on day 222 near OBS64. The whale traveled to the northeast and then to the south, with a total track length of only 1.5 km and swim speed generally below 10 km/hr.

To assess the uncertainties in these tracks, we analyzed in further detail the track determined on day 175 (Fig. 12.5). A Monte Carlo analysis (White et al., 2006) applied to localizations using OBS13 as the reference instrument yielded 95% confidence ellipsoids with mean x and y dimensions of 39 and 43 m, respectively (Fig. 12.5a). We also tested the sensitivity of inferred positions to the choice of reference instrument by recomputing the tracks using OBSs/OBHs 18, 35, 51, 52, and 54 as the reference instruments. Although the resulting tracks were all visually similar, Monte Carlo analysis of the combined results yielded 95% confidence ellipsoids with mean x and y dimensions of 325 and 355 m, respectively (Fig. 12.5b). Ellipsoid dimensions in the vertical plane were >100 m for both approaches, with some calls appearing to be above the ocean surface (perhaps due to our choice of sound-speed profile), so it is clear that network geometries were not suitable for determining call depth.

One should note that when using multiple reference sensors, the accuracy of the estimates degrades significantly. This may appear counterintuitive since using multiple reference sensors employs more sensor pairs so that one might anticipate a subsequent performance enhancement. Part of this location uncertainty comes from our simplification of the sound-speed profile, which can result in range errors of up to a few hundred meters at the longest ranges that are mitigated by the broad azimuthal distribution of OBSs used in localization. In addition, it should be noted that because the reference sensor used in the single-reference-sensor case is the sensor closest to the whale, all TDoA estimates are formed using at least one measurement with a high SNR. When using multiple reference sensors, some TDoA estimates are formed when both sensors collect data with a relatively poor SNR. In those cases, the TDoA estimates are unreliable and add considerable uncertainty to the location estimates, as seen in Fig. 12.5b.

12.3.3. Tracking

Because of the poor depth control, the Kalman filter was applied only in the horizontal plane, with the whale assumed to be at the ocean surface, consistent with previous suggestions that these animals dive no deeper than 50–100 m (Croll et al., 2001; Watkins et al., 1987). The filter yielded tracks very similar to the localization method for all tracks. A representative example is shown in Fig. 12.6. However, it resulted in significant smoothing of calculated speeds, which are unrealistic if successive locations are used without filtering (Fig. 12.6b).

12.4. DISCUSSION

We found no clear correlation between detection rate and the presence or absence of shooting. Other whale species have been found to increase the source level and frequency of their calling patterns to compensate for loud ambient noise (Clark et al., 2002), but we cannot infer such a pattern from our data. No reliable whale tracks were recovered during the shooting period, probably because of the low SNR during these periods and the corresponding likelihood of spurious cross-correlation peaks.

Localizations were possible on six days with a range of success over periods of 1–2 hours. Horizontal coordinates were recovered well, but we could not recover whale

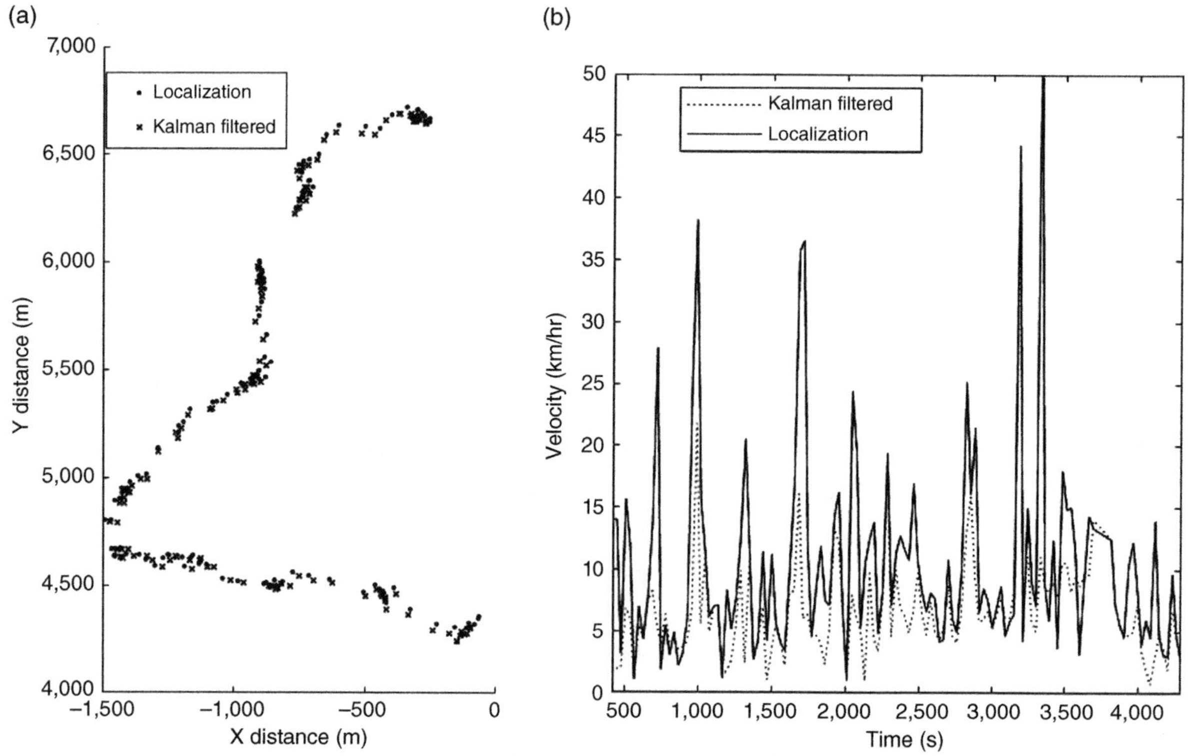

Figure 12.6 (a) Comparison between tracks from the localization algorithm (filled circles) and those generated by the Kalman filter (crosses) for the track on day 175. (b) Comparison of speed estimates based on localization tracks and those from the Kalman filter for the track shown in (a).

depths. The most continuous tracks were found on Julian days 175, 182, 187, and 222. The maximum error associated with these localizations was on the order of hundreds of meters, significantly larger than the spacing between adjacent calls. Based on visual observations, swim speeds as high as 20 km/hr have been reported for 20 mins (Watkins, 1981). Using acoustic methods, Dunn and Hernandez (2009) estimated speeds of 3–7 km/hr. Using tags tracked by satellite, Silva et al. (2013) calculated mean swim speeds of 5.7 km/hr during foraging and 7.7 km/hr during migration, with large variations around these means. Recent work has shown that the swim speed of males is related to whether or not they are singing (Clark et al., 2019), suggesting that singing animals mainly swim at slower speeds. In that study, the mean swimming speed of all tracked whales was found to be 6.7 km/hr, with a standard deviation of 3.7 km/hr. The mean swim speeds observed in this study correspond well with this data. However, some point-to-point estimates produce significantly higher values due to remaining outliers in the localizations that the Kalman filter is unable to smooth sufficiently.

The tracks here cover periods of up to 2 hours. The slow speeds estimated on days 175, 182, and 222 produce more meandering paths consistent with a singing male, whereas the direction changes could indicate foraging behavior (Croll et al., 2001). The track on day 187 is considerably faster and unidirectional, suggesting that the whale is showing a transiting behavior. These two track types, transiting and meandering, are reported elsewhere (McDonald et al., 1995; Rebull et al., 2006; Wilcock, 2012), albeit in most of these instances, the tracks are of longer duration.

12.5. CONCLUSIONS

We have used serendipitous recording of fin whale vocalizations during a deployment of 72 ocean bottom hydrophones/seismometers on the seabed with typical spacings of 4–6.5 km in ~5 km water depth west of Iberia to locate their sources and ultimately track several animals. From our analysis, we conclude the following:

1. Our call-detection algorithm based on the ratios of signal strengths in different frequency bands appears to be highly successful in detecting the broadband classic calls but failed to detect a larger proportion of the narrower-band backbeats.

2. Analysis of a ~two-month time series for several representative instruments revealed up to ~200 calls per hour;

there was no evidence that call-detection rates were affected by the presence of airgun shooting.

3. In a few cases, a series of calls lasting up to two hours was detected on enough seabed instruments to allow reliable estimation of time delays between instruments and thus localization and tracking. The use of a Kalman filter resulted in smoother tracks. Swim speeds along these tracks were highly variable, with means in the range of 2–15 km/hr, and these variations may be linked to various whale behaviors.

4. Long deployments of networks of fixed sound detectors on the seabed can provide rich datasets for the study of fin whale behavior, but the detector spacing of 4–6 km used in our study limited our ability to locate and track the animals.

ACKNOWLEDGMENTS

Seabed instrument deployments were funded by the UK Natural Environment Research Council (NERC; grant NE/E016502/1) and by the Geomar Helmholtz Centre for Ocean Research Kiel. OBSs were provided and operated by the NERC Ocean Bottom Instrumentation Facility. We thank Dirk Klaeschen and the officers, crew, and scientific parties of the research vessel *Poseidon* for their support at sea. We thank Jamie Burford, Masa Oue, and Laura Gilling for their contributions to the development of the techniques used in this study and Richard Davy for assistance with data access. We thank Luis Matias and an anonymous reviewer for their detailed and constructive comments. This work formed part of the University of Southampton Master's thesis of J. Fisher.

REFERENCES

Arulampalam, M. S., Maskell, S., Gordon, N., & Clapp, T. (2002). A tutorial on particle filters for online nonlinear/non-Gaussian Bayesian tracking. *IEEE Transactions on Signal Processing*, *50*(2), 174–188. https://doi.org/10.1109/78.978374

Au, W. W., & Lammers, M. O. (2016). *Listening in the ocean*. Springer. https://doi.org/10.1007/978-1-4939-3176-7

Baggenstoss, P. M. (2011). An algorithm for the localization of multiple interfering sperm whales using multi-sensor time difference of arrival. *Journal of the Acoustical Society of America*, *130*(1), 102–112. https://doi.org/10.1121/1.3598454

Bayrakci, G., Minshull, T. A., Sawyer, D. S., Reston, T. J., Klaeschen, D., Papenberg, C., et al. (2016). Fault-controlled hydration of the upper mantle during continental rifting. *Nature Geoscience*, *9*(5), 384–388. https://doi.org/10.1038/ngeo2671

Bibby, J. M., & Toutenburg, H. (1977). *Prediction and improved estimation in linear models. Wiley*. https://doi.org/10.1002/bimj.197800029

Castellote, M., Clark, W., & Lammers, M. O. (2012). Fin whale (*Balaenoptera physalus*) population identity in the western Mediterranean Sea. *Marine Mammal Science*, *28*(2), 325–344. https://doi.org/10.1111/j.1748-7692.2011.00491.x

Clark, W. (1995). Application of US Navy underwater hydrophone arrays for scientific research on whales. *Reports of the International Whaling Commission*, *45*, 210–212.

Clark, W., Borsani, J. F., & Notarbartolo-di-Sciara, G. (2002). Vocal activity of fin whales, *Balaenoptera physalus*, in the Ligurian Sea. *Marine Mammal Science*, *18*(1), 286-295. https://doi.org/10.1111/j.1748-7692.2002.tb01035.x

Clark, W., Gagnon, G. J., & Frankel, A. S. (2019). Fin whale singing decreases with increased swimming speed. *Royal Society Open Science*, *6*(6), 180525. https://doi.org/10.1098/rsos.180525

Croll, D. A., Acevedo-Gutiérrez, A., Tershy, B. R., & Urbán-Ramírez, J. (2001). The diving behavior of blue and fin whales: Is dive duration shorter than expected based on oxygen stores? *Comparative Biochemistry and Physiology - Part A: Molecular & Integrative Physiology*, *129*(4), 797–809. https://doi.org/10.1016/S1095-6433(01)00348-8

Croll, D. A., Clark, W., Acevedo, A., Tershy, B., Flores, S., Gedamke, J., & Urban, J. (2002). Bioacoustics: only male fin whales sing loud songs: these mammals need to call long-distance when it comes to attracting females. *Nature*, *417*(6891), 809–809. https://doi.org/10.1038/417809a

Davy, R. G., Morgan, J. V., Minshull, T. A., Bayrakci, G., Bull, J. M., Klaeschen, D., et al. (2018). Resolving the fine-scale velocity structure of continental hyperextension at the Deep Galicia Margin using full-waveform inversion. *Geophysical Journal International*, *212*(1), 244–263. https://doi.org/10.1093/gji/ggx415

Dunn, R. A., & Hernandez, O. (2009). Tracking blue whales in the eastern tropical Pacific with an ocean-bottom seismometer and hydrophone array. *Journal of the Acoustical Society of America*, *126*(3), 1084–1094. https://doi.org/10.1121/1.3158929

Edds, P. L. (1988). Characteristics of finback (*Balaenoptera physalus*) vocalizations in the St. Lawrence estuary. *Bioacoustics*, *1* (2–3), 131–149. https://doi.org/10.1080/09524622.1988.9753087

Edds-Walton, P. L. (1997). Acoustic communication signals of mysticete whales. *Bioacoustics*, *8*(1–2), 47–60. https://doi.org/10.1080/09524622.1997.9753353

Etter, P. (1995). *Underwater acoustic modeling: principles, techniques, and applications*. CRC Press.

Geijer, K. A., Notarbartolo di Sciara, G., & Panigada, S. (2016). Mysticete migration revisited: are Mediterranean fin whales an anomaly? *Mammal Review*, *46*(4), 284–296. https://doi.org/10.1111/mam.12069

Harris, D. V., Matias, L., Thomas, L., Harwood, J., & Geissler, W. H. (2013). Applying distance sampling to fin whale calls recorded by single seismic instruments in the northeast Atlantic. *The Journal of the Acoustical Society of America*, *134*(5), 3522–3535. https://doi.org/10.1121/1.4821207

Harris, D. V., Miksis-Olds, J. L., Vernon, J. A., & Thomas, L. (2018). Fin whale density and distribution estimation using acoustic bearings derived from sparse arrays. *The Journal of the Acoustical Society of America*, *143*(5), 2980–2993. https://doi.org/10.1121/1.5031111

Hatch, L. T., & Clark, W. (2004). *Acoustic differentiation between fin whales in both the North Atlantic and North Pacific*

Oceans. Paper presented at International Whaling Commission Scientific Committee, Sorrento, Italy.

Iwase, R. (2015). Fin whale vocalizations observed with ocean bottom seismometers of cabled observatories off east Japan Pacific Ocean. *Japanese Journal of Applied Physics*, *54*(7S1), 07HG03. https://doi.org/10.7567/jjap.54.07hg03

Kellogg, R. (1929). What is known of the migrations of some of the whale-bone whales. In *Annual report of the Board of Regents of the Smithsonian Institution 1928* (pp. 467–494). United States Government Printing Office.

Knapp, H., & Carter, G. (1976). The generalized correlation method for estimation of time delay. *IEEE Transactions on Acoustics, Speech, and Signal Processing*, *24*(4), 320–327. https://doi.org/10.1109/TASSP.1976.1162830

Leung, T. S., & White, P. R. (1998). Robust estimation of oceanic background spectrum. In J. G. McWhirter & I. K. Proudler (Eds.), *Mathematics in signal processing IV* (pp. 369–382). Clarendon Press. https://doi.org/10.1007/BF00142586

Li, X. R., & Jilkov, V. P. (2003). Survey of maneuvering target tracking. Part I. Dynamic models. *IEEE Transactions on Aerospace and Electronic Systems*, *39*(4), 1333–1364. https://doi.org/10.1109/TAES.2003.1261132

Locke, J., & White, P. R. (2011). The performance of methods based on the fractional Fourier transform for detecting marine mammal vocalizations. *The Journal of the Acoustical Society of America*, *130*(4), 1974–1984. https://doi.org/10.1121/1.3631664

Mahler, R. P. S. (2004). "Statistics 101" for multisensor, multitarget data fusion. *IEEE Aerospace and Electronic Systems Magazine*, *19*(1), 53–64. https://doi.org/10.1109/maes.2004.1263231

Marple, S. L. (1999). Computing the discrete-time analytic signal via FFT. *IEEE Transactions on Signal Processing*, *47*(9), 2600–2603. https://doi.org/10.1109/78.782222

Matias, L., & Harris, D. (2015). A single-station method for the detection, classification, and location of fin whale calls using ocean-bottom seismic stations. *The Journal of the Acoustical Society of America*, *138*(1), 504–520. https://doi.org/10.1121/1.4922706

McAulay, R. J., & Malpass, M. L. (1980). Speech enhancement using a soft-decision noise suppression filter. *IEEE Transactions on Acoustics, Speech, and Signal Processing*, *28*(2), 137–145. https://doi.org/10.1109/TASSP.1980.1163394

McCarthy, E., Moretti, D., Thomas, L., DiMarzio, N., Morrissey, R., Jarvis, S., et al. (2011). Changes in spatial and temporal distribution and vocal behavior of Blainville's beaked whales (*Mesoplodon densirostris*) during multiship exercises with mid-frequency sonar. *Marine Mammal Science*, *27*(3), E206-E226. https://doi.org/10.1111/j.1748-7692.2010.00457.x

McDonald, M. A., Hildebrand, J. A., & Webb, S. (1995). Blue and fin whales observed on a sea-floor array in the northeast Pacific. *The Journal of the Acoustical Society of America*, *98* (2), 712–721. https://doi.org/10.1121/1.413565

Mellinger, D. K., Stafford, K. M., Moore, S. E., Dziak, R. P., & Matsumoto, H. (2007). An overview of fixed passive acoustic observation methods for cetaceans. *Oceanography*, *20*(4), 36–45. https://doi.org/10.5670/oceanog.2007.03

Minshull, T. A., Sinha, M., & Peirce. (2005). Multi-disciplinary, sub-seabed geophysical imaging—a new pool of 28 seafloor instruments in use by the United Kingdom Ocean Bottom Instrument Consortium. *Sea Technology*, *46*(10), 27–31.

Moore, S. E., Stafford, K. M., Mellinger, D. K., & Hildebrand, J. A. (2006). Listening for large whales in the offshore waters of Alaska. *Bioscience*, *56*(1), 49–55. https://doi.org/10.1641/0006-3568(2006)056(0049:Lflwit)2.0.Co;2

Nguyen Hong Duc, P., Cazau, D., White, P. R., Gérard, O., Detcheverry, J., Urtizberea, F., & Adam, O. (2021). Use of ecoacoustics to characterize the marine acoustic environment off the North Atlantic French Saint-Pierre-et-Miquelon archipelago. *Journal of Marine Science and Engineering*, *9*(2), 177. https://doi.org/10.3390/jmse9020177

Nowacek, D. P., Christiansen, F., Bejder, L., Goldbogen, J. A., & Friedlaender, A. S. (2016). Studying cetacean behavior: new technological approaches and conservation applications. *Animal Behaviour*, *120*, 235–244. https://doi.org/10.1016/j.anbehav.2016.07.019

Oleson, E. M., Slrovic, A., Bayless, A. R., & Hildebrand, J. A. (2014). Synchronous seasonal change in fin whale song in the North Pacific. *PLoS ONE*, *9*(12). https://doi.org/10.1371/journal.pone.0115678

Pereira, A., Harris, D., Tyack, P., & Matias, L. (2020a). Fin whale acoustic presence and song characteristics in seas to the southwest of Portugal. *The Journal of the Acoustical Society of America*, *147*(4), 2235–2249. https://doi.org/10.1121/10.0001066

Pereira, A., Harris, D., Tyack, P., & Matias, L. (2020b). On the use of the Lloyd's Mirror effect to infer the depth of vocalizing fin whales. *The Journal of the Acoustical Society of America*, *148*(5), 3086–3101. https://doi.org/10.1121/10.0002426

Perry, S. L., DeMaster, D. P., & Silber, G. K. (1999). The great whales: history and status of six species listed as endangered under the U.S. Endangered Species Act of 1973. *Marine Fisheries Review*, *61*(1), 1–74.

Rebull, O., Cusí, J. D., Ruiz Fernández, M., & Muset, J. G. (2006). Tracking fin whale calls offshore the Galicia Margin, North East Atlantic Ocean. *The Journal of the Acoustical Society of America*, *120*(4), 2077–2085. https://doi.org/10.1121/1.2336751

Schuba, N., Gray, G. G., Morgan, J. K., Sawyer, D. S., Shillington, D. J., Reston, T. J., et al. (2018). A low-angle detachment fault revealed: three-dimensional images of the S-reflector fault zone along the Galicia passive margin. *Earth and Planetary Science Letters*, *492*, 232–238. https://doi.org/10.1016/j.epsl.2018.04.012

Silva, M. A., Prieto, R., Jonsen, I., Baumgartner, M. F., & Santos, R. S. (2013). North Atlantic blue and fin whales suspend their spring migration to forage in middle latitudes: building up energy reserves for the journey? *PLoS ONE*, *8*(10). https://doi.org/10.1371/journal.pone.0076507

Širović, A., Hildebrand, J. A., & Wiggins, S. M. (2007). Blue and fin whale call source levels and propagation range in the Southern Ocean. *The Journal of the Acoustical Society of America*, *122*(2), 1208–1215. https://doi.org/10.1121/1.2749452

Širović, A., Williams, L. N., Kerosky, S. M., Wiggins, S. M., & Hildebrand, J. A. (2013). Temporal separation of two fin whale call types across the eastern North Pacific. *Marine Biology*, *160* (1), 47–57. https://doi.org/10.1007/s00227-012-2061-z

Stafford, K. M., Mellinger, D. K., Moore, S. E., & Fox, G. (2007). Seasonal variability and detection range modeling of baleen whale calls in the Gulf of Alaska, 1999–2002. *The Journal of the Acoustical Society of America, 122*(6), 3378–3390. https://doi.org/10.1121/1.2799905

Stimpert, A. K., et al. (2015). Sound production and associated behavior of tagged fin whales (*Balaenoptera physalus*) in the Southern California Bight. *Animal Biotelemetry, 3*(1). https://doi.org/10.1186/s40317-015-0058-3

Thompson, P. O., Findley, L. T., & Vidal, O. (1992). 20-Hz pulses and other vocalizations of fin whales, *Balaenoptera physalus*, in the Gulf of Mexico. *The Journal of the Acoustical Society of America, 92*(6), 3051–3057. https://doi.org/10.1121/1.404201

Watkins, W. A. (1981). Activities and underwater sounds of fin whales. *Scientific Reports of the Whales Research Institute, 33*, 83–117.

Watkins, W. A., Tyack, P., Moore, K. E., & Bird, J. E. (1987). The 20-Hz signals of finback whales (*Balaenoptera physalus*). *The Journal of the Acoustical Society of America, 82*(6), 1901–1912. https://doi.org/10.1121/1.395685

Weirathmueller, M. J., Wilcock, W. S. D., & Soule, D. (2013). Source levels of fin whale 20 Hz pulses measured in the Northeast Pacific Ocean. *The Journal of the Acoustical Society of America, 133*(2), 741–749. https://doi.org/10.1121/1.4773277

White, P. R., Leighton, T. G., Finfer, D., Powles, & Baumann, O. N. (2006). Localisation of sperm whales using bottom-mounted sensors. *Applied Acoustics, 67*(11–12), 1074–1090. https://doi.org/10.1016/j.apacoust.2006.05.002

Wiggins, S. M., & Hildebrand, J. A. (2007). High-frequency acoustic recording package (HARP) for broad-band, long-term marine mammal monitoring. In *2007 symposium on underwater technology and workshop on scientific use of submarine cables and related technologies, vols 1 and 2* (pp. 551–557). IEEE. https://doi.org/10.1109/UT.2007.370760

Wilcock, W. S. D. (2012). Tracking fin whales in the northeast Pacific Ocean with a seafloor seismic network. *The Journal of the Acoustical Society of America, 132*(4), 2408–2419. https://doi.org/10.1121/1.4747017

Zimmer, W. M. X. (2011). *Passive acoustic monitoring of cetaceans. Cambridge University Press.* https://doi.org/10.1017/CBO9780511977107

13

Noise From Marine Traffic

David Dellong, Bazile G. Kinda, and Florent Le Courtois

ABSTRACT

Shipping activities are diverse, extend almost globally, and have increased in intensity over the past half-century. Motorized ships radiate underwater noise through numerous and complex mechanisms. The generated sound is radiated in the ocean and can be propagated over large distances. In acoustic recordings, individual vessel signatures can be characterized, and an increase in overall ambient noise levels can be observed near shipping routes. During the twentieth century, defense programs have developed their knowledge of acoustic radiation, propagation, and underwater signal analysis. Over the last decade, research and development in these fields has been driven primarily by the requirement to regulate underwater noise. Growing awareness of the environmental impacts of pollution has driven regulatory requirements. This chapter provides an overview of past and current research on underwater shipping noise at two levels: signature analysis of individual vessels and estimating and modeling ambient noise levels at larger scales.

13.1. INTRODUCTION

Underwater noise of human origin is receiving increasing attention from the scientific community because it can interfere with key activities of noise-sensitive marine animals. In the mid-latitudes, low-frequency ambient noise levels attributed to shipping are known to increase with sustained growth in maritime traffic, which is the primary means of transporting goods in a globalized world (Simard et al., 2016). In the Northeast Pacific Ocean, the 20–80 Hz ambient noise level appeared to increase by 10 dB between 1963 and 2001 (Amailland et al., 2018). This hypothesis is reinforced by other measurements in the Pacific Ocean, supporting the idea that shipping noise may have doubled every decade since 1963's early measurements (Viola et al., 2017; Vakili et al., 2020; Wagstaff, 1973; Amailland et al., 2018; Breeding Jr. et al., 1996). Several studies in the world's oceans have shown positive correlations between low-frequency ambient noise levels and shipping (Etter, 2018; Tollefsen & Dosso, 2008; ISO 18405, 2017; MacGillivray & de Jong, 2021; Al-Badrawi et al., 2020).

Climate changes, which can already be observed in the polar environments, are accompanied by new practices (e.g., new shipping routes and fishing areas), and a possible rise in ambient noise levels related to shipping is expected in the near future in pristine ocean environments. Recent studies showed an increase of about 9 dB in the average recorded noise in the Kongsfjorden fjord, Svalbard, after a heavy maritime traffic event during the summer of 2015 (ANSI/ASA S12.64-2009/Part 1, 2009). Similarly, predictive models of ambient noise in the Canadian Arctic have shown that noise from marine traffic could increase significantly in the coming decades (Farcas et al., 2016; Audoly et al., 2014).

Département Acoustique Sous-marine, Service Hydrographique et Oceanographique de la Marine, Brest, France

Noisy Oceans: Monitoring Seismic and Acoustic Signals in the Marine Environment, Geophysical Monograph 284, First Edition.
Edited by Gaye Bayrakci and Frauke Klingelhoefer.

DOI:10.1002/9781119750925.ch13

Increased ambient noise levels alter the acoustic environment of all compartments of marine organisms. Studies observing the biological impacts of marine traffic-induced noise are concerned with marine mammals, which use acoustics to perform key functions: feeding, communicating, maintaining social links, and finding partners. At the individual level, Rolland et al. (2012) have shown a correlation between chronic exposure to low-frequency shipping noise and stress-related hormones in North Atlantic right whales. Auditory masking was defined as the mechanism of underwater noise interfering with how marine mammals receive acoustic signals (EMODnet Bathymetry Consortium et al., 2018). In response to increasing ambient noise levels, individuals of the same species can adapt their vocal behavior by increasing their call amplitude to compensate for the reduction in communication space (McIntyre et al., 2021), while bowhead whales tend to lower their call frequency (Rolland et al., 2012) under similar conditions. The presence of shipping noise could reduce their communication space up to 60% (Fischer, 2008). Similar results have been observed for pilot whales (ISO 17208-1, 2016). In general, avoidance, foraging, and interrupted socializing were observed for several marine mammal species in the presence of shipping noise (Trabattoni et al., 2020; Blake, 1984; Dekeling et al., 2014). All these negative effects at the individual level can have consequences at the population level, such as habitat abandonment, changes of migratory routes, and changes in breeding success (Ren & Hermand, 2013), that need further investigation.

The impact of noise on marine fauna can affect the local ecological balance and cause societal aftereffects (e.g., perturbed presence of emblematic species). Consequently, underwater noise is recognized as pollution in many international texts (Le Gall & Bonnel, 2013; Urick, 1983), and regulations are emerging (Thode et al., 2017). In the European Union, the Marine Strategy Framework Directive (MSFD) (Collins, 1993) requires the member states to ensure good ecological status and that underwater noise is addressed through descriptor 11 (D11). Noise related to marine traffic at the ocean-basin scales must be reported in two adjacent third-octave bands (63 and 125 Hz) (Rivet et al., 2021).

Knowledge of shipping noise has greatly increased in recent decades. It was initially driven by defense purposes to asses sonar performance and detection (Tasker et al., 2010). Since the early alerts about the risk of underwater noise pollution, increased public interest has led to improved knowledge. Acoustic measurements were commonly used to monitor underwater soundscapes and thus provided relevant amounts of data on radiated shipping noise. Relatively recent management and regulatory needs involve monitoring large and remote areas. The high operating costs associated with offshore field measurements require noise-mapping strategies that rely on models combining shipping data and ocean environmental data (Johnson & Dudgeon, 1992; Erbe et al., 2016; Audoly & Rizzuto, 2015). Recent research has two goals (Porter & Liu, 1994): (1) controlling radiated noise by identifying onboard sources, and (2) understanding shipping noise in the marine environment using propagation models. This chapter presents the primary developments and research on the acoustic contribution of marine traffic to the soundscape at both scales.

The chapter is organized as follows: Section 13.2 presents the primary sound sources of a ship and its related source level as measured or modeled. Section 13.3 provides details on noise mapping methodologies and applications, and Section 13.4 provides insight into the evolution of maritime traffic noise regulations, followed by some concluding remarks.

13.2. UNDERWATER RADIATED NOISE

The noise radiated by a ship results from the complex acoustic contributions of individual sources (Antoni, 2009). These sources are of various natures and are usually classified based on their sound-generation mechanism. Depending on the ship, its construction, its associated activity, and its operating condition, the sources and their properties and positions vary greatly (Lighthill, 1952). Their location on the ship significantly affects their acoustic radiation and directivity pattern. Since 1950, to tackle the largest contribution to ambient noise, acoustic radiation has been primarily studied for commercial shipping at nominal transit speed (Tasker et al., 2010). More recently, studies have been conducted on radiated noise from smaller ships (McKenna et al., 2012).

The purpose of characterizing radiated noise is twofold. On one hand, it identifies the individual acoustic signatures of ships and allows classification (Antoni, 2009). On the other hand, it generalizes statistical noise modeling to assess shipping noise at coarse scales (Johnson & Dudgeon, 1992).

13.2.1. Sources of Shipping Noise

Sources are usually characterized by their emission level, spectrum, and radiation pattern and the temporal property of the emitted signal (Porter, 1992; McCarthy, 2007). They are classified based on the physical processes that drive sound generation, as shown in Fig. 13.1. Two primary physical processes are responsible for producing ship noise (Antoni, 2009; McCarthy, 2007): the propagation of structural vibrations from machinery and hydrodynamic noise from fluid displacement. The following subsections summarize current knowledge of the primary sources of noise.

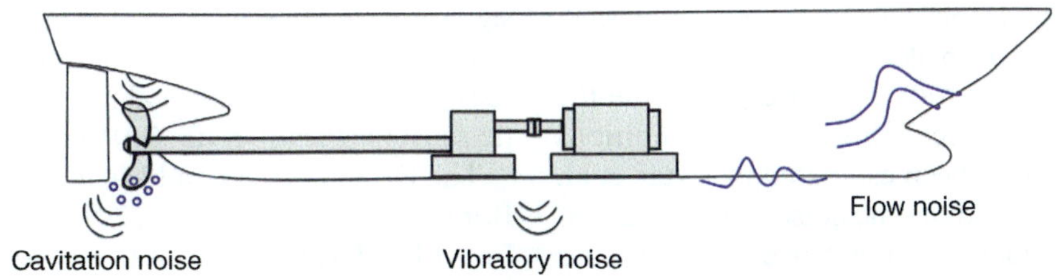

Figure 13.1 Schematic disposition of primary noise sources onboard a ship. Adapted from Amailland (2017).

Radiated Noise from Vibrations

Internal machinery on ships generates sound and vibrations. Sound propagates in air and water, whereas vibrations follow a solid path. Both are related because of the continuity of the displacement at the solid–fluid interface: the interaction of acoustic waves with solids generates a combination of longitudinal and transversal vibrations, and vibrations radiate sound waves in the fluid. The propagation through the hull of the ship generates its underwater vibroacoustic radiation.

Rotating machinery, including motors and drive shafts, is the naval industry's primary source of acoustic and vibratory noises. The ship's motors are often the most powerful sources on board and can be located close to and tightly fixed to the hull structure. They are the major contributor to the underwater radiation of the vibroacoustic field (e.g., applied study from Arveson and Vendittis [2000]). The motor rotates to activate the propellers. The vibrations and sounds generated by the mechanical interactions of rotating and nonrotating parts produce a spectrum of harmonic frequencies. The frequency content is proportional to the rotation speed frequency; when the speed increases, the frequencies change to a higher pitch.

The emission level depends on the number of motors and their rotation speed. Usually, the number of motors and their mechanical power depend on the boat's size and activity (Aumond et al., 2018). The larger it is, the more likely it is to generate noise.

The acoustic radiation of these sources is modulated by the transfer function of the structure. The acoustic properties of the hull are often assimilated into plates or shells between air and water. Depending on the ship's dimensions, the amplitude of radiation at particular frequencies can be increased by resonance of the hull structure. Thus, the amplitude of underwater radiated noise also depends on the direction. The hull radiation shows the favored direction of propagation in the prow–stern axis as well as in the depth dimension.

Hydrodynamic Noise and Cavitation

The relative motion of a ship, its components, and the water can generate motion of the fluid. The motion can become turbulent when particular conditions of speed and viscosity are met. Turbulence is a complex physical phenomenon that can generate acoustic radiation (Kotilainen et al., n.d.). Acoustic radiation from flow turbulence propagates over short distances compared to machine and vibration noise. The spectrum of emission covers a wide band of frequencies, usually below 1,000 Hz (Amailland, 2017).

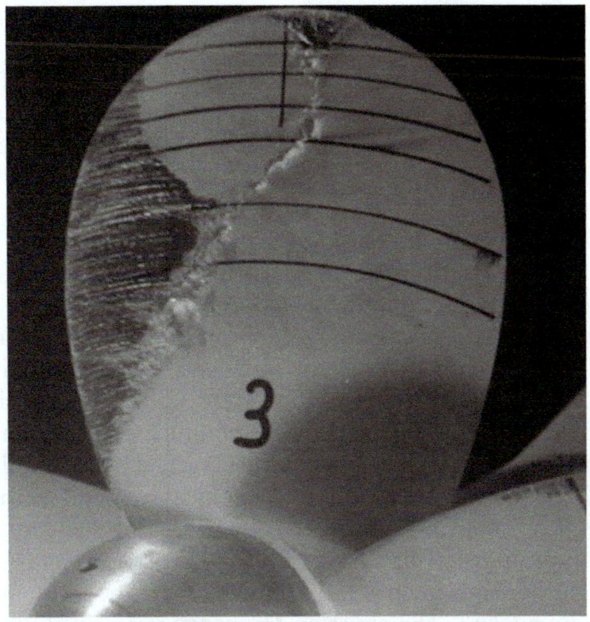

Figure 13.2 Photograph of cavitation. The bubbles are generated at the attack side of the propeller blade and collapse when the pressure increases again (Kuiper, 1997).

The interaction of motion and fluid is the origin of an additional major source of noise underwater. While rotating, the propeller's blades generate volumes where the hydrostatic pressure decreases. Water can evaporate into vapor bubbles under lower pressure (Aulanier et al., 2017); this phenomenon is called cavitation. An example from Kuiper (1997) is shown in Fig. 13.2. The oscillation and collapse of bubbles when the surrounding water pressure increases radiates acoustic waves and generates propeller vibrations. This effect is commonly referred to as "propeller singing" (Erbe et al., 2019).

The resulting combination of radiation bubbles and propeller singing is cavitation noise. Its occurrence requires particular conditions such as rotation speed, angle of attack of the propeller blades, and sea state (Leroy et al., 2008; Jensen, 2008), but it can become the ship's primary radiation source if unfavorable conditions are met. Source levels have been estimated in Aktas et al. (2016). As with other hydrodynamic noise sources, cavitation noise has low-frequency content over a wide band below 300 Hz. The radiation spectrum can show tonal components depending on the configuration. The source level is estimated between 130 and 150 dB re 1 μPa (Wales & Heitmeyer, 2002; Aktas et al., 2016). The radiation pattern is omnidirectional but can be shadowed by the hull structure in the prow direction depending on the propeller position (Lighthill, 1952) Assessing directivity pattern remains a major research topic (Arveson & Vendittis, 2000).

Acoustic Signatures of Ships in the Environment

A ship's acoustic signature is defined by the radiation of all the sources onboard. A recording of the signature in far-field conditions contains the propagation of radiation of all the individual sources. Because of the relative motion between the ship and the sensor, the received signal presents a modulation of its amplitude and frequency content: the Doppler effect (Tasker et al., 2010). This effect can be observed in acoustic measurements with a frequency shift of the sound wave caused by the emitter motion: as the ship moves toward the sensor, the frequency is high-pitched; as it moves away, the frequency is low-pitched. The emission frequency is received when the ship is at its closest point of approach (CPA) to the sensor. The amplitude of the signal is modulated in a similar way; it increases when the ship approaches and decreases when the ship moves away, depending on transmission losses in the environment. The magnitude of the frequency shift of the Doppler effect depends on the speed and emitted frequency, but it is usually around tenths of hertz for ship noise. Doppler frequency shift can provide information about speed and CPA distance (Southall et al., 2019).

The ship's acoustic signature contains a great deal of information about ship characteristics: the number of blades is related to individual low frequencies, and cavitation noise (Wales & Heitmeyer, 2002), range, and speed can be inferred from analysis of the Doppler effect. Moreover, the frequency-modulated content of the signature can be determined using cyclostationarity (Magnier & Gervaise, 2020) or Cepstrum analysis (Southall et al., 2019), allowing better identification and classification of vessels in complex soundscapes. Such representations are relevant to demodulate cyclic information in rotating machinery (Andrew et al., 2002). Thus, several blades, their rotation speed, and the number of motors can be inferred to identify ship activity. Detailed analysis of ship signatures is of great interest in passive monitoring. Similar signatures have been monitored using ocean-bottom seismometers (Southall et al., 2019) and distributed acoustic sensors (Picaut et al., 2019).

The Doppler effect and Lloyd's mirror (Bazile Kinda & Bonnel, 2015; Kinda et al., 2017) can both be observed in Fig. 13.3. The signature was recorded in September 2019 by the SHOM in the Celtic Sea near a shipping route. It highlights the strong constructive and destructive interference pattern (V-shaped striation pattern) related to the shallow area (120 m) and interaction with the seabed and propagation effects. Unfortunately, the frequency shift related to the Doppler effect is not easily seen in Fig. 13.3. The horizontal tonal lines are attributed to internal mechanisms of noise generators and can also be identified in Fig. 13.4. Wideband components also seen in this figure can be attributed to propeller blade cavitation (other examples are shown in McKenna et al. [2012]).

At low frequencies and in shallow areas (e.g., below 100 Hz for depth around 100 m), seabed properties greatly affect acoustic propagation. As ships radiate loud and continuous noise and are globally distributed, they can offer convenient sources for passive geoacoustic inversion (Etter, 2009; Parsons et al., 2021; Scorzolini et al., 2010). Acoustic propagation under such conditions is characterized by a V-shape striation pattern in the time-frequency representation (Fig. 13.4 and (Kinda et al., 2017; Aulanier et al., 2016)). Using dedicated signal processing and an inversion scheme, sound speed and density in the seabed can be estimated (Etter, 2009; Parsons et al., 2021). Later research topics aimed to find the number of layers and their thickness (Scorzolini et al., 2010; Ross & Kuperman, 1989). For this purpose, Bayesian inference has been found to be a suitable framework, particularly regarding uncertainties. This topic remains of high scientific interest with the increase in monitoring program coverage.

13.2.2. Measuring Radiated Noise

Ship noise defines the radiation of several sources of various natures that can have very different properties. The radiated noise properties are usually characterized by reversing propagation effects from acoustic measurements to infer source levels (or radiated level) for an equivalent point source whose amplitude is measured at 1 m (GEBCO Bathymetric Compilation Group, 2020). This task is challenging because the environment greatly affects propagation, and the naval industry produces a wide range of ship sizes and designs.

Providing a robust assessment of radiated noise is important for shipbuilders, given emerging requirements from environmental policies. Technical groups of the ISO and ANSI standardization committees proposed a measurement protocol that can be applied to ships of

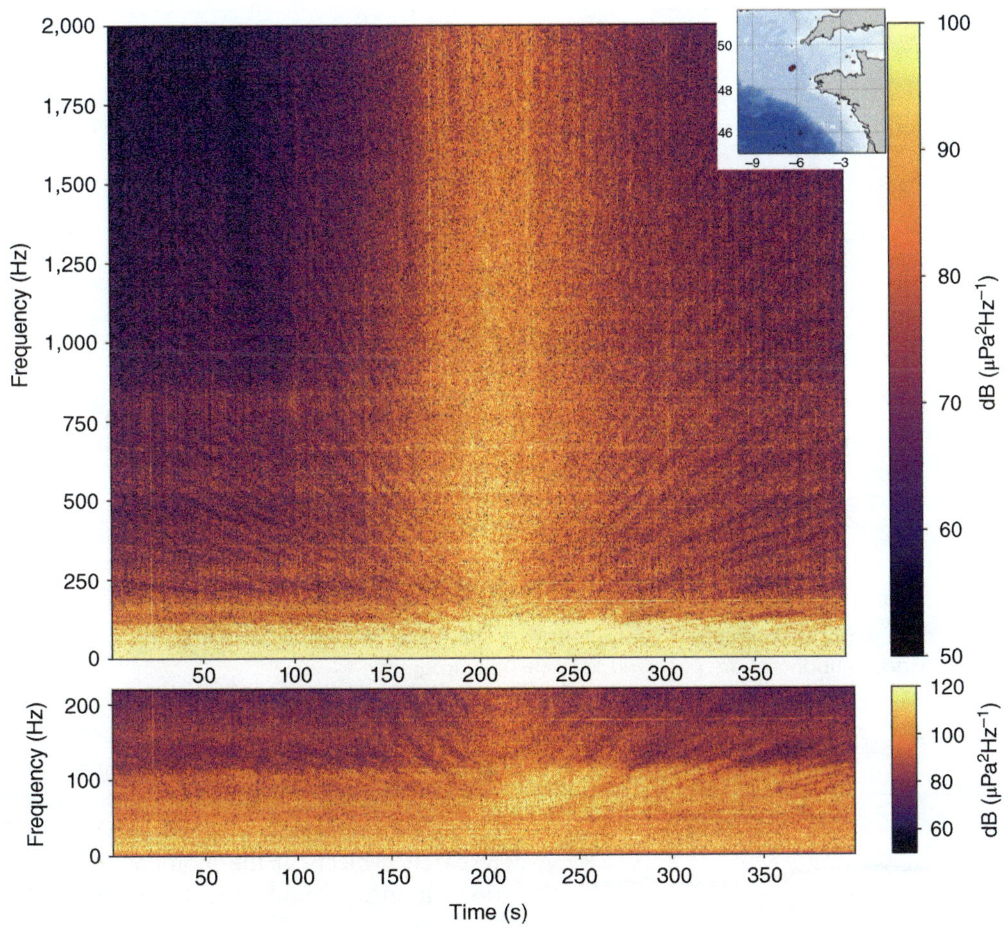

Figure 13.3 Example of a recorded vessel noise spectrogram with Lloyd's mirror interference patterns. The vessel (Maritime Mobile Service Identity: 636092253, identified from Automatic Identification System data, Exact Earth catalog) is passing close to a single hydrophone (CPA 2,400 m). Acoustic data from a SHOM station localized at longitude = −6.284999, latitude = 48.95027, instrument depth = 83 m below the surface, seabed depth = 120 m. The starting time is 17 September 2019 at 05:29:59. Top-right corner: bathymetry map (SHOM data) showing the hydrophone position. The subplot at the bottom shows a zoom of the lower frequencies of the spectrogram.

any size (Parks et al., 2011; GEBCO Bathymetric Compilation Group, 2020; Gervaise et al., 2012a), as well as a terminology standard (Gervaise et al., 2012b). The primary objective is to provide a general framework in which uncertainty in estimating radiated noise can be reduced or controlled. The methodology considers three grades of quality measurement methods: C, B, and A, from lowest to highest quality. Grades A and B require three hydrophones, whereas grade C requires only one hydrophone. The location should be as quiet as possible, away from general shipping and wind noise. The quality of the hydrophone in terms of frequency range and accuracy also affects the method grade. The frequency band defined by grade C should be between 50 Hz and 10 kHz. The measurement configuration must be in a free field propagation, and the ship is assumed to be a single source located at a theoretical acoustic center (i.e., the most energetic location). The depth should be at least 1, 1.5, or 3 times the ship length for grade C, B, and A, respectively. The radiated level is estimated by correcting only the received level by the distance between the receiver and the ship for the CPA of the trajectory. For grades A and B, estimated radiated levels are averaged over hydrophones to reduce the impact of surface reflections. An experimental configuration is shown in Fig. 13.5. Depending on the grade, dedicated processing methods are proposed to reduce uncertainty.

Although the standards aim to define a common methodology to measure radiated noise with confidence, additional research is needed to achieve greater accuracy in source characterization. Array processing is a common way to estimate source levels by reversing the propagation effects over several sensors. It allows filtering the direction of arrival of acoustic waves and then identifying and

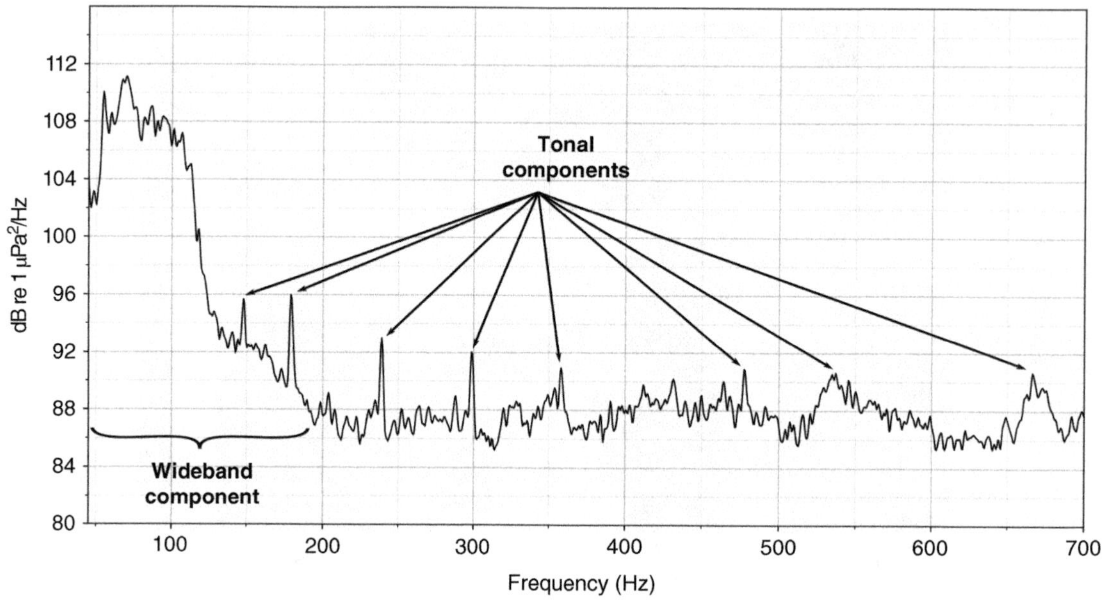

Figure 13.4 Periodogram representation computed from the same recording as in Fig. 13.3. Tonal and low-frequency wideband components are observed as in the spectrogram.

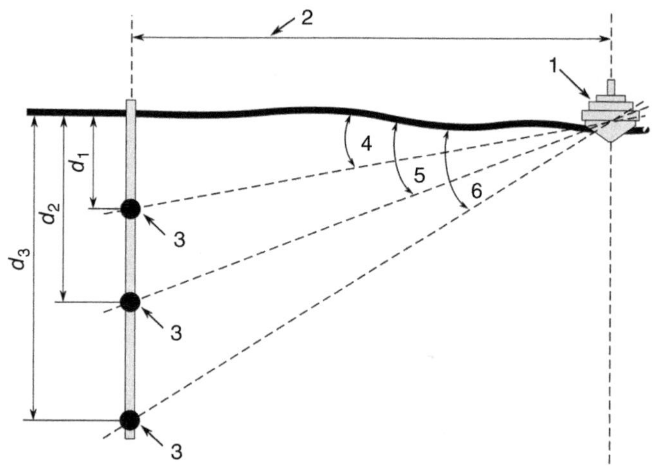

Figure 13.5 Grade A method to assess radiated noise from a ship. d1 = $d_{CPA} \cdot tan$ (15°); d2 = $d_{CPA} \cdot tan$ (30°); d3 = $d_{CPA} \cdot tan$ (45°); d_{CPA} = 100*m* or one overall ship length, whichever is greater (ISO 17208-2, 2019).

quantifying the contributions of primary sources. Beamforming and its improvements are the best array-processing techniques for simple propagation problems. They have been widely used in airborne acoustics for cars, planes, and trains (ISO 17208-2, 2019; Jensen et al., 2009). Most applications of array processing for imaging underwater radiated noise have been conducted in model-scale experiments (Commission, 2008; McCarthy, 2007; Amailland, 2017). The primary objective is to investigate solutions for an optimization strategy in noise reduction, particularly regarding cavitation noise. Array processing for underwater noise is mostly developed and applied to research projects at the moment, but the transportation noise industry could benefit from its application as well.

13.2.3. Modeling Underwater Radiated Noise

Research on developing models of radiated shipping noise has followed the progress of the methodology to measure shipping noise. Initial research aimed to provide the global contribution of shipping activity to ambient noise (Viola et al., 2017) without requiring additional information. The empirical model proposed by Wenz provided a reliable prediction of shipping noise for deep-water environments. Later updates were proposed for shallower areas (ISO 18405, 2017).

Modeling radiated noise matches levels of acoustic measurements to the actual parameters of a ship. Radiated noise modeling initially investigated the relation between average measured noise levels and ship speed. Ross proposed the first model using recordings during World War II (Porter, 1992) for frequencies above 100 Hz and speeds of 8–24 kt. This model gives a general formulation of the dependence of noise level on frequency and speed in the form of addition and subtraction of logarithmic terms. With the availability of new measurements from ships, including a larger variety of activities, additional logarithmic terms were added to include the effects of length (Tollefsen et al., 2018) and tonnage

Table 13.1 Commonly used source spectrum models for vessels, with their associated parameters.

Model	Required parameters	Frequency range (Hz)
Ross (Porter, 1992)	Frequency (Hz), speed (knots), tonnage (tons)	100–10,000
RANDI-3.1 (Aumond et al., 2018) (Tollefsen et al., 2018)	Frequency (Hz), length (feet), speed (knots)	10–10,000
W & H (Tournadre, 2014)	Frequency (Hz)	30–1,200
AQUO project (Ashokan et al., 2020)	Frequency (Hz), categorized length (feet), speed (knots)	10–10,000
Wittekind (Wales & Heitmeyer, 2002)	Frequency (Hz), displacement (t), speed (knots), cavitation inception speed (knots), hull block coefficient, engine mass (tons), number of operating engines	10–10,000

(Tasker et al., 2010) and allow extension to frequencies below 100 Hz and above 1 kHz. The model reflects the basic idea that the larger and the faster the ship, the noisier the radiated level.

The U.S. Naval Research Laboratory developed the RANDI-2 and RANDI-3 models (Research Ambient Noise Directionality) between the 1970s and 1990s to update the Ross model (Aumond et al., 2018). In particular, the RANDI-3 model introduced five categories of ships based on the associated activity: fishing, merchant, tanker, large tanker, and superlarge tanker. The radiated level is then expressed as the sum of a base spectrum that depends on the category, logarithmic values of speed and length, and a coefficient that depends on the frequency.

Wales and Heitmeyer (2002) proposed a model of ship-radiated noise by assuming that the emission spectra could be modeled as a bandpass filter. The parameters depend only on the frequency. It covers a wider frequency band than the Ross approach, particularly at low frequencies.

With the development of the ISO standard method, new modeling paradigms were investigated to estimate radiated noise as a function of the physical parameters of the source onboard. Most models rely on a statistical representation, so they may misestimate the actual physical phenomenon, particularly for the most modern ships. In Wittekind (2014), a new approach was proposed, with particular attention on the mechanism of cavitation. The radiation of cavitation noise can be classified according to three frequency bands. These frequency bands depend on several factors, including speed and ship dimensions. Radiated noise follows an addition of power laws of the three frequency bands. This modeling concept was improved in the AQUO project (Antoni, 2009), which added power-law terms for machinery noise and noncavitation propeller noise. Models parameters and examples of resulting spectra for a given vessel are shown in Table 13.1 and Fig. 13.6.

While these new approaches are relevant, they rely on the capacity to assess individual radiated noise. This may not be easily generalized to traffic situations covering a large area. On the other hand, the RANDI and Ross models require constant updates to accurately match ship construction changes, particularly at low frequencies (Porter & Liu, 1994). To tackle these limitations, matching traffic situations to acoustic measurements has been investigated as a relevant convenient model for source levels (Lighthill, 1952; Porter & Liu, 1994; Scorzolini et al., 2010; Kuiper, 1997). Traffic is then characterized then by the Automatic Identification System (AIS) signal sent by ships. AIS is more accurately described later in the chapter. Losses between receiver and ship locations are computed using physical model propagation. Computing accurate transmission losses is a major component in estimating source levels (Porter & Liu, 1994). Such approaches that combine AIS data and acoustic measurements to estimate radiated noise offer a way to obtain up-to-date noise models that can be built on convenient databases and will benefit from modern data processing.

There is currently a lack of consensus about the use of a radiated noise model (Lighthill, 1952). Significant mismatches between prediction and measurements continue to occur, particularly at lower frequencies. Global modeling approaches proposed in McKenna et al. (2012) and Simard et al. (2016) offer a suitable way to use large amounts of AIS and acoustic data. Such approaches will be strengthened as more data becomes available for wider configuration of ships and more environmental configurations, such as the ECHO program in the port of

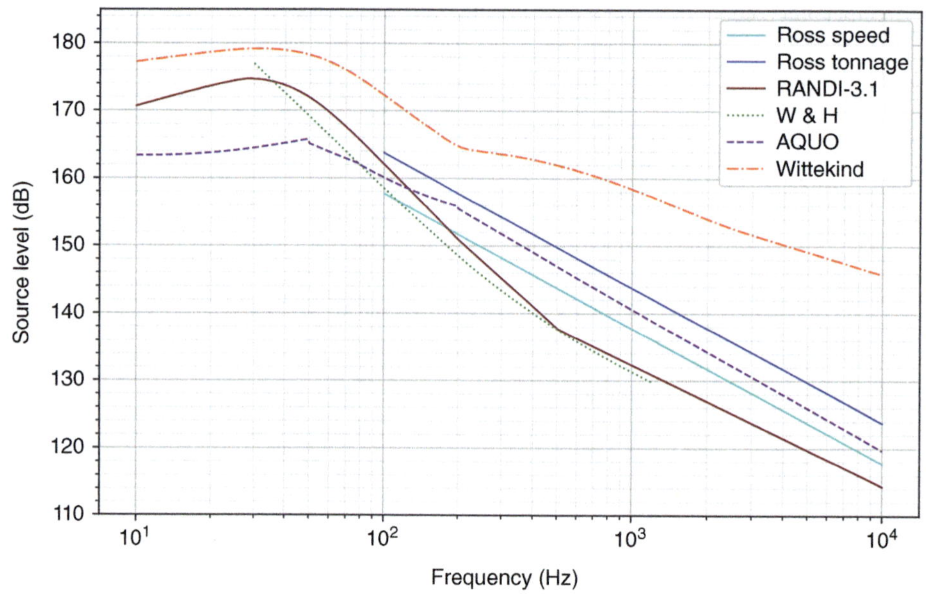

Figure 13.6 Example of source-level spectra produced by different models (from Table 13.1) for the same vessel. The hypothetical vessel has these parameters: length, 606 ft; speed, 14 kts; cavitation inception speed, 12 kts; hull block coefficient, 0.78; displacement, 45,574 t; displacement reference, 10,000 t; single engine of mass 40 t.

Vancouver. However, this must be supported by greater modeling capacity to improve transmission loss estimates.

13.3. NOISE MAPPING

Noise mapping refers to the computation of the sound levels over a given geographic area. In environmental sciences, noise maps can be based on measurements, models, or a combination. Noise mapping has been used extensively in urban areas to predict or measure noise levels from various sources (e.g., in (Audoly et al., 2015; Mustonen et al., 2019)). For an oceanic context, noise mapping based on measurements often leads to undersampling issues (especially of the temporal dynamics) and may represent static conditions only. This outcome derives from technical difficulties in setting up long-term continuous recorders at sea. Furthermore, sound propagation is more efficient underwater, so fewer recorders are needed to monitor large areas. For these reasons, mapping of ship traffic noise is mostly performed through modeling and includes acoustic recording as a validation step (De Jong et al., 2009; Cominelli et al., 2018; Chapman & Price, 2011).

To model ambient noise, multiple types of sources must be considered. The local acoustic environment is composed of three components: biophony (biologically produced sounds), geophony (geological and physical ocean sounds such as waves, wind, and currents), and anthrophony (human-produced sounds) (ISO 18405, 2017). Some models estimate the geophony component by modeling the wind and rain noises that are assumed to be related to a reference condition of noise levels (Erbe et al., 2016). In this section, the mapping of ambient noise levels is focused on soundscape modeling based on its anthropophony component.

13.3.1. Modeling Shipping Contributions

Underwater noise mapping performed through modeling is achieved by computing the SONAR equation for several receiving positions. For a position m, three steps are followed: (1) compute the source levels of the surrounding ships, (2) compute the transmission losses (TL) between the ships and the receiving position at several depths, and (3) compute the received levels (RL) (De Jong et al., 2009; Audoly & Rizzuto, 2015) following the SONAR equation in its passive form (Gervaise et al., 2012a):

$$RL_m = \sum_{n}^{N_s^m} SL_n - TL_{mn} + W_m \qquad (13.1)$$

where SL_n and TL_{mn} are, respectively, the emitted level and the transmission losses for the nth source. W_m is a local acoustic contribution that depends only on depth (e.g., wind noise). RL is expressed in dB ref. 1 μPa (later written as dB). The active and passive sonar equations infer detection ranges and performance (Tasker et al., 2010). Equation 13.1 is the Fourier-transform and log-transform of the convolution product of the source signal by the impulse response of the environment in the time domain: that is, the Green's function (Hatch et al., 2012)). Figure 13.7 illustrates the noise-mapping process. Each component of the SONAR equation is explored in the following section.

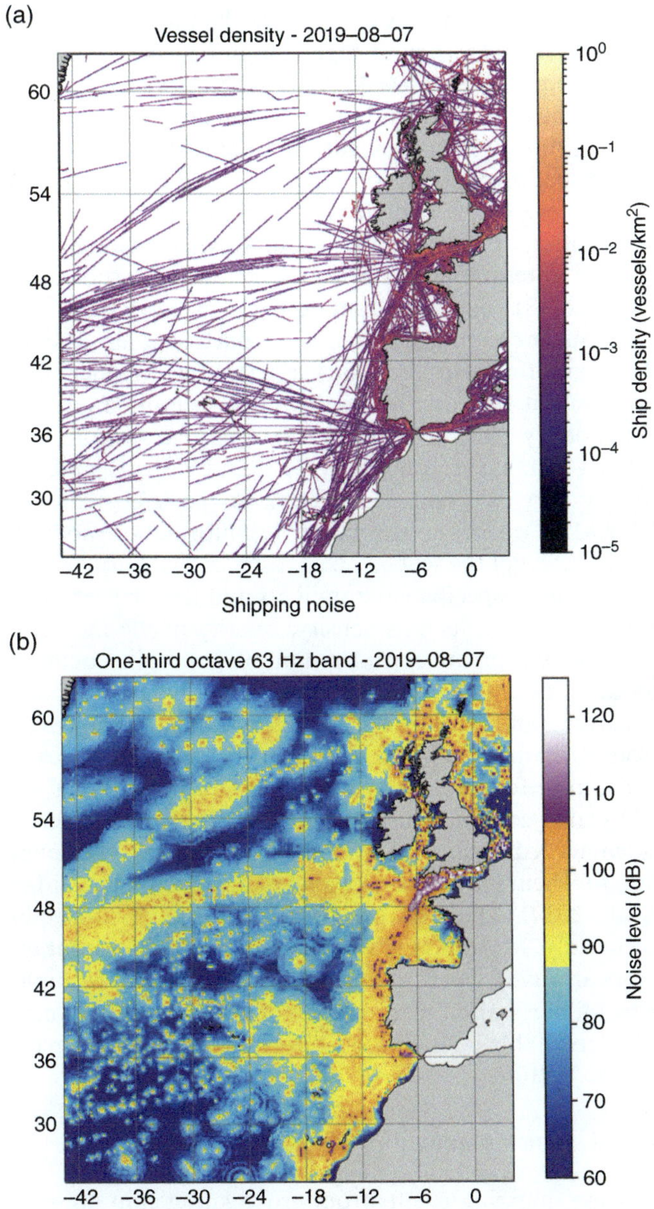

Figure 13.7 Example of the noise-mapping process. Data provided by the JONAS project for the day 08 July 2019 and for the one-third octave frequency band centered on 63 Hz. In this project, the Mediterranean Sea computation was omitted. (a) AIS vessel density map (provided by Exact Earth for the JONAS project). (b) Statistical shipping noise map modeled by SHOM software (Johnson & Dudgeon, 1992).

13.3.2. Source Properties

Shipping noise mapping performed with modeling requires an accurate trajectory of the ships using snapshots or statistical quantities over time (e.g., ship density). Hence, the spatial distribution of the vessels around the receivers is needed to accurately model *RL*. *SL* modeling and computation were explored earlier. Most modern shipping-noise models use AIS as a convenient dataset to determine the source's position. AIS is a vessel-tracking system primarily used for vessel security purposes. The tracking system is based on very-high-frequency (VHF) transceivers emitting from the vessels. Receivers are other vessels, land stations, or satellites (allowing larger areas to be covered since 2010 (Porter, 2011)). Numerous companies record this data continuously and can provide AIS catalogs.

AIS signals contain several fields of data. To model shipping noise, the positions, dates of emission, length, and speed are mandatory. Other parameters, such as tonnage, can be useful depending on the chosen source model (Table 13.1). Speed can be computed as well from the time derivative of ship positions. Those fields allow a source level to be assigned to each vessel over its trajectory (Johnson & Dudgeon, 1992; Mackenzie, 1981) using the models presented above.

However, the AIS catalog's comprehensiveness has a limit (Erbe et al., 2012). Land-based stations are coastal, and satellites are not always in position to track vessels, so "grey" areas and periods exist. Moreover, AIS transceivers are not required on some ships, especially recreational and small vessels. So, the vessel's route is interpolated to compensate for intermittent data coverage. In very coastal areas, noise maps may underestimate the noise produced by vessels without AIS. This last point may not greatly affect the noise-map levels because coastal areas tend to have a shallow environment, limiting acoustic propagation at low frequencies. Solutions for ships without AIS are currently being investigated (Le Courtois et al., 2016b).

AIS data can be used to model shipping noise directly using "snapshots" of the vessel's positions in a given area over a short period (e.g., 10 min, as in JOMOPANS project (discussed shortly; also see (De Jong, 2021)). This technique provides noise maps with high temporal and spatial resolution. Another approach involves using AIS densities, defined as the time spent by vessels in a given area over a given period. This method improves computational efficiency by allowing the *TL* computation to be performed over a high- or low-resolution spatial grid over very large areas. The resulting noise maps are assumed to be more robust to data gaps, but temporal dynamics can be missed (Johnson & Dudgeon, 1992).

13.3.3. Acoustic Propagation

Accurate acoustic propagation models are needed to determine the values of the losses between a source and a receiver. An acoustic signal propagated underwater becomes distorted due to multipath effects and attenuated by several loss mechanisms. They are considered the sum of a loss due to geometrical spreading and a loss due to attenuation in the medium (Hatch et al., 2012). Computing the *TL* due to attenuation is challenging, as propagation physics and wave equation simulation need to be considered and depend on specific environmental conditions.

Propagation Models

Underwater acoustic propagation is computed using numerical models. Most models consider 2D propagation along range and depth dimensions between the emitter and the receiver. More complex 3D models have been developed but require much longer computation times. Different models must be considered depending on the depth, seabed, water-column properties, and frequency. Models that allow variation of the environment with the range are called *range dependent*, in contrast to range-independent models. Four types of numerical models are commonly used to find a solution to the wave equation in the frequency domain: spectral (or fast-field program), normal mode, ray, and parabolic equation models. In the time domain, two models are used for full wave equation computation: direct finite-difference and finite-element (Hatch et al., 2012).

Most propagation models used for underwater noise mapping use normal-mode propagation, parabolic equations, ray methodologies, or a combination of several according to the frequency. These methods are considered appropriate for modeling shipping noise as they cover a sufficient range of geometries and frequencies. Each has already been developed to be integrated into the noise-mapping process (Ollivier et al., 2021; Carey, 2009; Ollivier et al., 2019; Castellote et al., 2012).

Choosing a specific model depends on the validity of the model assumptions, the number of computations required, and the existence of instabilities in the model algorithm (Hatch et al., 2012; Erbe et al., 2016). For example, parabolic equation models with a numerical implementation of RAM (range-dependent acoustic model) (Castellote et al., 2012) are applied on a 2D grid of ranges and depths. The cell sizes (or the model resolution) depend on the frequency. Particularly at low frequencies, this resolution can be low, so the required computation time is low (and vice versa). This method is commonly applied to evaluate the *TL* of ships at low frequencies (classically 10–400 Hz). Similar to geometric optics, *TL*s at higher frequencies are usually modeled using a ray approximation. BELLHOP is the best-known numerical implementation (Oudompheng et al., 2017). The informal validation domain of the primary models is defined in Etter (2009) and Farcas et al. (2016) based on depth, frequency, and range dependencies.

Environmental Data

Range-dependent models require three data types: bathymetry, seabed geoacoustic parameters, and sound speed profiles (SSPs) of the water column. Today, high-quality compilations of bathymetry data and models are easily accessible, even at high resolution and worldwide extent (e.g., (Eriksen et al., 2018; Collins, 1995)).

SSPs have a significant influence on acoustic propagation. Although measurements at large scales of SSPs are expensive, several empirical models have been explored and developed to exploit the relationship between the sea temperature, salinity, and sound speed. The long history of determining empirical equations is described in Etter (2018). The nine-term equation is cited here as it is the most commonly found (Le Courtois et al., 2016a). This empirical equation has a restrained applicability domain but was later refined by implementing a latitude dependency (Jensen et al., 2011). The temperature and salinity parameters can easily be found globally and are mostly derived from models (e.g., in the E.U. with the CNEMS-Copernicus Marine Service Information models).

The effect of the seabed on acoustic propagation can be significant in specific environments and frequencies. Seabed parameters such as density, celerity of the medium, and attenuation are considered in the parabolic equation model (Castellote et al., 2012). This data type is the most uncertain, and knowledge of the seabed must be improved from an acoustic point of view. Worldwide databases exist and can be found globally at https://data.shom.fr; for the EU scale, see Kotilainen et al. (2014). The seabed is classified according to its density, sound speed, and attenuation for shear and pressure waves in the medium(Halliday et al., 2017). These properties can be difficult to obtain accurately. Most models do not consider the propagation of shear waves, which can lead to a significant overestimation of the seabed attenuation for rocky-type bottoms. Most available data considers only an infinite half-space for the bottom.

13.3.4. Noise-Mapping Applications

Noise mapping results from propagating ship noise in the environment, as explained by equation 1. Noise levels are computed for a frequency and several depths. Frequency and depth combinations can be computed to achieve a noise map. Figure 13.7 shows a noise map at the one-third octave band centered on 63 Hz for the loudest level over the depth. Because of the low frequency and the range dependence, parabolic RAM was used for the propagation (Castellote et al., 2012). AIS and environmental data corresponding to 7 August 2019 was used for the North East Atlantic area. Based on this map, most European coastal areas are affected by anthropogenic shipping noise. The highest noise values are found along the main shipping lanes near the coasts. In the open seas of the North Atlantic, three main shipping lanes present high noise values. The lowest values are observed in the areas between routes in the open seas but also closest to the coasts. This can be explained by either low vessel density or limited propagation in shallow water at low frequencies.

Underwater noise maps are developed and produced by several research projects and over various marine regions. These projects allow us to improve our knowledge of underwater noise pollution in several domains, from vessel noise generation mechanics to acoustic propagation, recording, and vessel noise classification. A spatial approach was chosen to define noise thresholds based on noise maps (including models and recordings). In this context, the projects mentioned in Fig. 13.8 develop tools and methodologies to evaluate this impact, set relevant thresholds, and suggest mitigation measures. From 2008 to 2014, most of the conducted studies were used to evaluate feasibility; later projects (2014–2019) were more focused on methodology and standardization of processes. Since then, recent projects have been directed toward evaluation and feasibility of mitigation measures.

Over time, more projects are delivering results on underwater noise pollution and its impact, increasing the awareness of the general public and local actors about the blue economy. Knowledge of the environment will be crucial for the final steps in setting thresholds and mitigation measures.

13.4. CONCLUSION

Vessel-radiated noise is a combination of several sources onboard. Primary generating mechanisms can be divided into vibroacoustic fields and hydrodynamic interactions. Both mechanisms generate sounds with properties in terms of level, directivity, spectrum, and rhythm. These radiation properties depend on several ship specifications. The most-considered parameters are the activity associated with the ship and its size, speed, load, build year, and maintenance condition.

Shipping noise is a complex mixture of harmonic content and wideband noise over various ranges of frequencies. At short propagation ranges, underwater measurements have shown that acoustic signatures of ships can carry

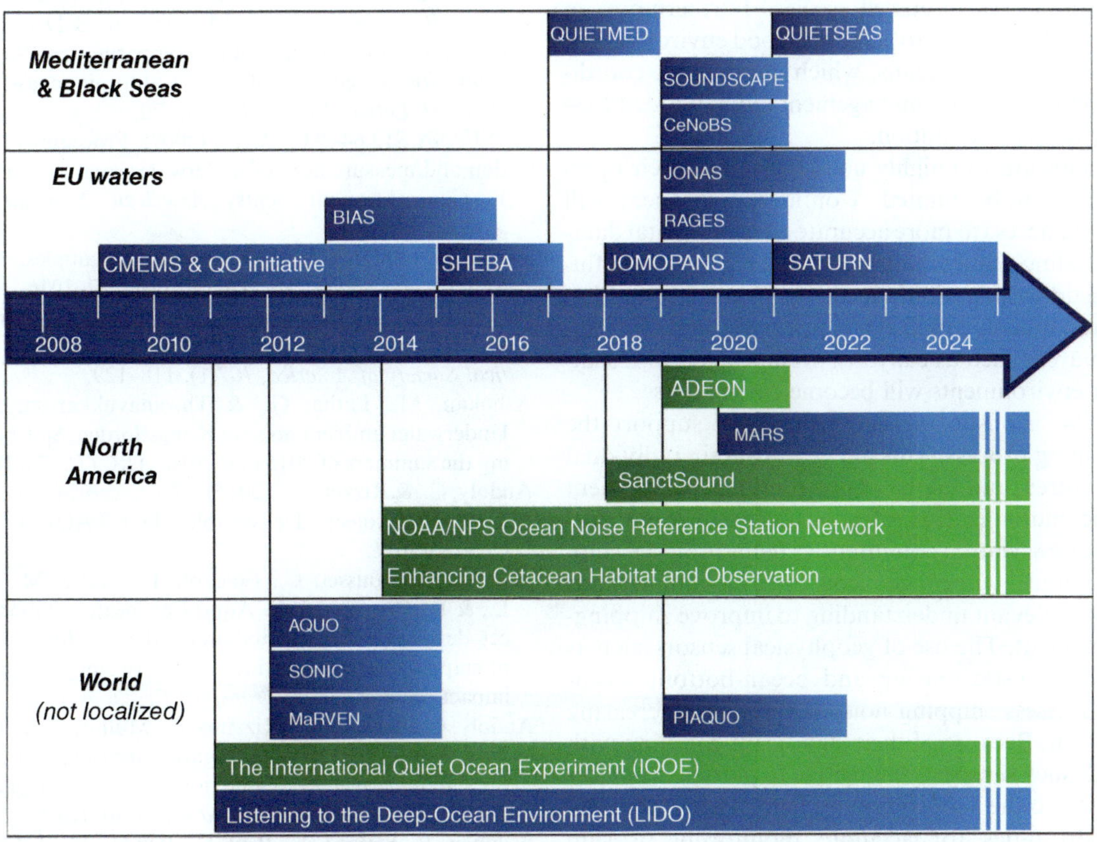

Figure 13.8 Examples of research projects studying shipping noise since 2008. International and European projects are classified vertically over their area of investigation.

information about ship motors and propellers. Considering the global contribution of shipping to the soundscape, it has been identified as increasing ambient noise levels at low frequency in recent decades.

Parametric models were initially proposed to match ambient noise levels to the speeds and lengths of a fleet. With the increase in acoustic recordings and AIS data, the capacity for parametrization was extended to categories of ship activities. Currently, improvements in the primary models are supported by dedicated and controlled measurements of radiated noise, as well as local inversion of shipping-noise contributions. Early research supported the naval industry, and later research monitored shipping noise at larger scales.

Knowledge about shipping noise has greatly increased in recent decades, primarily driven by acknowledging its adverse impact on ecosystems. Regulatory frameworks, environmental impact assessments, and incentives have helped with the estimation of underwater noise levels related to shipping activities. To support stakeholder and regulatory needs, shipping-noise maps are produced. Hot spots of noise can be identified to investigate mitigation measures. To reduce noise pollution, such programs must be accomplished jointly by regulators, ship manufacturers, and users. Reducing the noise produced by newly constructed vessels is currently being explored by several projects, as discussed earlier. Furthermore, in Europe, the MSFD aims to achieve a good environmental status in its seas and oceans, which necessitates coordinated planning for noise management along with addressing other forms of pollution.

Noise maps are still highly uncertain; thus, their operational use can be limited. Confidence in them will increase with access to more accurate environmental databases, including seabed parameters, which greatly influence acoustic propagation, especially at low frequencies. Propagation models can also be improved, and 3D propagation in areas such as canyons, archipelagos, and shallow-water environments will become easier to use.

Long-term acoustic measurements also support the noise-mapping process. This data can contain individual ship signatures and global contributions to ambient noise. The mutualization of acoustic monitoring programs among scientific communities helps improve shipping-noise models. Local geoacoustic inversion methods have shown relevant understanding to improve shipping-noise assessment. The use of geophysical sensors such as distributed acoustic sensing and ocean-bottom seismometers to assess shipping noise is receiving increasing consideration. Because of their significant coupling with the seabed, such sensors are sensitive to particle displacement related to the acoustic sound field. They offer relevant opportunities for accurate monitoring of ship acoustic radiation.

ACKNOWLEDGMENTS

This work has been supported by the JONAS project and co-funded by the European Regional Development Fund (ERDF) under the INTERREG Atlantic Area program and by the grant agreement between SHOM and the French Ministry of Ecological Transition. The authors would like to thank their colleagues for their careful reading, the reviewers for their helpful comments, and the editors for their invitation.

REFERENCES

Aktas, B., Atlar, M., Turkmen, S., Shi, W., Sampson, R., Korkut, E., & Fitzsimmons, P. (2016). Propeller cavitation noise investigations of a research vessel using medium size cavitation tunnel tests and full-scale trials. *Ocean Engineering, 120*, 122–135.

Al-Badrawi, M. H., Miksis-Olds, J. L., & Bradley, D. L. (2020). Bearing stake 1977 revisited: an understanding of ambient sound sources in the Indian Ocean. *The Journal of the Acoustical Society of America, 148*(4), EL320–EL325.

Amailland, S. (2017). *Caractérisation de sources acoustiques par imagerie en écoulement d'eau confinée* (PhD thesis). Le Mans.

Amailland, S., Thomas, J.-H., Pézerat, C., & Boucheron, R. (2018). Boundary layer noise subtraction in hydrodynamic tunnel using robust principal component analysis. *The Journal of the Acoustical Society of America, 143*(4), 2152–2163.

Andrew, R. K., Howe, B. M., Mercer, J. A., & Dzieciuch, M. A. (2002). Ocean ambient sound: comparing the 1960s with the 1990s for a receiver off the California coast. *Acoustics Research Letters Online, 3*(2), 65–70.

ANSI/ASA S12.64-2009/Part 1 (2009). Procedures for description and measurement of underwater sound from ships–Part 1: General requirements. *American National Standard ANSI/ASAS.*

Antoni, J. (2009). Cyclostationarity by examples. *Mechanical Systems and Signal Processing, 23*(4), 987–1036.

Arveson, P. T., & Vendittis, D. J. (2000). Radiated noise characteristics of a modern cargo ship. *The Journal of the Acoustical Society of America, 107*(1), 118–129.

Ashokan, M., Latha, G., & Thirunavukkarasu, A. (2020). Underwater ambient noise in Kongsfjorden, Spitsbergen, during the summers of 2015 and 2016. *ARCTIC, 73*(3), 386–392.

Audoly, C., & Rizzuto, E. (2015). *Ship underwater radiated noise patterns* (Project Deliverable D2). AQUO European Collaborative.

Audoly, C., Rousset, C., Folegot, T., André, M., Benedetti, L., & Baudin, E. (2014). Aquo European collaborative project–development of methods and indicators for the assessment of shipping noise footprint on underwater environment and impact on marine life. *Transport Research Arena.*

Audoly, C., Rousset, C., Rizzuto, E., Mullor, R. S., Hallander, J., & Baudin, E. (2015). Mitigation measures for controlling the ship underwater radiated noise, in the scope of aquo project. In *OCEANS 2015-Genova* (pp. 1–6). IEEE.

Aulanier, F., Simard, Y., Roy, N., Bandet, M., & Gervaise, C. (2016). Groundtruthed probabilistic shipping noise modeling and mapping: Application to blue whale habitat in the Gulf of St. Lawrence. In *Proceedings of Meetings on Acoustics 4ENAL* (Vol. 27, p. 070006). Acoustical Society of America.

Aulanier, F., Simard, Y., Roy, N., Gervaise, C., & Bandet, M. (2017). Effects of shipping on marine acoustic habitats in Canadian Arctic estimated via probabilistic modeling and mapping. *Marine Pollution Bulletin, 125*(1–2), 115–131.

Aumond, P., Can, A., Mallet, V., De Coensel, B., Ribeiro, C., Botteldooren, D., & Lavandier, C. (2018). Kriging-based spatial interpolation from measurements for sound level mapping in urban areas. *The Journal of the Acoustical Society of America, 143*(5), 2847–2857.

Bazile Kinda, G., & Bonnel, J. (2015). Passive acoustic observations of tide height in the Iroise Sea using ambient noise. *The Journal of the Acoustical Society of America, 138*(3), EL299–EL304.

Blake, W. K. (1984). *Aero-hydroacoustics for ships* (Vol. 1, technical report). David W. Taylor Naval Ship Research and Development Center.

Breeding Jr., J., Pflug, L. A., Bradley, M., & Walrod, M. H. (1996). *Research ambient noise directionality (randi) 3.1 physics description* (technical report). Naval Research Lab, Stennis Space Center.

Carey, W. M. (2009). Lloyd's mirror-image interference effects. *Acoustics Today, 5*(2), 14–20.

Castellote, M., Clark, C. W., & Lammers, M. O. (2012). Acoustic and behavioral changes by fin whales (Balaenoptera physalus) in response to shipping and airgun noise. *Biological Conservation, 147*(1), 115–122.

Chapman, N. R., & Price, A. (2011). Low frequency deep ocean ambient noise trend in the northeast Pacific Ocean. *The Journal of the Acoustical Society of America, 129*(5), EL161–EL165.

Collins, M. D. (1993). A split-step Padé solution for the parabolic equation method. *The Journal of the Acoustical Society of America, 93*(4), 1736–1742.

Collins, M. D. (1995). *User's guide for RAM versions 1.0 and 1.0P*. Naval Research Lab.

Cominelli, S., Devillers, R., Yurk, H., MacGillivray, A., McWhinnie, L., & Canessa, R. (2018). Noise exposure from commercial shipping for the southern resident killer whale population. *Marine Pollution Bulletin, 136*, 177–200.

Commission. (2008). Directive 2008/56/EC of the European Parliament and of the Council of 17 June 2008 establishing a framework for community action in the field of marine environmental policy (marine strategy framework directive). *Official Journal of the European Communities, 164*, 19–40.

De Jong, C. (2021). WP4 final report. Modelling. Report of the EU INTERREG Joint Monitoring Programme for Ambient Noise North Sea (JOMOPANS).

De Jong, C., Bosschers, J., & Hasenpflug, H. (2009). Model scale measurements of surface ship radiated flow noise. *NAG/DAGA, 5*.

Dekeling, R., Tasker, M., Van der Graaf, A., Ainslie, M., Andersson, M., André, M., ... et al. (2014). Monitoring guidance for underwater noise in European seas, part I: Executive summary. A guidance document within the common implementation strategy for the marine strategy framework directive by MSFD technical subgroup on underwater noise.

EMODnet Bathymetry Consortium et al. (2018). Emodnet digital bathymetry (DTM 2018).

Erbe, C., MacGillivray, A., & Williams, R. (2012). Mapping cumulative noise from shipping to inform marine spatial planning. *The Journal of the Acoustical Society of America, 132* (5), EL423–EL428.

Erbe, C., Reichmuth, C., Cunningham, K., Lucke, K., & Dooling, R. (2016). Communication masking in marine mammals: A review and research strategy. *Marine Pollution Bulletin, 103* (1–2), 15–38.

Erbe, C., Marley, S. A., Schoeman, R. P., Smith, J. N., Trigg, L. E., & Embling, C. B. (2019). The effects of ship noise on marine mammals—a review. *Frontiers in Marine Science, 6*, 606.

Eriksen, T., Greidanus, H., & Delaney, C. (2018). Metrics and provider-based results for completeness and temporal resolution of satellite-based AIS services. *Marine Policy, 93*, 80–92.

Etter, P. C. (2009). Review of ocean-acoustic models. In *OCEANS 2009* (pp. 1–6). IEEE.

Etter, P. C. (2018). *Underwater acoustic modeling and simulation*. CRC Press.

Farcas, A., Thompson, P. M., & Merchant, N. D. (2016). Underwater noise modeling for environmental impact assessment. *Environmental Impact Assessment Review, 57*, 114–122.

Fischer, R. (2008). Singing propellers—solutions and case histories. *Marine Technology and SNAME News, 45*(04), 221–227.

GEBCO Bathymetric Compilation Group (2020). GEBCO 2020 grid.

Gervaise, C., Kinda, B. G., Bonnel, J., St´ephan, Y., & Vallez, S. (2012a). Passive geoacoustic inversion with a single hydrophone using broadband ship noise. *The Journal of the Acoustical Society of America, 131*(3), 1999–2010.

Gervaise, C., Simard, Y., Roy, N., Kinda, B., & Menard, N. (2012b). Shipping noise in whale habitat: Characteristics, sources, budget, and impact on belugas in Saguenay–St. Lawrence Marine Park hub. *The Journal of the Acoustical Society of America, 132*(1), 76–89.

Halliday, W. D., Insley, S. J., Hilliard, R. C., de Jong, T., & Pine, M. K. (2017). Potential impacts of shipping noise on marine mammals in the western Canadian Arctic. *Marine Pollution Bulletin, 123*(1–2), 73–82.

Hatch, L. T., Clark, C. W., Van Parijs, S. M., Frankel, A. S., & Ponirakis, D. W. (2012). Quantifying loss of acoustic communication space for right whales in and around a US national marine sanctuary. *Conservation Biology, 26*(6), 983–994.

ISO 17208-1 (2016). *Underwater acoustics—quantities and procedures for description and measurement of underwater sound from ships—part 1: Requirements for precision measurements in deep water used for comparison purposes.*

ISO 17208-2 (2019). *Underwater acoustics—quantities and procedures for description and measurement of underwater sound from ships—part 2: Determination of source levels from deep water.*

ISO 18405 (2017). *Underwater acoustics—terminology.*

Jensen, F. (2008). *Sonar acoustics handbook*. NURC.

Jensen, F. B., Kuperman, W. A., Porter, M. B., & Schmidt, H. (2011). *Computational ocean acoustics*. Springer Science & Business Media.

Jensen, F. H., Bejder, L., Wahlberg, M., Soto, N. A., Johnson, M., & Madsen, P. T. (2009). Vessel noise effects on delphinid communication. *Marine Ecology Progress Series*, *395*, 161–175.

Johnson, D. H., & Dudgeon, D. E. (1992). *Array signal processing: concepts and techniques*. Simon & Schuster.

Kinda, G. B., Le Courtois, F., & Stéphan, Y. (2017). Ambient noise dynamics in a heavy shipping area. *Marine Pollution Bulletin*, *124*(1), 535–546.

Kotilainen, A. T., Kaskela, A., & Stevenson, A. (n.d.). EMODnet-Geology.

Kuiper, G. (1997). Cavitation research and ship propeller design. *Applied Scientific Research*, *58*(1), 33–50.

Le Courtois, F., Kinda, G. B., Boutonnier, J.-M., Stéphan, Y., & Sarzeaud, O. (2016a). Statistical ambient noise maps from traffic at world and basin scales. In *Acoustic and Environmental Variability, Fluctuations and Coherence*. Cambridge, England.

Le Courtois, F., Thomas, J.-H., Poisson, F., & Pascal, J.-C. (2016b). Genetic optimisation of a plane array geometry for beamforming. Application to source localisation in a high speed train. *Journal of Sound and Vibration*, *371*, 78–93.

Le Gall, Y., & Bonnel, J. (2013). Passive estimation of the waveguide invariant per pair of modes. *The Journal of the Acoustical Society of America*, *134*(2), EL230–EL236.

Leroy, C. C., Robinson, S. P., & Goldsmith, M. J. (2008). A new equation for the accurate calculation of sound speed in all oceans. *The Journal of the Acoustical Society of America*, *124*(5), 2774–2782.

Lighthill, M. J. (1952). On sound generated aerodynamically I. General theory. *Proceedings of the Royal Society of London. Series A. Mathematical and Physical Sciences*, *211*(1107), 564–587.

MacGillivray, A., & de Jong, C. (2021). A reference spectrum model for estimating source levels of marine shipping based on automated identification system data. *Journal of Marine Science and Engineering*, *9*(4), 369.

Mackenzie, K. V. (1981). Nine-term equation for sound speed in the oceans. *The Journal of the Acoustical Society of America*, *70*(3), 807–812.

Magnier, C., & Gervaise, C. (2020). Acoustic and photographic monitoring of coastal maritime traffic: Influence on the soundscape. *The Journal of the Acoustical Society of America*, *147*(6), 3749–3757.

McCarthy, E. (2007). *International regulation of underwater sound: establishing rules and standards to address ocean noise pollution*. Springer Science & Business Media.

McIntyre, D., Lee, W., Frouin-Mouy, H., Hannay, D., & Oshkai, P. (2021). Influence of propellers and operating conditions on underwater radiated noise from coastal ferry vessels. *Ocean Engineering*, *232*, 109075.

McKenna, M. F., Ross, D., Wiggins, S. M., & Hildebrand, J. A. (2012). Underwater radiated noise from modern commercial ships. *The Journal of the Acoustical Society of America*, *131*(1), 92–103.

Mustonen, M., Klauson, A., Andersson, M., Clorennec, D., Folegot, T., Koza, R., et al. (2019). Spatial and temporal variability of ambient underwater sound in the Baltic Sea. *Scientific Reports*, *9*(1), 1–13.

Ollivier, B., Le Courtois, F., Kinda, G. B., Ratsivalaka, C., Sarzeaud, O., & Boutonnier, J.-M. (2019). Analysis of the comprehensiveness of AIS data sets: Application to the underwater noise modelling at basin scale. In *OCEANS 2019–Marseille* (pp. 1–6). IEEE.

Ollivier, B., Shute, P., & Kinda, G. B. (2021). Underwater soundscape description from cyclostationarity point of view. *The Journal of the Acoustical Society of America*, *150*(3), 2245–2255.

Oudompheng, B., Nicolas, B., & Lamotte, L. (2017). Localization and contribution of underwater acoustical sources of a moving surface ship. *IEEE Journal of Oceanic Engineering*, *43*(2), 536–546.

Parks, S. E., Johnson, M., Nowacek, D., & Tyack, P. L. (2011). Individual right whales call louder in increased environmental noise. *Biology Letters*, *7*(1), 33–35.

Parsons, M. J., Erbe, C., Meekan, M. G., & Parsons, S. K. (2021). A review and meta-analysis of underwater noise radiated by small (¡ 25 m length) vessels. *Journal of Marine Science and Engineering*, *9*(8), 827.

Picaut, J., Fortin, N., Bocher, E., Petit, G., Aumond, P., & Guillaume, G. (2019). An open-science crowdsourcing approach for producing community noise maps using smartphones. *Building and Environment*, *148*, 20–33.

Porter, M. B. (1992). *The Kraken normal mode program* (technical report). Naval Research Lab Washington DC.

Porter, M. B. (2011). *The Bellhop manual and user's guide: preliminary draft*. (Tech. Rep. 260). Heat, Light, and Sound Research, Inc.

Porter, M. B., & Liu, Y.-C. (1994). Finite-element ray tracing. *Theoretical and Computational Acoustics*, *2*, 947–956.

Ren, Q.-Y., & Hermand, J.-P. (2013). Acoustic interferometry for geoacoustic characterization in a soft-layered sediment environment. *The Journal of the Acoustical Society of America*, *133*(1), 82–93.

Rivet, D., de Cacqueray, B., Sladen, A., Roques, A., & Calbris, G. (2021). Preliminary assessment of shipping noise monitoring using distributed acoustic sensing on an optical fiber telecom cable. https://doi.org/10.31223/X5SS3M

Rolland, R. M., Parks, S. E., Hunt, K. E., Castellote, M., Corkeron, P. J., Nowacek, D. P., et al. (2012). Evidence that ship noise increases stress in right whales. *Proceedings of the Royal Society B: Biological Sciences*, *279*(1737), 2363–2368.

Ross, D., & Kuperman, W. (1989). *Mechanics of underwater noise*. American Institute of Physics.

Scorzolini, A., De Perini, V., Razzano, E., Colavolpe, G., Mendes, S., Fiori, P., & Sorbo, A. (2010). European enhanced space-based AIS system study. In *2010 5th Advanced Satellite Multimedia Systems Conference and the 11th Signal Processing for Space Communications Workshop* (pp. 9–16). IEEE.

Simard, Y., Roy, N., Gervaise, C., & Giard, S. (2016). Analysis and modeling of 255 source levels of merchant ships from an acoustic observatory along St. Lawrence Seaway. *The Journal of the Acoustical Society of America*, *140*(3), 2002–2018.

Southall, B. L., Finneran, J. J., Reichmuth, C., Nachtigall, P. E., Ketten, D. R., Bowles, A. E., et al. (2019). Marine mammal noise exposure criteria: updated scientific recommendations for residual hearing effects. *Aquatic Mammals*, *45*(2).

Tasker, M., Amundin, M., Andre, M., Hawkins, A., Lang, W., Merck, T., et al. (2010). *Marine strategy framework directive—Task Group 11 report underwater noise and other forms of energy* (Report No. EUR, 24341). Publications Office of the European Union.

Thode, A. M., Blackwell, S. B., Conrad, A. S., Kim, K. H., & Michael Macrander, A. (2017). Decadal-scale frequency shift of migrating bowhead whale calls in the shallow Beaufort Sea. *The Journal of the Acoustical Society of America, 142*(3), 1482–1502.

Tollefsen, D., & Dosso, S. E. (2008). Bayesian geoacoustic inversion of ship noise on a horizontal array. *The Journal of the Acoustical Society of America, 124*(2), 788–795.

Tollefsen, D., Dosso, S. E., & Knobles, D. P. (2018). Ship of opportunity noise inversions for geoacoustic parameters of a layered mud-sand seabed. *The Journal of the Acoustical Society of America, 144*(3), 1972–1972.

Tournadre, J. (2014). Anthropogenic pressure on the open ocean: the growth of ship traffic revealed by altimeter data analysis. *Geophysical Research Letters, 41*(22), 7924–7932.

Trabattoni, A., Barruol, G., Dreo, R., Boudraa, A., & Fontaine, F. R. (2020). Orienting and locating ocean-bottom seismometers from ship noise analysis. *Geophysical Journal International, 220*(3), 1774–1790.

Urick, R. J. (1983). *Principles of underwater sound.* Peninsula Publishing.

Vakili, S., Ölçer, A., & Ballini, F. (2020). The development of a policy framework to mitigate underwater noise pollution from commercial vessels: the role of ports. *Marine Policy, 120*, 104132.

Viola, S., Grammauta, R., Sciacca, V., Bellia, G., Beranzoli, L., Buscaino, G., et al. (2017). Continuous monitoring of noise levels in the Gulf of Catania (Ionian Sea). Study of correlation with ship traffic. *Marine Pollution Bulletin, 121*(1–2), 97–103.

Wagstaff, R. (1973). Randi: Research ambient noise directionality model (technical report). Naval Undersea Center.

Wales, S. C., & Heitmeyer, R. M. (2002). An ensemble source spectra model for merchant ship-radiated noise. *The Journal of the Acoustical Society of America, 111*(3), 1211–1231.

Weilgart, L. (2019). Keeping the noise down: approaches to the mitigation and regulation of human-caused ocean noise. In *The Future of Ocean Governance and Capacity Development* (pp. 298–302). Brill Nijhoff.

Weilgart, L. S. (2007). A brief review of known effects of noise on marine mammals. *International Journal of Comparative Psychology, 20*(2).

Wenz, G. (1969). Low-frequency deep-water ambient noise along the Pacific coast of the United States. *US Navy Journal of Underwater Acoustics, 19*, 423–444.

Wenz, G. M. (1962). Acoustic ambient noise in the ocean: spectra and sources. *The Journal of the Acoustical Society of America, 34*(12), 1936–1956.

Wenz, G. M. (1972). Review of underwater acoustics research: noise. *The Journal of the Acoustical Society of America, 51*(3B), 1010–1024.

Wittekind, D. K. (2014). A simple model for the underwater noise source level of ships. *Journal of Ship Production & Design, 30*(1).

14

Tracking Multiple Underwater Vessels With Passive Sonar Using Beamforming and a Trajectory PHD Filter

Wei Yi[1], Ángel F. García-Fernández[2], and Boxiang Zhang[1]

ABSTRACT

This chapter proposes a signal-processing algorithm to detect and track an unknown and possibly variable number of underwater vessels using passive sonar array systems, particularly a towed linear array. We first model the narrow-band signal the array receives from several potential vessels, each emitting an acoustic signal. Then we explain beamforming and detection methods to obtain angle-of-arrival measurements from the detected vessels. Based on these angle-of-arrival detections, we use a multitarget tracker, the trajectory probability hypothesis density (PHD) filter, to estimate the number of vessels and their trajectories in the angle-of-arrival domain.

14.1. INTRODUCTION

Sonar (sound navigation and ranging) is a technique that uses underwater sound to detect, classify, localize, and track underwater targets, which may be animals, such as whales, or underwater vessels, such as submarines (Waite, 2002; Ainslie, 2010; Urick, 1983; Baumgartner et al., 2008; Nosal & Neil Frazer, 2007; Ferri et al., 2017). Acoustic source tracking has numerous applications: for example, underwater vehicle navigation, underwater search and rescue, military defense, ocean biological signal analysis (Luo et al., 2018; Isbitiren & Akan, 2011; Dalberg et al., 2006; Picard et al., 2018), and 3D underwater scene reconstruction (Murino & Trucco, 2000; Coiras et al., 2007; Saucan et al., 2015).

There are two main types of sonar systems: active and passive. Active sonar emits an acoustic signal that travels through the water and reflects on a target. Its echo is processed by a receiver: a hydrophone (Mellema, 2020; Coraluppi & Carthel, 2006; Ben Liu et al., 2020). In contrast, passive sonar measures and processes the underwater acoustic signal emitted by a target (Ziomek, 2017; Yocom et al., 2011). For example, an underwater vessel may generate an acoustic signal due to the propulsion machinery, the propellers, and the flow noise (Waite, 2002).

Basic passive sonar systems only provide bearings information on the target and do not provide information on the distance, as the power of the transmitted signal by the target is unknown (Waite, 2002). Nevertheless, passive sonar has the benefit of being nonintrusive in the environment and covertness (Carter, 1981; Luo et al., 2018). Passive sonar is the focus of this chapter.

Standard types of passive sonar are hull-mounted and towed linear arrays (Waite, 2002; Gershman et al., 2000). Towed linear arrays consist of a line of hydrophones towed behind a vessel on a cable that may be kilometers long. Based on the acoustic signal measured by a towed linear

[1]*School of Information and Communication Engineering, University of Electronic Science and Technology of China, Chengdu, China*

[2]*Department of Electrical Engineering and Electronics, University of Liverpool, Liverpool, UK*

Noisy Oceans: Monitoring Seismic and Acoustic Signals in the Marine Environment, Geophysical Monograph 284, First Edition.
Edited by Gaye Bayrakci and Frauke Klingelhoefer.

DOI:10.1002/9781119750925.ch14

array, we can apply beamforming techniques to transform this signal to the bearings domain (Krim & Viberg, 1996; Lorenz & Boyd, 2005). Here, there is a grid of possible angles of arrival, which are selected according to the angular resolution of the array, and each cell of the grid contains the incoming power in this direction. We can then apply a threshold to each cell in the grid to transform the original acoustic signal into a set of bearing detections.

A multitarget tracking algorithm can then be used to process the set of detections at each time step to estimate the trajectories of each target. In this chapter, we focus on detecting the number of underwater vessels and their trajectories in the bearings domain without transforming these into 3D coordinates. Performing this step would require additional information, such as prior data about the target location or additional sensors to perform triangulation (Waite, 2002; Heidemann et al., 2012; Georgy & Noureldin, 2014; Braca et al., 2014). Examples of tracking algorithms in the bearings domain are given in (Orton & Fitzgerald, 2002; Yifeng Zhou & Yip, and H. Leung., 1999; Goldberg, 1998; Northardt & Nardone, 2019; Yi et al., 2019; Saucan et al., 2015; Zhao et al., 2021; Saucan et al., 2015; Masnadi-Shirazi & Rao, 2018).

Multitracking algorithms can be formulated in the frameworks of multihypothesis tracking (MHT) (Reid, 1979; Mori et al., 1986; Blackman & Popoli, 1999), joint probabilistic data association (Bar-Shalom & Tse, 1975; Challa et al., 2011), joint multitarget probability density (Kreucher et al., 2005; Morelande et al., 2007), and random finite sets (RFSs) (Mahler, 2003; Mahler, 2014). In this chapter, we use the RFS framework due to its versatility and fully Bayesian nature. In this setting, we use probabilistic models for target birth, dynamics, and death and probabilistic models for target-generated measurements and clutter. Our objective is to estimate the set of trajectories of the targets in our surveillance area, where each target trajectory consists of the target birth time and its sequence of states at consecutive time steps (García-Fernández et al., 2020).

In a Bayesian setting, all information of interest about the trajectories is contained in the density of the set of trajectories given the sequence of measurements up to the current time step. We refer to this density as the posterior (density) on the trajectories (Please keep: on the set of trajectories). Under the standard multitarget motion and measurement models with a Poisson RFS birth model, the predicted and posterior densities on the current sets of targets are Poisson multi-Bernoulli mixtures (PMBMs) (Williams, 2015; García-Fernández et al., 2018). Analogously, the posterior on the trajectories is also a PMBM (Xia et al., 2019). The PMBM is a state-of-the-art MHT tracking algorithm that contains information about potentially undetected and detected targets. Information about undetected targets is useful for search-and-track operations (Boström-Rost et al., 2021).

A computationally lighter alternative to the PMBM filter is the Poisson multi-Bernoulli (PMB) filter, which is a PMBM with only one mixture component. The PMB filter can be applied to estimate the current set of targets (Williams, 2015; Brekke et al., 2021) and the current set of trajectories (García-Fernández et al., 2020). A computationally cheaper alternative to PMBM and PMB filters, though with a decrease in performance, is a probability hypothesis density (PHD) filter, which can also be used for targets (Mahler, 2003; Vo & Ma, 2006; Clark et al., 2007) and trajectories (García-Fernández & Svensson, 2019). The PHD filter for sets of trajectories is called the trajectory PHD (TPHD) filter. The PHD and TPHD filters are computationally fast: they approximate the posterior density via a Poisson point process, also called Poisson RFS, in which the number of objects is Poisson distributed and object states are independent and identically distributed.

In this work, we first process the raw signal data of the hydrophone. Then we obtain the set of detections using conventional beamforming (CBF) (Bartlett, 1948) and a cell-averaging constant false alarm rate (CA-CFAR) (Watts, 2000) detector, which is an adaptive algorithm to detect the presence of a target against unknown background noise. Finally, we apply a TPHD filter to estimate the number of targets and their trajectories. The flowchart of this procedure is shown in Fig. 14.1. We also show how

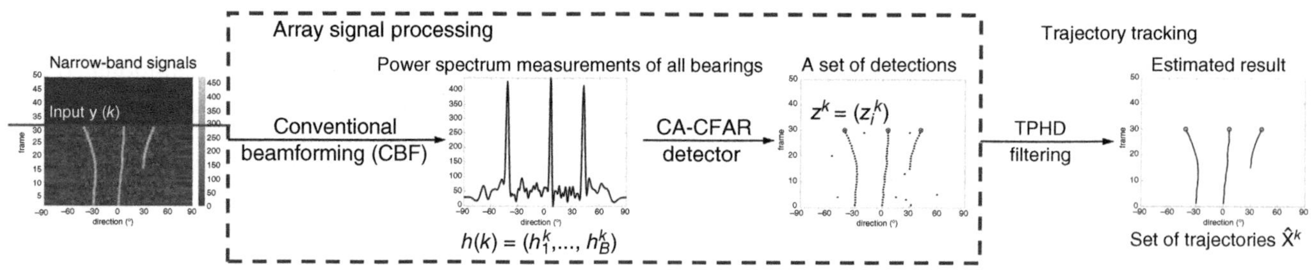

Figure 14.1 Architecture of the proposed procedure. The narrow-band input signal of the receiver is first processed using beamforming. Beamforming generates a power spectrum measurement for each bearing cell. We then apply a CA-CFAR detector to generate a set of detections at each time step. Finally, the TPHD filter estimates the current set of trajectories of the underwater vessel.

the proposed processing chain works via experimental results in the underwater environment.

The rest of the chapter is organized as follows. The signal models are given in Section 14.2. In Section 14.3, we explain how to obtain target detections from the received acoustic signal using CBF and CA-CFAR. Then we elaborate on the tracking system model of multiple vessels and the TPHD filter in Section 14.1. In Section 14.5, the efficacy of the proposed method is illustrated both in simulation and on collected real-world data. Finally, we draw conclusions in Section 14.6.

14.2. NARROW-BAND SIGNAL MODEL

In this section, we explain the mathematical model of the acoustic signal received by a uniform linear array (ULA). We show a typical scenario in Fig. 14.2. The ULA, together with the beamforming technique explained in Section 14.3, will be used to measure the power of incoming signals at different bearings, which will then be used to detect the targets.

We assume that at time step t, there are $L_t \geq 0$ targets. Each target moves in the area of interest and transmits an acoustic signal that is narrow band (Yi et al., 2019). Each target is indexed by l, where $l = 1, 2, \ldots, L_t$. Then the lth target has bearing θ_l^t (in radians) and bearing velocity $\dot{\theta}_l^t$ (in radians/s). The lth target state is therefore denoted as $x_l^t = \left[\theta_l^t \, \dot{\theta}_l^t\right]'$, where $'$ denotes transpose. The set of targets is then $\mathbf{x}^t = \left\{x_1^t, \ldots, x_{L_t}^t\right\}$.

We measure the sound emitted by the targets using a ULA system with M hydrophone elements, assuming that the emitters are far from the array. The narrow-band signal received by the first element of the array from the lth acoustic emitter is $\tilde{y}_l^1(t) = s_l(t)e^{jw_l t} + v^1(t)$. In this equation, $s_l(t)$ is the complex envelope of the received signal, ω_l is the carrier angular frequency, and $v^1(t)$ denotes the noise at the receiver. The signal measured by the mth element of the ULA is

$$\tilde{y}_l^m(t) = s_l\left(t - \tau^m\left(\theta_l^t\right)\right)e^{j\omega_l\left(t - \tau^m\left(\theta_l^t\right)\right)} + v^m(t), \tag{14.1}$$

where $\tau^m\left(\theta_l^t\right)$ is the time delay between the received signal at the first and mth elements, and $v^m(t)$ is the receiver noise for the mth element. In addition, we should note that $m = 1, \ldots, M$. For a ULA in which the distance between two consecutive array elements is d, $\tau^m\left(\theta_l^t\right) = (m-1)d \sin\theta_l^t/c$, where c is the speed of propagation of the acoustic wave and $\tau\left(\theta_l^t\right) = d \sin\theta_l^t/c$ is the propagation time delay between two consecutive elements of the array. As we have considered that the acoustic signal is narrow band, the complex envelope $s_l(t)$ changes slowly in time

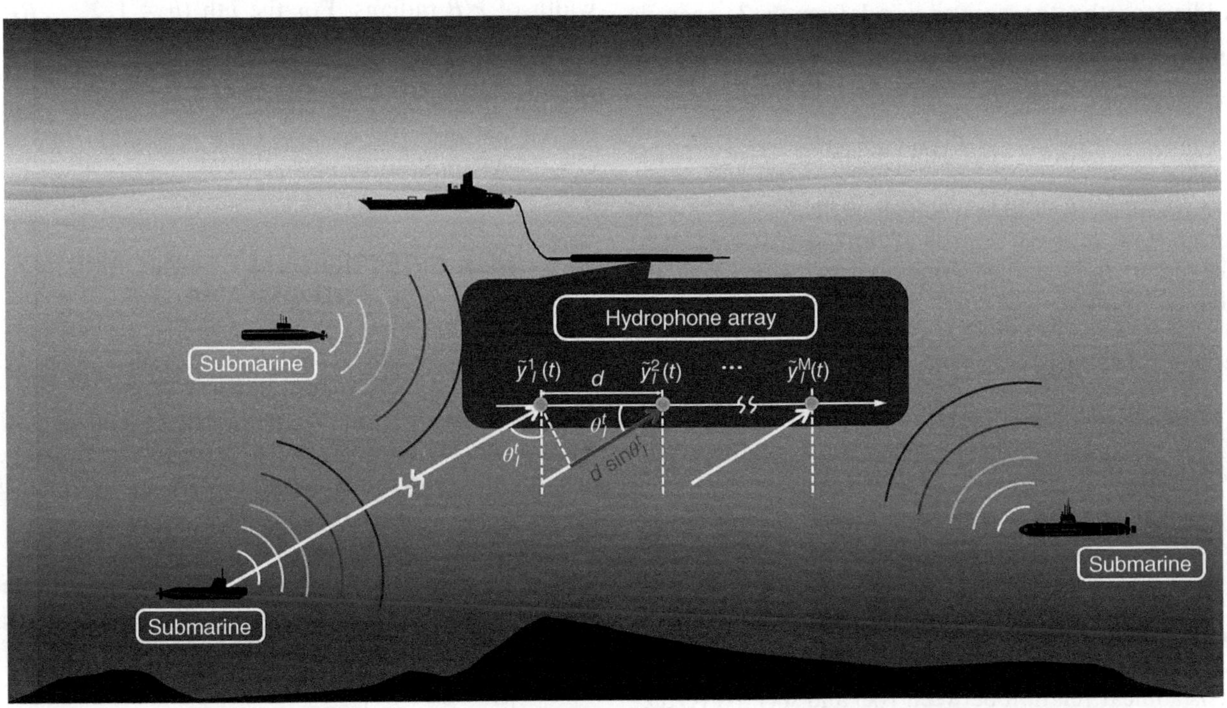

Figure 14.2 Scenario that shows the use of a passive sonar array to detect multiple underwater vessels. Each underwater vessel emits an acoustic signal. This signal is received by a ULA in a towed passive sonar system. The ULA has M elements separated by d. One of the acoustic emitters is located at a bearing θ_l^t with respect to the ULA coordinate system. The ULA can sense the power of the incoming signal at this bearing.

compared to the carrier signal $e^{jw_l t}$. Then we can make the approximation (Krim & Viberg, 1996)

$$s_l(t) \approx s_l\left(t - \tau^m\left(\theta_l^t\right)\right), m = 1, 2, \cdots, M \tag{14.2}$$

Then, equation (1) becomes

$$\tilde{y}_l^m(t) \approx s_l(t) e^{j\omega_l t} e^{-j\omega_l \tau^m\left(\theta_l^t\right)} + v^m(t). \tag{14.3}$$

The array received signal vector $\tilde{y}_l(t) = \left[\tilde{y}_l^1(t)\, \tilde{y}_l^2(t) \ldots \tilde{y}_l^M(t)\right]'$ can then be written as

$$\tilde{y}_l(t) = a\left(\theta_l^t\right) s_l(t) e^{j\omega_l t} + v(t), \tag{14.4}$$

where $a\left(\theta_l^t\right) = \left[1\, e^{-j\phi_l^t} \cdots e^{-j(M-1)\phi_l^t}\right]'$ is the steering vector, with $\phi_l^t = \omega_l \tau\left(\theta_l^t\right)$, and $v(t) = \left[v^1(t) \cdots v^M(t)\right]'$ is the vector that includes the noise. This noise is independent of the signal and is a simple way to represent the effect of undesired signals, such as thermal noise or interference from the surroundings, on the received signal.

In general, the complex carrier $e^{j\omega_l t}$ does not carry meaningful information, and only the complex received baseband signal is considered (Krim & Viberg, 1996). Then the discretized complex baseband signal of equation (??) using a sampling time T is given by

$$y_l(k) = a\left(\theta_l^k\right) s_l(k) + v(k), \tag{14.5}$$

where k is an integer that indicates the time step. That is, time step k corresponds to time $t = k \cdot T$.

Adding together the received signals from the L_k acoustic emitters with incident angles $\theta^k = \left[\theta_1^k, \cdots, \theta_{L_k}^k\right]'$, we can write equation (??) as

$$y(k) = A\left(\theta^k\right) s(k) + v(k), \tag{14.6}$$

where $s(k) = \left[s_1(k)\, s_2(k) \cdots s_{L_k}(k)\right]'$ is the vector with the L_k discretized received signals at time step k, and $A(\theta^k)$ is the $M \times L_k$ signal steering matrix, which has the analytical formula

$$A\left(\theta^k\right) = \left[a\left(\theta_1^k\right) a\left(\theta_2^k\right) \cdots a\left(\theta_{L_k}^k\right)\right] = \begin{bmatrix} 1 & 1 & \cdots & 1 \\ e^{-j\theta_1^k} & e^{-j\theta_2^k} & \cdots & e^{-j\theta_L^k} \\ \vdots & \vdots & \ddots & \vdots \\ e^{-j(M-1)\theta_1^k} & e^{-j(M-1)\theta_2^k} & \cdots & e^{-j(M-1)\theta_{L_k}^k} \end{bmatrix}. \tag{14.7}$$

We can then see that for a given steering matrix $A(\theta^k)$, there is a linear relation between $y(k)$ and $s(k)$. Nevertheless, $A(\theta^k)$ is a nonlinear transformation of the bearing vector θ^k, which implies that the relation between $y(k)$ and θ^k is nonlinear.

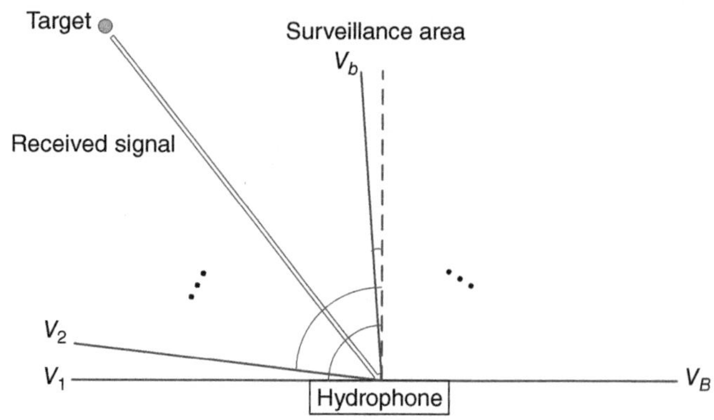

Figure 14.3 The surveillance area $-\pi/2$ to $\pi/2$ is equally divided into B bearing units.

To track multiple vessels using detection-based multi-target tracking (MTT) algorithms, we need to preprocess the received acoustic data and create a set of detections. To this end, we use beamforming (Krim & Viberg, 1996) and CFAR (Watts, 2000). The beamformer linearly combines the received array data $y(k)$ with the beamforming weights, which can enhance the signal from the desired spatial bearing and decrease the signals from other bearings.

As shown in Fig. 14.3 (Yi et al., 2019), the surveillance area of the hydrophone is equally divided into B bearing cells as V_1, V_2, $\cdots$, V_B: that is, each bearing cell has a width of π/B radians. For the bth ($b = 1, 2, \ldots, B$) cell, we obtain the output of the beamformer at this bearing after processing the received array signal $y(k)$. This output is then thresholded to create a detection by using a CFAR detector. We explain CBF in Section 14.3.1 and CA-CFAR in Section 14.3.2.

14.3. DETECTION VIA BEAMFORMING AND CA-CFAR

14.3.1. CBF

As a widely used beamforming algorithm, CBF (Bartlett, 1948) performs well in isotropic and homogeneous noise. Given the collected narrow-band signal $y(k)$, the output of CBF at bearing cell V_b is denoted as

$$h_b^k = \left(a(V_b)\right)^H \cdot R^k \cdot a(V_b) \tag{14.8}$$

where the superscript $(\cdot)^H$ denotes the conjugate transpose, $a(V_b)$ is the weight coefficient of the bearing V_b, $R^k = \mathbb{E}\left[y(k)\left(y(k)\right)^H\right]$ is the correlation matrix of the received signal, and $\mathbb{E}(\cdot)$ is the statistical expectation.

In the CBF algorithm, the correlation matrix can be approximately estimated by statistical average as

$$\hat{R}^k = \frac{1}{N}\sum_{n=1}^{N} y_n(k)(y_n(k))^H, \tag{14.9}$$

where $y_n(k)$ is the nth snapshot (frequency domain sampling) of $y(k)$, and N is the number of snapshots.

Given the raw input signal, we apply beamforming for all units $V_1, \ldots, V_B$ to generate the power-spectrum measurement frame at time step k: $h^k = \left[h_1^k, \ldots, h_B^k\right]$. Then, by sorting the spatial spectrum in chronological order, we can obtain the bearing-time record (BTR), $h^{1:k}$, which is the standard method to display sonar data.

14.3.2. CA-CFAR

After applying CBF at each time step, we adopt the CA-CFAR (Watts, 2000) detector to detect the spectrum measurement. The detected bearings constitute the set of measurements for the considered detection-based MTT algorithm, as explained in Section 14.4.

In CFAR, we go through all bearings cells and adaptively select the detection threshold based on the neighboring cells. Figure 14.4 shows a diagram of the CA-CFAR technique, which determines whether a target (vessel) is present in a test bearing cell. The reference cells at time step k include the square of the output of the beamforming algorithm: $H_b^k = \left(h_b^k\right)^2$. The cells are assumed to be evenly separated on both sides of the test cell. Then the leading I samples and lagging I samples are processed to obtain a statistic S, which depends on the CFAR processor. With a scaling factor α, the statistic S is used to determine the threshold $T = \alpha S$, which is then compared with the value of the test cell H_0 to judge the presence ($H_0 \geq T$) or absence (H_0 T) of a target in the test cell.

As shown in Fig. 14.4, the average power in the leading and lagging I cells around the test cell is denoted by U_1 and U_2, respectively. The choice of the statistic S as a function of U_1 and U_2 depends on the CFAR processor, some of which have been studied in the literature. Considering a uniform clutter background, the CA-CFAR processor chooses S as the average value of all the reference cells

$$S = (U_1 + U_2)/2. \tag{14.10}$$

At each time step k, the CA-CFAR processor is used for all spectral measurements $\left(h_1^k, \ldots, h_B^k\right)$. The corresponding bearings of the test cells that have been hypothesized to originate from a target constitute the set of detections $z^k = \left\{z_b^k : z_b^k = \theta_b^k \text{ with } \left(h_b^k\right)^2 \geq T\right\}$.

14.4. TRAJECTORY PHD FILTER FOR MULTIPLE UNDERWATER VESSELS

Based on the set of detections, the TPHD filter (García-Fernández & Svensson, 2019) can estimate the number of vessels present at the current time step and their trajectories. In this section, we first explain the models used in the tracker in Section 14.4.1. Then, in Section 14.4.2, we present the TPHD filter equations.

14.4.1. System Model

The target state $x \in \mathbb{R}^{n_x}$ is a vector that contains kinematic information about the target, such as position and velocity at the current time. In this chapter, each target state is a 2D vector that contains its bearings and bearings rate as we perform tracking in the bearings domain. Each target appears, moves, and disappears from the scene of interest. Therefore, the trajectory state of the lth target can be expressed as a variable $X_l = \left(\beta_l, x_l^{1:i}\right)$, where β_l is the birth time step, i is the trajectory length, and $x_l^{1:i} = \left(x_l^1, \cdots, x_l^i\right)$ denotes the state sequence of the lth target at consecutive time steps. At time step k, we consider trajectories that started before time step k and may last up to time step k. That is, one of these trajectories belongs to the trajectory state space at time k, which is described mathematically as

$$\mathcal{T}^k = \biguplus_{(\beta,i)\in J_k} \{\beta\} \times \mathbb{R}^{in_x}, \tag{14.11}$$

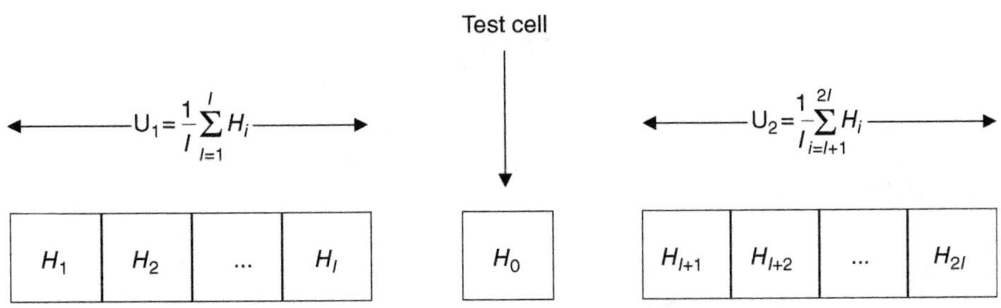

Figure 14.4 Diagram of CA-CFAR detector. It adaptively sets the detection threshold in a test cell based on average power in the I cells before and after the test cell.

where $J_k = \{(\beta, i) : 0 \leq \beta \leq k, 1 \leq i \leq k - \beta + 1\}$, and $\uplus$ denotes disjoint set union (Mahler, 2014), which is used to highlight that these sets are disjoint.

In MTT, we should account for a variable and an unknown number of trajectories so we can write the variable representing the MTT state as a set of trajectories (García-Fernández et al., 2020). The set of trajectories at time k including all the trajectories that exist between the interval $1 \sim k$ can be expressed as

$$X^k = \{(X_1, \cdots, X_L) : X_n \in \mathcal{T}_k\}, \tag{14.12}$$

where $X^k \in \mathcal{F}(\mathcal{T}^k)$ and $\mathcal{F}(\mathcal{T}^k)$ denotes the set of all finite subsets of $\mathcal{T}^k$.

The ground truth set of trajectories, which represents the trajectories of the underwater vessels, is modeled as a random variable, which gives rise to an RFS of trajectories. In the standard MTT models, a target with state x survives to the next time step with a probability $P_S(x)$ with a transition density $t(\cdot | x)$, and new targets are born independently of the surviving targets. At each time step, a target with state x is detected with a probability $P_D(x)$ and generates a measurement with density $s(\cdot | x)$. There are also clutter measurements distributed as a Poisson point process with intensity $\kappa(z)$.

Next, we describe the single-target motion model and the single-measurement model.

Vessel Motion Model

Conditional on the fact that that the bearing motion of the underwater vessel is usually slow and the maneuvering is small, we assume that the evolution of each target state $t(x_l^k | x_l^{k-1})$ follows the nearly constant velocity (CV) motion model (Bar-Shalom et al., 2001):

$$x_l^k = F x_l^{k-1} + w_l^k, \tag{14.13}$$

where F is the state-transition matrix

$$F = \begin{bmatrix} 1 & T \\ 0 & 1 \end{bmatrix}, \tag{14.14}$$

where T is the time interval between adjacent time steps and w_l^k is the process noise with covariance matrix

$$Q = q \times \begin{bmatrix} T^3/3 & T^2/2 \\ T^2/2 & T \end{bmatrix}, \tag{14.15}$$

where q is the process-noise intensity describing the uncertainty of the target moving velocity in the tracking process. Therefore, the single-target transition density is

$$t(x_l^k | x_l^{k-1}) = \mathcal{N}(x_l^k; F x_l^{k-1}, Q). \tag{14.16}$$

It should be noted that a small value of q models vessels that move predictably, roughly following a straight line. On the other hand, as we increase q, the trajectories become less predictable due to higher uncertainty in the dynamics.

Measurement Model

Since each measurement denotes the bearing detection of a target, the individual measurement generation according to (Bar-Shalom et al., 2001)

$$z_l^k = H x_l^k + v_l^k, \tag{14.17}$$

where H is the measurement generation matrix given by

$$H = [1 \;\; 0], \tag{14.18}$$

v_l^k is the measurement noise with covariance matrix

$$C = \sigma_v^2, \tag{14.19}$$

where σ_v is the measurement-noise standard deviation describing the uncertainty of the detected measurement position. Therefore, the single-measurement density is

$$s(z_l^k | x_l^k) = \mathcal{N}(z_l^k; H x_l^k, C) \tag{14.20}$$

14.4.2. TPHD Filter

Given the multitrajectory density $\pi(X)$, which perfectly characterizes an RFS of trajectories, its PHD, which is also called intensity in the point-process literature, is defined as (Ronald, 2007; Mahler, 2014)

$$D_\pi(X) = \int \pi(\{X\} \cup X) \delta X. \tag{14.21}$$

It should be noted that the PHD is a function of a single trajectory X and provides a summary of the full information of the RFS of trajectories, which is contained in the multitrajectory density $\pi(X)$. That is, the PHD is the first-order multitarget moment, and it can be used to calculate the expected number $\hat{N}_A$ of trajectories in a certain spatiotemporal region $A \subseteq \mathcal{T}^k$ using

$$\hat{N}_A = \int_A D_\pi(X) dX = \sum_{(t,i) \in J_{(k)}} \int 1_A(t, x^{1:i}) D_\pi(t, x^{1:i}) dx^{1:i} \tag{14.22}$$

where $1_A(\cdot)$ is the indicator function of a subset A: $1_A(z) = 1$ if $z \in A$ and $1_A(z) = 0$ otherwise. Therefore, the expected number of trajectories in the entire surveillance area can be obtained by substituting $A = \mathcal{T}^k$ in equation (22).

A PHD uniquely characterizes a Poisson RFS, which is a type of RFS in which the cardinality distribution is Poisson and the elements are independently identically distributed (IID) for any cardinality. Next, we provide an example showcasing how a Poisson RFS encodes probabilistic information on a set of trajectories.

Example

Let us consider a Poisson RFS of trajectories with a one-dimensional target state x that up to time step 2 has the following PHD:

$$D_\pi(2, x^1) = \mathcal{N}(x^1; 50, 1) \tag{14.23}$$

$$D_\pi(1, x^{1:2}) = \mathcal{N}\left(x^{1:2}; (10,10.1), \begin{bmatrix} 1 & 1 \\ 1 & 2 \end{bmatrix}\right), + \mathcal{N}\left(x^{1:2}; (100,100.1), \begin{bmatrix} 1 & 1 \\ 1 & 2 \end{bmatrix}\right), \tag{14.24}$$

where it is assumed that the trajectory status follows the Gaussian distribution $\mathcal{N}$ and $D_\pi(X) = 0$ for $X \neq (2, x^1)$ and $X \neq (1, x^{1:2})$. Integrating this PHD with $A = \mathcal{T}^2$ (see equation (22)), the expected number of trajectories is three. Integrating the PHD for trajectories starting at time step 1 and length 2, $A = \{1\} \times \mathbb{R}^{2n_x}$, we expect two trajectories being born at time step 1 and being alive at time step 2. The two peaks of the PHD for trajectories of the form $X = (1, x^{1:2})$ correspond to a trajectory with a path (10,10.1) and another trajectory with a path (100,100.1).[1] Integrating the PHD for trajectories starting at time step 2, $A = \{2\} \times \mathbb{R}^{n_x}$, there is one expected trajectory at time step 2 with length one. The peak of the PHD for a trajectory of the form $(2, x^1)$ corresponds to a target located at value 50. We can therefore have a PHD with two or more Gaussian mixture components with the same birth time and the same length, and estimate the trajectories of the expected targets.

The TPHD filter propagates the PHD of the RFS of trajectories present at the current time step, so it can be understood as a Poisson RFS approximation of the underlying posterior density on the set of present trajectories. Then the PHD of the Poisson multitrajectory density is propagated over time using the TPHD recursion, which consists of a prediction and an update step that are performed at each time step.

TPHD prediction: Given the posterior PHD $D_{k-1}(X) = D_{k-1}(\beta, x^{1:l-1})$ at time $k - 1$, the predicted PHD $D_{k|k-1}(X) = D_{k|k-1}(\beta, x^{1:l})$ at time k is given by

$$D_{k|k-1}(X) = D_{\gamma,k}(X) + D_{\zeta,k}(X), \tag{14.25}$$

where

$$D_{\gamma,k}(X) = D_{\gamma_\tau}(x^1) 1_{\{k\}}(\beta), \tag{14.26}$$

$$D_{\zeta,k}(X) = 1_{\mathbb{N}_{k-1}}(\beta) P_S(x^{l-1}) t(x^l | x^{l-1}) D_{k-1}(X), \tag{14.27}$$

where $\gamma_\tau(\cdot)$ is the Poisson multitarget density of new born trajectories, $P_S(x^{l-1})$ is the trajectory surviving probability, $t(\cdot | x^{l-1})$ is the state transition density of single surviving target and $\mathbb{N}_k = \{1, \ldots, k\}$.

TPHD update: Given the predicted PHD $D_{k|k-1}(X) = D_{k|k-1}(\beta, x^{1:l})$, the updated PHD $D_k(X) = D_k(\beta, x^{1:l})$ at time k is

$$D_k(X) = D_{k|k-1}(X)(1 - P_D(x^l)) + \sum_{z \in \mathbf{z}^k} \frac{P_D(x^l) s(z|x^l) D_{k|k-1}(X)}{\kappa(z) + \langle P_D(\cdot) s(z|\cdot), D_{\tau,k|k-1}(\cdot) \rangle}, \tag{14.28}$$

where $P_D(x^l)$ is the trajectory detection probability, $s(\cdot | \cdot)$ is the target-measurement likelihood function and $\kappa(z) = \lambda_c \breve{c}(z)$ is the intensity of Poisson clutter and $\breve{c}(z)$ is the clutter spatial density function, $\langle a, b \rangle = \int a(x)b(x)dx$ is the inner product, and $D_{\tau,k|k-1}(\cdot)$ denotes the predicted PHD of the targets at time k, defined as

$$D_{\tau,k|k-1}(x^l) = \sum_{\beta=1}^{k} \int D_{k|k-1}(\beta, x^{1:l}) dx^{1:l-1}. \tag{14.29}$$

After filtering, we can extract the number and trajectory states of all targets at the current time step.

Gaussian Mixture Implementation

The TPHD filtering recursion can be implemented in closed-form for linear/Gaussian dynamic and measurement models (such as those explained in Section 14.4.1), constant probability of detection and survival, and a birth intensity that is a Gaussian mixture. In this section, we summarize the primary characteristics of this implementation. All details can be found in (García-Fernández & Svensson, 2019).

We can define a Gaussian trajectory density as

$$\mathcal{N}(\beta, x^{1:i}; t^k, m^k, P^k) = \begin{cases} \mathcal{N}(x^{1:i}; m^k, P^k) & \beta = \beta^k, i = i^k \\ 0 & \text{otherwise} \end{cases} \tag{14.30}$$

where $i^k = \dim(m^k)/n_x$. Equation (30) represents a single-trajectory Gaussian density with start time β^k, duration i^k, mean $m^k \in \mathbb{R}^{i^k n_x}$, and covariance matrix $P^k \in \mathbb{R}^{i^k n_x \times i^k n_x}$ evaluated at $(\beta, x^{1:i})$.

Then the posterior PHD in the Gaussian mixture TPHD implementation is of the form

$$D_k(X) = \sum_{j=1}^{J^k} w_j^k \mathcal{N}\left(X; \beta_j^k, m_j^k, P_j^k\right). \tag{14.31}$$

[1] *The nonzero off-diagonal elements of the 2D covariance matrix in equation (24) incorporate correlations across time for the target trajectories.*

where J^k is the number of components and $w_j^k > 0$, β_j^k, m_j^k, and P_j^k are the weight, birth time, mean, and covariance matrix of the jth component of the PHD.

It should be noted that by applying equation (22), the expected number of trajectories in the entire surveillance area according to the PHD in equation (31) is the sum of the weights:

$$\hat{N}_{T^k} = \sum_{j=1}^{J^k} w_j^k. \tag{14.32}$$

The Gaussian mixture TPHD filter implementation requires several approximations to manage the computational complexity. In particular, we need to prune the number of components in the posterior PHD by discarding negligible weights. The implementation also uses absorption, in which a Gaussian PHD component absorbs another Gaussian PHD component, whose distribution for the current target state is sufficiently close and has a smaller weight. Measurement gating can also be used to reduce the number of data associations in the update step (Challa et al., 2011). Finally, as time goes on, single-trajectory Gaussian densities represent trajectories over longer time steps, so the covariance matrices become increasingly large. To make this approach computationally tractable, we consider the L-scan approximation, in which past states before the most recent L time steps are considered independent in the single-trajectory Gaussian densities (García-Fernández & Svensson, 2019).

14.5. EXPERIMENTS

In this section, we first show through simulation experiments the processing results of the CBF and CA-CFAR algorithms to produce a set of bearings detections. Then, also via simulations, we evaluate the trajectory-tracking performance of the TPHD filter (Section 14.5.1) based on the detections provided by CBF and CA-CFAR. Finally, a set of real-world experimental data using a ULA in the underwater environment is used to verify the proposed processing chain in Section 14.5.2.

14.5.1. Testing Using Numerical Simulations

We proceed to analyze a standard underwater bearing-only tracking scenario. The parameters of the ULA are as follows: $M = 40$, $d = 0.5$ m with a working frequency interval from 100 to 1,000 Hz for the wide-band signal. The propagation speed of the acoustic wave is c = 1531 m/s. The number of time steps (frames) considered in the numerical simulation is K = 50. The parameters of frequency segmenting are N = 2,000 and $\tilde{N}$ = 100, where N is the number of snapshots and $\tilde{N}$ is the number of frequency segments in wide-band signal processing (Yi et al., 2019). The sampling frequency is 4,000 Hz. To analyze the performance of the proposed processing chain, we perform a Monte Carlo (MC) simulation with 200 runs.

Results of CBF and CA-CFAR

We consider a surveillance area in the bearings domain from −90° to 90°. As shown in Fig. 14.5, we analyze a simple scenario where three underwater vessels move in the

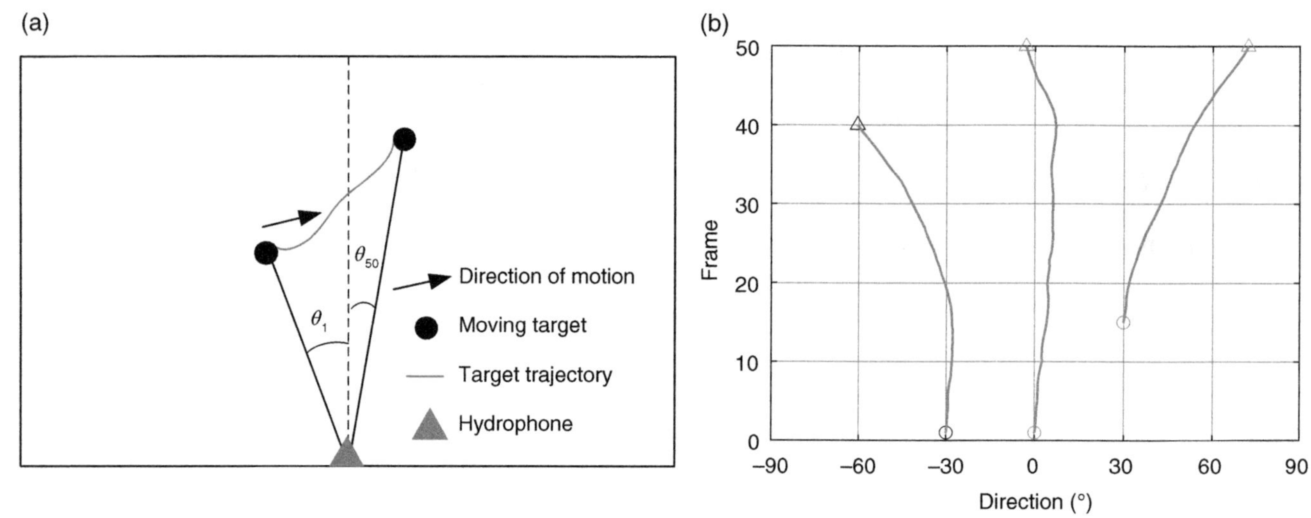

Figure 14.5 (a) Diagram of the scenario used in the numerical simulations. (b) The ground truth set of trajectories in the bearings domain. The start and end points for each trajectory are marked by ○ and △, respectively.

area of interest. In the complex underwater environment, sonar signals are often submerged in noise, so the target signal-to-noise ratio (SNR) is very low, where $SNR = 10\log_{10}\left(\mathbb{E}\left[s_l(k)(s_l(k))^H\right]/\mathbb{E}\left[v(k)(v(k))^H\right]\right)$. Setting the SNR at each acoustic emitter of the narrow-band input signal to −17 dB and −21 dB, the BTR of the target and the measurement points passing through CA-CFAR are shown in Fig. 14.6, in which the scaling factors for SNR = −17 dB and SNR = −21 dB are $\alpha = 10$ and $\alpha = 7.5$, respectively. It can be seen from Fig. 14.6a,c that CBF enhances the signal and suppress the noise, which improves the input SNR. Meanwhile, CA-CFAR processing can accurately detect underwater vessels using passive sonar array systems from the thresholded measurement. Moreover, the set of detections also contains clutter and missed detections. With the decrease in SNR, the number of clutter measurements increases, and the phenomenon of missed detection becomes more prominent, as shown in Fig. 14.6b,d.

Considering a single target scene with 50 frames, we estimate the root mean square error (RMSE), the probability of detection (P_D), and the clutter rate (average clutter in a single frame) to evaluate the detector performance of the CBF and CA-CFAR algorithms. These quantities will then be used in the TPHD filter to characterize detection performance for a given SNR. At each frame, the target is deemed to be successfully detected if the estimated

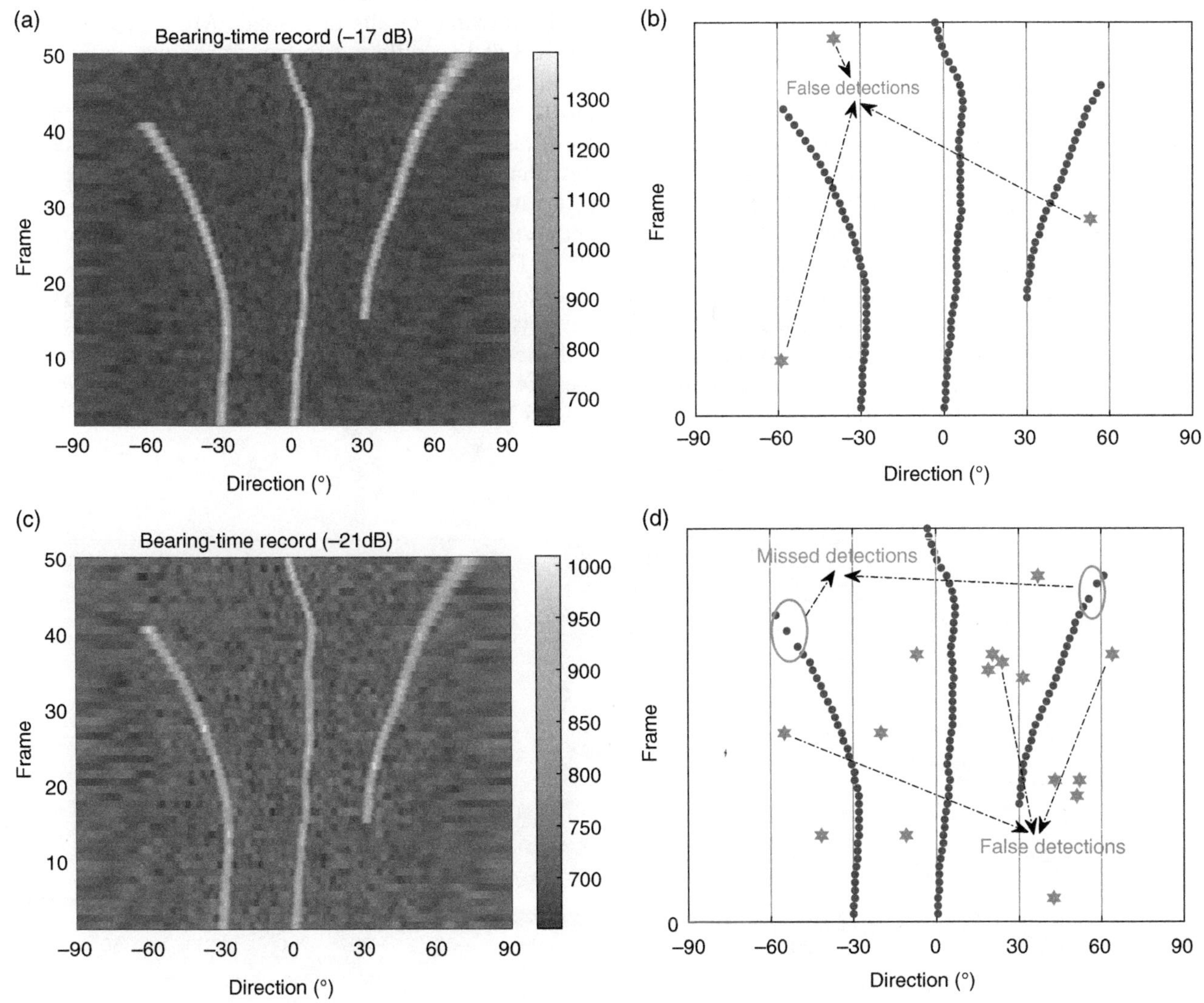

Figure 14.6 (a) Bearing-time record for SNR = −17 dB. (b) Detected bearing measurement through CA-CFAR for SNR = −17 dB. (c) Bearing-time record for SNR = −21 dB. (d) Detected bearing measurement through CA-CFAR for SNR = −21 dB.

Table 14.1 Detector performance parameters of 200 MC simulations for the CBF and CA-CFAR algorithms.

Parameter \ SNR	−23 dB	−21 dB	−19 dB	−17 dB	−15 dB
RMSE (°)	2.7318	1.9636	1.4089	0.9894	0.9728
P_D	0.8391	0.8970	0.9490	0.9893	0.9964
Clutter rate	1.6002	0.7138	0.4335	0.1967	0.0027

direction error is less than 4°. On the other hand, if it is not less than 4, the detected point is clutter, and the clutter rate is the average of the clutter number over time. With 200 MC simulations, the detector performance parameters are shown in Table 14.1. As expected, with increased SNR, detection performance improves.

TPHD Filter Performance

In this subsection, we consider a scenario with three moving vessels. Targets 1, 2, and 3 appear at k = 1, 1, 15 frames, and target 2 disappears at k = 36 frames. The parameter q of the covariance matrix of the dynamic model (15) is set to $q = 1°^2/s^3$.

At each time step, the TPHD filter processes a set of bearings detections. As explained in Section 14.4, the measurement model is linear and Gaussian. The probability of survival of the targets is a constant set to $P^S = 0.98$. As we analyze two different SNRs, the probability of detection P_D, the clutter rate λ_c, and the measurement noise standard deviation σ_v (detection error) are different in each scenario. For SNR = −21 dB, $P_D = 0.90$, $\lambda_c = 1$, and $\sigma_v = 2$. For SNR= − 17 dB, $P_D = 0.99$, $\lambda_c = 0$, and $\sigma_v = 1$. Similar to (Yi et al., 2020), the TPHD filter adopts the L-scan Gaussian mixture implementation with parameter L set to $L = 2$, and the Poisson birth intensity is a Gaussian mixture (equation (31)) with the following parameters: $J_\gamma^k = 3$, $\omega_{\gamma,j}^k = 0.03$, $m_{\gamma,1}^k = [-30; 0]$, $m_{\gamma,2}^k = [47.5; 0]$, $m_{\gamma,3}^k = [30; 0]$, and $P_{\gamma,j}^k = \text{diag}([5; 5])^2$. The pruning parameter for PHD components is $T_p = 1 \times 10^{-5}$, the absorption threshold is $T_a = 4$, and the maximum number of targets the TPHD filter handles at one time is $N_{max} = 100$.

The tracking results of a single MC run with SNR= −17 dB at two different frames are shown in Fig. 14.7. It can be observed that the set of trajectories estimated by the TPHD filter is in good agreement with the real set of target trajectories at each time step. We should also note that the TPHD filter estimates the set of alive trajectories at each time step: that is, trajectories that are present at the current time step. Three targets are alive at time step 30, so Fig. 14.7a shows the estimated trajectories of these three targets. In contrast, at time step 40, two targets are alive, as one of the previous targets is no longer in the surveillance area, so 14.7b shows the estimated trajectories of these two targets. The estimated trajectories are accurate from a localization perspective. In this case, the TPHD

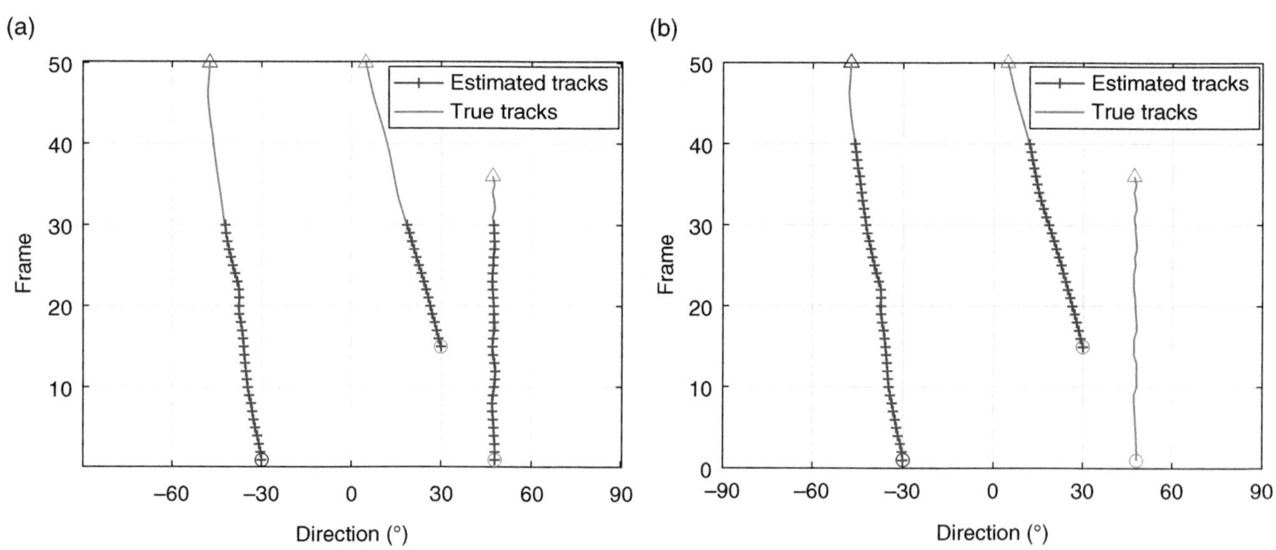

Figure 14.7 Ttracking results of a single MC run where the SNR of the received array signal is −17 dB at different frames: (a) k = 30; (b) k = 40. The TPHD filter properly estimates the set of alive trajectories at each time step. The start and end points for each true trajectory are marked by ○ and △, respectively.

filter estimates the time step when the targets entered the surveillance area with no error.

In the following, we use a metric for sets of trajectories to evaluate the error between the estimated set of trajectories and the ground truth set of trajectories at each time step. We use the metric for sets of trajectories in (Garcí-Fernández et al., 2020), which we refer to as trajectory metric (TM). At each time step, this metric penalizes localization errors for properly detected targets, costs for false and missed targets, and costs for track switching. The metric is obtained by solving a multidimensional assignment problem in which, at each time step, estimated trajectories are assigned to ground-truth trajectories or remain unassigned. We use the linear programming implementation of the TM with parameters $p = 2$, $c = 10°$, $\gamma = 1$. Parameter c indicates the maximum distance (in this case, in degrees) so we can consider a target properly detected by a target estimate. Parameter γ indicates how much track switching is penalized in the metric, and $p = 2$ indicates that the metric penalizes the squared costs.

The TM errors to estimate the set of alive trajectories for each time step are shown in Fig. 14.8b. It can be seen in Fig. 14.8 that the TPHD filter has better tracking performance for both target-number estimation and TM error with the increase in SNR, especially missed-detection error and false-detection error. The analysis is consistent with Fig. 14.6. The decomposition of the trajectory metric into the different costs is shown in Fig. 14.9. Here, we can see that the missed-detection error and false-detection error are larger than the localization error and trajectory-switching error; the problems in this underwater environment are primarily false interference and missed detection.

14.5.2. Testing Using Real-World Experimental Data

In this section, we analyze the performance of the proposed beamforming-TPHD processing chain using an experimental data set. The experimental data has a duration of 186 frames. The data set was collected using a towed passive ULA in a shallow sea. The ULA measures the acoustic noise produced by moving ships in the surroundings and the noise generated by the ship in which the array is mounted. The received raw array signals are processed according to the CBF algorithm explained in Section 14.3, and the resulting BTR is shown in Fig. 14.10. In this figure, we do not take into account the signal that lies in the directions in the interval between −90° and −60°. The high-power signal generated in this interval is generated by the ship carrying the ULA and is therefore irrelevant to help us detect other vessels. Five targets are moving in the surveillance area. The trajectories of two of these targets cross at around frame 90 (at a direction of approximately 60°). At around frame 110, the emitted signal from one of the targets is too low, and the mounted sensing system cannot detect its presence. Therefore, it is impossible to continue tracking this target from this time frame, and the multitarget tracking algorithms consider that this target has disappeared from the scene of interest, as it may have moved sufficiently far away.

The processing results with SNR = −17 dB are depicted in Fig. 14.11, in which the P_D, clutter rate, σ_v, and other

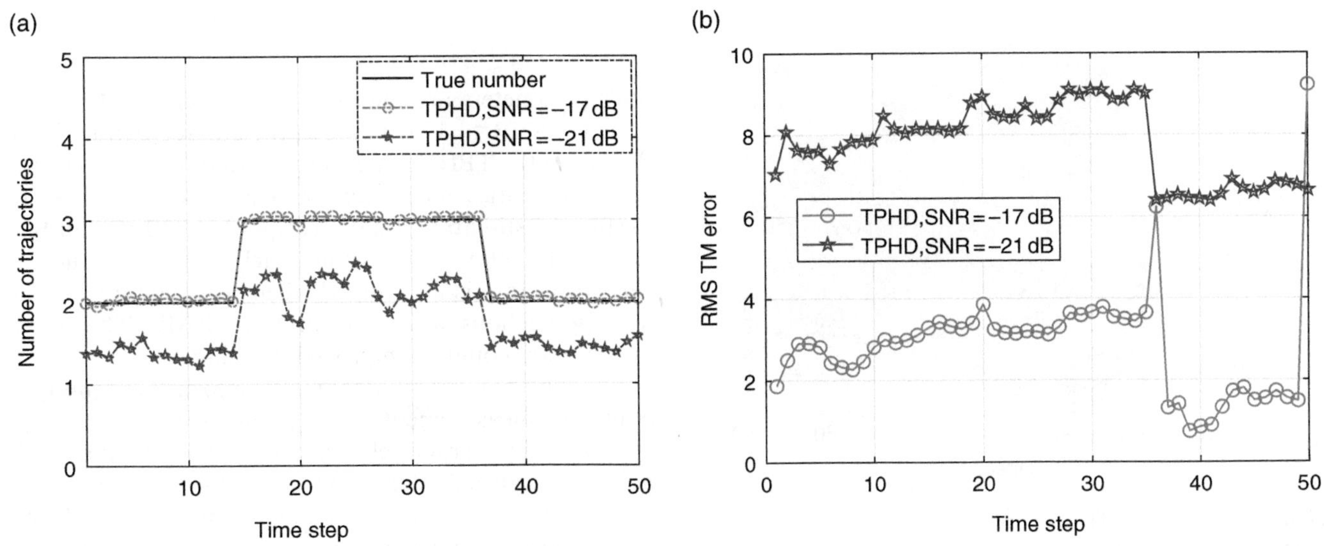

Figure 14.8 (a) Number of estimated trajectories at each time frame for two different SNRs. (b) TM error to estimate the set of alive trajectories at each time frame.

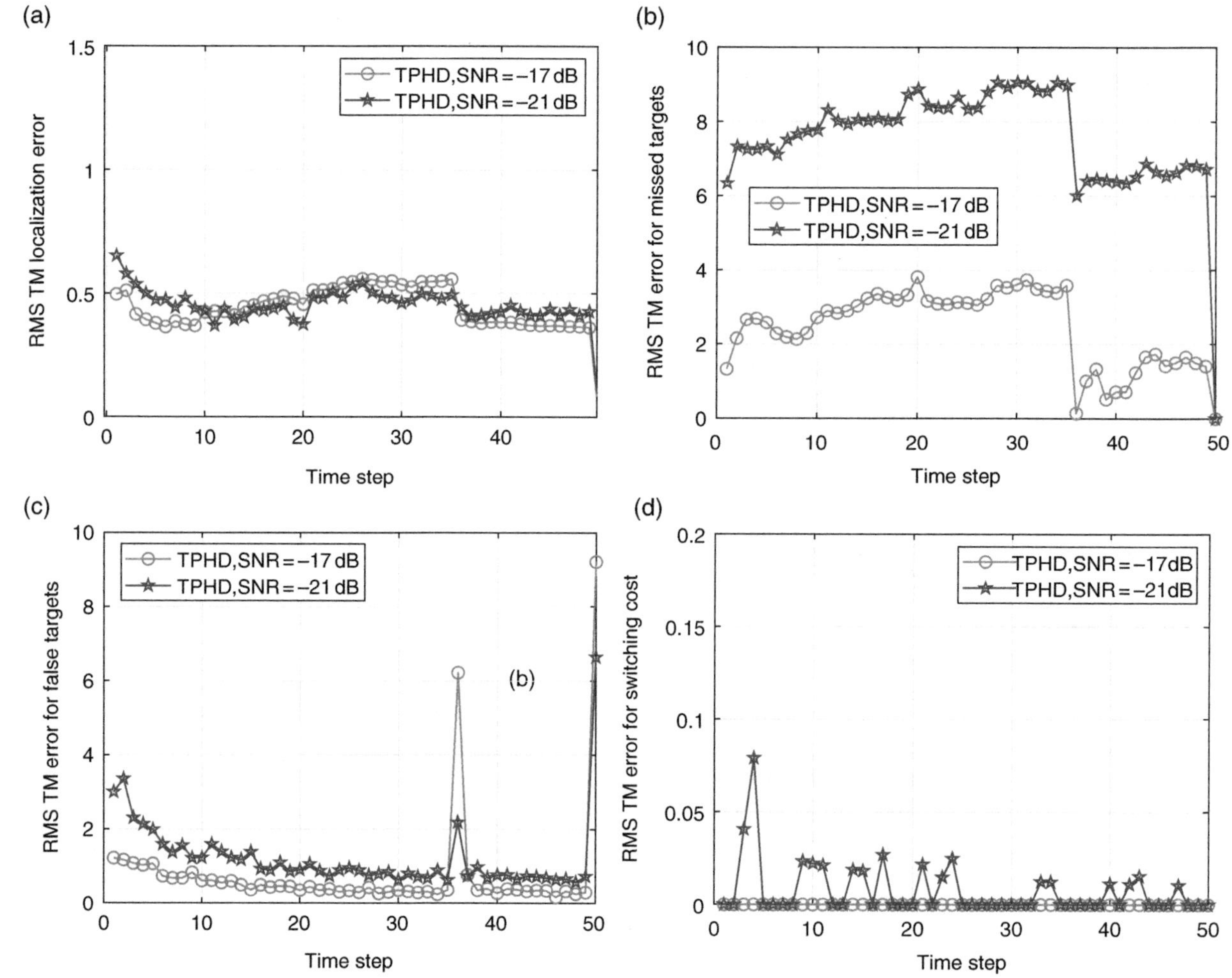

Figure 14.9 Decomposition of the trajectory metric error against time frame: (a) localization error; (b) missed-detection error; (c) false-detection error; (d) trajectory-switching error. The error reduction after increasing the SNR is primarily due to fewer missed targets.

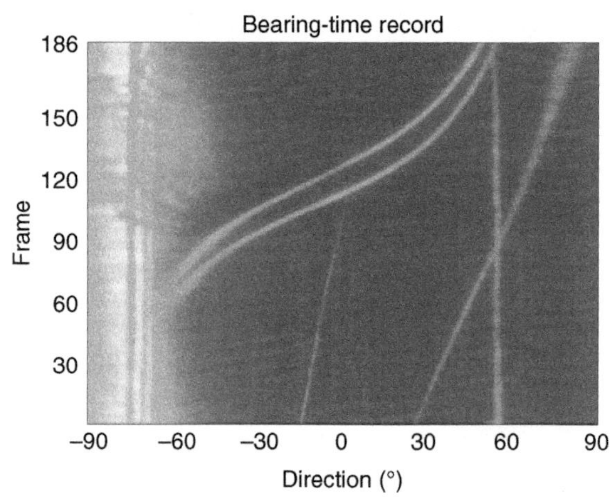

Figure 14.10 Raw BTR map of the recorded real data. The high-power signal in the directions between −90° and −60° is generated by the ship that carries the passive sonar and is discarded by the proposed detection-and-tracking system.

filter parameters are the same as the numerical simulations for SNR= −17 dB in the previous subsection. Note that since the TPHD filter only provides estimates of the alive trajectories at each time step, Fig. 14.11b consists of the filter's estimation results at steps 81, 107, 174, and 186. The five targets are successfully tracked, although the target 2 trajectory is interrupted when crossing target 3 due to misdetections arising from CA-CFAR. This track interruption could be improved by using a higher probability of survival or a filter that takes into account this type of merged measurement in which only one target can create a detection in each cell. As explained before, target 1 is no longer detected after frame 110, so the tracking system indicates that it is no longer in the surveillance area. Overall, the proposed processing chain can track multiple underwater vessels by estimating the set of alive trajectories at each time step.

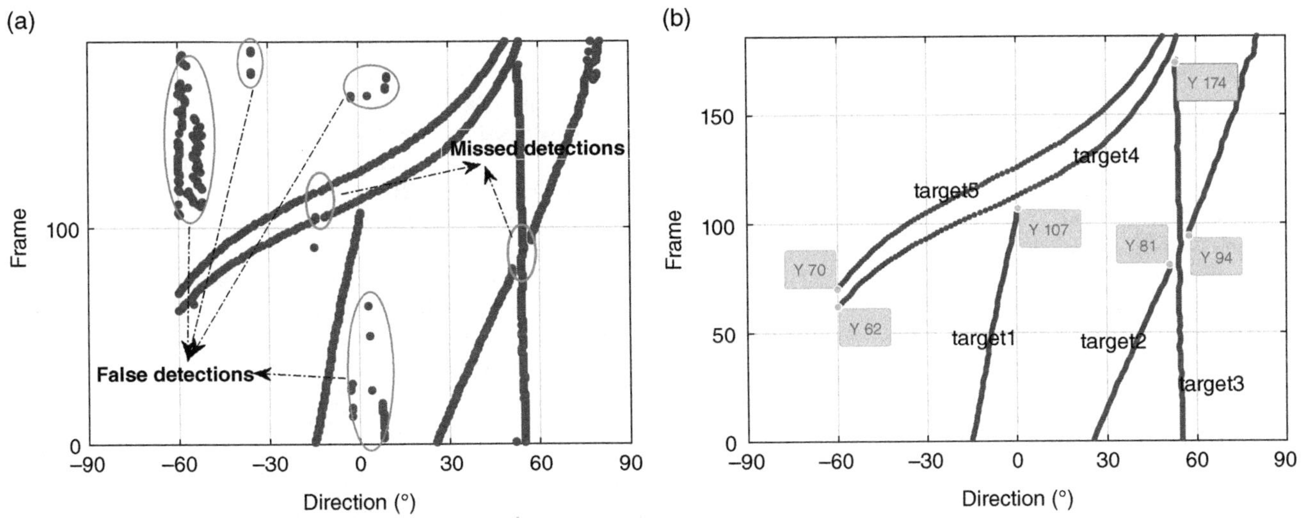

Figure 14.11 (a) Detected bearing measurement through CA-CFAR with SNR = −17 dB. (b) Tracking results of the TPHD filter at time steps 81, 107, 174, and 186.

14.6. CONCLUSIONS

This chapter proposes a complete processing chain from received acoustic signals to tracked trajectories for the underwater tracking of an unknown and time-varying number of moving vessels using passive array sonar systems. First we introduced the narrow-band signal model received by passive-array sonar systems. Then the BTR of narrow-band signals can be obtained by CBF. To track multiple vessels using detection-based MTT algorithms, a set of detections is generated by processing the BTR using the CA-CFAR detector. Given the set of detections at each time step, the TPHD filter estimates the number and individual trajectories of vessels in real time. The performance of the proposed processing chain was demonstrated both in simulation and on collected real-world data.

As part of future work, the proposed processing chain can be modified by using different CFAR detectors tailored to the underwater environment or more complex multitarget trackers. In addition, it is possible to design a processing chain based on track-before-detect methods (Li et al., 2018; García-Fernández, 2016) that can track multiple targets directly using raw BTR data rather than the detection sets after CA-CFAR. Information fusion among multiple passive-array sonar systems can also be considered in the processing chain (Yi et al., 2020; Li et al., 2019; Li et al., 2018). An improved processing chain for active sonar (Lo & Ferguson, 2004) is also a relevant line of future work. In active sonar, the sonar measurement includes the range information, and the tracker can accurately estimate the target's position, not just its bearing.

REFERENCES

Ainslie, M. A. (2010). *Principles of Sonar Performance Modeling*. Springer.

Bar-Shalom, Y., & Tse, E. (1975). Tracking in a cluttered environment with probabilistic data association. *Automatica*, *11*(5), 451–460. https://doi.org/10.1016/0005-1098(75)90021-7

Bar-Shalom, Y., Kirubarajan, T., & Li, X. R. (2001). *Estimation with Applications to Tracking and Navigation*. John Wiley & Sons, Inc.

Bartlett, M. (1948). Smoothing periodograms from time–series with continuous spectra. *Nature*, *161*, 686–687.

Baumgartner, M. F., Freitag, L., Partan, J., Ball, K. R., & Prada, K. E. (2008). Tracking large marine predators in three dimensions: the real-time acoustic tracking system. *IEEE Journal of Oceanic Engineering*, *33*(2), 146–157.

Blackman, S., & Popoli, R. (1999). *Design and Analysis of Modern Tracking Systems*. Artech House.

Boström-Rost, P., Axehill, D., & Hendeby, G. (2021). Sensor management for search and track using the Poisson multi-Bernoulli mixture filter. *IEEE Transactions on Aerospace and Electronic Systems*, *57*(5), 2771–2783.

Braca, P., Willett, P., LePage, K., Marano, S., & Matta, V. (2014). Bayesian tracking in underwater wireless sensor networks with port-starboard ambiguity. *IEEE Transactions on Signal Processing*, *62*(7), 1864–1878.

Brekke, E. F., Hem, A. G., & Tokle, L.-C. N. (2021). Multitarget tracking with multiple models and visibility: Derivation and verification on maritime radar data. *IEEE Journal of Oceanic Engineering*, *46*(4), 1272–1287.

Carter, G. (1981). Time delay estimation for passive sonar signal processing. *IEEE Transactions on Acoustics, Speech, and Signal Processing*, *29*(3), 463–470.

Challa, S., Morelande, M. R., Musicki, D., & Evans, R. J. (2011). *Fundamentals of Object Tracking*. Cambridge University Press.

Clark, D., Tena-Ruiz, I., Petillot, Y., & Bell, J. (2007). Particle PHD filter multiple target tracking in sonar image. *IEEE Transactions on Aerospace and Electronic Systems*, *43*(1), 409–416.

Coiras, E., Petillot, Y., & Lane, D. M. (2007). Multiresolution 3-D reconstruction from side-scan sonar images. *IEEE Transactions on Image Processing*, *16*(2), 382–390.

Coraluppi, S., & Carthel, C. (2006). Advances in active sonar tracking. In *14th European Signal Processing Conference*, 1–5.

Dalberg, E., Lauberts, A., Lennartsson, R., Levonen, M., & Persson, L. (2006). Underwater target tracking by means of acoustic and electromagnetic data fusion. In *9th International Conference on Information Fusion*, 1–7.

Ferri, G., Munafo, A., Tesei, A., Braca, P., Meyer, F., Pelekanakis, K., et al. (2017). Cooperative robotic networks for underwater surveillance: an overview. *IET Radar, Sonar and Navigation*, *11*(12), 1740–1761.

García-Fernández, A. F. (2016). A track-before-detect labeled multi-Bernoulli particle filter with label switching. *IEEE Transactions on Aerospace and Electronic Systems*, *52*(5), 2123–2138.

García-Fernández, A. F., Rahmathullah, A. S., & Svensson, L. (2020). A metric on the space of finite sets of trajectories for evaluation of multi-target tracking algorithms. *IEEE Transactions on Signal Processing*, *68*, 3917–3928.

García-Fernández, A. F., & Svensson, L. (2019). Trajectory PHD and CPHD filters. *IEEE Transactions on Signal Processing*, *67*(22), 5702–5714.

García-Fernández, A. F., Svensson, L., & Morelande, M. R. (2020a). Multiple target tracking based on sets of trajectories. *IEEE Transactions on Aerospace and Electronic Systems*, *56*(3), 1685–1707.

García-Fernández, A. F., Svensson, L., Williams, J. L., Xia, Y., & Granström, K. (2020b). Trajectory Poisson multi-Bernoulli filters. *IEEE Transactions on Signal Processing*, *68*, 4933–4945.

García-Fernández, A. F., Williams, J. L., Granström, K., & Svensson, L. (2018). Poisson multi-Bernoulli mixture filter: direct derivation and implementation. *IEEE Transactions on Aerospace and Electronic Systems*, *54*(4), 1883–1901.

Georgy, J., & Noureldin, A. (2014). Unconstrained underwater multi-target tracking in passive sonar systems using two-stage PF-based technique. *International Journal of Systems Science*, *45*(3), 439–455.

Gershman, A., Nemeth, E., & Bohme, J. (2000). Experimental performance of adaptive beamforming in a sonar environment with a towed array and moving interfering sources. *IEEE Transactions on Signal Processing*, *48*(1), 246–250.

Goldberg, J. (1998). Joint direction-of-arrival and array-shape tracking for multiple moving targets. *IEEE Journal of Oceanic Engineering*, *23*(2), 118–126.

Heidemann, J., Stojanovic, M., & Zorzi, M. (2012). Underwater sensor networks: applications, advances and challenges. *Philosophical Transactions of the Royal Society A*, *370*, 158–175.

Isbitiren, G., & Akan, O. B. (2011). Three-dimensional underwater target tracking with acoustic sensor networks. *IEEE Transactions on Vehicular Technology*, *60*(8), 3897–3906.

Kreucher, C., Kastella, K., & Hero, A. O. III. (2005). Multitarget tracking using the joint multitarget probability density. *IEEE Transactions on Aerospace and Electronic Systems*, *41* (4), 1396–1414. https://doi.org/10.1109/TAES.2005.1561892

Krim, H., & Viberg, M. (1996). Two decades of array signal processing research: the parametric approach. *IEEE Signal Processing Magazine*, *13*(4), 67–94.

Li, S., Yi, W., Hoseinnezhad, R., Battistelli, G., Wang, B., & Kong, L. (2018a). Robust distributed fusion with labeled random finite sets. *IEEE Transactions on Signal Processing*, *66*(2), 278–293.

Li, S., Yi, W., Hoseinnezhad, R., Wang, B., & Kong, L. (2018b). Multiobject tracking for generic observation model using labeled random finite sets. *IEEE Transactions on Signal Processing*, *66*(2), 368–383.

Li, S., Battistelli, G., Chisci, L., Yi, W., Wang, B., & Kong, L. (2019). Computationally efficient multi-agent multi-object tracking with labeled random finite sets. *IEEE Transactions on Signal Processing*, *67*(1), 260–275.

Liu, B., Tang, X., Tharmarasa, R., Kirubarajan, T., Jassemi, R., & Halle, S. (2020). Underwater target tracking in uncertain multipath ocean environments. *IEEE Transactions on Aerospace and Electronic Systems*, *56*(6), 4899–4915.

Lo, K., & Ferguson, B. (2004). Automatic detection and tracking of a small surface watercraft in shallow water using a high-frequency active sonar. *IEEE Transactions on Aerospace and Electronic Systems*, *40*(4), 1377–1388.

Lorenz, R., & Boyd, S. (2005). Robust minimum variance beamforming. *IEEE Transactions on Signal Processing*, *53*(5), 1684–1696.

Luo, J., Han, Y., & Fan, L. (2018). Underwater acoustic target tracking: a review. *Sensors*, *18*(1).

Mahler, R. P. S. (2003). Multitarget Bayes filtering via first-order multitarget moments. *IEEE Transactions on Aerospace and Electronic Systems*, *39*(4), 1152–1178. https://doi.org/10.1109/TAES.2003.1261119

Mahler, R. P. S. (2007). *Statistical Multisource–Multitarget Information Fusion*. Artech House.

Mahler, R. P. S. (2014). *Advances in Statistical Multisource–Multitarget Information Fusion*. Artech House.

Masnadi-Shirazi, A., & Rao, B. D. (2018). A covariance-based superpositional CPHD filter for multisource DOA tracking. *IEEE Transactions on Signal Processing*, *66*(2), 309–323.

Mellema, G. R. (2020). Improved active sonar tracking in clutter using integrated feature data. *IEEE Journal of Oceanic Engineering*, *45*(1), 304–318.

Morelande, M. R., Kreucher, C. M., & Kastella, K. (2007). A Bayesian approach to multiple target detection and tracking. *IEEE Transactions on Signal Processing*, *55*(5), 1589–1604. https://doi.org/10.1109/TSP.2006.889470.

Mori, S., Chong, C.-Y., Tse, E., & Wishner, R. (1986). Tracking and classifying multiple targets without a priori identification. *IEEE Transactions on Automatic Control*, *31*(5), 401–409. https://doi.org/10.1109/TAC.1986.1104306

Murino, V., & Trucco, A. (2000). Three-dimensional image generation and processing in underwater acoustic vision. *Proceedings of the IEEE*, *88*(12), 1903–1948.

Northardt, T., & Nardone, S. C. (2019). Track-before-detect bearings-only localization performance in complex passive sonar scenarios: a case study. *IEEE Journal of Oceanic Engineering*, *44*(2), 482–491.

Nosal, E.-M., & Frazer, L. N. (2007). Modified pairwise spectrogram processing for localization of unknown broadband sources. *IEEE Journal of Oceanic Engineering*, *32*(3), 721–728.

Orton, M., & Fitzgerald, W. (2002). A Bayesian approach to tracking multiple targets using sensor arrays and particle filters. *IEEE Transactions on Signal Processing*, *50*(2), 216–223. https://doi.org/10.1109/78.978377

Picard, K., Brooke, B. P., Harris, P. T., Siwabessy, P. J., Coffin, M. F., Tran, M., Spinoccia, M., Weales, J., Macmillan-Lawler, M., & Sullivan, J. (2018). Malaysia airlines flight MH370 search data reveal geomorphology and seafloor processes in the remote southeast Indian ocean. *Marine Geology*, *395*, 301–319.

Reid, D. (1979). An algorithm for tracking multiple targets. *IEEE Transactions on Automatic Control*, *24*(6), 843–854. https://doi.org/10.1109/TAC.1979.1102177

Saucan, A.-A., Chonavel, T., Sintes, C., & Le Caillec, J.-M. (2015a). Marked Poisson point process PHD filter for DOA tracking. In *23rd European Signal Processing Conference (EUSIPCO)* (pp. 2621–2625). IEEE.

Saucan, A.-A., Sintes, C., Chonavel, T., & Le Caillec, J.-M. (2015b). Model-based adaptive 3D sonar reconstruction in reverberating environments. *IEEE Transactions on Image Processing*, *24*(10), 2928–2940.

Urick, R. J. (1983). *Principles of Underwater Sound.* Peninsula Publishing.

Vo, B.-N., & Ma, W.-K. (Nov. 2006). The Gaussian mixture probability hypothesis density filter. *IEEE Transactions on Signal Processing*, *54*(11), 4091–4104. https://doi.org/10.1109/TSP.2006.881190

Waite, A. D. (2002). *Sonar for Practising Engineers*. Wiley.

Watts, S. (2000). The performance of cell-averaging cfar systems in sea clutter. *Record of the IEEE 2000 International Radar Conference* (pp. 398–403). IEEE.

Williams, J. L. (2015a). An efficient, variational approximation of the best fitting multi-Bernoulli filter. *IEEE Transactions on Signal Processing*, *63*(1), 258–273. https://doi.org/10.1109/TSP.2014.2370946

Williams, J. L. (July 2015b). Marginal multi-Bernoulli filters: RFS derivation of MHT, JIPDA and association-based MeMBer. *IEEE Transactions on Aerospace and Electronic Systems*, *51*(3), 1664–1687.

Xia, Y., Granström, K., Svensson, L., García-Fernández, A. F., & Williams, J. L. (2019). Multi-scan implementation of the trajectory Poisson multi-Bernoulli mixture filter. *Journal of Advances in Information Fusion*, *14*(2), 213–235.

Yi, W., Fu, L., García-Fernández, A. F., Xu, L., & Kong, L. (2019). Particle filtering based track-before-detect method for passive array sonar systems. *Signal Processing*, *165*, 303–314.

Yi, W., Li, S., Wang, B., Hoseinnezhad, R., & Kong, L. (2020). Computationally efficient distributed multi-sensor fusion with multi-Bernoulli filter. *IEEE Transactions on Signal Processing*, *68*, 241–256.

Yocom, B. A., La Cour, B. R., & Yudichak, T. W. (2011). A Bayesian approach to passive sonar detection and tracking in the presence of interferers. *IEEE Journal of Oceanic Engineering*, *36*(3), 386–405.

Zhao, J., Gui, R., Dong, X., & Wu, S. (2021). Time-varying DOA tracking algorithm based on generalized labeled multi-Bernoulli. *IEEE Access*, *9*, 5943–5950.

Zhou, Y., Yip, P., & Leung, H. (1999). Tracking the direction-of-arrival of multiple moving targets by passive arrays: algorithm. *IEEE Transactions on Signal Processing*, *47*(10), 2655–2666.

Ziomek, L. J. (2017). *An Introduction to Sonar Systems Engineering*. Taylor and Francis Group.

15

Deciphering the Submarine Soundscape: New Insights, Broader Implications, Future Directions

Frauke Klingelhoefer[1] and Gaye Bayrakci[2]

ABSTRACT

This book gives an overview of different sources of signals and noise in the oceans, including human-generated, biological, and geological noise. In this chapter, we synthesize the previous chapters into larger cross-cutting themes. We present some sources of further information. Then the subject of this book is shown in its broader context scientifically and from society's point of view. Finally, we propose future actions that could be taken to enhance the understanding of noise in the oceans.

15.1. INTRODUCTION

Noise is present in the oceans, originating from natural sources all around: for example, from surface rain, wind, waves, and, more locally, icebergs. But ships and planes also introduce noise (Erbe et al., 2018). Natural noise created inside the Earth, including volcanoes, earthquakes, landslides, and gas escaping, is transferred to the ocean, along with human-made noise from drilling and other seafloor activities. Animals living in the ocean, such as fish and marine mammals, also create sounds to communicate.

In water, sound travels about 4.5 times faster than in the atmosphere. It can reach very long distances, creating an ocean soundscape rich in different signals and noise, contrary to the previously prevailing opinion that oceans are a "silent world" (Cousteau & Dumas, 1953). In these soundscapes, we can distinguish biophony (biologically produced sounds), geophony (geological and physical ocean sounds such as waves, wind, and currents), and anthrophony (human-produced sounds). As no light can penetrate the deep ocean, sounds are used by marine mammals to communicate, navigate, and find food. These vocalizations can travel over 100 km under average noise conditions (e.g., Munger et al., 2011). However, if noise pollution grows, animals cannot communicate through the noise and may become disoriented (e.g., Wellgart, 2007). On the other hand, the fact that sound travels long distances is used to monitor potential hazards, submarine landslides, volcanoes, and earthquakes (e.g., Clare et al., 2021). In the water, ship or submarine tracking with active sonar from very long distances is used for military applications, also creating noise.

In this book, we have compiled an overview of the different sources of noise in the ocean, their potential uses, and their impact on humans and nature. Methodological chapters, case studies, and reviews were incorporated to give a broad insight into modern studies of marine noise. The book remains inexhaustive because the field is broad and rapidly developing as politicians and society realize the importance of the ocean for marine life and human society, especially geohazards in coastal areas, and its influence on the climate.

[1]*IFREMER, Centre Bretagne, Plouzané, France*

[2]*National Oceanography Centre, Southampton, UK*

Noisy Oceans: Monitoring Seismic and Acoustic Signals in the Marine Environment, Geophysical Monograph 284, First Edition.
Edited by Gaye Bayrakci and Frauke Klingelhoefer.

DOI:10.1002/9781119750925.ch15

Submarine and emergent volcanoes pose a direct threat to coastal communities. One example is the volcano of Mayotte Island, which began to show an increase in activity from 2018 on and is now closely surveyed by a marine seismological array. Active volcanoes and earthquakes frequently trigger marine landslides, which damage seafloor infrastructure and can also be the origin of tsunamis arriving at coasts. Additionally, marine landslides travel further than terrestrial ones, with run-outs up to hundreds of kilometers: for example, the Storegga slide offshore Norway (Bugge et al., 1988). Thus, the study of noise from marine volcanoes (Chapter 4) and landslides (Chapter 5) can be used to monitor their activity and send alerts in case of danger. However, monitoring these dangers in the marine environment remains challenging, although progress is being made using cabled observatories, such as the one planned to monitor the offshore Mayotte submarine volcano system (Aiken et al., 2021) and the European Multidisciplinary Seafloor and Water Column Observatory (EMSO) monitoring submarine landslides in the Ligurian sea offshore Nice (Garziglia et al., 2018).

Climate change is heavily influenced by processes in the oceans, which often remain poorly understood today. Greenhouse gases are trapped in the form of gas hydrates in the seafloor but may be partly released as the oceans warm (Mestdagh et al., 2017). This melting of gas hydrates may induce seafloor instability at continental margins, and therefore trigger landslides and tsunamis (Brown et al., 2006). Vertical gas migration within the seafloor creates typical short waveforms (< 1s) called *short-duration events* (SDEs) (Tary et al., 2012). The frequency of recorded SDEs provides information about fluid pathways in the subsurface, slope stability, and the integrity of future CO_2 storage locations. Recording and quantifying these SDEs can help us quantify global greenhouse gas cycles and understand the influence of marine processes on climate change. A review of the characteristics of SDEs is compiled in Chapter 8, and an additional study case of SDE recordings is given in Chapter 9.

Icebergs contribute considerably to the submarine soundscape of the Southern Ocean, and with an increase in ice-sheet breakup and iceberg calving due to global warming, this noise may increase significantly in the future (Dziak et al., 2013). Icebergs may also contribute directly to climate evolution, as massive iceberg discharges have been proposed as triggers for global climate change in the past (Broecker, 1994). Chapter 6 gives an overview of the characteristics and past studies of iceberg noise. Monitoring iceberg noise also has a potential application for climate change studies and future climate models.

Situated in the deep ocean, near mid-ocean ridges and volcanoes, hydrothermal vents, discussed in Chapter 7, discharge geothermally heated seawater. They are natural laboratories to study the interaction between the earth and the water column and are the principal mechanism for lithospheric cooling (e.g., Kardell et al., 2020). Studying the sounds created by these discharges is complicated by the high temperature of the discharged water rapidly degrading the instruments used for recording (Crone et al. 2006). The sound of hydrothermal vents attracts larval populations from up to 300 km distance (Mullineaux et al. 2010). The expelled fluids are rich in dissolved minerals that form deposits of mineral ores when they cool. They are therefore an important resource of marine minerals; however, detailed studies are necessary to mitigate the risk posed by the interference of human-made noise from mineral mining with the highly specialized and fragile populations living at hydrothermal vents.

Whales and other marine mammals can be detected and tracked by their vocalizations (e.g., Malinka et al., 2020). However, the increasing ocean soundscape poses a direct danger to their health and behavior. Noise can interrupt their interaction, which is primarily based on sounds, affect their navigation skills, and lead to the abandoning of habitats. Therefore, it is important to closely study and measure possible sources of human noise in the ocean. For example, to mitigate the risk of marine mammals being injured by ships, ships can perform passive acoustic monitoring of noise using a towed hydrophone. Knowledge of noise radiated from ships, originally developed for military defense systems, is also used today to calculate ocean noise pollution. Chapter 13 gives an overview of noise from marine traffic, and Chapter 11 discusses vocalizations from baleen whales. A case study of tracking fin whales is described in Chapter 12, and a second case study of tracking underwater vessels is presented in Chapter 14.

Ambient, evenly distributed noise is present in all oceans. It originates from a wide variety of sources: for example meteorological (wind and storms), oceanographic (turbulent currents and internal waves), and gas bubbles generated in the water (gas seeps and rainfall). Although their impact seems smaller than that of volcanoes or ships, they contribute to the marine soundscape and therefore have to be taken into account and quantified. This type of ambient marine noise can also be used to study the subsurface structure of the Earth (e.g., Harrison et al., 2019). In this book, Chapter 10 presents ambient noise from general bubble formation, and Chapters 8 and 9 deal with gas forming bubbles when migrating within sediments. A study of the different origins of ambient noise recorded on land and offshore Taiwan is presented in Chapter 2, and a specific part of the ambient noise—microseisms—and their correlation to significant wave heights is presented in Chapter 3.

15.2. WHAT WAS NOT INCLUDED

This book does not present noise in the ocean from seismic surveys, as an abundance of literature exists that primarily treats the processing of seismic data and also studies of the influence of seismic noise on marine mammals and other life in the sea. For similar reasons, we did not include a chapter on tectonic earthquakes; the subject is well researched, and numerous studies exist. This book instead focuses on non-seismic noise in the oceans, meaning noise unrelated to seismic shots and tectonic earthquakes.

We were unsuccessful in soliciting chapters on more regional human-made noises, including noise from marine wind turbines, marine construction, mining, drilling, and dredging. These sources have less impact than ship noise—for example, because they are used less frequently. However, the first calculations of mining noise predict that a single operation will exceed ambient noise levels experienced in gentle weather conditions (~94 dB re 1 μPa) to a range of ~500 km (Williams et al., 2022).

A promising new technique still in the testing stage is distributed acoustic sensing (DAS). Using an optoelectronic interrogator allows fiber-optic cables to be used as densely spaced geophone arrays with repeated laser signals. As this is a very young technology, we did not succeed in soliciting a contribution. At the moment, the application of DAS is rare in the marine environment (Lindsey et al., 2019; Sladen et al., 2019; Williams et al., 2019). However, in the future, sounds in the ocean may be monitored in real time using existing or retired telecommunications cables.

15.3. FURTHER INFORMATION

Further sources for studies of ocean noise can be readily found in the following non-exhaustive list:

- A Joint Task Force on SMART cables was founded in 2012 to investigate the use of submarine telecommunications cables for ocean and climate monitoring and disaster warning by the International Telecommunication Union (ITU), the Intergovernmental Oceanographic Commission of the United Nations Educational, Scientific and Cultural Organization (UNESCO/IOC), and the World Meteorological Organization (WMO). This task force studies and evaluates scientific and engineering challenges and risks associated with the use of submarine telecommunications cables and its societal benefits (Howe et al., 2019) (https://www.itu.int/en/ITU-T/climatechange/task-force-sc/Documents/JTF-flyer.pdf).
- A future vehicle without an engine (POLARPOD) is constructed to passively drift in the Southern Ocean, circling Antarctica. As the drift is slow, it will generate low acoustic noise and be able to record accurate soundscapes along its trajectory using permanently mounted hydrophones on the part of it immersed in water. The construction is organized by Jean-Louis Etienne and IFREMER (French National Institute for Marine Research), CNRS (French National Centre for Scientific Research), and CNES (Centre National d'Etudes Spatiales, CNES) (https://www.polarpod.fr/fr).
- Rapid progress has been made on cabled seafloor observatories. A well-known example is the NEPTUNE observatory in Canada, which has been recording since 2009. It includes 900 km of submarine telecommunication cables and two instrumented branches, one to the Axial volcano and the other to a major gas-hydrate ridge (https://io.ocean.washington.edu/story/NEPTUNE_Canada).
- The GEOSCOPE network is a global seismological network that aims to fill gaps in isolated sites. It includes 33 broadband seismometers providing real-time data, which is freely available (http://geoscope.ipgp.fr/index.php/en/geoscope-network/).
- A future observatory called MARMOR (Marine Advanced Geophysical Research Equipment and Mayotte Multidisciplinary Observatory for Research and Response) will be deployed offshore Mayotte and record sound as well as chemical parameters of the seawater along with biogeophysical parameters (https://www.marmor-project.org/).

15.4. BROADER CONTEXT

The first studies of marine noise (Cato, 1978) recorded only for short periods and at narrow frequency bands. In comparison, terrestrial soundscapes have been regularly studied since the 1970s, as they directly impact human life and are easily recorded using microphones in the air. The number and length of marine studies then increased rapidly as data storage became more affordable and instrumentation more powerful (e.g., Howe et al., 2019b). Increasingly, simple acoustic studies are being replaced by multipurpose observatories, including, for example, zooplankton fish profilers and physiochemical sensors, which are implemented in the ADEON (Atlantic Deepwater Ecosystem Observatory network) (e.g., Miksis-Olds et al., 2021). The ADEON and other observatories also used the relative silence in the oceans at the beginning of the COVID pandemic to better describe the human influence on ocean soundscapes (Bates et al., 2021; Rutz et al., 2020; Miksis-Olds et al., 2021b).

Other marine geophysical approaches, such as seismic profiling to image the Earth's interior, developed quickly as industrial projects provided funding opportunities. Another example is seismology, where the impact of earthquakes on human life is clearly visible, leading to more abundant funding. In comparison, exploring the submarine soundscape trailed because no commercial use was recognized. However, in recent years, legislation has changed in many countries; they now require studies of the impact of noise on marine life for scientific and industrial applications (e.g., Ritts, 2017). These legislation changes may increase the number of marine noise studies.

Since the invention of submarines, military research of ocean noise has been strongly supported, as it potentially allows the identification of intruders, characterizing their vessels, and distinguishing friendly vessels from intruders. The military also uses this knowledge to build silent ships and submarines that are nearly undetectable by sonar. These measures include acoustic quieting using propellers that produce reduced cavitation, fitting anechoic tiles, and reducing engine vibrations. For example, modern diesel submarines have a radiated noise level of 90 dB (Moore, 2019), while large tankers can reach 188 dB (e.g., McKenna, 2012). It would certainly be interesting to use quieting measures developed by the military in modern civil ships to reduce their noise impact.

Today, society has become aware of the stakes of marine sciences and the importance of reducing human-made ocean noise and mitigating risk from this noise. Monitoring geohazards has become more powerful, and real-time sea observation has become possible. Thus, another goal is to implement up-to-date observatories in hazardous regions. This book is situated in the context of the "Decade of Ocean Science for Sustainable Development (2021–2030)" proclaimed by the United Nations to support efforts to reverse the cycle of decline in ocean health and ensure that ocean science can fully support countries in creating improved conditions for sustainable development of the ocean. Twenty-six ocean decade programs have been created, including many aspects discussed in this book, such as geohazard monitoring, the interaction between the maritime acoustic environment and life, and building a digital twin of the ocean.

15.5. FUTURE IMPACT AND IMPLICATIONS

In this book, we have attempted to propose answers to some questions regarding the origin of different waveforms typically recorded in the oceanic environment. This association is necessary for reducing the overall soundscape and thereby improving the quality of the environment for marine life.

Some chapters of this book include reviews of current knowledge about sounds from specific features, like icebergs and volcanoes. They will interest readers working on marine seismic and acoustic data and enable them to improve their identification of sounds and lead to a better knowledge of marine soundscapes.

We also presented examples of using sound recording in the ocean, such as tracking marine mammals or surveillance of marine geohazards. Similar experiments can, for example, be used to study the marine influence on our climate.

15.6. GAPS IN CURRENT SCIENTIFIC KNOWLEDGE AND FUTURE DIRECTIONS FOR RESEARCH

One reason for gaps in knowledge about marine noise is the relatively small size of networks. Additionally, global seismological networks consist primarily of stations installed on land, as they are easier to use and maintain. An additional difficulty is that the exact location of some targets, such as landslides, is unknown and therefore requires large networks. Other sources are very weak and will need very dense networks: for example, fluid-expulsion sites on the seafloor. In the future, more and better-adapted networks must be installed to optimize monitoring noise in the oceans.

Due to the difficulty in accessing deep oceans, there are gaps in our current knowledge about their soundscapes compared to coastal areas. However, the deep oceans host the most specialized and fragile biotopes, so research should focus on these deep areas: for example, along mid-ocean ridges and in the deep trenches of subduction zones.

There are still gaps in knowledge to be able to associate a recorded waveform with its origin, such as SDEs. It will be important to better define which waveform corresponds to a certain family of sounds. For example, tank experiments can help to clearly assign a sound to its source; however, the tank's size and signal wavelength limit the possible observations, and reverberations may interfere with signals. Thus, there is a need for a waveform database. Once established, the nomenclature should be unified between acoustic and seismic/seismologic researchers and between researchers and industry.

Recording and monitoring the oceans using improved resolution in terms of stations and acoustic bandwidth will lead to much larger quantities of data, as is already the case for DAS recordings. This will require applying big data processing techniques using automated methods, including the use of artificial intelligence to detect recurring signals such as SDEs, marine mammal vocalization, volcanic tremor, and human-made noise. However, DAS measurements depend on the distance separating two positions along the cable (the so-called gauge length) and limit the recorded noise frequency range. This will allow more extensive monitoring of geohazards and exchanges between the Earth's crust and water.

New developments in ocean and earth acoustics include using natural noises in the ocean to image the subsurface and detect the presence of human vehicles. Therefore, they do not increase the sound level. For example, fin whale songs have been successfully used to study sedimentary and upper crustal layers, thereby avoiding the need for seismic shots (Kuna and Nábělek, 2021). As another example, the U.S. military is trying to use shrimp clicks for surveillance of human activity (Wilkinson, 2019).

Further research is needed to implement scientific instrumentation in existing and retired telecommunication cables. Efficient means of transmitting information hundreds of kilometers must be developed, as well as methods to keep the telecommunication information confidential (protected from any potential "listening" by scientific instruments).

15.7. PROPOSED FUTURE ACTIONS

Although geophysicists record soundscapes in the ocean, and some sounds are straightforward to interpret, many open questions remain about the origin of some of the noise. To monitor underwater hazards, marine volcanoes and landslides must be equipped with modern observatories, such as the future observatory of Mayotte, to keep coastal communities as safe as possible. These should not only record sound and movement of the seafloor but also monitor other parameters, such as particle load and chemistry of the water, to detect active sites and record changes in this activity. Finally, to better understand the interactions between sounds of all origins (human and natural) and marine life forms, more interdisciplinary research is needed, including geophysics, water acoustics, and biology.

In the future, comparable noise maps should be created for all oceanic regions. State-of-the-art risk assessment and mitigation tools should be further developed. In addition, legislation can be unified and changed. For example, in the European context, the Marine Strategy Framework Directive (MSFD 2008/56/EC), with its descriptor 11, focuses on underwater noise and clearly states that the spatial distribution, temporal extent, and levels of anthropogenic impulsive sound sources and anthropogenic continuous low-frequency should not exceed levels that adversely affect populations of marine animals. However, these policies should be internationally discussed and implemented in all countries to the same degree.

REFERENCES

Aiken, C., Saurel, J. M., & Foix, O. (2021). Earthquake location and detection modeling for a future seafloor observatory along Mayotte's volcanic ridge. *Journal of Volcanology and Geothermal Research, 418*, 107322.

Bates, A. E., Primack, R. B., Biggar, B. S., Bird, T. J., Clinton, M. E., Command, R. J., et al. (2021). Global COVID-19 lockdown highlights humans as both threats and custodians of the environment. *Biological Conservation, 263*, 109175. https://doi.org/10.1016/j.biocon.2021.109175

Brown, H. E., Holbrook, W. S., Hornbach, M. J., & Nealon, J. (2006). Slide structure and role of gas hydrate at the northern boundary of the Storegga Slide, offshore Norway. *Marine Geology, 229*(3–4), 179–186.

Bugge, T., Belderson, R. H., & Kenyon, N. H. (1988). The Storegga slide. *Philosophical Transactions of the Royal Society of London. Series A, Mathematical and Physical Sciences, 325* (1586), 357–388.

Cato, D. H. (1978). Marine biological choruses observed in tropical waters near Australia. *Journal of the Acoustical Society of America, 64*, 736–743. https://doi.org/10.1121/1.382038

Clare, M., Lintern, D. G., Pope, E., Baker, M., Ruffell, S., Zulkifli, M., et al. (2021). Seismic and acoustic monitoring of submarine landslides: ongoing challenges, recent successes and future opportunities. https://eartharxiv.org/repository/view/2736/

Cousteau, J.-Y., & Dumas, F. (1953). *The silent world.* Harper & Brothers Publishers.

Crone, T. J., Wilcock, W. S., Barclay, A. H., & Parsons, J. D. (2006). The sound generated by mid-ocean ridge black smoker hydrothermal vents. *PLoS One, 1*(1), e133.

Erbe, C., Williams, R., Parsons, M., Parsons, S. K., Hendrawan, I. G., & Dewantama, I. M. I. (2018). Underwater noise from airplanes: an overlooked source of ocean noise. *Marine Pollution Bulletin, 137*, 656–661.

Garziglia, S., Apprioual, R., Roudaut, M., Bompais, X., & Barbot, S. (2018). Pore pressure and temperature data from piezometer ST4-PZ2L-01B on EMSO-LIGURE NICE observatory since September 2016. *SEANOE.* https://doi.org/10.17882/51490

Harrison, C. H., & Simons, D. G. (2002). Geoacoustic inversion of ambient noise: A simple method. *The Journal of the Acoustical Society of America, 112*(4), 1377–1389.

Howe, B. M., Arbic, B. K., Aucan, J., Barnes, C. R., Bayliff, N., Becker, N., et al. (2019). SMART cables for observing the global ocean: science and implementation. *Frontiers in Marine Science, 6*, 424.

Howe, B. M., Miksis-Olds, J., Rehm, E., Sagen, H., Worcester, P. F., & Haralabus, G. (2019b). Observing the oceans acoustically. *Frontiers in Marine Science, 6*, 426.

Kardell, D. A., Zhao, Z., Ramos, E. J., Estep, J., Christeson, G. L., Reece, R. S., & Hesse, M. A. (2021). Hydrothermal models constrained by fine-scale seismic velocities confirm hydrothermal cooling of 7–63 Ma South Atlantic Crust. *Journal of Geophysical Research: Solid Earth, 126*(6), e2020JB021612.

Kuna, V. M., & Nábělek, J. L. (2021). Seismic crustal imaging using fin whale songs. *Science, 371*(6530), 731–735.

Lindsey, N. J., Dawe, T. C., & Ajo-Franklin, J. B. (2019). Illuminating seafloor faults and ocean dynamics with dark fiber distributed acoustic sensing. *Science*, *366*(6469), 1103–1107.

Malinka, C. E., Atkins, J., Johnson, M. P., Tønnesen, P., Dunn, C. A., Claridge, D. E., et al. (2020). An autonomous hydrophone array to study the acoustic ecology of deep-water toothed whales. *Deep Sea Research Part I: Oceanographic Research Papers*, *158*, 103233.

McKenna, M. F., Ross, D., Wiggins, S. M., & Hildebrand, J. A. (2012). Underwater radiated noise from modern commercial ships. *The Journal of the Acoustical Society of America*, *131* (1), 92–103.

Mestdagh, T., Poort, J., & De Batist, M. (2017). The sensitivity of gas hydrate reservoirs to climate change: Perspectives from a new combined model for permafrost-related and marine settings. *Earth-Science Reviews*, *169*, 104–131.

Miksis-Olds, J., Warren, J., Blair, H., & Lucca, B. M. (2021). Atlantic deepwater ecosystem observatory network: Patterns of acoustic backscatter and community structure of the US Outer Continental Shelf. *The Journal of the Acoustical Society of America*, *149*(4), A62–A62.

Miksis-Olds, J., Martin, S. B., Lowell, K., Verlinden, C., & Heaney, K. D. (2021b). Minimal COVID-19 quieting measured in the deep, offshore waters of the US Outer Continental Shelf. *The Journal of the Acoustical Society of America*, *150*(4), A51–A52.

Moore, S. (2019). Acoustic signature control on maritime platforms. In *Proceedings of acoustics* (Vol. 10, No. 13). Australian Acoustical Society.

Mullineaux, L. S., Adams, D. K., Mills, S. W., & Beaulieu, S. E. (2010). Larvae from afar colonize deep-sea hydrothermal vents after a catastrophic eruption. *Proceedings of the National Academy of Sciences*, *107*(17), 7829–7834.

Munger, L. M., Wiggins, S. M., & Hildebrand, J. A. (2011). North Pacific right whale up-call source levels and propagation distance on the southeastern Bering Sea shelf. *The Journal of the Acoustical Society of America*, *129*(6), 4047–4054.

Rutz, C., Loretto, M.-C., Bates, A. E., Davidson, S. C., Duarte, C. M., Jetz, W., et al. (2020). COVID-19 lockdown allows researchers to quantify the effects of human activity on wildlife. *Nature Ecology and Evolution*, *4*, 1156–1159. https://doi.org/10.1038/s41559-020-1237-z

Ritts, M. (2017). Amplifying environmental politics: ocean noise. *Antipode*, *49*(5), 1406–1426.

Sladen, A., Rivet, D., Ampuero, J. P., De Barros, L., Hello, Y., Calbris, G., & Lamare, P. (2019). Distributed sensing of earthquakes and ocean-solid Earth interactions on seafloor telecom cables. *Nature Communications*, *10*(1), 5777.

Tary, J. B., Geli, L., Guennou, C., Henry, P., Sultan, N., Çağatay, N., & Vidal, V. (2012). Microevents produced by gas migration and expulsion at the seabed: a study based on sea bottom recordings from the Sea of Marmara. *Geophysical Journal International*, *190*(2), 993–1007.

Weilgart, L. S. (2007). A brief review of known effects of noise on marine mammals. *International Journal of Comparative Psychology*, *20*(2).

Wilkinson, A. (2019). Military tries out fish as underwater spies. *Scientific American* (29 March).

Williams, E. F., Fernández-Ruiz, M. R., Magalhaes, R., Vanthillo, R., Zhan, Z., González-Herráez, M., & Martins, H. F. (2019). Distributed sensing of microseisms and teleseisms with submarine dark fibers. *Nature Communications*, *10*(1), 5778.

Williams, R., Erbe, C., Duncan, A., Nielsen, K., Washburn, T., & Smith, C. (2022). Noise from deep-sea mining may span vast ocean areas. *Science*, *377*(6602), 157–158.

INDEX

Noisy Oceans: Monitoring Seismic and Acoustic Signals in the Marine Environment, Geophysical Monograph 284, First Edition.
Edited by Gaye Bayrakci and Frauke Klingelhoefer.

DOI:10.1002/9781119750925.index

Printed in the USA/Agawam, MA
September 16, 2026

910661.210